Reagents for Organic Synthesis

ADVISORS

JEFF ROHDE   MARK BILODEAU

STRUCTURE COMPOSITION

TIMOTHY GRINSTEINER   TIM JAMISON
MIHAI AZIMIOARA   TAE BUM SHIN
MARK RUSSETT   ORLANDO SCHARER
MARGARET HSU   SEPEHR SARSHAR

PROOFREADERS

MICHAEL H. KRESS   ALEXANDER R. MUCI
REBECCA M. MARSH   DAVID H. RIPIN
RICHARD M. GOODMAN   BRYAN E. ROBERTS
JAY F. LARROW   DEBORAH EVRARD
SUSAN L. VANDERVELDE   KARLENE A. CIMPRICH
DEBORAH T. HUNG

Fieser and Fieser's
Reagents for Organic Synthesis

VOLUME SEVENTEEN

Mary Fieser
Harvard University

A WILEY-INTERSCIENCE PUBLICATION
John Wiley & Sons, Inc.
NEW YORK / CHICHESTER / BRISBANE / TORONTO / SINGAPORE

Library of Congress Cataloging in Publication Data:
ISBN 0-471-00074-4
ISSN 0271-616X

Printed in the United States of America

10 9 8 7 6 5 4 3 2 1

PREFACE

This volume of Reagents includes material published in late 1990 to early 1993. As usual, this volume has received tremendous help from Harvard graduate students. Mark Bilodeau and Jeff Rohde have read and corrected large portions of the manuscript. Other advisers have provided invaluable correction of the page proofs. Another group, for the first time, was responsible for composition of the chemical formulas in ChemDraw.

MARY FIESER

Cambridge, Massachusetts
June 1994

CONTENTS

Reagents for Organic Synthesis

A

Acylsilanes.

Review. Cirillo and Panek[1] have reviewed progress in the chemistry of these compounds during the last two to three years (48 references). A new synthesis involves homologation of aldehydes with methoxybis(trimethylsilyl)methyllithium (**1**)[2] to form enol ethers of an acylsilane (equation I). Acylsilanes can be prepared in high yield by

(I) $\underset{\textbf{1}}{Li\overset{\displaystyle OCH_3}{\overset{|}{C}}[Si(CH_3)_3]_2}$ $\xrightarrow[\text{THF}]{\text{RCHO,}}$ $\underset{H}{\overset{R}{>}}C=C\underset{Si(CH_3)_3}{\overset{OCH_3}{<}}$ $\xrightarrow[51\text{-}92\%]{H_3O^+}$ $RCH_2\overset{\displaystyle O}{\underset{}{\overset{||}{C}}}Si(CH_3)_3$

(II) $R\overset{\displaystyle O}{\underset{}{\overset{||}{C}}}SPy$ $+\ Al[Si(CH_3)_3]_3$ $\xrightarrow[83\text{-}98\%]{\begin{array}{c}CuCN\\THF\end{array}}$ $R\overset{\displaystyle O}{\underset{}{\overset{||}{C}}}Si(CH_3)_3$

(III) $\underset{R\quad Si(CH_3)_3}{\overset{S\quad S}{\diagdown\diagup}}$ $\xrightarrow[70\text{-}96\%]{e,\ CH_3CN\ /\ H_2O}$ $R\overset{\displaystyle O}{\underset{}{\overset{||}{C}}}Si(CH_3)_3$

reaction of S-2-pyridyl esters with aluminum tris(trimethylsilane) and CuCN (1 equiv.), equation II.[3] A third general synthesis of acylsilanes involves anodic oxidation of 2-alkyl-2-trimethylsilyl-1,3-dithianes (equation III)[4].

Acyldimethylphenylsilanes undergo highly selective hydrogenation (Pd/C) to aldehydes without effect on benzyl or BOM ether groups, acetonides, or *t*-butyldimethylsilyl groups (equation IV).

(IV) $C_6H_5\!-\!\underset{\underset{O}{|}}{\overset{\overset{(CH_3)_2}{|}}{Si}}\underset{OSi(CH_3)_3}{\overset{OMOM}{\diagdown}}CH_3$ $\xrightarrow[75\%]{\begin{array}{c}H_2,\ Pd\ /\ C\\C_2H_5OH\end{array}}$ $H\underset{\underset{O}{|}}{\diagdown}\underset{OSi(CH_3)_3}{\overset{OMOM}{\diagdown}}CH_3$

Acylsilanes undergo highly diastereoselective reactions with RLi and RMgX to afford *syn*-adducts, particularly when the chiral center is substituted by a phenyl group.[5]

> 100:1

89% | Bu$_4$N$^+$F$^-$

Acylsilanes are reduced to α-alkoxysilanes by a variety of metal hydrides. Reduction with (−)-chlorodiisopinocamphenylborane provides (R)-alcohols in >80% ee and in 30−65% yield.[6]

R, 98% ee

[1] P. F. Cirillo and J. S. Panek, *Org. Prep. Proc. Intl.*, **24**, 553 (1992).

[2] J. Yoshida, S. Matsunaga, Y. Ishichi, T. Makawa, and S. Isoe, *J. Org.*, **56**, 1307 (1991).

[3] M. Nakada, S. Nakamura, S. Kobayashi, and M. Ohno, *Tetrahedron Letters*, **32**, 4929 (1991).

[4] K. Suda, J. Watanabe, and T. Takanami, *ibid.*, **33**, 1355 (1992).

[5] P. F. Cirillo and J. S. Panek, *ibid.*, **32**, 457 (1991).

[6] J. A. Soderquist, C. L. Anderson, E. I. Miranda, and I. Rivera, *Tetrahedron Letters*, **31**, 4677 (1990); J. D. Breynak, J. B. Strickland, G. W. Lamb, D. Khasnis, S. Modi, D. Williams, and H. Zhang, *J. Org.*, **56**, 7076 (1991).

Alanine, CH$_3$CH(NH$_2$)COOH.

α-Methyl α-amino acids.[1]   These amino acids are of general interest because such substituents in peptides can restrict the rotational freedom of amide bonds, and hence the secondary and tertiary structures of proteins. A general synthesis is an extension of Seebach's method for α-methylation of α-amino acids, but involves alkylation of the α-methyl α-amino acid D- or L-alanine. Thus the Schiff base formed from this D-amino acid and benzaldehyde is cyclized by benzyl chloroformate to the (2S,4R)-oxazolidinone **1**. The enolate of **1** is alkylated by various electrophiles from the face opposite to that of the phenyl group. The products, such as (2S,4S)-**2**, can be hydrolyzed by brief treatment with LiOH or NaOH in CH$_3$OH/H$_2$O to a protected derivative of the α-methyl α-amino acid, such as α-methylaspartic acid (**3**).

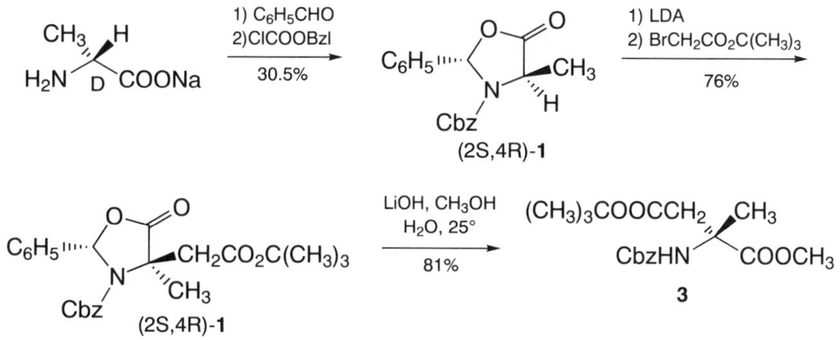

The enolate of (2S,4R)-**1** can also undergo Michael addition to *t*-butyl acrylate. The yield is low, but the adduct is obtained in 98% ee. This reaction provides a route to a derivative of 2-methylglutamic acid (equation I).

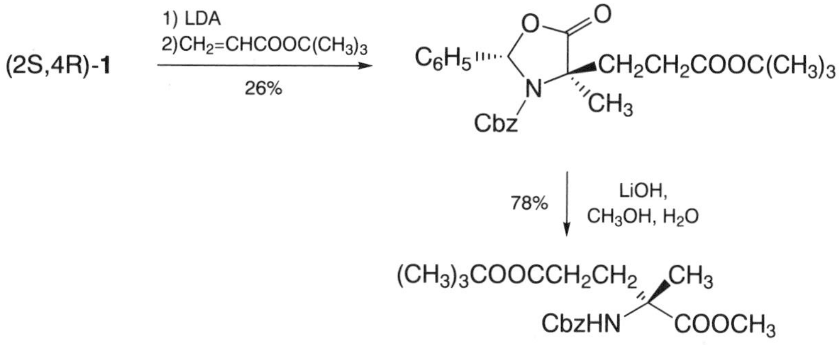

[1]E. Altmann, K. Nebel, and M. Mutter, *Helv.*, **74**, 800 (1991).

β-Alanine, $H_2NCH_2CH_2COOH$.

Stereoselective alkylation. As an extension of a method for diastereoselective alkylation of α-amino acids (**14**, 263–264), Seebach *et al.*[1] have converted this simple β-amino acid into the 2-*t*-butylperhydropyrimidinone **2** via an imine with pivaldehyde.

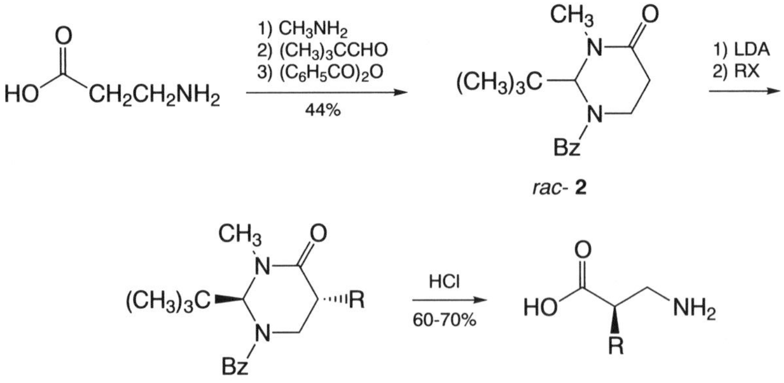

rac- **2**

3   **4** (86-97% ds)

Surprisingly, an X-ray analysis of **2** shows that the *t*-butyl group has the axial configuration. The enolate of **2** reacts with alkyl halides with high diastereoselectivity to give the *trans*-product (**3**) (also axial). These products (**3**) can be hydrolyzed to α-substituted β-amino acids (**4**) with 6*N* HCl at 160–180°. This overall process should be applicable to a synthesis of emantiomerically pure β-amino acids.

[1] E. Juaristi, D. Quintana, B. Lamatsch, and D. Seebach, *J. Org.,* **56**, 2553 (1991).

Alkylaluminum halides.

   Reaction of RAlCl$_2$ with β-formylpropionamides.[1]   Reaction of $(CH_3)_3Al$ with the β-formylpropionamides **1** shows no stereoselectivity, but the reaction with CH_3AlCl_2 shows high *threo*-selectivity. In contrast the reaction with CH_3TiCl_3 is *erythro*-selective.

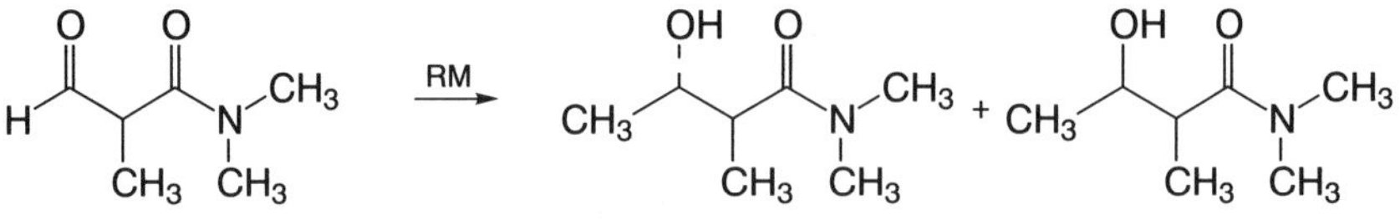

1

CH_3AlCl_2	63%		97 : 3
CH_3TiCl_3	38%		2 : 8

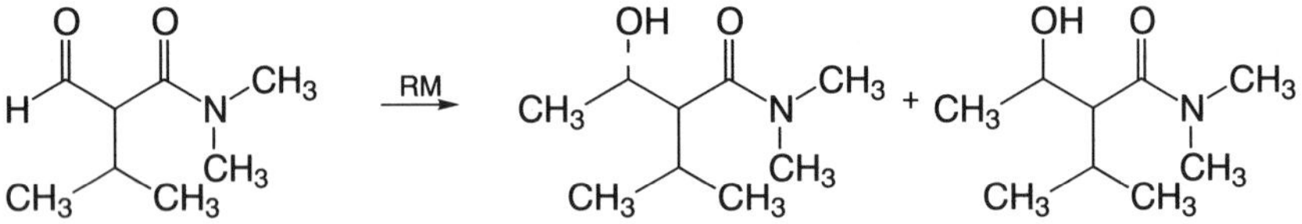

CH_3AlCl_2	62%		99 : 1
$C_2H_5AlCl_2$	72%		99 : 1

Cyclization of epoxydienes. Corey and Sodeoka[2] have examined in detail the optimum conditions for cyclization of the epoxydiene **1**. Of several Lewis acids, CH_3AlCl_2 was found to increase the rate of cyclization more than $(CH_3)_2AlCl$ or $(CH_3)_2AlOTf$, but the yield was comparable in all cases. Methylene chloride is superior to $ClCH_2CH_2Cl$, CH_3NO_2, or toluene as solvent. Under optimal conditions the cyclization was carried out at $-78°$ for 1 hour followed by silylation. The principal products are the silyloxy ketone **2** and the monocyclic silyloxy methyl ketone **3**, formed under all conditions as a minor product

Deoxygenation of ketones. Reductive deoxygenation of ketones can be effected with an aluminum source and a Lewis acidic aluminum reagent. In the case of diaryl ketones *i*-BuAlCl₂ can be used as the hydride source as well as the Lewis acid.

Reductive deoxygenation of alkyl aryl ketones is best effected in two steps: reduction with *i*-Bu₂AlH followed by treatment with AlBr₃. The yield can be increased by addition of Cp₂TiCl₂ in the second step.

$$C_6H_5COCH_3 \xrightarrow[\substack{2) \text{ AlBr}_3, \text{ Cp}_2\text{TiCl}_2 \\ 75\%}]{1) \text{ } i\text{- BuAlH}} C_6H_5CH_2CH_3$$

Reductive deoxygenation of dialkyl ketones is best effected in three steps: sequential treatment with *i*-Bu₂AlH, AlBr₃, and finally *i*-Bu₂AlH and a Ni(acac)₂ catalyst.

$$CH_3(CH_2)_4 \overset{O}{\underset{}{\|}} Bu \xrightarrow[\text{3) } i\text{ - Bu}_2\text{AlH, Ni(acac)}_2]{\substack{\text{1) } i\text{ - Bu}_2\text{AlH} \\ \text{2) AlBr}_3}} \quad n\text{ - nonane} \; + \; \text{nonenes}$$

n - nonane (48%) + nonenes (32%)

These reactions can be used for reduction of the corresponding secondary alcohols to alkanes, but do not reduce primary alcohols or phenols.

Cyclization of alkynyltitanium complexes.[4] Reaction of an alkynylmagnesium bromide with dichlorobis(cyclopentadienyl)titanium results in a Ti complex **1** in essentially quantitative yield. This complex is more stable to air oxidation and to water than the Grignard reagent. On treatment with an alkylaluminum halide, these complexes ($n = 1-3$) undergo cyclization to an exocyclic trisubstituted alkene attached to a 4-, 5-, or 6-membered ring (**2**).

This cyclization is regioselective; thus **3** undergoes intramolecular *syn*-insertion to give **4** as the exclusive cyclic product.

Diels–Alder catalyst.[5] $C_2H_5AlCl_2$ is the most effective catalyst for Diels–Alder reactions of (S)-(+)-carvone **1** with silyloxy-1,3-dienes. This reaction is a key step in a total synthesis of (+)-α-cyperone (**5**) in 40% overall yield from (S)-carvone.

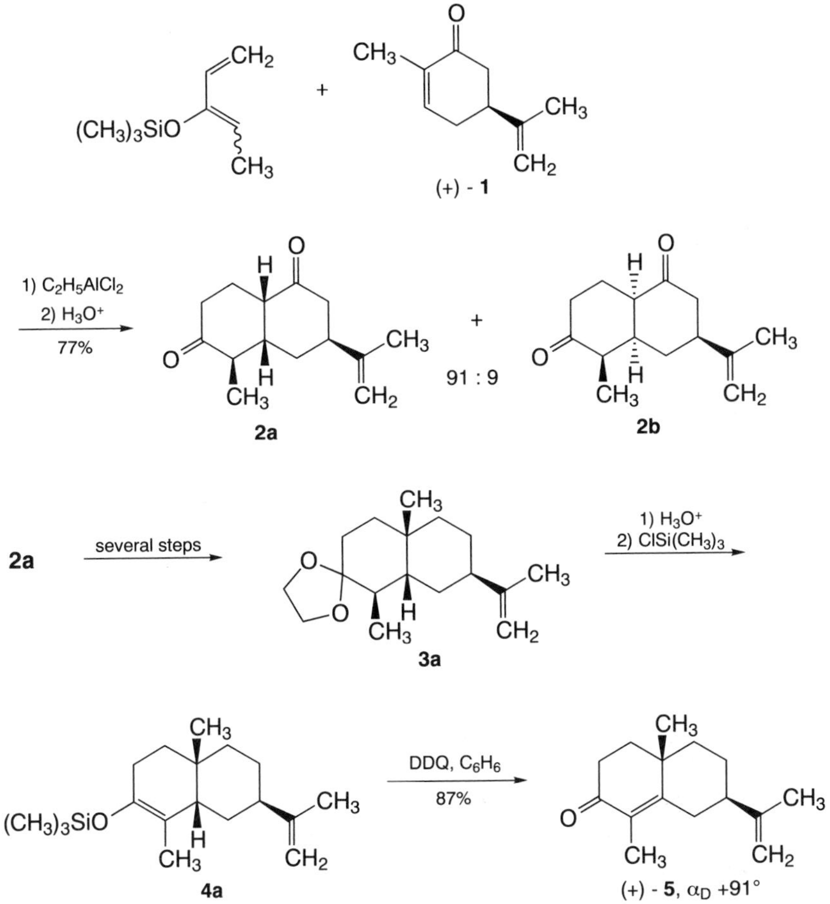

[1] H. Fujii, M. Taniguchi, K. Oshima, and K. Utimoto, *Tetrahedron Letters,* **33**, 4579 (1992).
[2] E. J. Corey and M. Sodeoka, *Tetrahedron Letters,* **32**, 7005 (1991).
[3] J. L. Eisch, Z.-R. Liu, and M. P. Boleslawski, *J. Org.,* **57**, 2143 (1992).
[4] A. E. Harms and J. R. Stille, *Tetrahedron Letters,* **33**, 6565 (1992).
[5] A. A. Haaksma, B. J. M. Jansen, and A. de Groot, *Tetrahedron,* **48**, 3121 (1992).

Alkyllithiums.

2,4-Dialkylphenols.[1]   Reaction of *p*-benzoquinone with an alkyllithium (3 equiv.) followed by an acidic hydrolysis provides 2,4-dialkylphenols in 60–70% isolated yield.

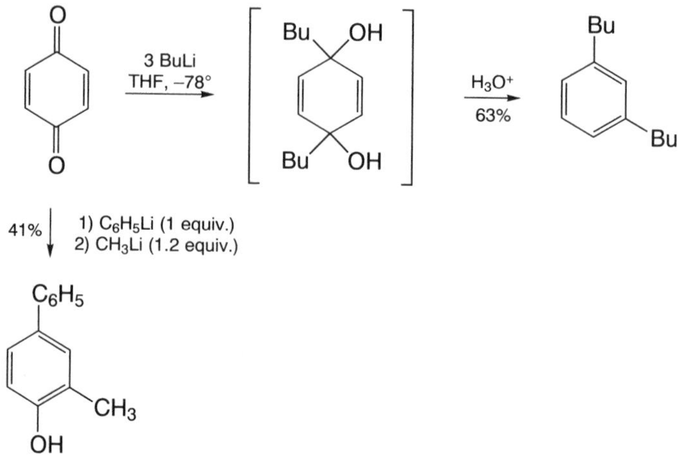

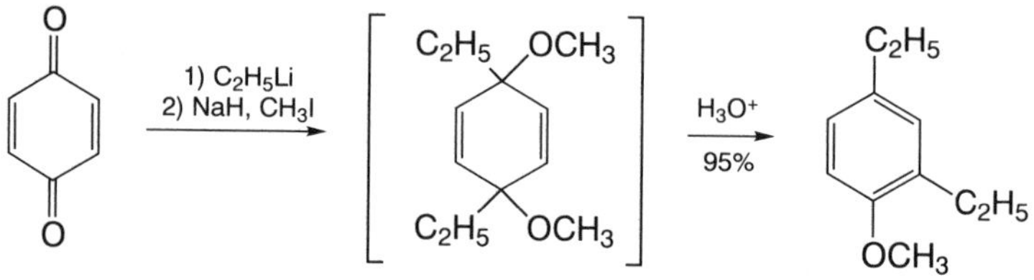

The alkyl groups can also be derived from different alkyllithiums. 2,4-Dialkyl-anisoles can also be prepared by this reaction.

Reaction with CO_2.[2]   Reaction of R^1Li with CO_2 (solid) in ether, pentane, or THF at $-40 \rightarrow -10°$ provides a product **a** that on treatment with lithium hydride forms an aldehyde (43–72% yield). Reaction of **a** with R^1Li or R^2Li gives a ketone after hydrolysis. Addition of a second RLi to **b** before work-up provides a tertiary alcohol.

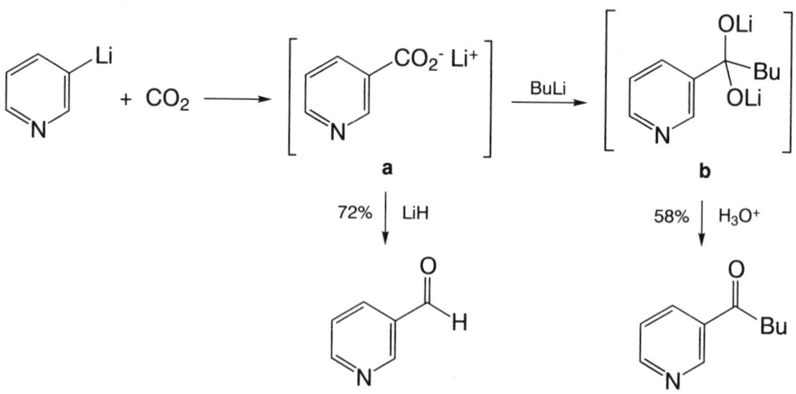

This protocol provides a ready access to a range of aldehydes or ketones with moderate to high yield.

   (Aminosilyl)lithiums.[3] Silyl anions are unstable and are usually generated *in situ* in the presence of a quenching agent. Aminosilyl anions are more stable both to substitution and α-elimination. They can be prepared by reaction of aminochlorosilanes with lithium dispersed in THF or with lithium wire. Yields are generally in the range 80–100% and the products are stable at 0° for 1–6 days.

   The reagents couple with chlorosilanes to form disilanes. They undergo conjugate addition to unsaturated esters.

$$(CH_3)SiCl \quad + \quad (C_2H_5)_2N(C_6H_5)_2SiLi \quad \xrightarrow[78\%]{AcCl}$$

$$(CH_3)_3SiSi(C_6H_5)_2Cl \quad \xrightarrow[47\%]{\substack{(C_2H_5)_2NCH_3C_6H_5SiLi \\ TMEDA}} \quad (CH_3)_3Si\overset{\overset{\displaystyle (C_6H_5)_2}{|}}{Si}SiC_6H_5CH_3N(C_2H_5)_2$$

[1] F. Alonso and M. Yus, *Tetrahedron,* **48**, 2709 (1992).
[2] G. Zadel and E. Breitmaier, *Angew. Chem. Int. Ed.,* **31**, 1035 (1992).
[3] K. Tamao, A. Kawachi, and Y. Ito, *Am. Soc.,* **114**, 3989 (1992).

Allylbarium reagents, $CH_2\!=\!CHCH_2BaCl$.
These reagents can be generated *in situ* by reaction of allylic chlorides with barium(0) with retention of the double bond geometry.

   β,γ-Unsaturated carboxylic acids.[1] Carboxylation of allylbarium reagents, generated *in situ,* proceeds with high α-selectivity. Note that carboxylation of allylmagnesium proceeds with high γ-selectivity.

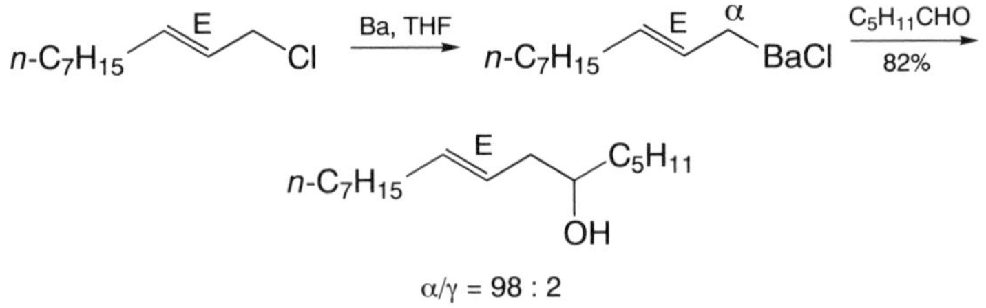

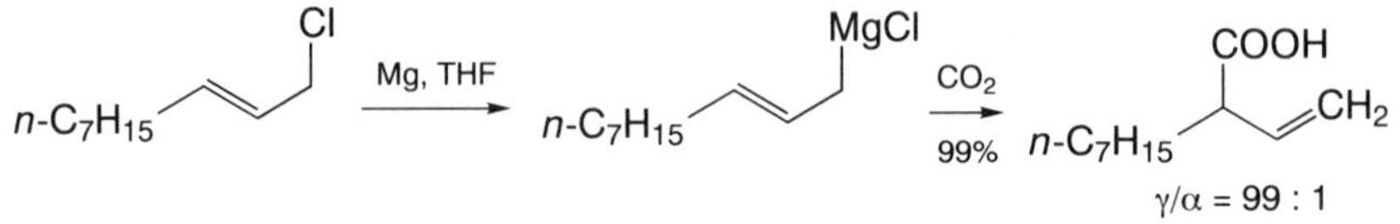

$\alpha/\gamma = 98 : 2$

$\gamma/\alpha = 99 : 1$

[1] A. Yanagisawa, S. Habaue, and H. Yamamoto, *Am. Soc.*, **113**, 8955 (1991); A. Yanagisawa, K. Yasue, and H. Yamamoto, *Synlett*, 593 (1992).

4-Allylcyclobutenones.

Bicyclo[3.2.0]heptenones.[1] Allyl Grignard reagents react with cyclobutenediones to give 1,2-adducts (**1**) in moderate yield. The products when heated in refluxing toluene give bicyclo[3.2.0]heptenones (**2**) in high yield. The reaction evidently involves ring opening followed by an intramolecular [2 + 2]cycloaddition of a vinyl ketene and a methylene group.

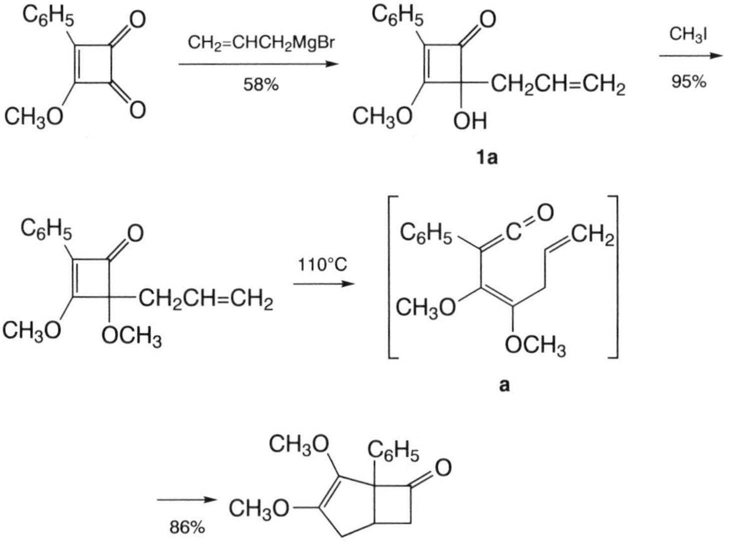

This methodology has been extended to 4-alkynyl-4-(propargyloxy)cyclobutenones,[2] formed by addition of lithium acetylides to dimethyl squarate (**3**) followed by treatment of the 1,2-adducts with propargyl iodides. These products (**4**) when heated at 135° rearrange to methylenebenzofurans (**5**), which isomerize in an acidic medium to benzofurans (**6**).

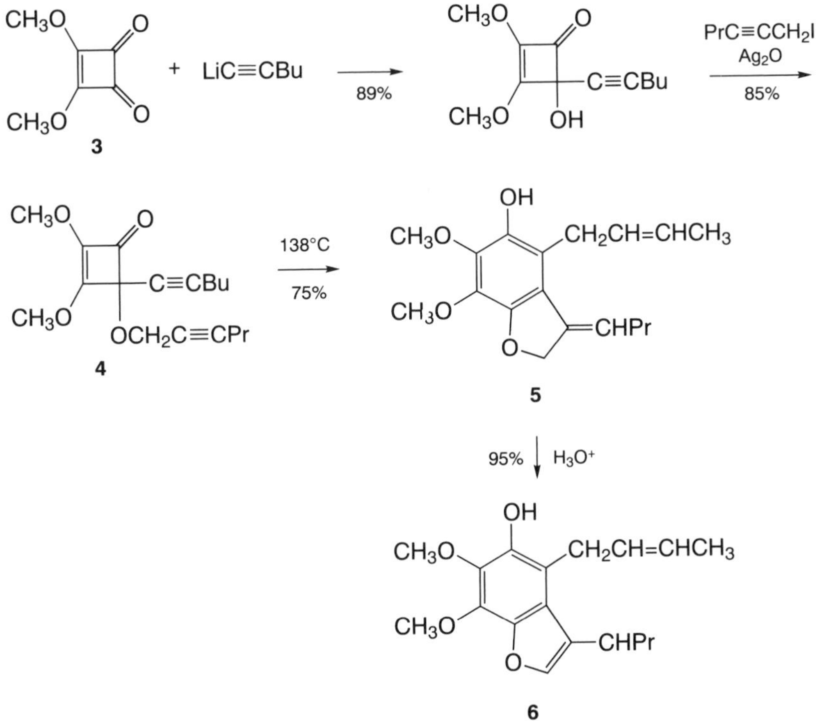

[1] S. L. Xu, H. Xia, and H. W. Moore, *J. Org.*, **56**, 6094 (1991).
[2] S. L. Xu, M. Taing, and H. W. Moore, *ibid.*, **56**, 6104 (1991).

Allyl(diisopinocampheyl)borane,

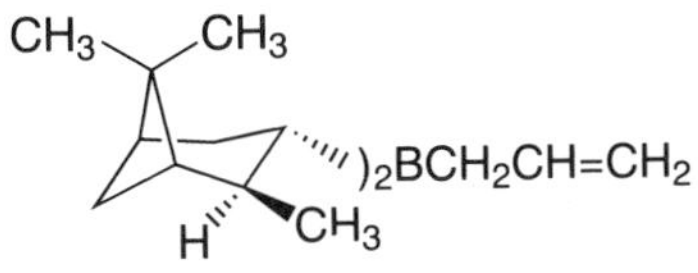

1

Enantioselective allylboration; (R)-γ-amino-β-hydroxybutyric acid.[1] Reaction of bromoacetaldehyde with **1** in ether at $-100°$ provides the (R)-bromohydrin **2** in 89% ee (~50% yield), which is converted to the (R)-epoxide **3** on treatment with KOH. This epoxide is a precursor to **4**, (R)-(-) γ-amino-β-hydroxybutyric acid (GABOB), a neuromodulator.

$$BrCH_2CHO \xrightarrow[\sim 50\%]{1} $$

(R) - **2**, 89% ee

(R) - **3**

1) $NaIO_4\cdot RuCl_3$
2) NH_3
34%

(R) - **4**, 89% ee

[1] Y. N. Bubnov, L. I. Lavrinovich, A. G. Zykov, and A. V. Ignatenko, *Mendeleev Comm.*, 86 (1992).

Allyltributyltin reagents, $CH_2=CHCH_2SnBu_3$ (**1**).

-    *Allylation of dioxane acetals*[1]   (cf. **12**, 375–378). Allylation of chiral dioxane acetals such as **2** has been shown to proceed with marked stereoselectivity with allyltrimethylsilane promoted by $TiCl_4/Ti(O\text{-}i\text{-}Pr)_4$. A more recent study, indicates that

(±)-**2**

$CH_2=CHCH_2MR_3$
$TiCl_4$, $Ti(O\text{-}i\text{-}Pr)_4$

lk-**3**

ul-**3**

lk : *ul*

		lk : *ul*
M = Si, R = CH₃	100%	58:1
M = Sn, R = Bu	100%	270:1
M = Sn, R = C₆H₅	100%	90:1

allyltributylstannane (**1**) is significantly more selective than allyltrimethylsilane and that phenyl groups on the metal decrease the selectivity. The same selectivity is shown in the case of dioxane acetals bearing other alkyl, aryl, and alkynyl groups.

β,γ-Enones.[2] Organic halides are known to undergo tin hydride mediated carbonylation. This reaction can be extended to a three-component coupling of RI, CO, and allyltins to form *β,γ*-enones.

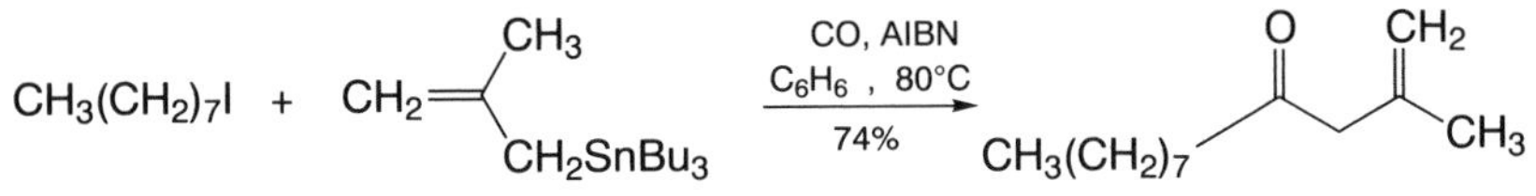

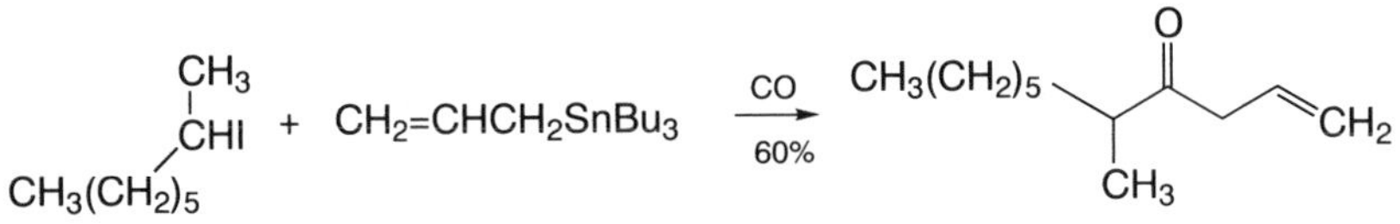

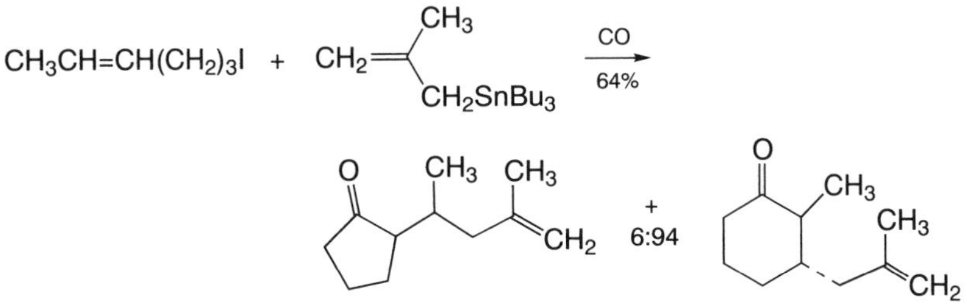

$CH_3(CH_2)_7I$ + $\xrightarrow[74\%]{\begin{array}{c}CO, AIBN \\ C_6H_6 , 80°C\end{array}}$

+ $CH_2=CHCH_2SnBu_3$ $\xrightarrow[60\%]{CO}$

$CH_3CH=CH(CH_2)_3I$ + $\xrightarrow[64\%]{CO}$

+ 6:94

Asymmetric addition of allyltins to RCHO.[3] Addition of γ-substituted allylstannanes to aldehydes to form homoallylic alcohols can proceed in high diastereo- and enantioselectivity when carried out in the presence of 1 equiv. of the Yamamoto chiral acyloxyborane **1** (**16**, 314). If trifluoroacetic anhydride is used as promotor, then only catalytic amounts of **1** are required. Virtually quantitative yields are obtained if **1** and $(TFA)_2O$ are used in the ratio 1 eq:1.2 eq.

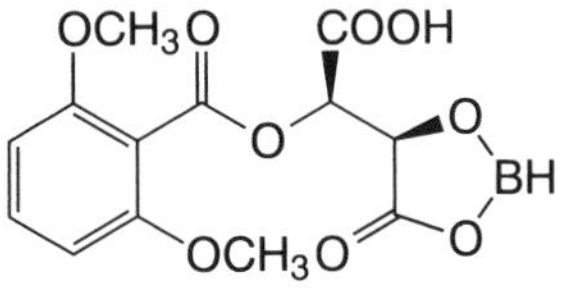

1

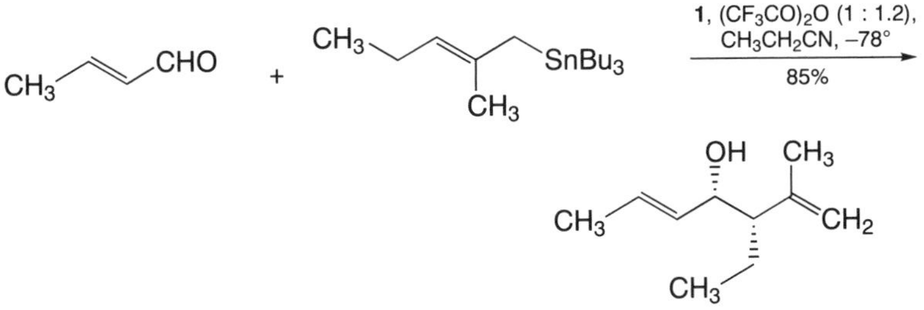

+ $\xrightarrow[85\%]{\begin{array}{c}\textbf{1}, (CF_3CO)_2O \ (1:1.2), \\ CH_3CH_2CN, -78°\end{array}}$

95% ee, syn/anti = 92 : 8

[1] S. E. Denmark and N. G. Almstead, *J. Org.*, **56**, 6485 (1991).

[2] I. Ryu, H. Yamazaki, K. Kusano, A. Ogawa, and N. Sonoda, *Am. Soc.*, **113**, 8558 (1991).

[3] J. A. Marshall and Y. Tang, *Synlett,* 653 (1992).

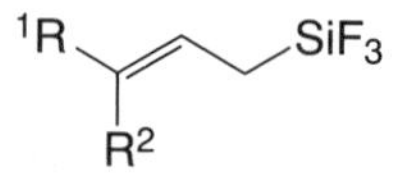

Allyltrifluorosilanes,

γ-Selective coupling with electrophilies.[1]   In the presence of a Pd(0) catalyst and Bu$_4$NF (1 equiv.), aryl, vinyl, or allylic halides, triflates, or acetates couple with allyltrifluorosilanes at the γ-position, with migration of the double bond. Pd(OAc)$_2$ and dppb in combination with TASF (tris(diethylamino)sulfonium difluorotrimethylsilicate)

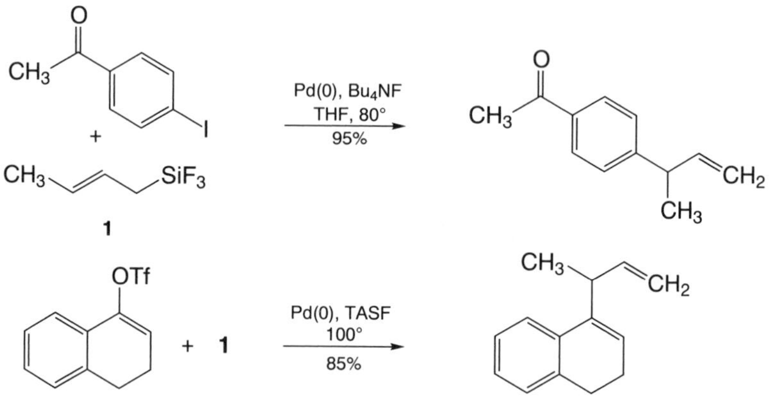

is particularly effective for coupling of enol triflates. Only one coupling of an allylic acetate in low yield was reported. This coupling provides a practical route to the antiinflammatory agent ibuprofen (equation I).

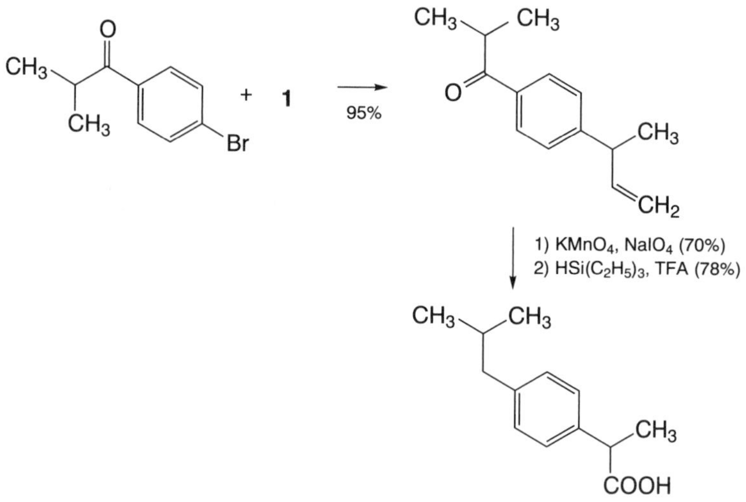

[1] Y. Hatanaka, Y. Ebina, and T. Hiyama, *Am. Soc.,* **113**, 7075 (1991).

Aluminum chloride.

Acylation.[1] Acylation of 2-methylbutane with the crotyl chloride **1** in CH_2Cl_2 in the presence of $AlCl_3$ and $CuSO_4/CH_3NO_2$ provides a synthesis of cyclopentenones (equation I). A similar reaction of alkenyl chlorides with methylcyclopentane or methyl-

cyclohexane provides a synthesis of tetrahydropentalenones or tetrahydroindenones.

Fries rearrangement, review. Martin[2] has reviewed the rearrangement of aryl esters to *o*- and *p*-hydroxyaryl ketones, usually effected with $AlCl_3$. The rearrangement can occur at room temperatures in polar solvents (nitroalkanes). Photo-induced rearrangement usually provides the same products as obtained by Lewis acid induced rearrangement. The review lists 281 references to literature mainly from 1964 to 1990.

[1] C. Morel-Fourrier, J.-P. Dulcère, and M. Santelli, *Am. Soc.*, **113**, 8062 (1991).
[2] R. Martin, *Org. Prep. Proc. Int.*, **24**, 373 (1992).

(R)- and (S)-1-Amino-2-methoxymethylpyrrolidine (1, Ramp, Samp).

Asymmetric Michael cyclization; trans-disubstituted cyclopentanecarboxylates.[1] The lithium enolates of SAMP hydrazones (**2**) of methyl ketones react with methyl (E)-6-bromohex-2-enoate (**3**) to form the Michael adducts (**4**), which are cleaved by ozone to give *trans*-disubstituted cyclopentanecarboxylates **5** with high diastereo- and entantioselectivity.

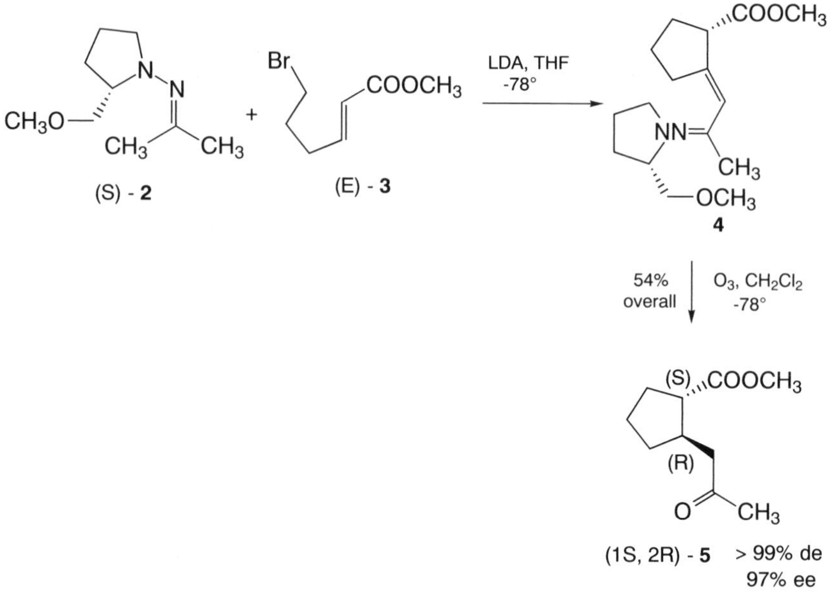
LDA, THF
-78°
54%
overall
O_3, CH_2Cl_2
-78°
(S) - 2
(E) - 3
4
(1S, 2R) - 5
> 99% de
97% ee

[1] D. Enders, H. J. Scherer, and G. Raabe, *Angew. Chem. Int. Ed.,* **30**, 1664 (1991).

Aniline and Nitrobenzene.

Direct coupling to 4-nitrosodiphenylamine and 4-nitrodiphenylamine.[1] Addition to tetramethylammonium hydroxide dihydrate, $(CH_3)_4NOH \cdot 2H_2O_6$ to a mixture of aniline and nitrobenzene results in a α-complex **1**, which is then converted to a mixture of 4-nitrosodiphenylamine **(2)** and 4-nitrodiphenylamine **(3)**. The ratio of **2** and **3**

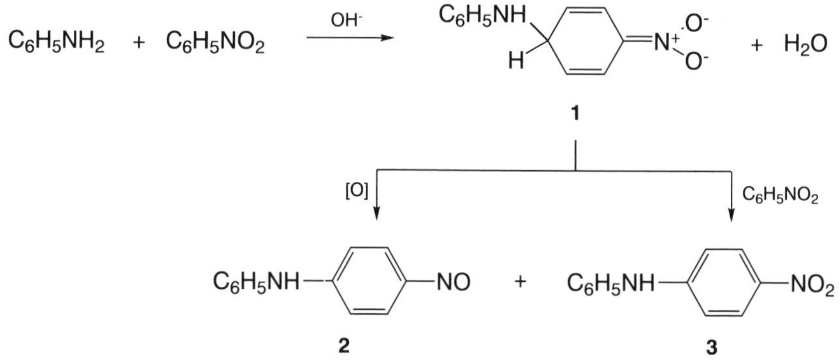
OH⁻
[O]
$C_6H_5NO_2$
1
2
3

depends on the ratio of aniline and nitrobenzene. As the ratio of nitrobenzene is increased, the yield of **3** is increased. Evidently nitrobenzene can act as an intermolecular oxidizing agent of **1** to give **3**.

[1] M. K. Stern, F. D. Hileman, and J. K. Bashkin, *Am. Soc.,* **114**, 9237 (1992).

Antimony(V) chloride–Silver hexafluoroantimonate.
This combination generates $SbCl_4^+ \cdot SbF_6^-$ (1).

Beckmann rearrangement.   The antimony(V) salt **1** is an effective catalyst for Beckmann rearrangement of ketoxime trimethylsilyl ethers to amides or lactams.[1]

Pinacol rearrangement.[2]   This reaction is generally effected with at least one equiv. of conc. H_2SO_4. It can also be effected with the salt (**1**), prepared from $SbCl_5$ and $AgSbF_3$, in high yield under mild conditions. In addition only 20 mol% of **1** is required.

[1] T. Mukaiyama and T. Harada, *Chem. Letters*, 1653 (1991).
[2] **Ibid,** *idem,* 81 (1992).

2-Arenesulfonyl-3-aryloxaziridines, 11, 108; **12,** 392–393; **13,** 23–24; **14,** 22.

Oxidative desulfonylation; α-diketones.[1]   The potassium enolate of an α-sulfonyl ketone on oxidation with a typical oxaziridine of this type, 2-[(p-chlorophenyl)sulfonyl]-3-(p-chlorophenyl)oxaziridine (**1**), is converted into an α-diketone in 78–85% yield. Use of lithium or sodium bases results in poor yields. Addition of HMPA or 18-crown-6 can be beneficial. The procedure is applicable to cyclic and acyclic systems and is compatible with olefins, even allenic groups.

1, Ar = 4-ClC₆H₄

1) KOC(CH₃)₃
2) **1**, THF, -78°

81%

¹ D. R. Williams, L. A. Robinson, G. S. Amato, and M. Y. Osterhout, *J. Org.*, **57**, 3740 (1992).

Arene(tricarbonyl)chromium complexes.

Nucleophilic substitution.[1] Carbon nucleophiles react with the Cr(CO)₃ complex of indane followed by oxidative decomplexation (iodine) to effect highly selective or exclusive substitution α to the ring junction. Similar but less pronounced selectivity obtains in reactions of the Cr(CO)₃ complex of 1,2,3,4-tetrahydronaphthalene. Good but

1) LiC(CH₃)₂CN
2) I₂

90%

91:9

80%

73:27

85%

20:80

opposite selectivity can be obtained even in substitution of *o*-xylene. Evidently some steric effects are operative as well as electronic effects.

Stereoselective benzylic substitution reactions.[2] A novel synthesis of the phenolic diterpene dihydroxyserrulatic acid (**6**) depends on highly regio- and stereoselective reactions of a tetralin(tricarbonyl)chromium complex with nucleophiles. Thus the Cr(CO)₃ complex **1** is converted in high yield to the *endo*-acetate **2** on hydride reduction followed by acetylation. This product reacts with an (E)-crotylsilane catalyzed by BF₃ etherate to give the *exo*-isobutenylated tetralin complex (**3**) with the desired *syn*-selectivity at

C_{11}. The carbonyl group of **3** is converted to an *endo*-methyl group by addition of CH_3Li followed by ionic hydrogenolysis to provide the complex **4**. Reaction of **4** with 2-lithio-1,3-dithiane results in substitution *meta* to the methyoxy group to provide **5** after demetalation. This product has the desired *trans*-arrangement of the groups at C_1 and C_4 as well as the precursor for the carboxyl group at C_6 for the complete synthesis of the diterpene **6**. The conversion of **5** into **6** was completed in four conventional steps.

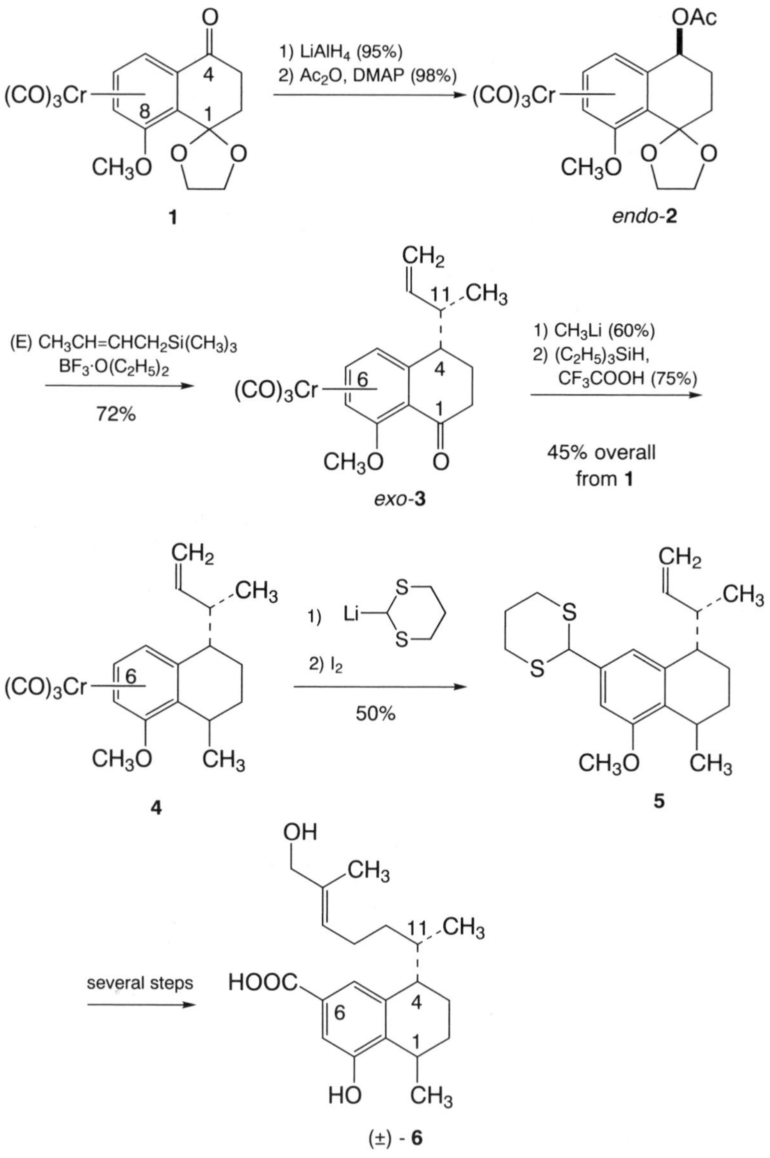

Asymmetric reaction of the (R)-(−)-tolualdehyde(tricarbonyl)chromium complex, (R)-1, with tosylmethyl isocyanide (TosMic).[3] This complex reacts with TosMic in K_2CO_3/CH_3OH at 0° to form only one (2) of the four possible products. Decomplexation followed by $LiAlH_4$ reduction provides (S)-3 in 100% de.

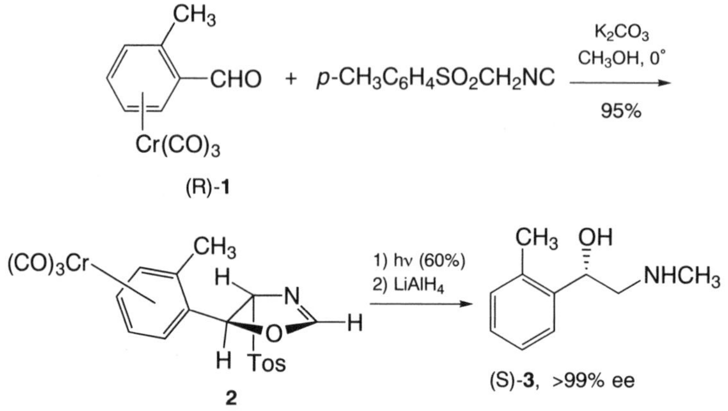

(R)-1

2

(S)-3, >99% ee

anti-Selective aldol reactions.[4] The racemic tricarbonyl[o-(trimethylsilyl)benz-aldehyde]chromium(0) complex 1 reacts with the acyclic o-silyl ketene acetal 2 in the presence of $TiCl_4$ in CH_2Cl_2 at −78° followed by decomplexation with CAN to form a mixture of *anti-* and *syn*-aldols in the ratio 71:29. The *anti*-selectivity is independent of the geometry of the silyl ketene acetal. Surprisingly, o-trimethylsilylbenzaldehyde itself also reacts with silyl ketene acetals with comparable *anti*-selectivity. The complexed aldehydes, however, have the advantage that they can be resolved to (+)- and (−)-1. Reactions of these complexed aldehydes can proceed in 92–98% ee and are also highly *anti*-selective.

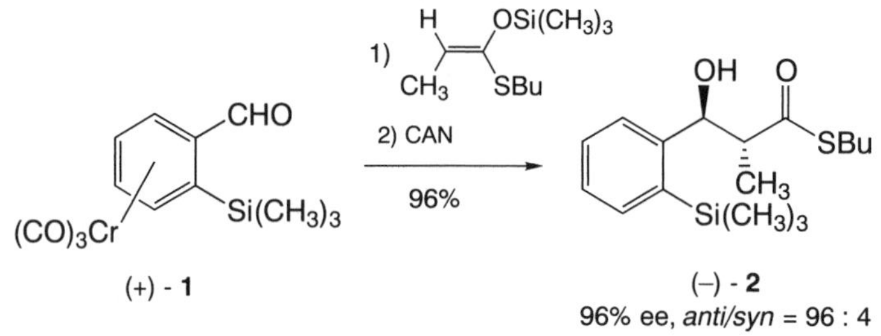

(+) - 1

(−) - 2
96% ee, *anti/syn* = 96 : 4

(−) - 1

(+) - 2
94% ee, anti/syn = 95 : 5

Enantioselective addition of $(C_2H_5)_2Zn$ *to RCHO.*[5]   Chiral complexes of the type **1** can be effective catalysts for this reaction.

+ **1a**	87%	S, 93% ee
+ **1b**	98%	S, 87% ee

Polycyclic arene(tricarbonyl)chromium complexes.[6]   These complexes are best prepared by treatment of polycyclic arenes with $(NH_3)_3Cr(CO)_3$[7] and BF_3 etherate. As in complexation with $Cr(CO)_6$, the terminal or most aromatic ring is complexed selectively. However, the lower temperatures used in the newer method are advantageous with thermally labile polycyclic arenes. These complexes are useful for substitution reactions at positions that are not available by electrophilic substitution of the arene directly. One such reaction is hydroxylation effected by simultaneous reaction with a base (BuLi or LDA) and tributoxyborane (excess) followed by H_2O_2/HOAc workup. Regioselective silylation is effected by reaction of the complex with LiTMP and $(CH_3)_3SiCl$ with

1) LiTMP, B(OBu)₃
2) H₂O₂, HOAc

59%

1

similar yield. Carboethoxylation is effected by reaction with ethyl chloroformate and a base. Since the carboethoxy group is smaller in size and electron-withdrawing, the reaction with the pyrene complex (1) gives a mixture of 2-carboethoxypyrene, 1,2-dicarboxyethoxypyrene, and 1,2,3-tricarboethoxypyrene.

2-Arylcarboxylic acids.[8] α-Lithiation of a tricarbonylchromium complex of an alkylarene followed by reaction with CO_2 provides complexes of α-arylcarboxylic acids.

[1] E. P. Kündig, C. Grivet, E. Wenger, G. Bernardinelli, and A. F. Williams, *Helv.*, **74**, 2009 (1991).
[2] M. Uemura, H. Nishimura, T. Minami, and Y. Hayashi, *Am. Soc.*, **113**, 5402 (1991).
[3] A. Solladié-Cavallo, S. Quazzotti, S. Colonna, A. Manfredi, J. Fischer, and A. DeCian, *Tetrahedron: Asymmetry*, **3**, 287 (1992).
[4] C. Mukai, M. Miyakawa, A. Mihira, and M. Hanaoka, *J. Org.*, **57**, 2034 (1992).
[5] M. Uemura, R. Miyake, and Y. Hayashi, *J. C. S. Chem. Comm.*, 1696 (1991).
[6] J. A. Morley and N. F. Woolsey, *J. Org.*, **57**, 6487 (1992).
[7] G. A. Razuvaev et al., *Organomet. Chem.*, **111**, 131 (1976).
[8] V. N. Kalinin, I. A. Cherepanov, and S. K. Moiseev, *Mendeleev Comm.*, 113 (1992).

5-Aza-semicorrins, chiral.

These ligands are generally prepared from D- or L- pyroglutamic acid. An attractive short synthesis is shown in equation I. Several 5 aza-semicorrins such as **1** are known and differ only in the nature of the substituent (R) at the stereogenic centers.

Enantioselective cyclopropanation and allylic alkylation.[1] These semicorrins, particularly when treated with CuOTf, are comparable to bis(oxazoline) ligands (16, 38–41) for cyclopropanation.

94% ee   75:25   68% ee

They appear to be more useful than bis(oxazolines) as ligands for palladium-catalyzed allylic alkylation with dimethyl malonate/BSA [N,O-bis(trimethylsilyl)acetamide].

95% ee

R=Si(CH$_3$)$_2$tBu

2

[1] U. Leutenegger, G. Umbricht, C. Fahrni, P. von Matt, and A. Pfaltz, *Tetrahedron,* **48**, 2143 (1992).

Azidotrimethylsilane–Chromic anhydride.

Acyl azides.[1] A mixture of N$_3$Si(CH$_3$)$_3$ and CrO$_3$ (1:1) in CH$_2$Cl$_2$ effects conversion of aldehydes to acyl azides in 70–90% isolated yields. In the case of aryl aldehydes, the reaction is conducted at 25°, but the conversion of aliphatic aldehydes is best conducted at −10° to prevent rearrangement of the acyl azide to the alkyl isocyanate.

$$C_6H_5CHO \xrightarrow[87\%]{\substack{N_3Si(CH_3)_3,\ CrO_3 \\ CH_2Cl_2,\ 25°}} C_6H_5CON_3$$

$$CH_3(CH_2)_6CHO \xrightarrow[74\%]{CH_2Cl_2,\ -10°} CH_3(CH_2)_6CON_3$$

[1] J. G. Lee and K. H. Kwak, *Tetrahedron Letters,* **33**, 3165 (1992).

Azidotrimethylsilane–Trifluoromethansulfonic acid.

Azidation of alkenes.[1] Trifluoromethanesulfonic acid is an efficient catalyst for addition of hydrazoic acid to alkenes. The reaction is also facilitated by silica gel or alumina. A further advantage is that these adsorbents can effect *in situ* hydrolysis of azidotrimethylsilane to hydrazoic acid, HN_3. This new procedure is particularly useful for preparation of tertiary, benzylic, and allylic azides.

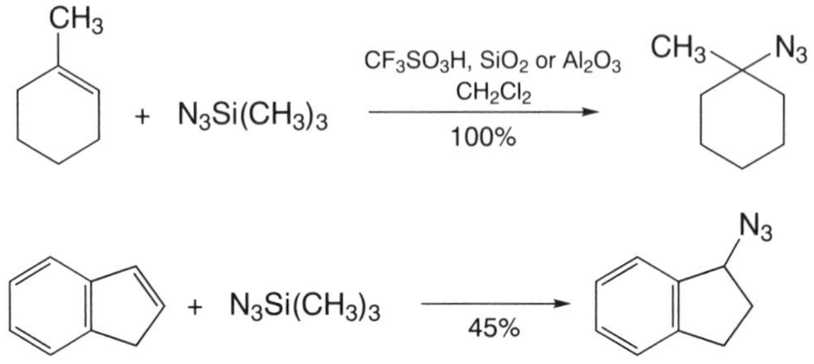

[1]2G. W. Breton, K. A. Daus, and P. J. Kropp, *J. Org.,* **57**, 6646 (1992).

B

Barium, activated (Ba*).

Coupling of allylic halides.[1] In the presence of Ba*, prepared *in situ* by reduction of BaI$_2$ by lithium biphenylide, *cis*- or *trans*-allylic halides undergo, α,α'-coupling with retention of the double bond geometry. Unsymmetrical dienes can be prepared by α,α'-cross coupling. Thus (E,E)-farnesyl chloride is converted into squalene in 64% yield.

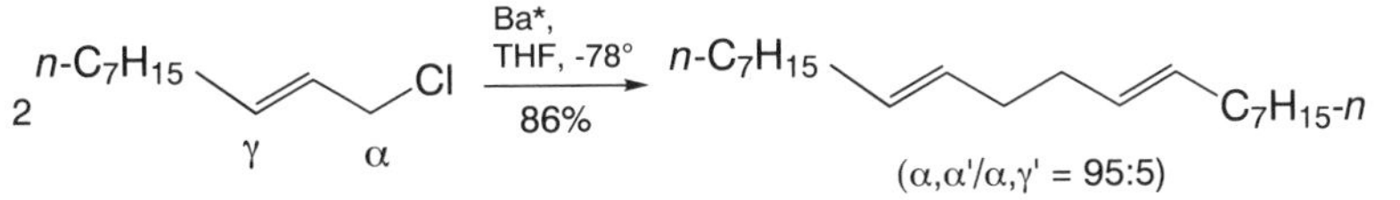

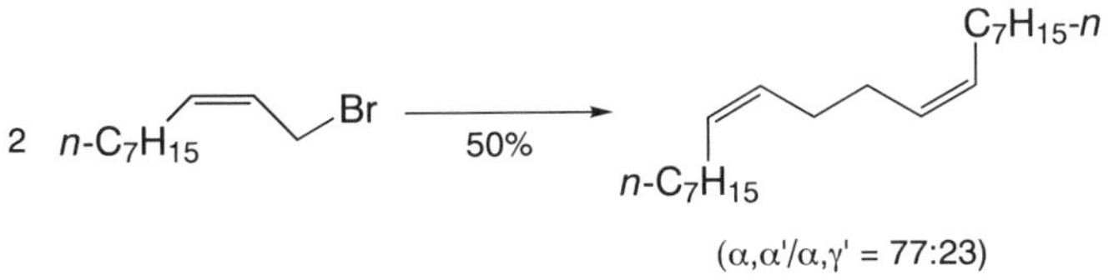

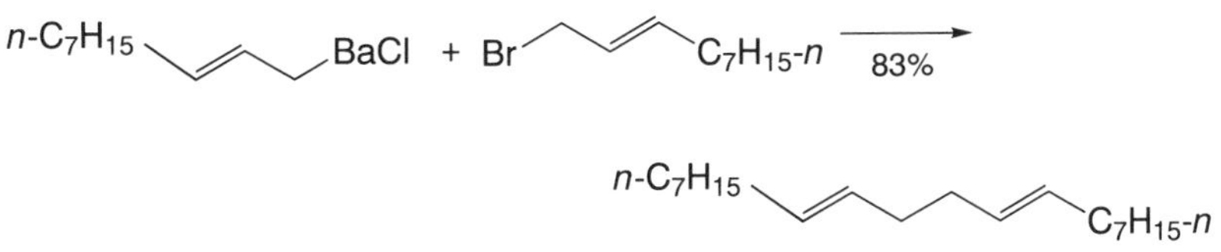

Allyl-allyl coupling.[2] The reaction of (E)-prenyl chlorides with BaI$_2 \cdot$ 2H$_2$O and lithium biphenylide (2 equiv.) to form allylbarium reagents has been used for an allyl/allyl coupling to form diisoprenoids. Thus the allylic alcohol **1**, prepared by allylic hydroxylation of geraniol, was converted into the corresponding bromide **2**. This bromide couples with geranylbarium (**3**) to give **4** (61% yield) with only primary/primary coupling with preservation of the geometry of both allyl units.

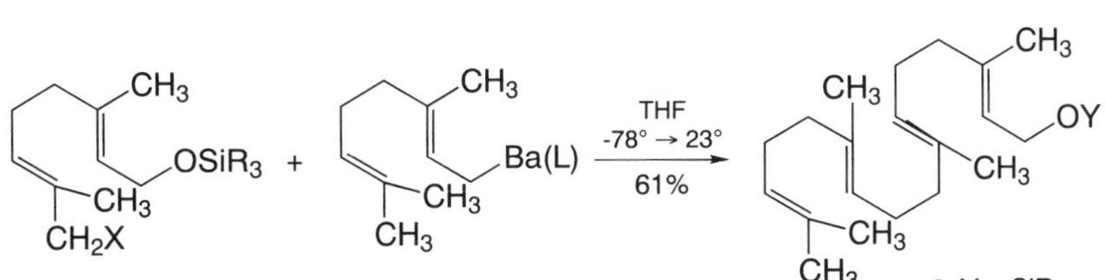

1, X = OH
2, X = Br

3

4, Y = SiR$_3$
5, Y = H

Coupling of (E,E)-farnesyl bromide with farnesylbarium provides all (E)-squalene (79% yield).

[1] A. Yanagisawa, S. Habaue, and H. Yamamoto, *Am. Soc.,* **113**, 5893, 8955 (1991); A. Yanagisawa, H. Hibino, S. Habane, Y. Hisada, and H. Yamamoto, *J. Org.,* **57**, 6386 (1992).
[2] E. J. Corey and W.-C. Shieh, *Tetrahedron Letters,* **33**, 6435 (1992).

1,2-[(Benzenediolato)-O,O']oxotitanium,

1 (16; 19)

Stereoselective 1,2-cisglycosylation of tribenzyl-D-arabinofuranose (2).[1] The reaction of **2** with various trimethylsilyl ethers when catalyzed by **1** and trimethylsilyl triflate results in stereoselective formation of 1,2-*cis*-arabinofuranosides. (**3**).

[1] T. Mukaiyama, M. Yamada, S. Suda, Y. Yokomizo, and S. Kobayashi, *Chem. Letters,* 1401 (1992).

Benzeneselenenyl chloride.

cis-2,5-Disubstituted tetrahydrofurans.[1]   Cyclization of triethylsilyl ethers of 3-hydroxyalkenes with C_6H_5SeCl in the presence of K_2CO_3 provides *cis*-2,5-disubstituted tetrahydrofurans with high stereoselectivity. Other protecting groups show generally low stereoselectivity.

		cis / trans
R = CH$_2$OTBDMS	92%	>100 : 1
R = CH(CH$_3$)$_2$	83%	>100 : 1
R = CH$_3$	91%	15 : 1

[1] S. H. Kang, T. S. Hwang, W. J. Kim, and J. K. Lim, *Tetrahedron Letters,* **32**, 4015 (1991).

Benzylchlorobis(triphenylphosphine)palladium(II).

Stille macrocyclization.   Several macrocylic antibiotics contain a γ-oxo-α,β-unsaturated ester group. Macrolides of this type can be obtained by Stille intramolecular coupling.[1] The substrates are available from Mitsunobu esterification of propriolic acid with an ω-hydroxy ester followed by hydrostannylation to give β-stannyl alkenoates **1** as a 1:1 mixture of (Z)- and (E)-isomers. Selective saponification of the methyl ester

1 (Z/E = 1 : 1)

| **2** | R = H | 53% |
| | R = CH$_3$ | 70% |

(E) - 3

with LiOH provides the free acid which is converted into the acid chloride to give **2** in >90% yield. Intramolecular coupling of **2** to form **3** is effected with a Pd(II) catalyst under 3 atms. of carbon monoxide at high dilution in toluene. This cross coupling can be used to obtain 12- to 20-membered rings. Dimeric macrolides are formed as the main or only products in the case of shorter precursors.

[1] J. E. Baldwin, R. M. Adlington, and S. H. Ramcharitar, *J. C. S. Chem. Comm.*, 940 (1991).

N-(Benzyldimethylsilyl)methyl-2-(+)-methoxymethyl)pyrrolidine.

(S)-(+)-1

Asymmetric synthesis of 1,3-diols.[1]   The anion (*sec*-BuLi) of (S)-**1** reacts with ethylene oxide to give a single product **2**. Oxidative cleavage of the C-Si bond gives the 1,3-diol (S)-(−)-**3** in >99% ee. Similar reaction of the anion of **1** with *t*-butyloxirane or 1,2-epoxyhexane gives a mixture of two diastereomers, which are separable by chromatography and converted to *syn*- and *anti*-1,3-diols on oxidative cleavage. In these reactions as in the reaction with aliphatic epoxides, the 1,3-diols have the (1S)-configuration.

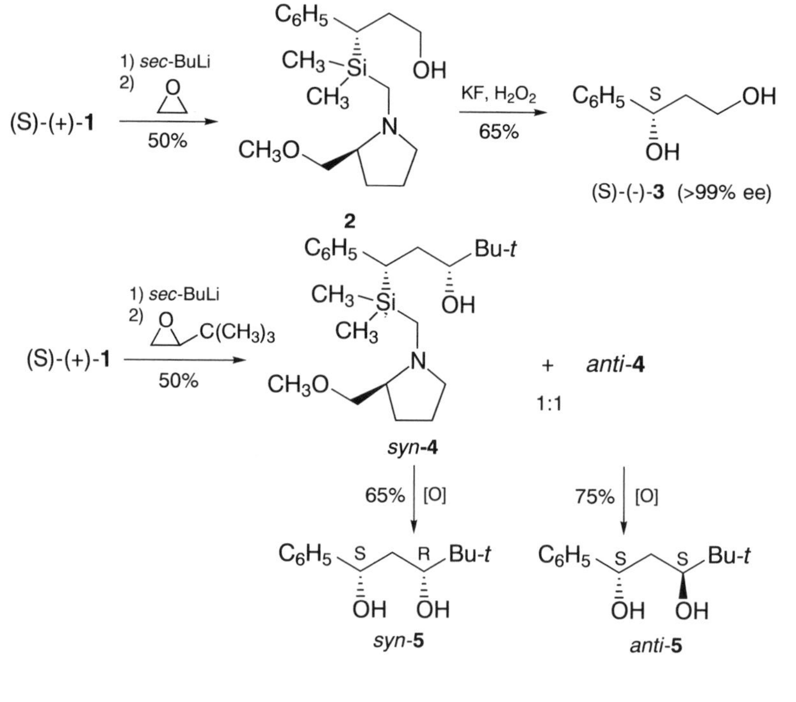

[1] T. H. Chan and K. T. Nwe, *J. Org.*, **57**, 6107 (1992).

1,1'-Bi-2,2'-naphthol–Dichlorodiisopropoxytitanium (1), **15**, 26–27; **16**, 24–25.

Enantioselective ene cyclization of unsaturated aldehydes.[1]   The titanium complex (R)-**1a** can effect enantioselective catalytic cyclization of **2** and **3**. Substitution of the chlorine atoms by perchlorate improves the enantioselectivity significantly.

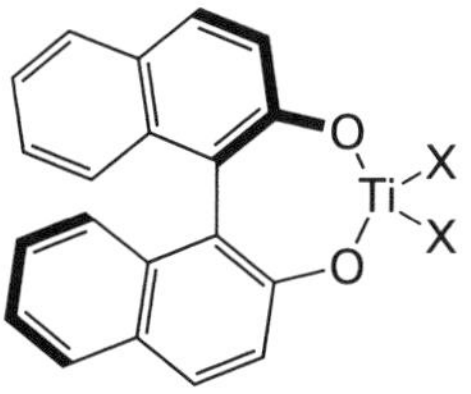

(R)-**1a**, X = Cl
(R)-**1b**, X = ClO$_4$

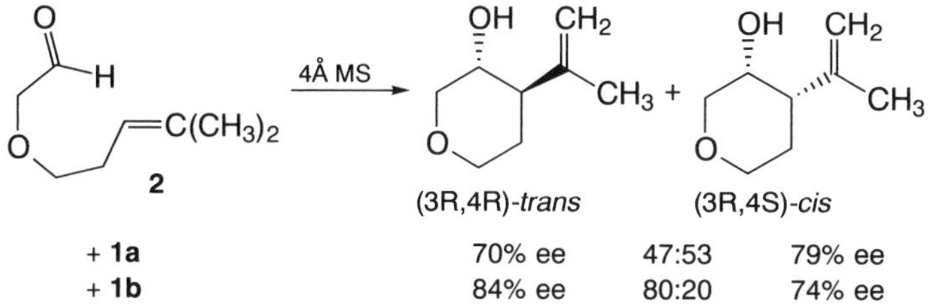

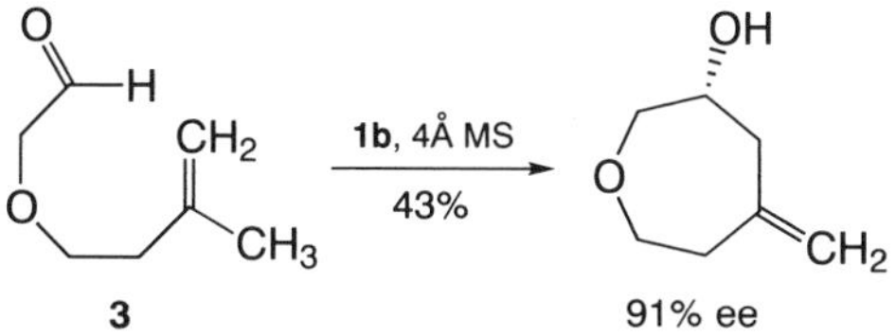

	(3R,4R)-*trans*	47:53	(3R,4S)-*cis*
+ **1a**	70% ee	47:53	79% ee
+ **1b**	84% ee	80:20	74% ee

3   91% ee

Asymmetric carbonyl-ene reaction with 1.[2]   The reaction of racemic allylic alcohols (**2**) with methyl glyoxalate catalyzed by (R)-**1** can proceed with remote asymmetric induction, which suggests that (R)-**1** can discriminate between the two ene components.

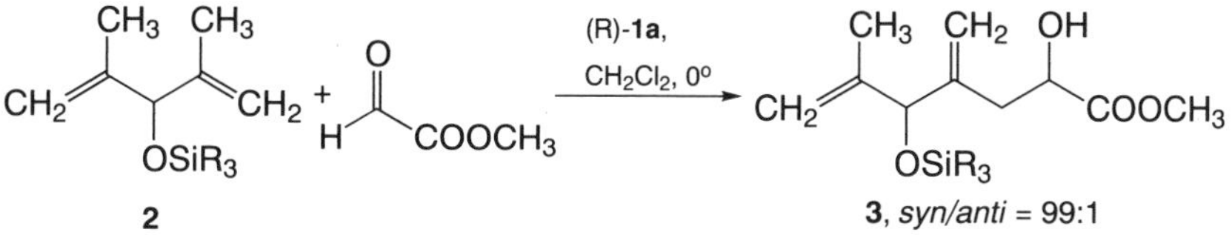

2   **3**, *syn/anti* = 99:1

Asymmetric glyoxylate ene reaction.[3]   The reaction of exocyclic alkenes (**2**) with methyl glyoxylate in the presence of (R)-**1** and molecular sieves proceeds in 63–99% ee.

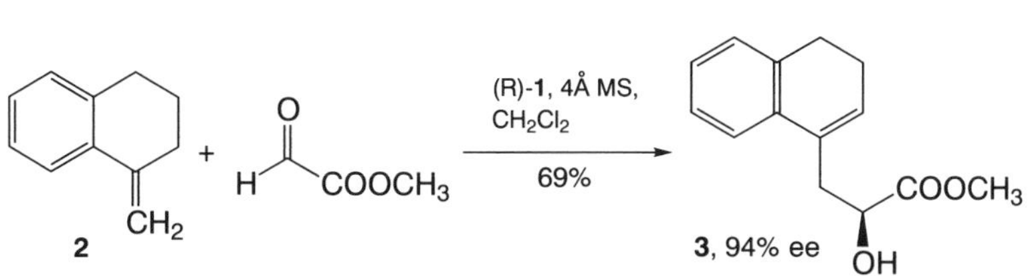

2   **3**, 94% ee

[1] K. Mikami, E. Sawa, M. Terada, and T. Nakai, *Tetrahedron Letters,* **32**, 6571 (1991); K. Mikami and M. Shimizu, *Chem. Rev.,* **92**, 1021 (1992).
[2] K. Mikami, S. Narisawa, M. Shimizu, and M. Terada, *Am. Soc.,* **114**, 6566 (1992).
[3] F. T. van der Meer and B. L. Feringa, *Tetrahedron Letters,* **33**, 6695 (1992).

Bis(acetonitrile)dichloropalladium(II).

Oxygenation of allylic amides.[1]   Oxygenation of allylic amides (or lactams) in the presence of a Pd(II)–CuCl catalyst in 1,2-dichloroethane containing HMPA (essential) results in an aldehyde. In the presence of water, the oxygenation results mainly in a methyl ketone.

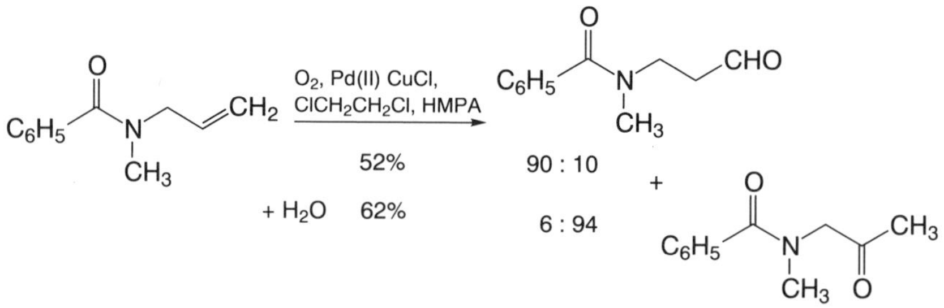

The allyl acetate **1** is oxidized under these anhydrous conditions to an aldehyde (**2**) in reasonable yields.

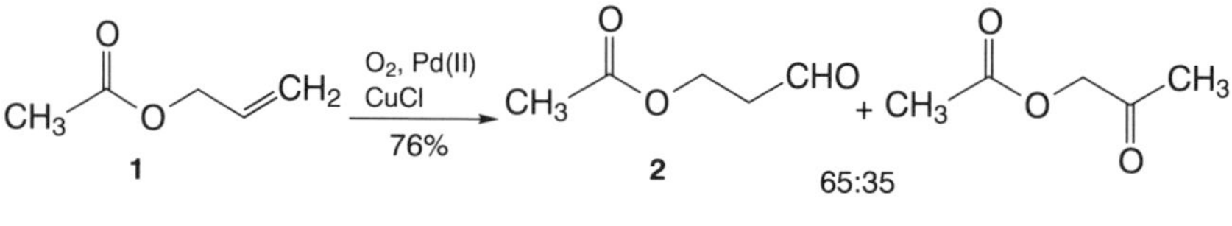

Amidation of electron-deficient alkenes.[2]   In the presence of $Cl_2(CH_3CN)_2Pd$ and CuCl (1:1) amides react with 1-alkenes bearing COOR, COR, CHO, or $CONR_2$ groups to form (E)-enamides in generally good yield. Cyclic amides are more reactive than open-chain amides. Carbamates are particularly reactive.

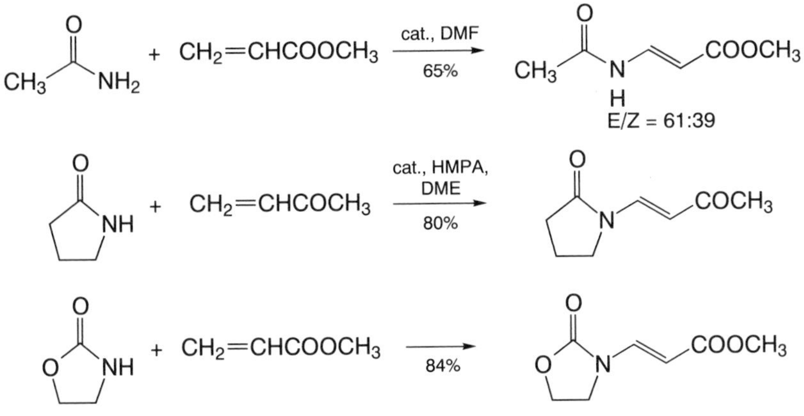

E/Z = 61:39

[1] T. Hosokawa, S. Aoki, M. Takano, T. Nakahira, Y. Yoshida, and S.-I. Murahashi, *J. C. S. Chem. Comm.*, 1559 (1991).
[2] T. Hosokawa, M. Takano, Y. Kuroki, and S.-I. Murahashi, *Tetrahedron Letters*, **33**, 6643 (1992).

Bis[(allyl)trifluoroacetato]nickel(II), 1.

Poly(p-phenylene), 5.[1]   This catalyst has been used to polymerize 1,3-butadiene with >97% 1,4-regioselectivity. Attempts to polymerize phenylenes directly results in mixtures of *para-*, *meta-*, and *ortho*-linked oligomers. In contrast polymerization of *cis*-5,6-bis(trimethylsilyloxy)-1,3-cyclohexadiene 2 with 1 in C_6H_5Cl results in exclusive formation of the 1,4-poly-1,3-cyclohexadiene 3. This polymer is converted into poly(*p*-phenylene) (5) by conversion of the trimethylsilyloxy groups to acetyl groups followed by pyrolysis.

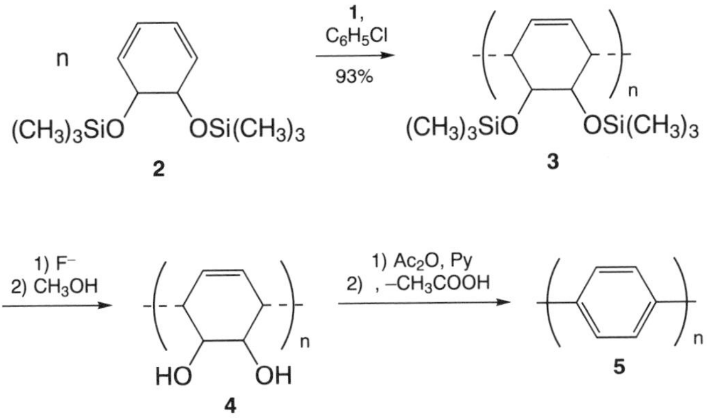

[1] D. L. Gin, V. P. Conticello, and R. H. Grubbs, *Am. Soc.*, **114**, 3167 (1992).

Bis(cyclooctadiene)nickel.

Phenol synthesis. In the presence of this Ni(0) catalyst cyclobutenones can couple with internal alkynes to form phenols in 50–80% yield. Two phenols are obtained from unsymmetrical alkynes with only slight regioselectivity based on the size of the alkyl substituents. Modest selectivity obtains with oxygen-substituted alkynes.

$R^1, R^2 = C_2H_5$	75%
$R^1 = CH_3, R^2 = (CH_2)_2OTHP$	47% 65 : 35

This phenol synthesis differs from the thermal reaction of cyclobutenones with alkynes (**15**, 160) in that activated alkynes are not required and the ultimate positions of the alkyne substituents in the phenol are different.

[1] M. A. Huffman and L. S. Liebeskind, *Am. Soc.*, **113**, 2771 (1991).

Bis(1,5-cyclooctadiene)nickel(0) – Chlorotrialkylsilane, Ni(COD)₂—ClSiR₃.

Conjugate addition of alkenyltributylstannanes to α, β-enals. The reaction can be effected with a chlorotrialkylsilane (1 equiv.) and a catalytic amount of Ni(COD)₂ in DMF and results in an (E)-silyl enol ether in 48–79% yield.[1] The reaction evidently involves a 1-silyloxyallylnickel(II) intermediate (**a**).

$$CH_2{=}CHCHO \ + \ CH_2{=}CHSnBu_3 \ + \ ClSiR_3 \xrightarrow[\text{DMF, 25°}]{\text{Ni(0)}}$$

a   E/Z = > 50:1

[1] B. A. Grisso, J. R. Johnson, and P. B. Mackenzie, *Am. Soc.*, **114**, 5160 (1992).

Bis(1,5-cyclooctadiene)nickel – Triphenylphosphine.

[2 + 2 + 2] Homo-Diels–Alder reactions of norbornadiene.[1] In the presence of Ni(COD)₂ and P(C₆H₅)₃ (1:2) norbornadienes can react with activated alkenes at 80° to give [2 + 2 + 2] adducts. Preliminary studies of the regioselectivity indicate that this homo-Diels–Alder reaction can be useful for synthesis of complex molecules. Thus reaction of methyl norbornadiene-2–carboxylate with acrylonitrile results in only two

products (*endo/exo* = 2.3:1), in which the electron-withdrawing groups are *para* to the diene substituent (equation I). Only two products are formed from reaction of the same

(I)

CH$_3$OOC

+

CH$_2$=CHCN

Ni(COD)$_2$
P(C$_6$H$_5$)$_3$
ClCH$_2$CH$_2$Cl, 80°

94%

CH$_3$OOC

`CN

endo-para

+

CH$_3$OOC

CN

exo-para

2.3 : 1

(II)

CH$_3$OOC

+

CH$_2$=CHSO$_2$C$_6$H$_5$

75%

CH$_3$OOC

SO$_2$C$_6$H$_5$

exo-para

+

CH$_3$OOC

C$_6$H$_5$SO$_2$

exo-meta

2 : 1

diene with phenyl vinyl sulfone, the *exo-para* and *exo-meta* isomers in a ratio of 2:1.

The presence of a methoxy group on the diene lowers the reactivity, but 2-methoxynorbornadiene can react with methyl vinyl ketone to provide a single *endo-ortho* adduct (equation III).

(III)

—OCH$_3$ +

CH$_2$=

CH$_3$

=O

51%

—OCH$_3$

`COCH$_3$

endo-ortho

These preliminary results indicate that substituent groups can exert a marked effect on the regioselectivity of this homo-Diels–Alder reaction with a general preference for *ortho-* and *para*-isomers.

[1] M. Lautens and L. G. Edwards, *J. Org.*, **56**, 3761 (1991).

Bis(cyclopentadienyl)chloro(hydrido)zirconium, Cp$_2$ZrHCl.

β-Keto esters → enoates.[1]   The enolates of β-keto esters, prepared with lithium hexamethyldisilazane in DME, on reaction with the Schwartz reagent (1 equiv.) at 0–20° are converted into enoates in 55–70% yield.

[1] A. G. Godfrey and B. Ganem, *Tetrahedron Letters*, **33**, 7461 (1992).

Bis(2,6-diphenylphenoxide)tris(4-methylbenzyl)niobium(V), 1.[1]

All-cis Hydrogenation of polynuclear arenes.[2]   This niobium aryloxide is an efficient catalyst for regioselective hydrogenation of polynuclear arenes. Thus phenanthrene is reduced to the 9,10-dihydro derivative and an octahydrophenanthrene. Anthracene is

converted to an octahydro derivative. Analysis of the ^{1}H NMR spectra indicates that the hydrogen atoms are all introduced on the same face in the case of naphthalene, anthracene, and acenaphthene.

[1] R. W. Chesnut, G. G. Jacob, J. S. Yu, P. E. Fanwick, I. P. Rothwell, *Organometallics*, **10**, 321 (1991).

[2] J. S. Yu, B. C. Ankianiec, M. T. Nguyen, and I. P. Rothwell, *Am. Soc.*, **114**, 1927 (1992).

2,2′-Bis(diphenylphosphino)-1,1′binaphthyl (BINAP, 1).

Asymmetric hydrogenation.[1]   Hydrogenation of the α-substituted β-keto ester **2** with a Ru(II) catalyst coupled with (R)-1 provides the desired β-hydroxy ester **3** in 90% ee but with only moderate *syn*-selectivity (74% de). Hydrogenation with high enantio- and diastereoselectivity can be achieved using a BINAP derivative in which the phenyl groups of the phosphine are replaced by 3,5-di-t-butyl phenyl groups. (2S,3R)-**3** is a useful percursor to the β-lactam **4**.

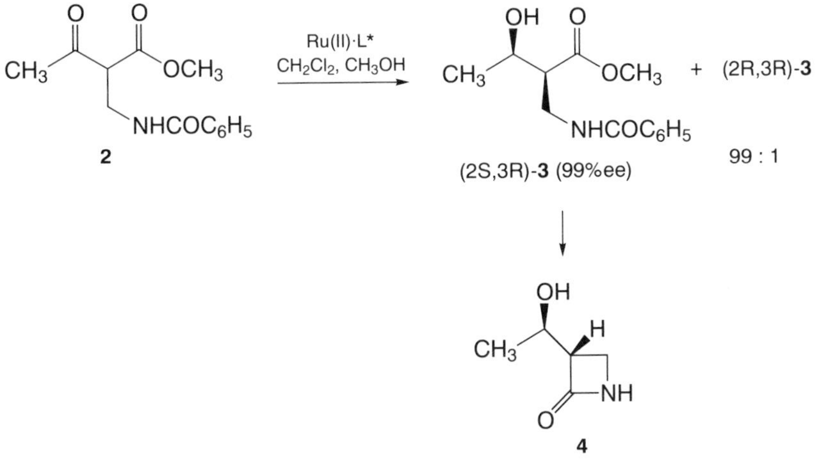

Asymmetric hydrogenation of 2- and 4-alkylidene-γ-butyro-lactones.[2] This reaction can be effected by catalysis with Ru(II)/(S)- or (R)-BINAP in 94–98% ee regardless of the geometry of the olefinic group. This method is also applicable to 2-alkylidenecyclopentanes.

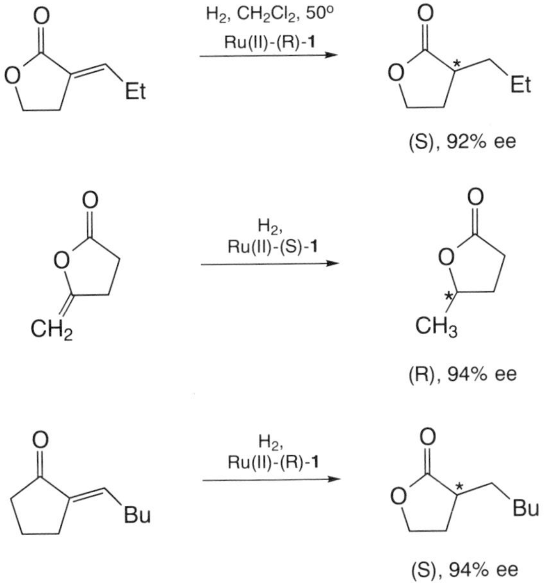

(R)- or (S)-2,2¹-Bis(diphenylphosphino)1,1′-binaphthyl (BINAP)

Asymmetric Heck reaction.[3]   Treatment of the vinyl triflate **1** with tetrabutylammonium acetate (1.7 equiv.) and a catalytic amount of Pd(OAc)$_2$ and (S)-BINAP in DMSO at 20° results in the fused bicyclic pentanoid **2** in 80% ee, formed by a Heck reaction followed by an anion capture process. This product can be converted into **3**, an

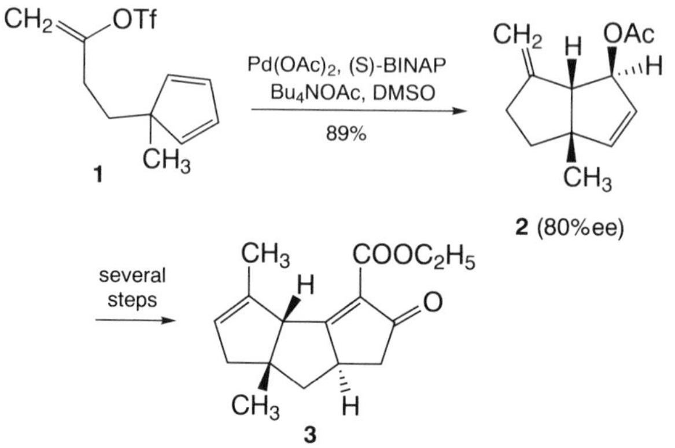

intermediate in the synthesis of several sesquiterpenes.

Catalytic asymmetric arylation of alkenes.[4]   This arylation can be effected by a Heck-type arylation of 2,3-dihydrofuran with an aryl triflate catalyzed by Pd(OAc)$_2$/(R)-BINAP in the presence of 1,8-bis(dimethylamine)naphthalene (proton sponge) as base. This reaction was used to prepare an antagonist (**4**) of platelet activating factor.

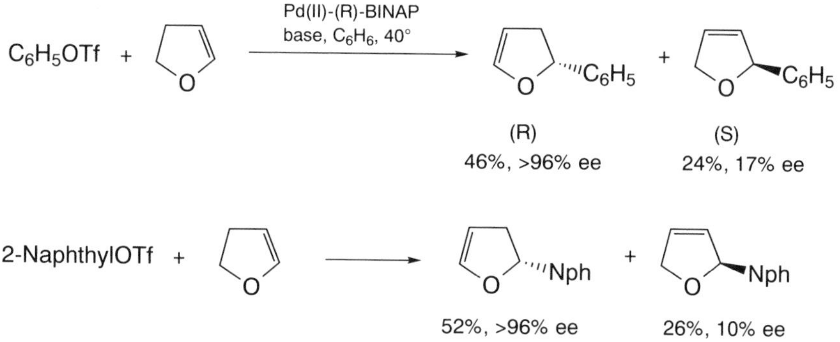

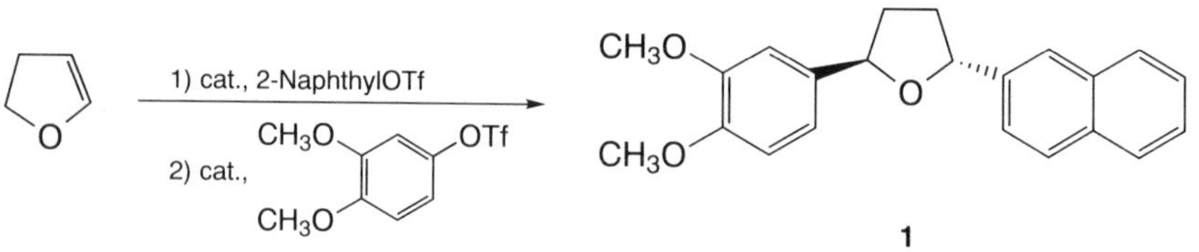

Chiral spirooxindoles.[5]   An asymmetric Heck reaction of the acryloyl 2^1-iodoaniline (**2**) effected with tris(dibenzylideneacetone)dipalladium and (R)-**1** as the chiral ligand can provide either enantiomer of the spiroindole **3** depending upon an added promotor. Thus the presence of a silver salt provides (S)-(+)-**3**, whereas addition of a basic tertiary amine can favor formation of (R)-(−)-**3**. The most effective amine is 1,2,2,6,6-pentamethylpiperidine (PMP).

Pd$_2$(dba)$_3$, (R)-**1**, 80°
Ag$_3$PO$_4$, CH$_3$CON(CH$_3$)$_2$
81%

(S)-(+)-**3**, 71% ee

2

Pd$_2$(dba)$_3$, (R)-**1**, 80°
PMP, CH$_3$CON(CH$_3$)$_2$
77%

(R)-(-)-**3**, 66% ee

Enantioselective cyclization of 4-disubstituted 4-pentenals.[6]   Cyclization of 4-pentenals such as **2** to 3-substituted cyclopentanones (**3**) can be effected in >99% ee by use of Rh[(S)- or (R)-BINAP]ClO$_4$ as catalyst.

Rh[(S)-BINAP]ClO$_4$
CH$_2$Cl$_2$, 20°

2, R = *t*-Bu   87%   **3**, S, >99% ee
R = C(CH$_3$)$_2$CH$_2$CH$_3$ 84%   S, >99% ee
R = Bu   90%   S, 91% ee

β-Amino acids.[7]   A useful route to chiral β-amino acids involves enantioselective hydrogenation of methyl β-(acylamino)acrylates such as **1**. Thus hydrogenation of (E)-**1** catalyzed by a complex of $Ru(OCOCH_3)_2$ with (R)-BINAP furnishes (S)-2 in 96% ee. Surprisingly (Z)-1 is converted under the same conditions to (R)-2 in 5% ee. Hydro-

genation of either (E)- or (Z)-1 catalyzed by $Rh^+(CH_3OH)_2ClO_4{}^-$ complexed with (R)-BINAP results in (R)-2 but with modest enantioselectivity (45–60% ee).

Asymmetric carbonylation of alkenyl halides; α-methylene lactones.[8]   Optically active α-methylene lactones can be prepared by carbonylation of alkenyl halides such as 1 catalyzed by $PdCl_2$–(R)-BINAP (5 mol %) in the presence of thallium acetate.

BINAP-Ru(II) complexes.[9]   A suitable BINAP-Ru(II) catalyst for asymmetric hydrogenation of ketones (14, 40–41) can be prepared most readily by heating (R)- or (S)-BINAP with $[RuCl_2$ (benzene)$]_2$ in DMF at 100° to effect ligand exchange. A suitable catalyst can also be obtained by ligand exchange with $RuCl_2[Sb(C_6H_5)_3]_3$, prepared from $RuCl_3$ and $Sb(C_6H_5)_3$, in o-dichlorobenzene at 160°. Either of these catalysts gives satisfactory results in hydrogenations at 100 atm. at 25° or at 4 atm. at 100°.

[1] K. Mashima, Y. Matsumura, K. Kusano, H. Kumobayashi, N. Sayo, Y. Hori, T. Ishizaki, S. Akutagawa, and H. Takaya, *J. C. S. Chem. Comm.*, 609 (1991).

[2] T. Ohta, T. Miyake, N. Siedo, H. Kumobayashi, S. Akutagawa, and H. Takaya, *Tetrahedron Letters*, 33, 635 (1992).

[3] K. Kagechika and M. Shibasaki, *J. Org.*, 56, 4093 (1991).

[4] T. Hayashi, A. Kubo, and F. Ozawa, *Pure Appl. Chem.*, 64, 421 (1992).

[5] A. Ashimori and L. E. Overman, *J. Org.*, 57, 4571 (1992).

[6] X.-M. Wu, K. Funakoshi, and K. Sakai, *Tetrahedron Letters*, 33, 6331 (1992).

[7] W. D. Lubell, M. Kitamura, and R. Noyori, *Tetrahedron Asymmetry*, 2, 543 (1991).

[8] T. Suzuki, Y. Uozumi, and M. Shibasaki, *J. C. S. Chem. Comm.*, 1593 (1991).

[9] M. Kitamura, M. Tokunaga, T. Ohkuma, and R. Noyori, *Tetrahedron Letters*, 32, 4163 (1991).

1,4-Bis(diphenylphosphine)butane(norbornadiene)rhodium(I) tetrafluoroborate, 1.[1]

Diastereoselective hydrogenation of γ-hydroxyvinylsilanes and -tins, Syn-1,3-diols.[1] This catalyst is known to effect diastereoselective hydrogenation of chiral allylic alcohols (**12**, 426–428). It can also effect diastereoselective hydrogenation of vinyltins and vinylsilanes having a γ-hydroxyl group (equations I and II).

(I)

syn (>100:1)

(II)

syn (>500:1)

68% KBr, HOOAc NaOAc, HOAc

syn - diol

[1] M. Lautens, C.-H. Zhang, and C. M. Crudden, *Angew. Chem. Int. Ed.*, **31**, 232 (1992).

[1,4-Bis(diphenylphosphino)butane]norbornadienerhodium trifluoromethanesulfonate, 1(14, 44–45).

Diastereoselective hydrogenation of vinyl sulfones and sulfoxides.[1] Hydrogenation of the vinyl sulfone **2** with the Rh catalyst **1** proceeds in 99% de to give **3**. Similar hydrogenation of the corresponding sulfoxide provides **5**, which on perborate oxidation gives the *threo*-isomer of **3**. Apparently the steric course of hydrogen-

2   anti -3, 99% de

(R,R) - **4**   syn - **5**, 99% de   syn - **3**

ation of sulfones is controlled by the hydroxy group, whereas the configuration at sulfur controls the course of hydrogenation of sulfoxides.

[1] D. Ando, C. Bevan, J. M. Brown, and D. W. Price, *J. C. S. Chem. Comm.*, 592 (1992).

2,2''-Bis[1-(diphenylphosphino)ethyl]-1,1''-biferrocene (TRAP).

(S,S)-(R,R)-TRAP

Asymmetric Michael addition.[1]   A catalyst prepared from $RhH(CO)[P(C_6H_5)_3]_3$ and TRAP can effect asymmetric addition of α-cyano carboxylates to vinyl ketones or acrolein. Note that isopropyl esters show the highest selectivity.

[1] M. Sawamura, H. Hamashima, and Y. Ito, *Am. Soc.*, **114**, 8295 (1992).

Bis(dipivaloylmethanato)nickel(II), $Ni(dpm)_2$.

Baeyer–Villiger oxidation.[1]   Complexes of nickel coordinated with several 1,3-diketones are effective catalysts for oxidations with molecular oxygen and an aldehyde or alcohol reductant (16, 30–32). Such oxygenations are much safer than those involving the usual oxidants such as hydrogen peroxide or a peracid. Such a catalyzed oxygenation can be used for Baeyer–Villiger oxidation of either cyclic or acyclic ketones. Highest yields are obtained with bis(dipivaloylmethanato)nickel(II) as catalyst (1 mol%) and isovaleraldehyde as reductant.

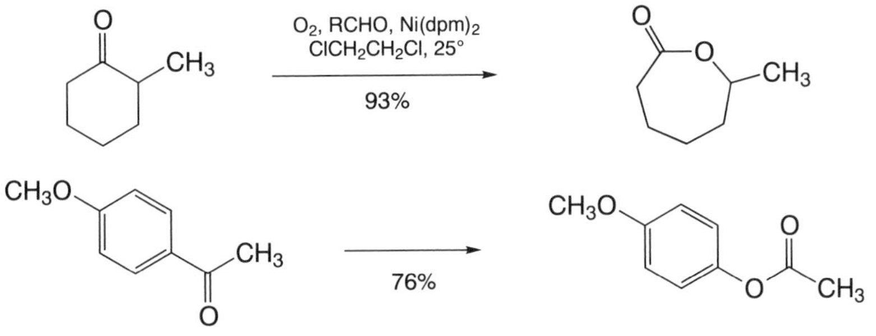

[1] T. Yamada, K. Takahashi, K. Kato, T. Takai, S. Inoki, and T. Mukaiyama, *Chem. Letters,* 641 (1991).

[*N,N'*-Bis(4-methoxysalicylidene)-*o*-phenylenediamine]cobalt(II),

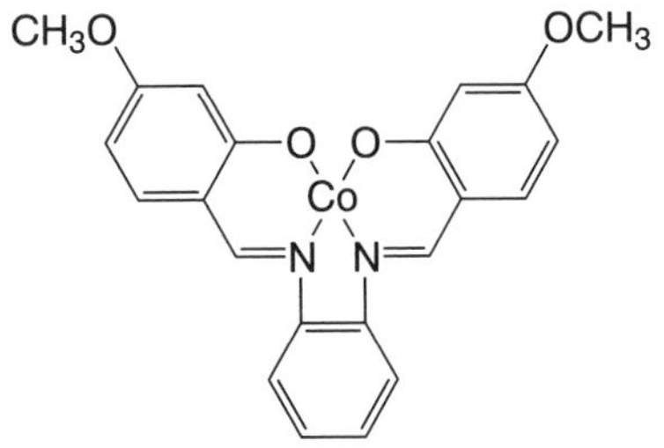

(1)

Expoxidation of alkenes.[1]   In the presence of this complex, alkenes undergo epoxidation with oxygen (oxidant) and cyclic ketones as reductant. The most useful reductant is 2-methylcyclohexanone. Yields are ~80% for trisubstituted or *exo*-terminal alkenes.

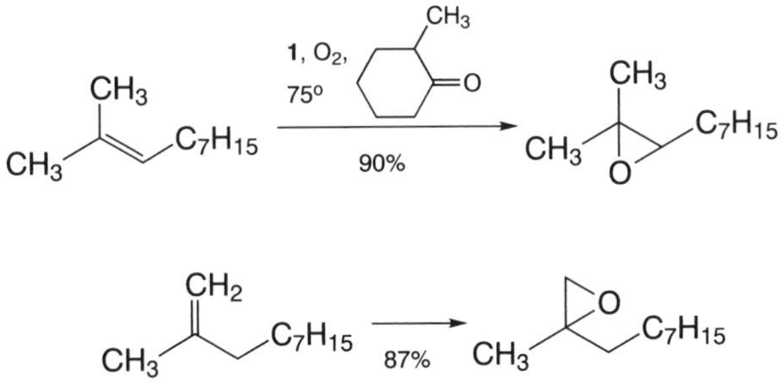

[1] T. Takai, E. Hata, K. Yorozu, and T. Mukaiyama, *Chem. Letters,* 2077 (1992).

1,2-Bis(methylene)cycloalkane–Magnesium, 16, 198.

β,γ-Enones.   Reaction of 1,2-bis(methylene)cyclohexane with activated magnesium provides a magnesium complex (1). This complex reacts with ethyl acetate at −10°

to form a spiro enol containing a cyclopropane ring (**a**). This intermediate when heated undergoes ring expansion and protonation to form **2**.

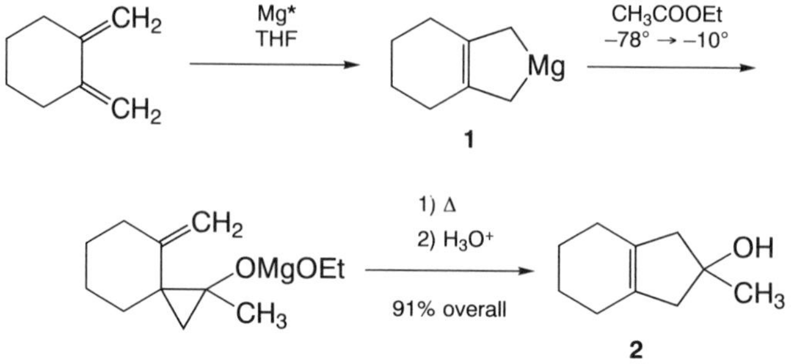

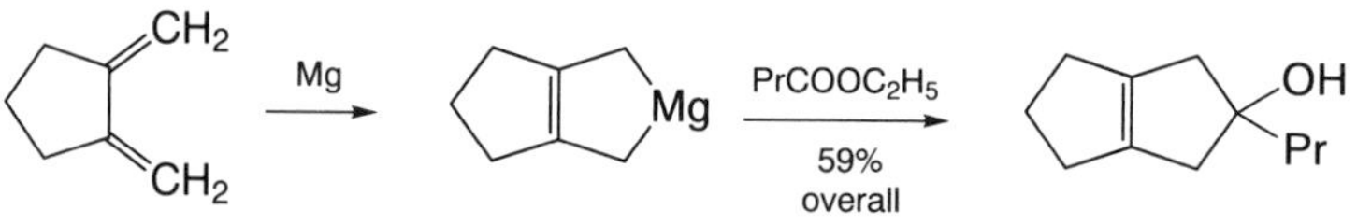

[1] H. Xiong and R. D. Rieke, *Am. Soc.*, **114**, 4415 (1992).

Bis(oxazolines), 16, 38–41.

Asymmetric hydrosilylation.[1]   Bis(oxazolines) (**1**) are easily prepared by reaction of (S)-amino alcohols with diethyl oxalate followed by cyclization. When present in excess, ligands of this type can effect enantioselective Rh(I)-catalyzed hydrosilylation

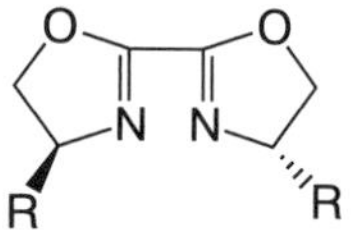

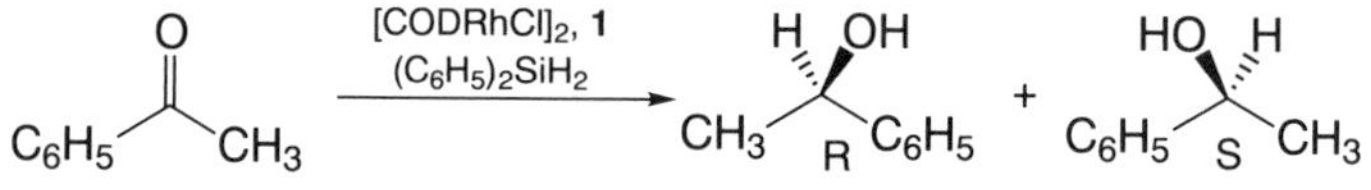

of acetophenone to give (R)-1-phenylethanol as the major product. The highest enantioselectivity, 84%, obtains with **1**(R=CH$_2$C$_6$H$_5$).[1] Similar results, but using chiral pyridyloxazolines, have been reported by Brunner.[2] In this case, highest enantioselectivity obtains with the ligand bearing *t*-butyl as the alkyl group.

These chiral bis(oxazolines) effect only slight enantioselectivity in a Michael or a Reformatsky reaction. The latter reaction actually is inhibited by this ligand.

Enantioselective Diels–Alder catalyst.[3] The chiral bis(oxazoline) **1**, prepared from (S)-phenylglycine, in combination with MgI_2 or magnesium tetraphenylborate (**2**) serves as a highly enantioselective catalyst for the Diels–Alder reaction of cyclopentadiene with 3-acryloyloxazolidine-2-one (equation I).

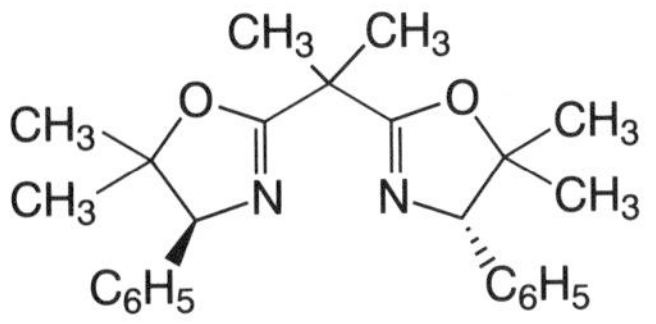

1·Mg[$(C_6H_5)_4B]_2$ (**2**)

1

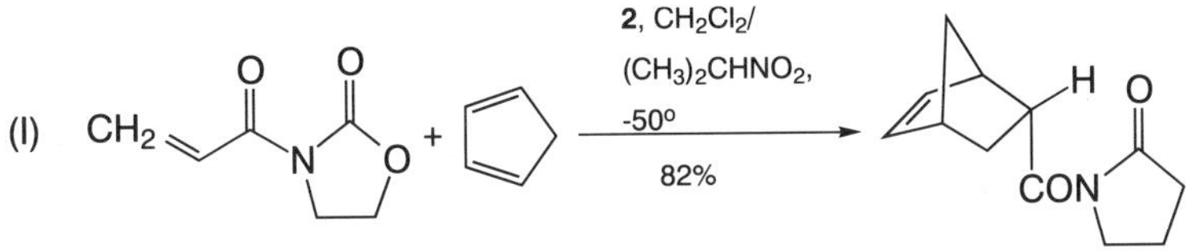

(I)

endo/exo = 97:3
90.4% ee

Enantioselective cyclopropanation (16, 38–39).[4] The bis(oxazoline) **1**, prepared from *t*-leucinol and 2,2-dimethylpropane-1,3-dioyl chloride), forms a white, crystalline complex (**2**) with CuOTf which is an effective catalyst for asymmetric cyclopropanation with ethyl diazoacetate.

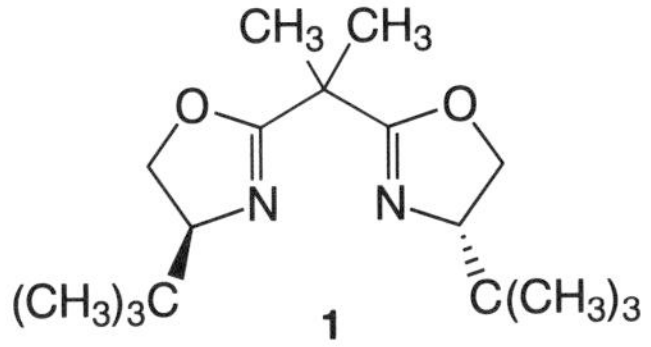

1

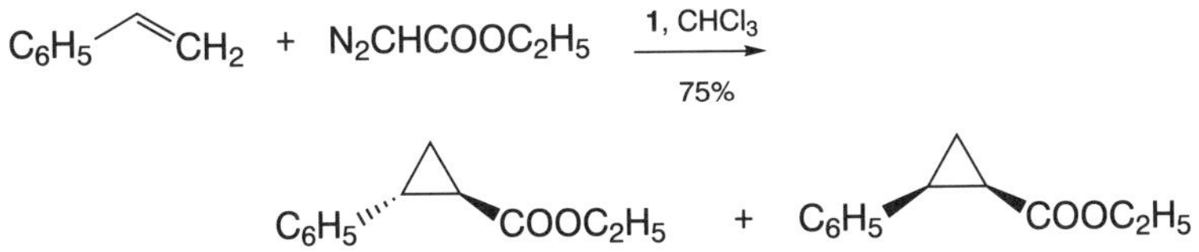

(1R,2R), 96% ee   73 : 27   (1R,2S), 94% ee

[1] G. Helmchen, A. Krotz, K.-T. Ganz, and D. Hansen, *Synlett,* 257 (1991).
[2] H. Brunner and U. Obermann, *Ber.,* **122**, 499 (1989).
[3] E. J. Corey and K. Ishihara, *Tetrahedron Letters,* **33**, 6807 (1992).
[4] D. A. Evans, K. A. Woerpel, and M. J. Scott, *Angew. Chem. Int. Ed.,* **31**, 430 (1992).

Bis(pentamethylcyclopentadienyl)[bis(trimethylsilyl)methyl]yttrium, $(C_5Me_5)_2YCH[Si(CH_3)_3]_2$ **(1).**

Hydrosilylation of alkenes.[1] This lanthanide is an effective catalyst for hydrosilylation of mono- and 1,1-disubstituted alkenes at 20° in C_6H_6 or $C_6H_5CH_3$. Hydrosilylation of dienes with a terminal and an internal olefin is highly selective.

$$CH_3CH{=}CH(CH_2)_2CH{=}CH_2 \xrightarrow[96\%]{\mathbf{1},\ C_6H_5SiH_3} CH_3CH{=}CH(CH_2)_3CH_2SiH_2C_6H_5$$

CH2
CH2
91%
CH2
(CH2)3SiH2C6H5

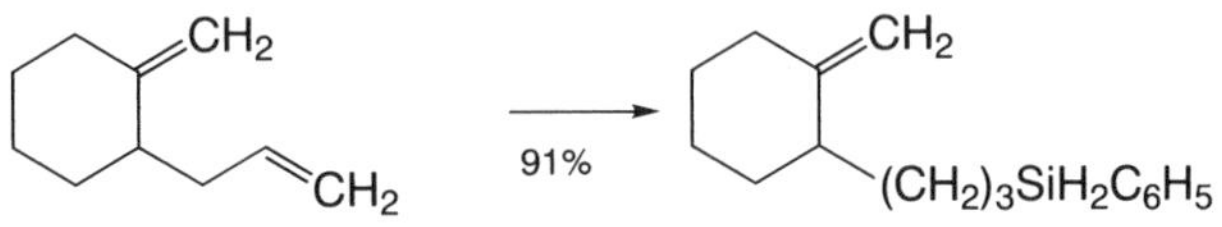

[1] G. A. Molander and M. Julius, *J. Org.*, **57**, 6347 (1992).

Bis(pentamethylcyclopentadienyl)methylyttrium, $Cp^*_2YCH_3(THF)$, **1.** Preparation.[1]

Cyclization of 1,5- and 1,6-dienes.[2] Treatment of **1** with hydrogen generates "Cp^*_2YH" (**2**) with release of CH_4. This hydride is an effective catalyst for cyclization of 1,5- and 1,6-dienes, even those with bulky substituents. No cyclization is observed when the substrate bears CN, $COOCH_3$, or $SO_2C_6H_5$ groups.

CH2
CH2
R
2
R
CH3
+
R
CH3
R = OBzl
= OC(C6H5)3
(6 : 1), 84%
(21 : 1), 99%
(5%)
—

CH2
CH2
OC(C6H5)3
OC(C6H5)3
CH3
+
OC(C6H5)3
Pr
Pr
100 : 1
(70%)
25%

CH2
CH2
CH3
+
C2H5
C2H5
53%
26%

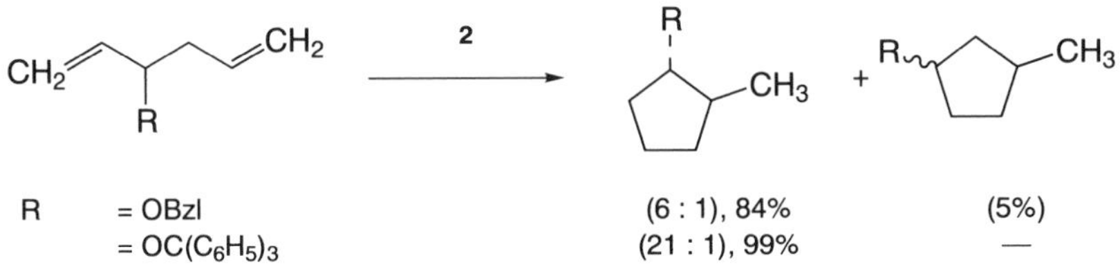

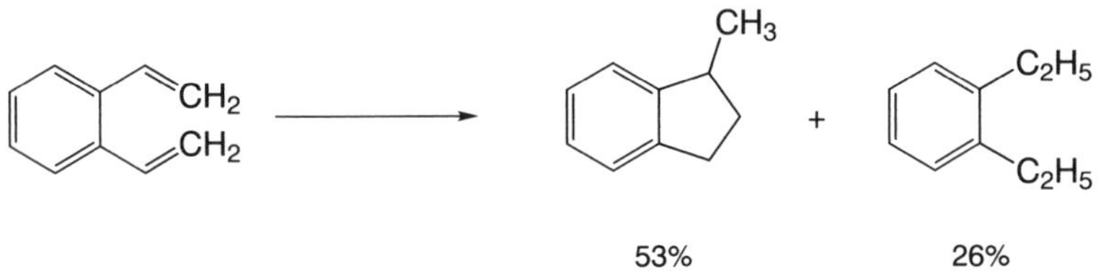

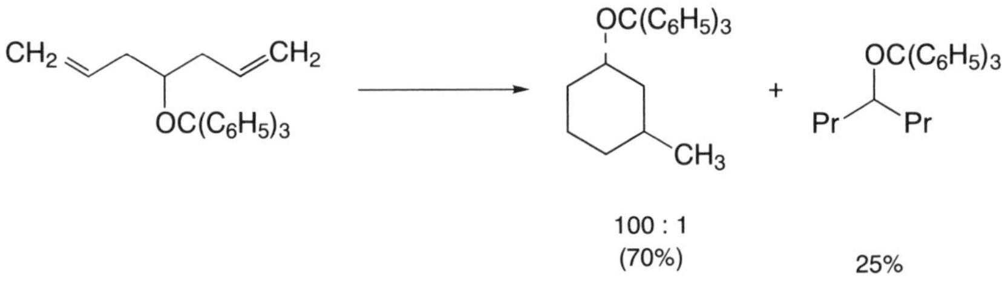

Hydrogenation of dienes.[3]   This organolanthanide when treated with H_2 loses methane to form catalyst "Cp_2YH," which can effect highly selective hydrogenation of terminal double bonds in the presence of other double bonds, even those of allylic ethers.

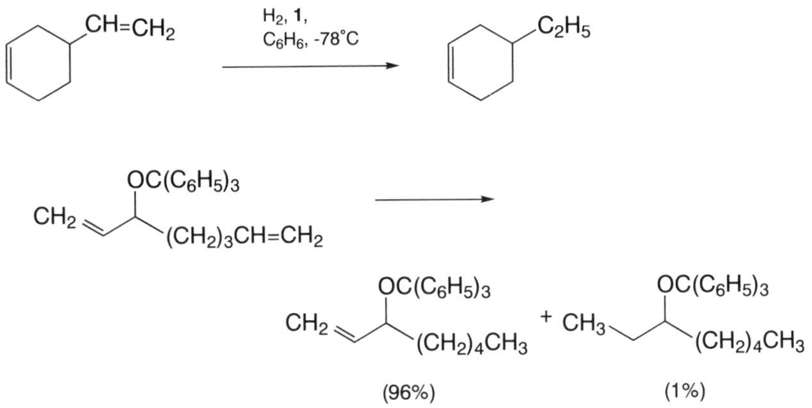

[1] K. H. den Haan, J. L. de Boer, J. H. Teuben, W. J. J. Smeets, and A. L. Spek, *J. Organomet. Chem.*, **327**, 31 (1987).
[2] G. A. Molander and J. O. Hoberg, *Am. Soc.*, **114**, 3123 (1992).
[3] *Idem., J. Org.*, **57**, 3266 (1992).

Bis(pentamethylcyclopentadienyl)samarium tetrahydrofuran,
$(C_5Me_5)_2Sm \cdot THF$ (1). Preparation.[1]

Hydroboration with catecholborane (CB).[2]   This lanthanide is an efficient catalyst for hydroboration of a wide range of alkenes with CB at 25°. Highest rates obtain in the order: monosubstituted terminal > disubstituted terminal > disubstituted internal > trisubstituted.

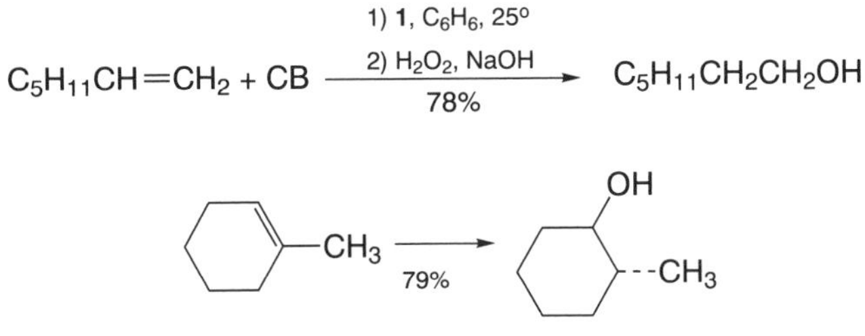

[1] W. J. Evans and T. A. Ulibarri, *Inorg. Synth.*, **28**, 297 (1990).
[2] K. N. Harrison and T. J. Marks, *Am. Soc.*, **114**, 9220 (1992).

Bis(phospholanes).

The C_2-symmetric bis(phospholanes) (**3** and **4**) can be prepared by reaction of a dilithium bis(phosphido)ethane (**1**) with the cyclic sulfate (**2**) of chiral 1,4-diols.

(S,S)-**3**, R = CH_3

(S,S)-**4**, R = CH_3

Enantioselective hydrogenation.[1]   Rhodium catalysts bearing these ligands effect asymmetric hydrogenation of acetamidoacrylates, $RCH=C(COOCH_3)NHAc$. In general, enantioselectivities of about 100% are obtained, particularly with ligands of type (S,S)-**4**, R=C_2H_5.

High enantiomeric excess can be obtained in hydrogenation of enol acetates catalyzed by a Rh catalyst with these chiral ligands. In this reaction, no one ligand is consistently superior, but the enantioselectivity varies with R in the bis(phospholanes) and the R' in the enol acetate.

R' = $C_2H_5O_2C$ >99% ee

[1] M. J. Burk, *Am. Soc.*, **113**, 8518 (1991).

Bispiperazines, chiral.

Asymmetric osmylation.   Chiral ligands of *N,N'*-dialkyl bispiperazines linked by two carbons (**1**, *n* = 2) can effect highly enantioselective dihydroxylation of *trans*-disubstituted alkenes.

1, n = 2, a) R = CH₃, b) CH₂C(CH₃)₃, or c)CH₂—(with dimethylphenyl)—CH₃

L* = **1a**	84%	94%ee
= **1b**	87%	91%ee
= **1c**	84%	89%ee

When $n = 3$ or **4**, these ligands suppress the reaction completely. If R is an acyl group, the enantioselectivity is depressed. These ligands (**1**) can also effect asymmetric osmylation of mono- and disubstituted terminal alkenes, but not of *cis*-disubstituted alkenes.

[1] K. Fuji, K. Tanaka, and H. Miyamoto, *Tetrahedron Letters*, **33**, 4021 (1992).

Bissteroidal pyrazines.

A new group of steroidal alkaloids, known as cephalostatins and shown to be cytotoxic, are composed of two identical steroidal groups connected by a pyrazine unit. Smith and Heathcock[1] have developed two routes to alkaloids of this type in which identical steroid groups are linked at C_2 and C_3 of the A ring to the pyrazine ring (Schemes I and II).

Scheme I

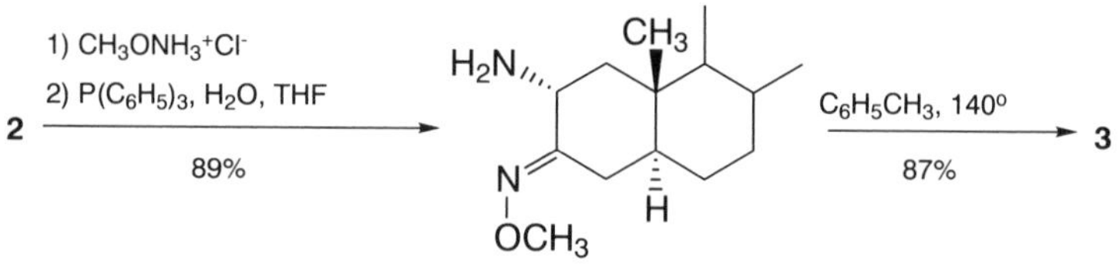

Scheme II

The report also includes a synthesis of an unsymmetrical cephalostatin **6**. In order to avoid dimerization of **4**, and **5**[1] are heated first at 90° for 24 hours to obtain an

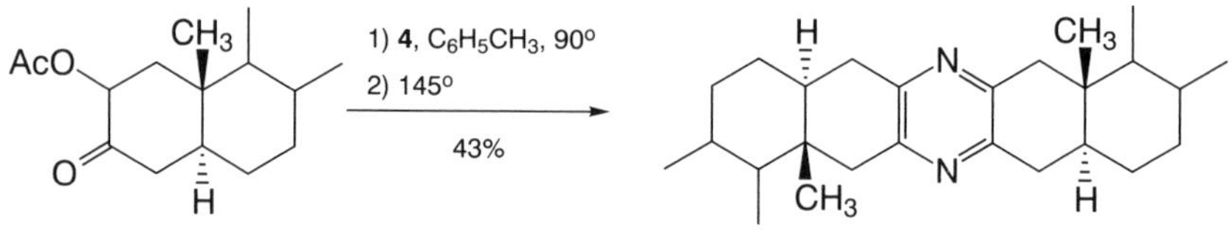

intermediate that forms the pyrazine **6** at 145°.

[1] S. C. Smith and C. H. Heathcock, *J. Org.*, **57**, 6379 (1992).

(S,S)-Bis-*p*-tolylsulfinylmethane (1).

Preparation from menthyl sulfinate and (R)-methyl-*p*-tolyl oxide.[1]

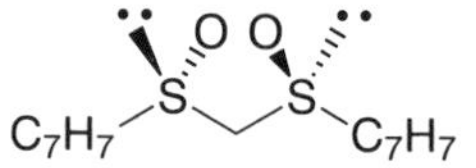

Asymmetric reactions with carbonyl compounds.[2]   The anion of this C_2-symmetric reagent (**1**) can react with aromatic aldehydes with high diastereoselectivity in contrast to the anion of methyl (R)-*p*-tolyl sulfoxide which reacts with low diastereoselectivity.

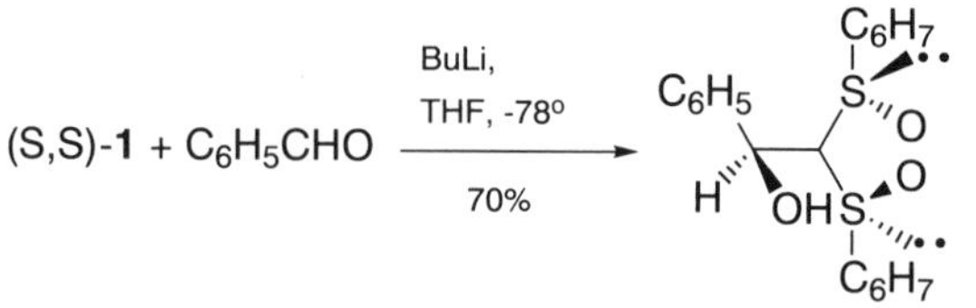

In contrast, **1** reacts with an α, β-enal to form a 4-substituted 1,1-bis(p-tolylsulfinyl)-1,3-butadiene (equation I).

(I) (S,S)-**1** + $CH_3CH=CHCHO$ $\xrightarrow[\text{64%}]{\text{BuLi, THF}}$

(S,S)-**3**, α_D -22°

[1] N. Kunieda, J. Nokami, and M. Kinoshita, *Bull. Chem. Soc. Jpn.*, **49**, 256 (1976).
[2] G. Solladié, F. Colobert, P. Ruiz, C. Hamdouchi, M. C. Carreño, and J.-L. Garcia Ruano, *Tetrahedron*, **32**, 3695 (1991).

(1R,2R)-1,2-N,N'-Bis(trifluoromethanesulfonylamino)cyclohexane – Titanium(IV) isopropoxide.

Enantioselective alkylation of RCHO.[1]   The reaction of diethylzinc with aldehydes proceeds in high enantioselectivity when catalyzed by the complex (**1**) obtained from the combination of the optically active C_2-symmetric disulfonamide and Ti(O-i-Pr)$_4$.

C_6H_5CHO + $(C_2H_5)_2Zn$ $\xrightarrow[\text{98%}]{\text{1, } C_6H_5CH_3 \text{ , -20°}}$

(S), 98%ee

High enantioselectivity can also be obtained with $(CH_3)_2Zn$, $(Bu)_2Zn$, and $(C_5H_{11})_2Zn$, and also with aliphatic aldehydes.

[1] H. Takahashi, T. Kawakita, M. Ohno, M. Yoshioka, and S. Kobayashi, *Tetrahedron*, **48**, 5691 (1992).

9-Borabicyclo[3.3.1]nonane (9-BBN).

1,4-Hydroborations of enones.[1]   Phenyl 1-alkenyl ketones are reduced by 9-BBN or catecholborane (CB) (1 equiv.) in >95% yield to boron (Z)-enolates (>99/1), irrespective of the substituents on the alkenyl group. The reaction is more rapid with

CB. The (Z)-boron enolates derived from 9-BBN undergo aldol reactions to provide aldols in >96% *syn*-selectivity. However, (Z)-boron enolates formed from CB show lower selectivity (syn/anti = 3:1).

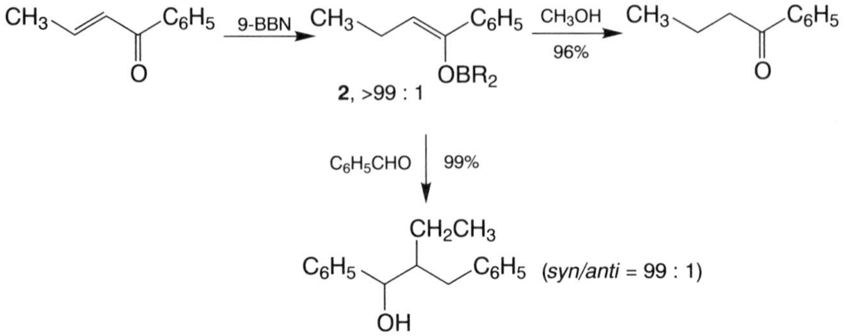

Alkenyl alkyl ketones undergo both 1,2- and 1,4-hydroboration with 9-BBN, with preference for 1,2-addition.

[1] Y. Matsumoto and T. Hayashi, *Synlett,* 349 (1991).

Borane–Dimethyl sulfide, $BH_3 \cdot S(CH_3)_2$.

C-Aryl glycosides.[1]   Reductive aromatization of quinone ketals by $BH_3 \cdot S(CH_3)_2$ provides a new route to C-aryl glycosides. Thus 2-lithiodihydropyran (**2**) and **3** react to form a quinol ketal (**4**). Reaction of **4** with $BH_3 \cdot S(CH_3)_2$ reduces the glycal double bond and provides the C-glycoside **5** derived from *p*-methylanisole.

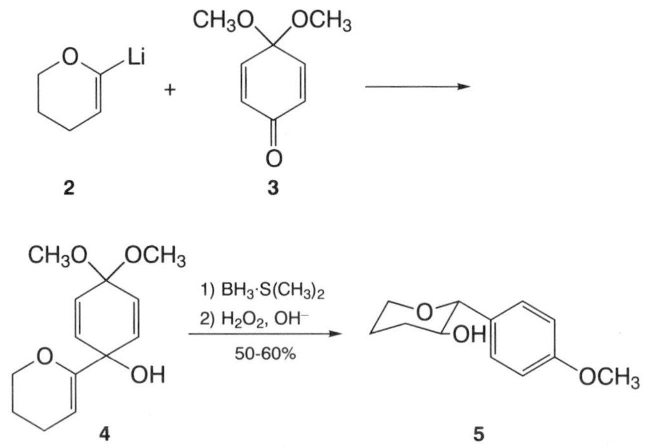

An alternative approach is formulated in equation II.

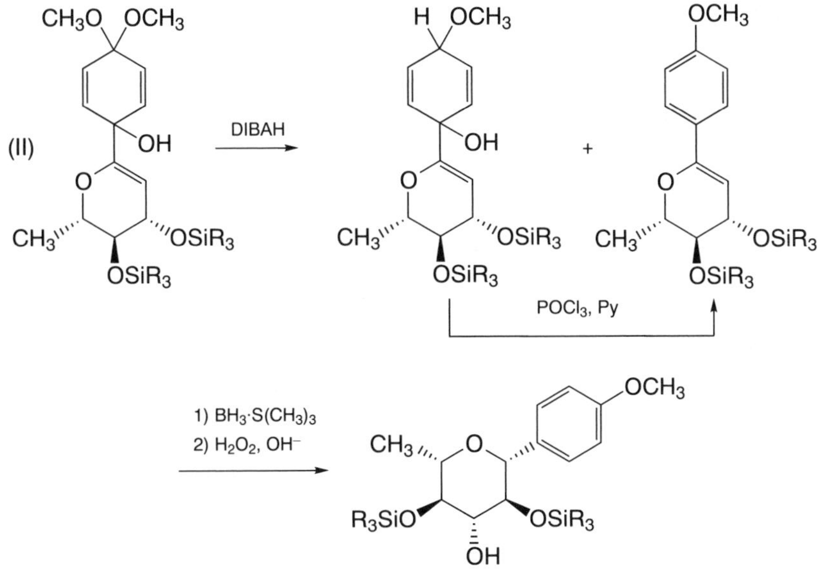

Diastereoselective hydroboration.[2]   Hydroboration of α-alkoxy-β, γ-unsaturated esters with $BH_3 \cdot S(CH_3)_2$ (1.05 equiv.) in THF at $0 \rightarrow 20°$ followed by standard oxidation affords 2-alkoxy-1,3-diols with significant diastereoselectivity.

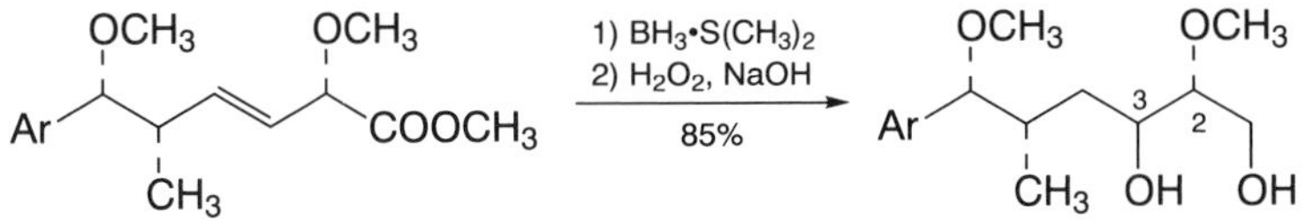

2,3-*anti* / 2,3-*syn* = 8:1

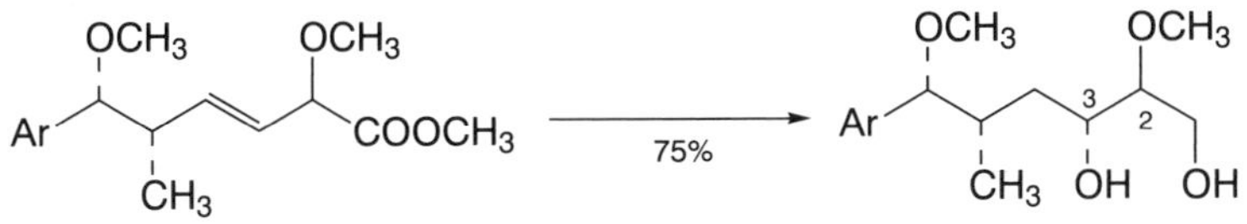

2,3-*anti* / 2,3-*syn* = 4:1

The diastereoselectivity is independent of the size of the alkoxy group and of the stereochemistry of the double bond.

[1] K. A. Parker and C. A. Coburn, *Am. Soc.*, **113**, 8516 (1991).
[2] J. S. Panek and F. Xu, *J. Org.*, **57**, 5288 (1992).

Boron trifluoride etherate.

Diastereoselective addition of chiral (E)-crotylsilanes to RCHO; tetrahydrofurans.[1]
BF$_3$ etherate is the most efficient Lewis acid promotor for diastereoselective addition
of α-substituted β-silyl-(E)-crotylsilanes (1) to α- and β-benzyloxy aldehydes to form

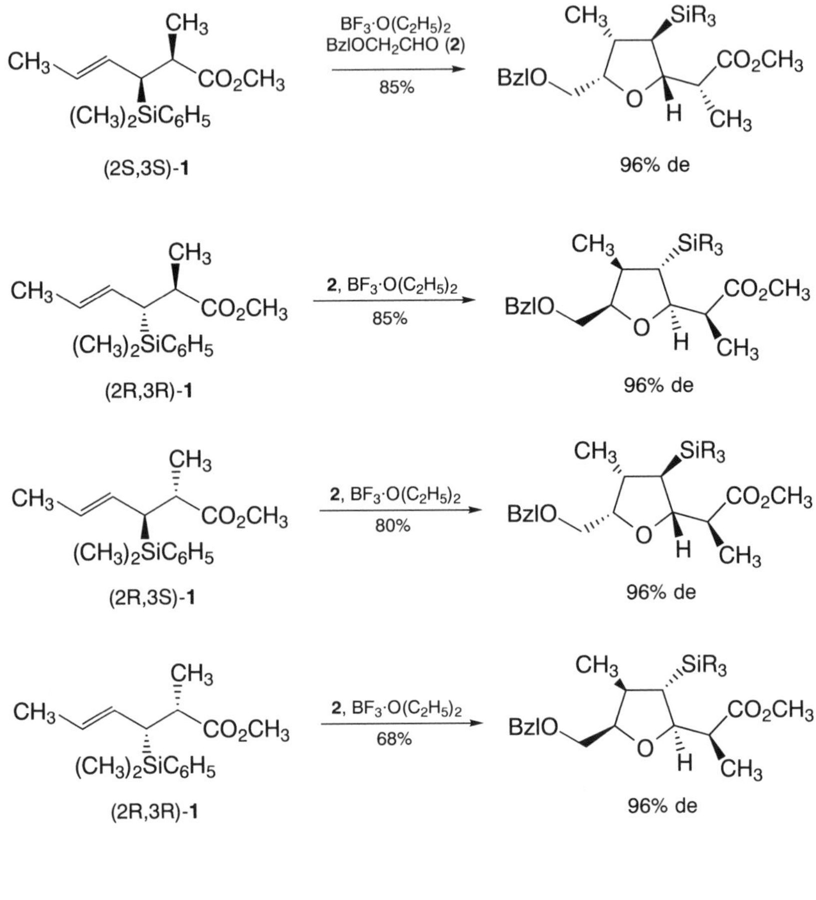

2,5-*cis*-furans. This heterocyclization involves a 1,2-silyl migration.

Dehydration of tert-alcohols.[2]   This reaction can be effected with BF$_3$ etherate (1–3
equiv.) in CH$_2$Cl$_2$ at 25°. Yields are usually higher than those obtained with the Burgess
reagent, thionyl chloride/amine, or *p*-toluenesulfonic acid. When dehydration could
result in two different alkenes, the more thermodynamically stable alkene predominates.
Noepentyl *tert*-alcohols result in mixtures of alkenes, some resulting from a carbocation
rearrangement.

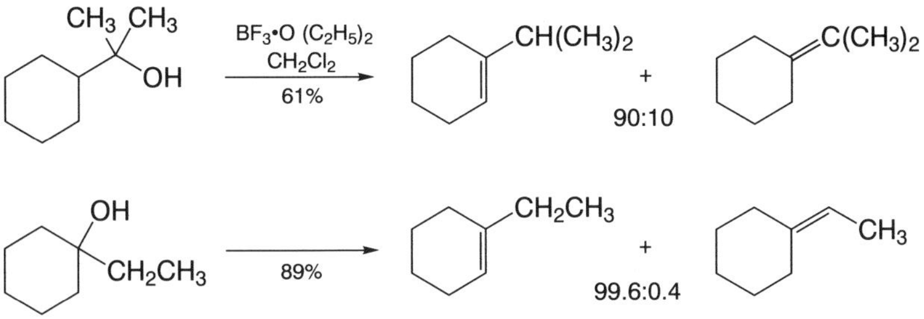

1 J. S. Panek and M. Yang, *Am. Soc.,* **113**, 9868 (1991).
2 G. H. Posner, E. M. Shulman-Roskes, C. H. Oh, J.-C. Carry, J. V. Green, A. B. Clark, and T. E. N. Anjeh, *Tetrahedron Letters,* **32**, 6489 (1991).

5-Bromo-2-pyrone,

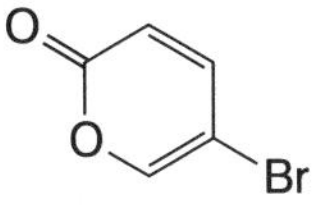

(1)

Preparation:

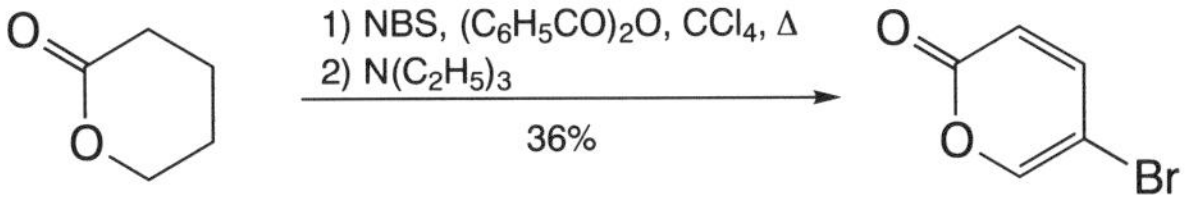

36%

1

Diels-Alder cycloadditions.[1]   This pyrone undergoes Diels–Alder reactions at 25–100° to form adducts that on radical debromination afford halogen-free bicyclic lactones. It is therefore a practical equivalent to 2-pyrone itself.

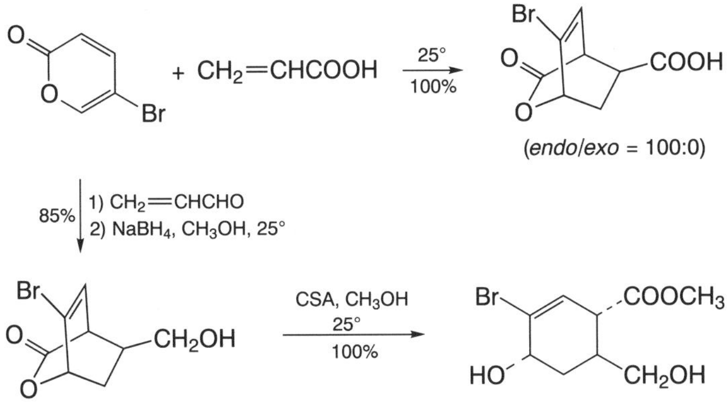

1 K. Afarinkra and G. H. Posner, *Tetrahedron Letters,* **33**, 7839 (1992).

(5S,6R)-4-*t*-Butoxycarbonyl-5,6-dipenyl-2,3,5,6-tetrahydro-4*H*-1,4-oxazin-2-one, 14, 58–59.

Asymmetric synthesis of arylglycines.[1]   These glycines are difficult to obtain in optically pure form because they readily undergo base-catalyzed racemization. One of the most useful routes involves the Williams glycinate (1). Bromination followed by alkylation with an arylcuprate or an arene ($ZnCl_2$ catalysis) provides 2. Removal of the Boc group ($ISiR_3$), hydrolysis of the lactone ring, and oxidative removal of benzaldehyde provides the free amino acid. Even a furanylglycine can be prepared (equation II).

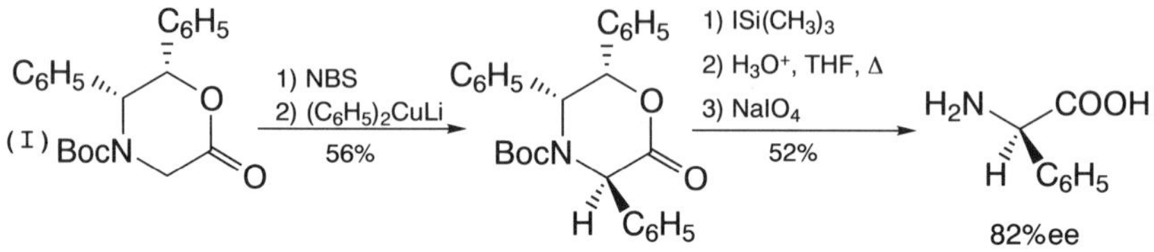

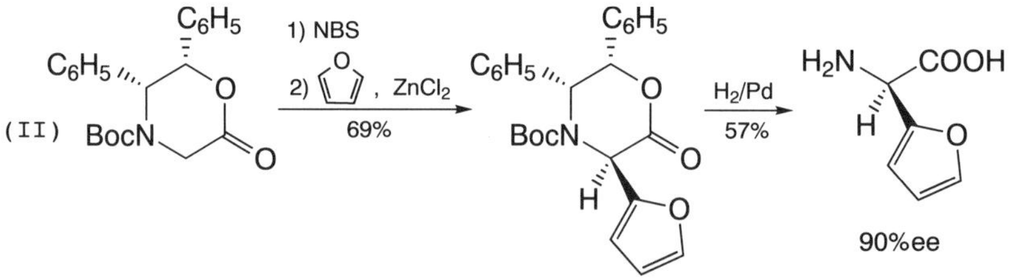

This oxazinone route can be used to obtain bisaryl glycines by a version of the Stille coupling reaction (equation III).

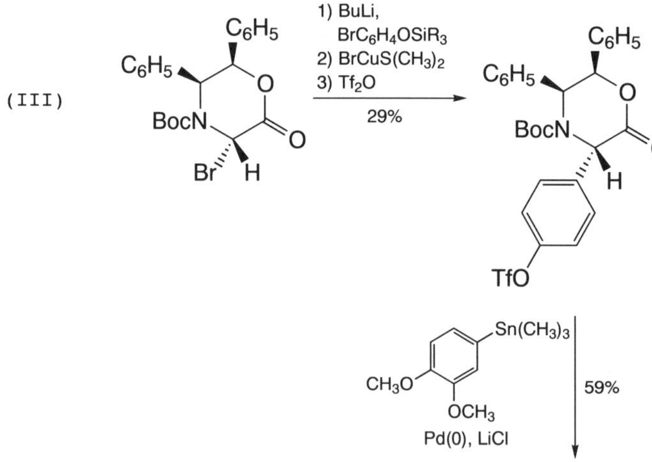

[1] R. M. Williams and J. Hendrix, *Chem. Rev.*, **92**, 889 (1992).

trans-(−)-**2**-*t*-**Butylcyclohexanol,**   (1R,2S)-(−)-**1**

This chiral alcohol (−)-**1** can be obtained from racemic **1** by conversion to the chloroacetate (reaction with ClCH$_2$COOH), which is saponified by enzyme PLAP (pig liver acetone powder) at 40% conversion to give (−)-**1** and the (+)-ester, which can then be saponified to provide (+)-**1**.

This alcohol when used as a chiral auxiliary in four test reactions shows higher asymmetric induction than *trans*-2-phenylcyclohexanol or menthol. It is only slightly less effective than 8-phenylmenthol as a chiral auxiliary.[1]

[1] P. Esser, H. Buschmann, M. Meyer-Stork, and H.-D. Scharf, *Angew. Chem. Int. Ed.*, **31**, 1190 (1992).

t-**Butyldimethylsilyl trifluoromethanesulfonate (1),** 12, 86; 15, 54–55.

Intramolecular Michael-aldol cyclization.[1]   This silyl triflate (**1**) in combination with NEt$_3$ can effect cyclization of α, β-enoates substituted by a keto group to a polycyclic system fused to a cyclobutane. Thus reaction of (E)-**2** with 4 equiv. of **1** and 7

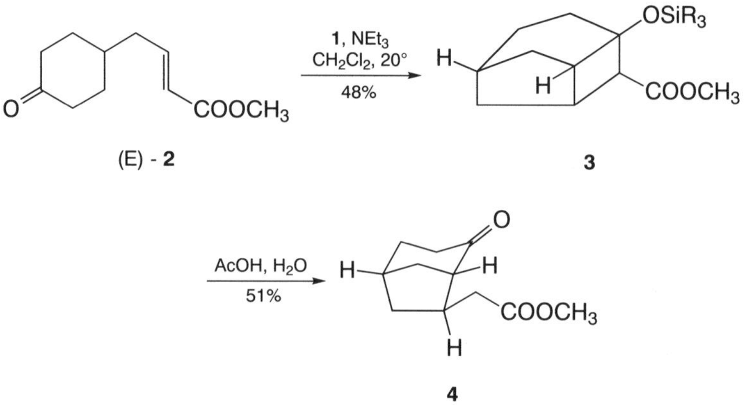

(E) - **2**   **3**

4

equiv. of NEt₃ provides **3**, which can be converted into **4** by a retro-aldol reaction. Another example:

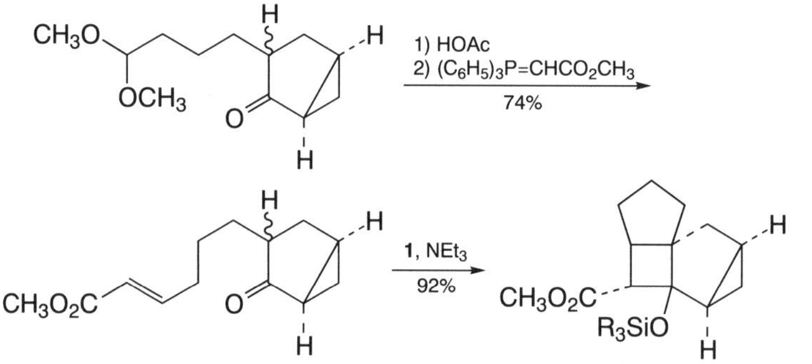

[1] M. Ihara, M. Ohnishi, M. Takano, K. Makita, N. Taniguchi, and K. Fukumoto, *Am. Soc.*, **114**, 4408 (1992).

t-Butyl hydroperoxide.

Stereoselective epoxidation of chiral γ-amino enoates.   Since γ-dibenzylamino-β-keto esters can be reduced with high stereoselectivity (**16**, 304) and since γ-dibenzylaminoenoates (**1**) undergo stereoselective cuprate addition, Reetz[1] has investigated epoxidation of these substrates (**1**). In this case, epoxidation with *t*-BuOOH requires the presence of potassium *t*-butoxide, which results in ester hydrolysis as well. Under these conditions epoxidation can proceed with high stereoselectivity. Essentially a single product (**2**) is obtained and this reacts with cuprate also to form a single product (**3**).

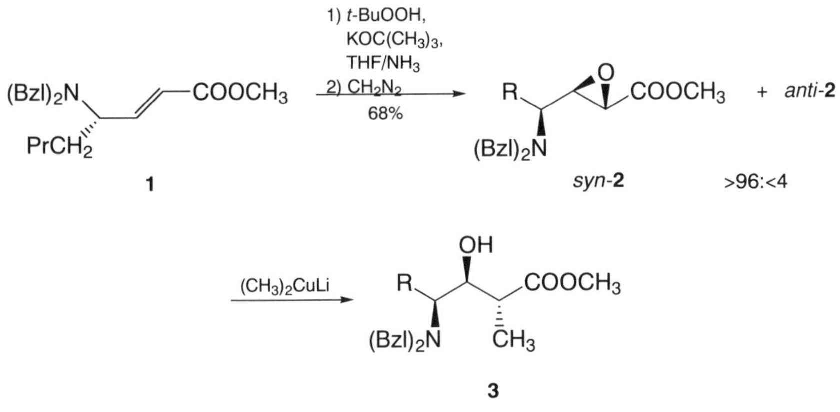

Use of *m*-chloroperbenzoic acid provides an amine oxide **4**, which rearranges with complete chirality transfer to a hydroxyamine (**5**).

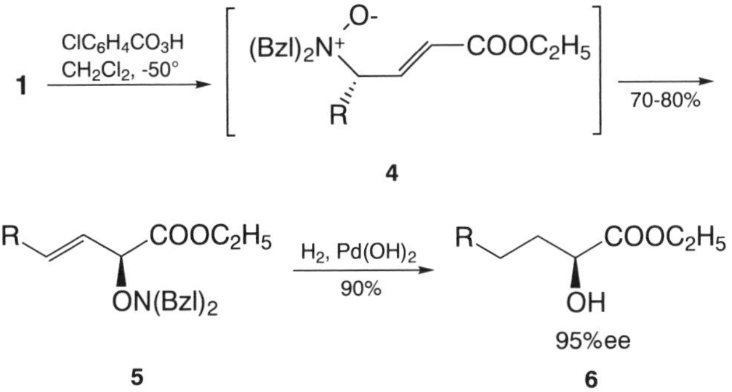

[1] M. T. Reetz and E. H. Lauterbach, *Tetrahedron Letters,* **32**, 4477, 4481 (1991).

t-Butyl hydroperoxide–Dialkyl tartrate–Titanium(IV) isopropoxide.

Sharpless asymmetric epoxidation of β-hydroxy acrylates.[1]   A β-hydroxy acrylate on classical epoxidation (alkaline hydrogen peroxide) gives a 1:1 mixture of *syn*- and *anti*-epoxides, but on Sharpless asymmetric epoxidation gives the *syn*-epoxide selectively (99:1). No reaction occurs in the absence of the hydroxyl group. The asymmetric Sharpless epoxidation also is possible with cyclic β-hydroxy ketones (second example).

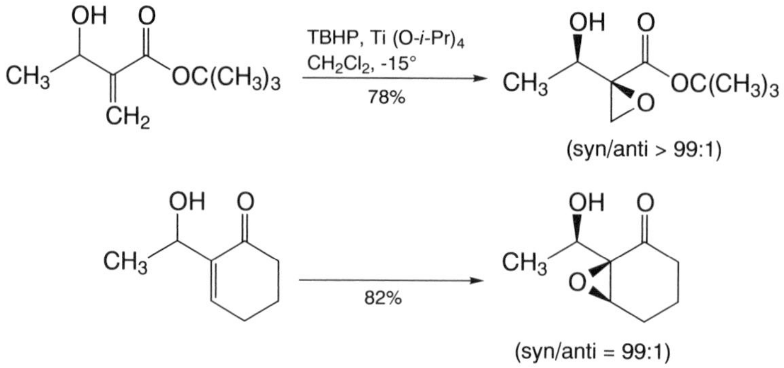

(syn/anti > 99:1)

(syn/anti = 99:1)

Double kinetic resolution.   Davies *et al.*[2] have noted that the enantiomeric selectivity of Sharpless asymmetric epoxidation of an allylic alcohol can be enhanced in some cases by use of two kinetic resolutions. Thus epoxidation of the allylic alcohol **1** with (+)-DiPT as the chiral component (58% conversion) provides the epoxide **2** and the less reactive enantiomer (R) of **1**, which can be recovered and epoxidized with (−)-DiPT. Using this technique, the epoxide **3** was obtained from **1** in 86% ee. This strategy is

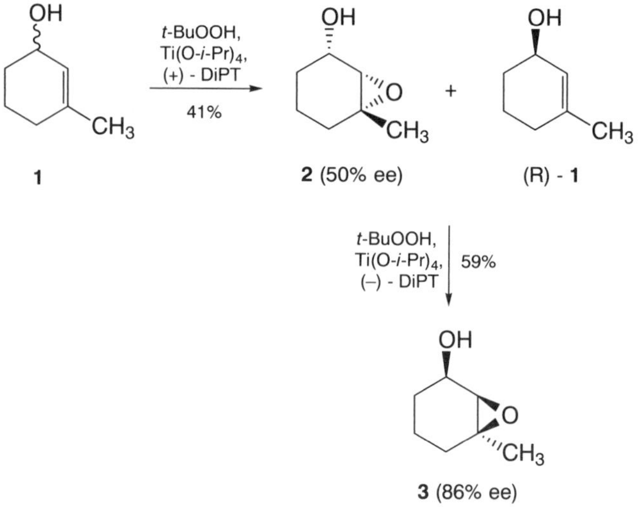

only useful when the racemates of the allylic alcohol do not differ significantly in reactivity and when both optical forms of the chiral reagent are available.

[1] M. Bailey, I. E. Markó, and W. D. Ollis, *Tetrahedron Letters,* **32**, 2687 (1991).
[2] S. M. Brown, S. G. Davies, and J. A. A. de Sousa, *Tetrahedron Asymmetry,* 511 (1991).

Butyllithium

Diastereoselective [2.3] Wittig rearrangement.[1]   Treatment of allylic propargylic ethers (**1**) with BuLi at −85° provides either *syn-* or *anti-β*–silyl alcohols (**2**) in high selectivity, depending on the R group. A variation of this Wittig rearrangement followed by Peterson elimination provides a stereocontrolled synthesis of terminal conjugated trienes.

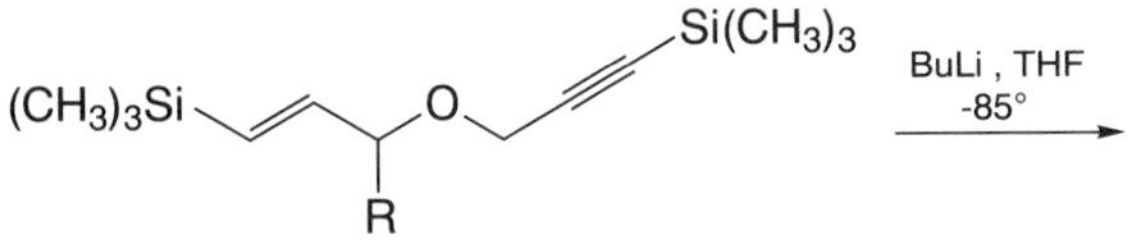

R = H	90%	3:97
R = CH₃	60%	96:4
R = C₂H₅	65%	97:3

Thus the ether **3** rearranges to **4** with high (E)- and *anti*-selectivity to give **4**. This product is reduced by sodium bis(2-methoxyethoxy)aluminum hydride to the (1E,5E)-diene **5**. Acidic elimination followed by protiodesilylation gives sarohornene B (**6**) in high selectivity. In contrast, basic elimination and protiodesilylation provides sarohornene C (**7**) with high selectivity.

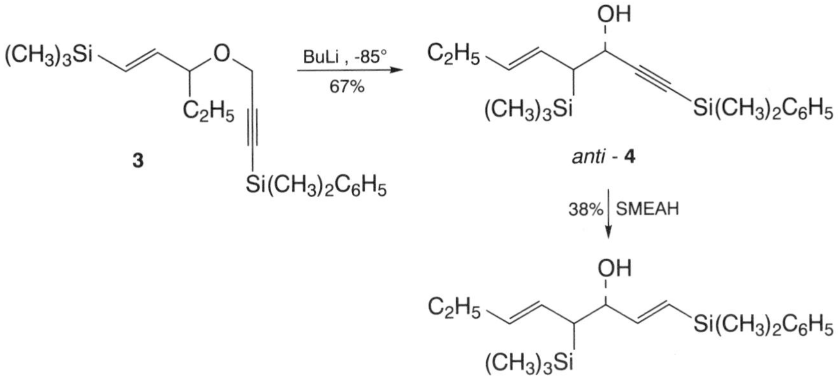

1) H₃O⁺ (99%)

2) F⁻ (99%)

$C_2H_5 \diagup\diagdown\diagup CH_2$

6

5

$C_2H_5 \diagup\diagdown\diagup CH_2$

1) KH , 0° (99%)

2) F⁻ (68%)

7

Scheme I

[1] N. Kishi, T. Maeda, K. Mikami, and T. Nakai, *Tetrahedron,* **48**, 4087 (1992).

Butyllithium–Potassium *t*-butoxide.

ortho-Metalation.[1]   Directed *ortho*-metalation of an anthracene derivative such as **1** with an alkyllithium presents a problem since attack occurs mainly at C_9 or C_{10} instead of desired attack at C_2. *ortho*-Metalation can be effected with *t*-BuOK/BuLi (5 equiv. of each) in THF at −90° followed by reaction with $(CH_3)_3SnCl$ to give the arylstannane **2** in 64° yield. This product can undergo tin–lithium exchange on treatment with methyllithium in toluene (−78 → 30°) without reaction at C_9 or C_{10}. The aryllithium thus generated reacts with aldehyde **3** to introduce the side chain required for synthesis of a precursor (**4**) to the anthraquinone antibiotic vineomycinone B_2 methyl ester (**5**).

1) BuLi/KOC(CH₃)₃

2) (CH₃)₃SnCl

THF, −90°

64%

1

2

1) CH₃Li, C₆H₅CH₃, −78° → 30°

2) OHC $\diagup\diagdown$ CH₂ ; CH₃ ''OBzl   (**3**)

50%

4

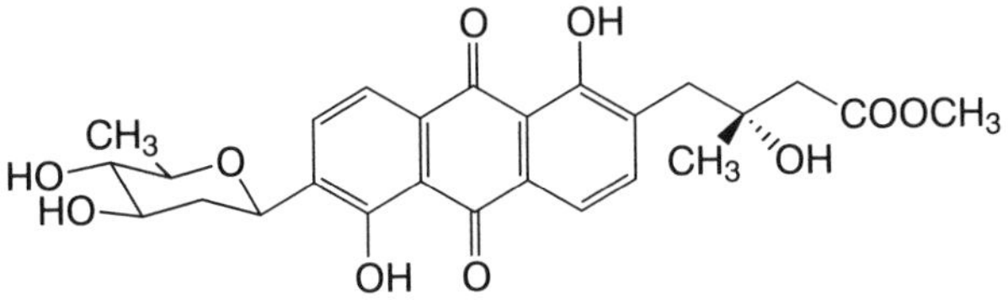

5

[1] T. Matsumoto, M. Katsuki, H. Jona, and K. Suzuki, *Am. Soc.,* **113**, 6982 (1991).

(2R)-2-*t*-Butyl-6-methyl-4*H*-1,3-dioxin-4-one (1) (14, 69–70).

An improved preparation of **1** from the acetal (**2**) obtained from pivaldehyde and (R)-3-hydroxybutanoic acid, permits preparation of this chiral derivative of acetylacetic acid on a large scale (13 g.).[1]

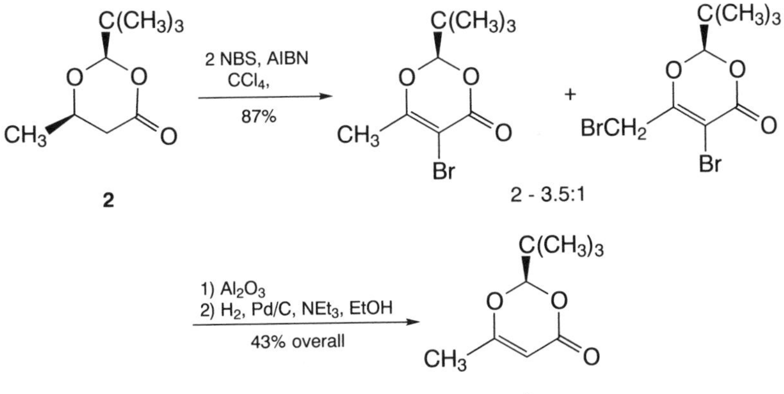

[1] D. Seebach, U. Gysel, K. Job, and A. K. Beck, *Synthesis,* 39 (1992).

1-(*t*-Butylperoxy)-1,2-benziodoxol-3(1*H*)-one, 1, stable at 25°.

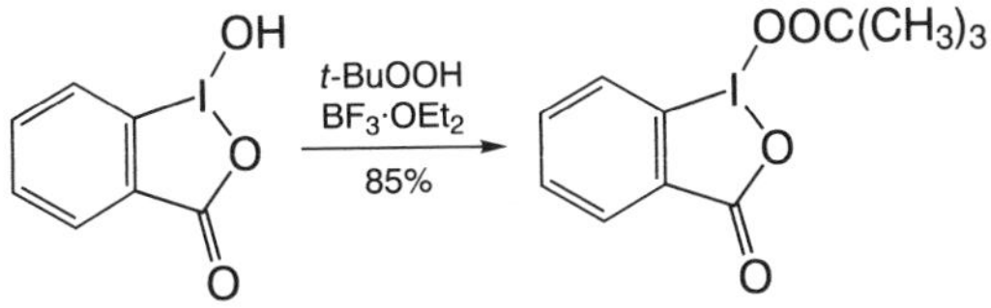

1

Oxidation. This new iodinane is a versatile oxidizing reagent.[1]

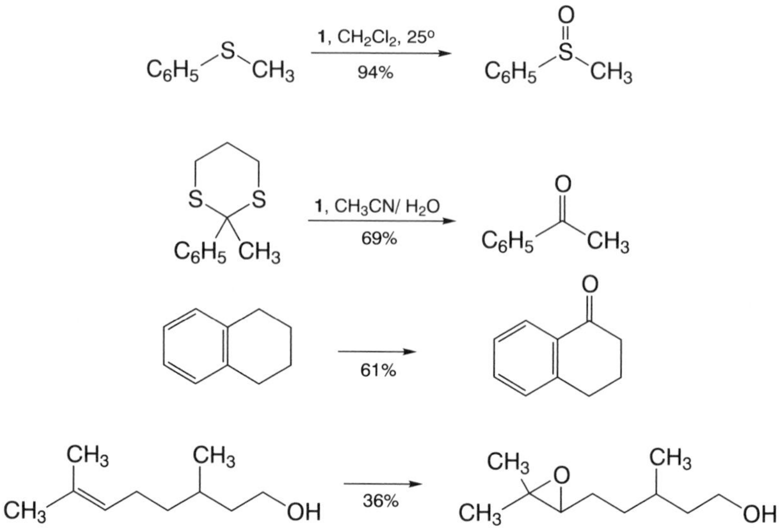

[1] M. Ochiai, T. Ito, and Y. Masaki, *Am. Soc.*, **114**, 6269 (1992).

C

D-Camphor dimethyl acetal (1).

Selective protection of myo-inositols.[1]   Inositol derivatives, particularly various phosphates, are involved in various biological processes, but are difficult to prepare in enantiomerically pure form. A useful expedient is to use D-camphor not only as a protective group but also as the chiral auxiliary. Thus D-camphor dimethyl acetal (**1**) reacts with *myo*-inositol (**2**) to give the 2,3-protected tetrol **3** in 31% yield. The 1-hydroxyl group of **3** reacts selectively with the bulky *t*-butyl-dimethylsilyl chloride to give **4** in 88% yield. This product reacts selectively with pivaloyl chloride to form the C_4-protected ester in 78% yield. The tetrol **3** reacts with the bifunctional silylating reagent, 1,3-dichloro-1,1,3,3-tetraisopropyldisiloxane, to give a protected 5,6-diol. By use of various acylating reagents it is possible to obtain derivatives of *myo*-inositols in which hydroxyl groups are free at C_1; C_1 and C_4; C_1, C_3, and C_4; C_1, C_4, and C_5; C_1, C_3, and C_4; C_1, C_4, and C_5; C_1, C_2, and C_6; C_1, C_2, C_4, and C_5; and C_1 and C_6.

[1] K. S. Bruzik and M.-D. Tsai, *Am. Soc.*, **114**, 6361 (1992).

Camphor-derived oxazinone (1).

The oxazinone **1** is prepared in several steps from ketopinic ethyl ester.

m.p. 242°, α_D -155°

Asymmetric aldol reactions.[1]   The enolate of the N-propionyl derivative of **1** can undergo highly *syn*-selective aldol reactions to provide the "non-Evans" *syn*-aldols (**16**, 48).

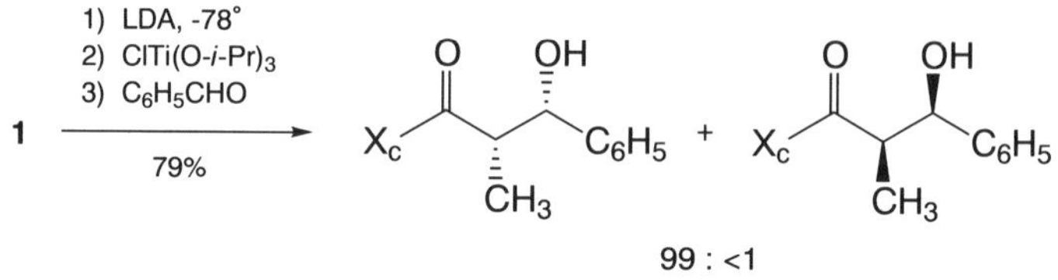

99 : <1

[1] K. H. Ahn, S. Lee, and A. Lim, *J. Org.*, **57**, 5065 (1992).

(1R)- or (1S)-10-Camphorsulfonamides.

Chiral α-nitroso reagents; anti-β-amino alcohols.   The α-chloro-α-nitroso reagent **2** is prepared from the 10-camphorsulfonamide **1** in 78% yield by oximation followed by chlorination. This reagent reacts with the zinc enolate (**a**) of an ethyl

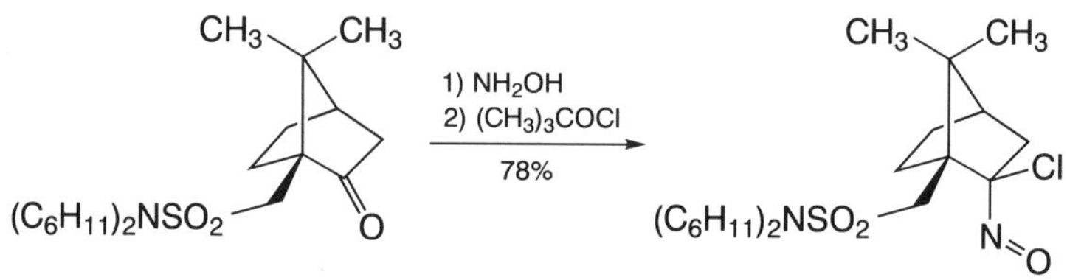

ketone **3** to form essentially a single nitrone (**4**). Hydrolysis (H_3O^+) of **4** provides **1** and a β-keto N-hydroxylamine hydrochloride **5**, which is converted by reduction and hydrogenolysis into (−)-norephedrin (**6**) in 96% ee. This sequence provides a general route to *anti*-β-amino alcohols in 96–99.0% ee.[1]

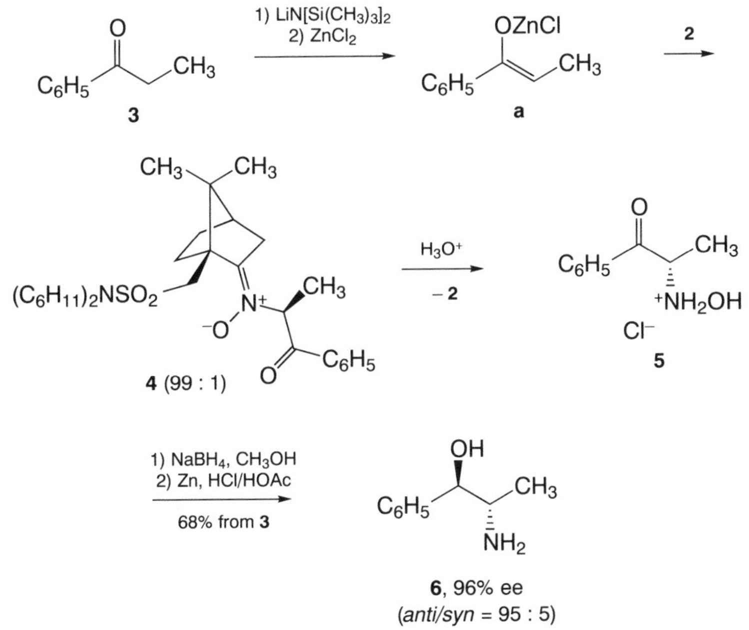

3

a

2

4 (99 : 1)

5

1) NaBH$_4$, CH$_3$OH
2) Zn, HCl/HOAc

68% from 3

6, 96% ee
(anti/syn = 95 : 5)

[1] W. Oppolzer, O. Tamura, G. Sundarababu, and M. Signer, *Am. Soc.,* **114**, 5900 (1992).

10-Camphorsultam (1), 13, 62; 14, 71–72.

Chiral β-hydroxy esters.[1]   The N-acetyl derivative **1** of Oppolzer's auxiliary can be converted into the O-silyl-N,O-ketene acetal **2**, which reacts with aldehydes in the presence of TiCl$_4$ to afford aldols **3** or **4**. These are hydrolyzed to β-hydroxy esters **5**. The stereoselectivity depends upon the steric demands of the aldehyde group R mainly, with

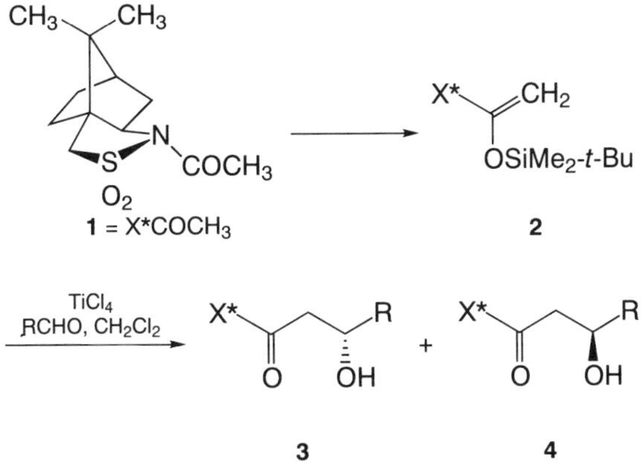

1 = X*COCH$_3$

2

TiCl$_4$
RCHO, CH$_2$Cl$_2$

3

4

R = C_6H_5 61%   91 : 9
= *i*-C_3H_7 75%   96.4 : 3.6
= *c*-C_6H_{11} 73%   95 : 5

LiOH, H_2O_2

5

slight dependence on the silyl substituents.

[1] W. Oppolzer and C. Starkemann, *Tetrahedron Letters*, **33**, 2439 (1992).

(10-Camphorylsulfonyl)oxaziridines, 13, 64–65; **14**, 72; **16**, 61–62.

Kinetic resolution of a lactone.[1]   Hydroxylation of the enolate (LDA) of (±)-3-methylalerolactone (**2**) with (−)-1 (0.5 equiv.) provides **3** in 58% yield and 60% ee.

(−) - **1**

1) LDA, TMEDA
2) (−) - **1**, THF, -78°

58%

(±) - **2**   **3**, (60% ee)

Flash chromatography followed by two crystallizations provides (2S,3R)-(−)-verrucarinolactone (**3**) in >95% ee.

[1] F. A. Davis and A. Kumar, *J. Org.*, **57**, 3337 (1992).

p-**Carboxybenzenesulfonyl azide**, *p*-HOOCC$_6$H$_4$SO$_2$N$_3$ (**1**), **2**, 62.

Azidation of N-hydroxy β-lactams.[1]   Reaction of **1** with the N-hydroxy β-lactam **2** does not provide the expected α-diazo-β-keto ester but compound **3**, formed by azide transfer to C$_3$ with cleavage of the N-hydroxy bond. Similar results obtain in reaction of the N-hydroxy β-lactam (4) with trisyl azide in the presence of 2 equiv. of tri-ethylamine.

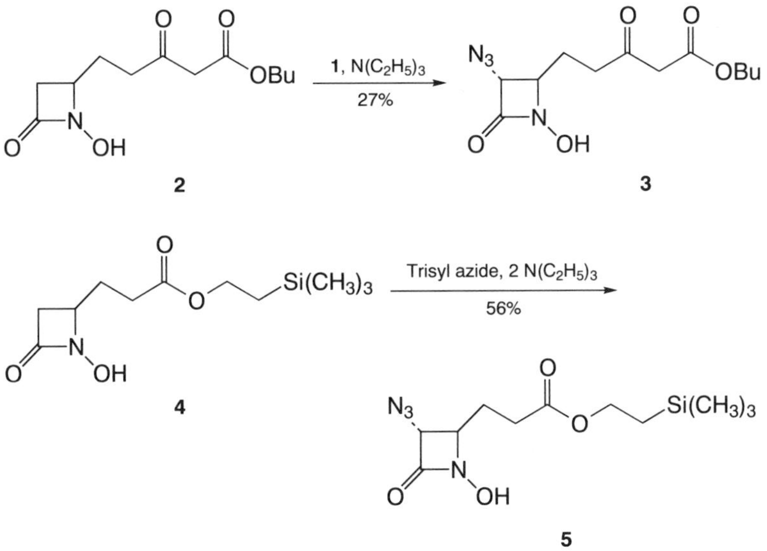

[1] C. M. Gasparski, M. Teng, and M. J. Miller, *Am. Soc.*, **114**, 2741 (1992).

Catecholborane (CB), 16, 65–66.

Catalyzed hydroboration.[1]   Hydroboration by CB can be catalyzed by a number of Rh(I) and Ir(I) catalysts, particularly RhCl[P(C_6H_5)$_3$]$_2$ and Ir(cod)(PCy$_3$)(py)PF$_6$. Advantages are that selective hydroboration owing to steric factors can be improved and the regioselectivity can be enhanced. Moreover, the Rh(I)-catalyzed hydroboration of acyclic and cyclic allylic alcohols proceeds with high diastereoselectivity, opposite to that observed with 9-BBN (equations I and II).

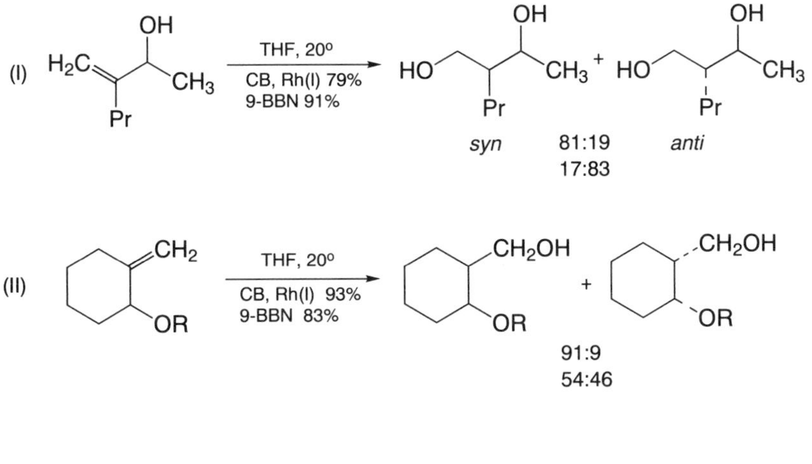

In addition a few groups can direct Rh(I)- or Ir(I)-catalyzed hydroboration. Thus an amide group can direct hydroboration of cyclohexene-derived alkenes (equation III).

syn/anti = 91:9

[1] D. A. Evans, G. C. Fu, and A. Hoveyda, *Am. Soc.,* **114**, 6671 (1992); D. A. Evans, G. C. Fu, and B. A. Anderson, *ibid.,* **114**, 6674 (1992).

Ceric ammonium nitrate–Sodium azide (CAN, NaN$_3$).

α-Azido ketones.[1]   Reaction of a triisopropylsilyl enol ether in CH$_3$CN with NaN$_3$ (4 equiv.) followed by CAN (3 equiv.) at $-20°$ results in an α-azido ketone and triisopropyl azide. Other trialkylsilyl enol ethers are less useful for this azidation because of their ready hydrolysis.

72%

[1] P. Magnus and L. Barth, *Tetrahedron Letters,* **33**, 2777 (1992).

Cesium fluoride, CsF.

Alkylation of RCOOH.[1]   In the presence of CsF (1.5 equiv.) carboxylic acids react with alkyl iodides in DMF at $10-15°$ to form esters in $70-95\%$ yield. DMSO can also be used as solvent, but less polar solvents are not effective.

This reaction can also convert Bu$_3$SnCOR into carboxylic esters (equation I).

(I) C_6H_5COOH $\xrightarrow{Bu_3SnOSnBu_3}$ $Bu_3SnCOOC_6H_5$

$\xrightarrow[94\%]{C_2H_5I, CsF}$ $C_6H_5COOC_2H_5$

[1] T. Sato, J. Otera, and H. Nozaki, *J. Org.,* **57**, 2166 (1992).

μ-Chlorobis(cyclopentadienyl)(dimethylaluminum)-μ-methylene titanium (Tebbe reagent).

1,3-Dienes.[1]   Reaction of the Tebbe reagent with 2-butyne provides the dimethylti-tanacyclobutene **1**. Aldehydes react with **1** to provide a complex (**a**) formed by insertion into the titanium−vinyl bond. This complex is unstable and at 60° decomposes to a 1,3-diene (**2**) and $Cp_2Ti=O$. Ketones react with **1** to give two insertion products. One is formed by insertion into the titanium−vinyl bond and corresponds to (**a**); the other (**b**) is formed by insertion into the titanium−alkyl bond. This latter complex does not decompose into a diene when heated gently. The trend to formation of the undesired

intermediate **b** increases as the size of the alkyl groups of **1** increases and as the size of alkyl groups on the aldehyde or ketone increases. Even so, this route is useful for preparation of 1,3-dienes from simple alkynes and unhindered aldehydes or ketones.

Ketone methylenation.[2]   A comparison of methylenation of about 10 ketones using the Tebbe reagent or the Wittig reagent (methylenetriphenylphosphorane) indicates that the yields are consistently higher with the former reagent, and that the yield improves with hindrance in the substrate. Another advantage is that the Tebbe reagent is available commercially (Strem and Alfa), but expensive.

[1] K. M. Doxsee and J. K. Mouser, *Tetrahedron Letters,* **32**, 1687 (1991).
[2] S. H. Pine, G. S. Shen, and H. Hoang, *Synthesis,* 165 (1991).

Chlorobis(cyclopentadienyl)hydridozirconium(IV)

(zirconium hydrochloride, Schwartz reagent).

Conjugate addition of alkenes.[1]   Hydrozirconation of alkenes followed by addition of a catalytic amount of $CuBr \cdot S(CH_3)_2$ results in *in situ* preparation of alkyl cuprates that undergo 1,4-addition to an enone.

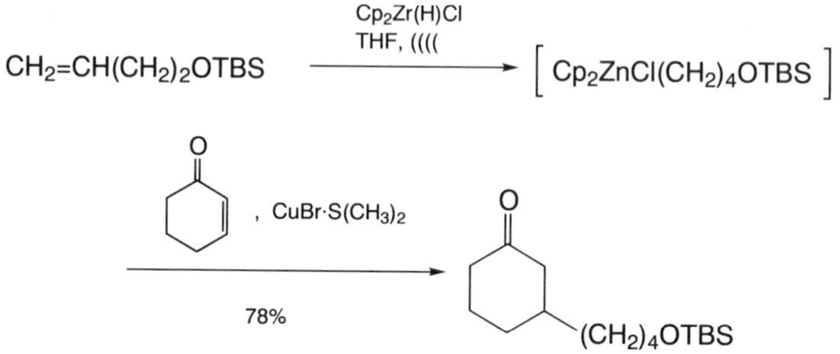

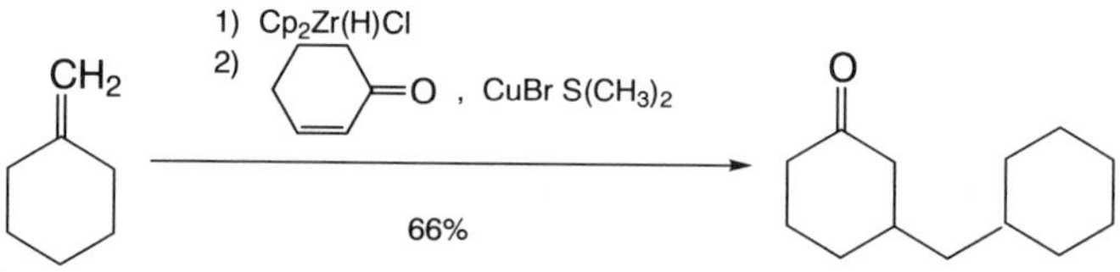

(Z)-Vinylstannes.[2]   (Z)-Vinylstannanes can be prepared by hydrozirconation of stannylacetylenes followed by proton quench.

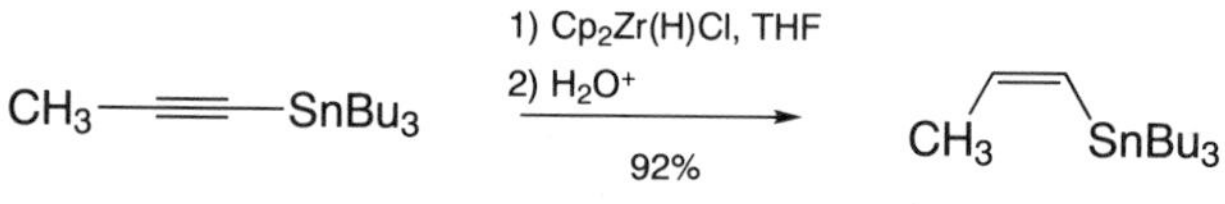

[1] P. Wipf and J. H. Smitrovich, *J. Org.*, **56**, 6494 (1991).
[2] B. H. Lipschutz, R. Keil, and J. C. Barton, *Tetrahedron Letters*, **33**, 5861 (1992).

Chlorobis(cyclopentadienyl)methylzirconium, $Cp_2Zr(Cl)CH_3$, 15, 81

Indole synthesis.[1]   A novel route to 4-iodoindoles involves reaction of N-allyl-N-benzyl-2-bromoaniline (**1**) with this zirconocene to form the zirconacycle **2**, which is converted into **3** by iodination and elimination. This product undergoes ene reactions with various active enophiles at 85° to form 4-iodoindole derivatives such as **4** and **5**.

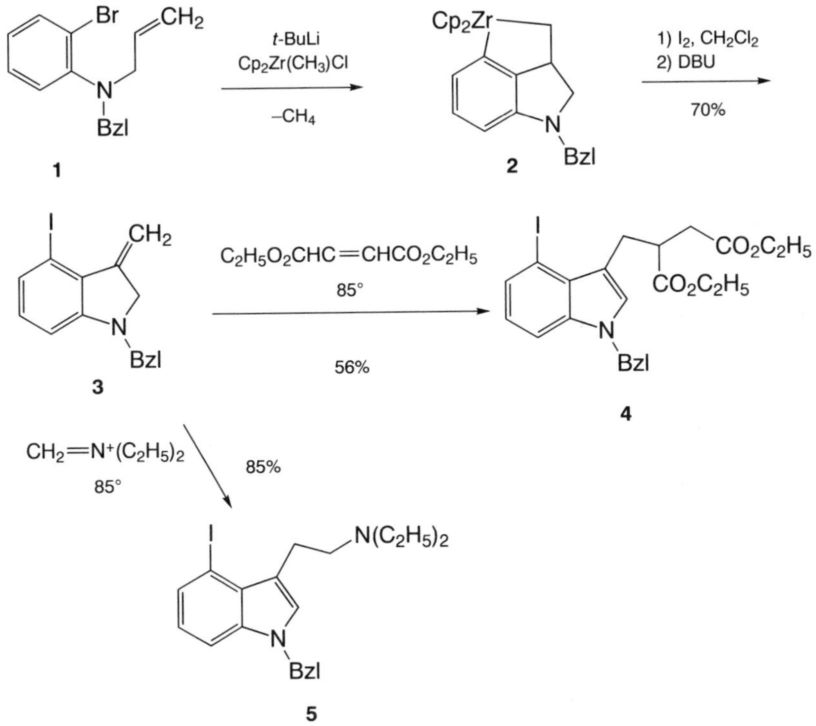

[1] J. H. Tidwell, D. R. Senn, and S. L. Buchwald, *Am. Soc.*, **113**, 4685 (1991).

B-Chlorobis(iso-2-ethylapinocampheyl)borane, Eap_2BCl **(1).**
    Preparation:[1]

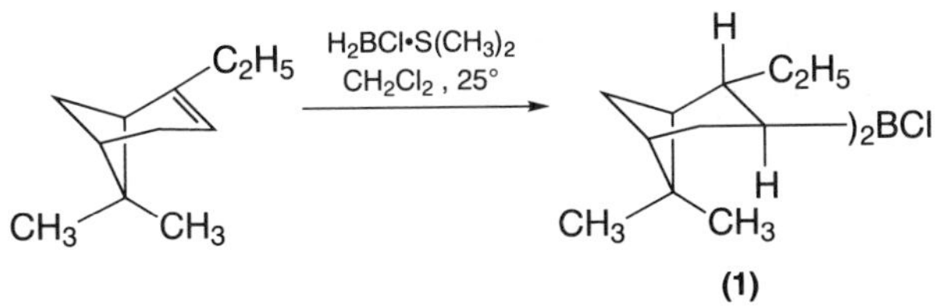

(1)

Asymmetric reduction of ketones.[2]   Both Alpine–Borane and B-chlorodiisopino-campheylborane (Ipc_2BCl) have one main defect for asymmetric reduction of prochiral ketones: they show little enantioselectivity in reduction of dialkyl ketones in which the alkyl groups are similar in size. This problem is now solved by reductions with Eap_2BCl (**1**). Thus acetylcyclohexane is reduced by **1** in 97% ee and isopropyl methyl ketone is reduced in 95% ee (65% yield).

[1] H. C. Brown, P. V. Ramachandran, A. V. Teodorovic, and S. Swaminathan, *Tetrahedron Letters*, **32**, 6691 (1991).

[2] H. C. Brown, P. V. Ramachandran, *Acc. Chem. Res.*, **25**, 16 (1992).

Chlorobis(triphenylphosphine)palladium.

C-Aryl glucals. 1-Iodo-3,4,5-tri-*o*-(triisopropylsilyl)-D-glucal **1** can be prepared in

high yield by stannylation followed by tin/iodine exchange of a protected D-glucal. If *t*-butyldimethylsilyloxy groups are used for protection of the D-glucal, the same conversion to an iodo glucal proceeds in low yield (12–30%).[1]

In the presence of a Pd catalyst, $ClPd[P(C_6H_5)_3]_2$, this vinyl iodide couples with metalated aromatics, particularly arylzinc chloride or arylboronic acids. Aryl Grignard reagents, however, couple in very low yield. The resulting C-aryl glucals can be converted to C-aryl glycosides by functionalization of the enol ether double bond.[2]

[1] R. W. Friessen and R. W. Loo, *J. Org.*, **56**, 4821 (1991).

[2] R. W. Friessen and C. F. Sturino, *ibid.*, **55**, 5808 (1990).

Chlorocyclopentadienylbis(triphenylphosphine)ruthenium, $CpClRu[P(C_6H_5)_3]_2$ (**1**).

Isomerization of allylic alcohols to saturated carbonyl compounds.[1] In the presence of this ruthenium complex, allylic alcohols rearrange to saturated aldehydes or ketones. The rate depends on the number of substituents on the double bond, with highest rates in the case of monosubstituted alkenes. 1,1-Disubstituted alkenes rearrange faster than 1,2-disubstituted alkenes.

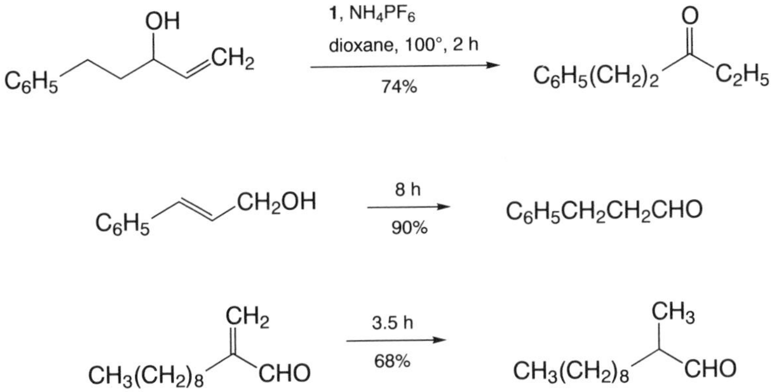

[1] B. M. Trost and R. J. Kulawiec, *Tetrahedron Letters*, **32**, 3039 (1991).

Chloro(cyclopentadienyl)dimethyltitanium, CpTiCl(CH$_3$)$_2$, **1**, prepared *in situ* from CpTiCl$_3$ and 2CH$_3$Li.

[2 + 2]Cycloaddition of alkynylamines; heteroannelations.[1]   Reaction of the alkynylamine **2** with **1** leads to a titanacycle (**a**) with evolution of methane (2 equiv.). On addition of CH$_3$OH the Δ^1-pyrroline **3** is formed in 96% yield. The intermediate **a** can undergo reactions at either carbon or nitrogen with appropriate electrophiles. Thus acylation occurs exclusively on nitrogen (**4**). Intramolecular [2 + 2]cycloaddition can also be used to obtain tetrahydropyridines.

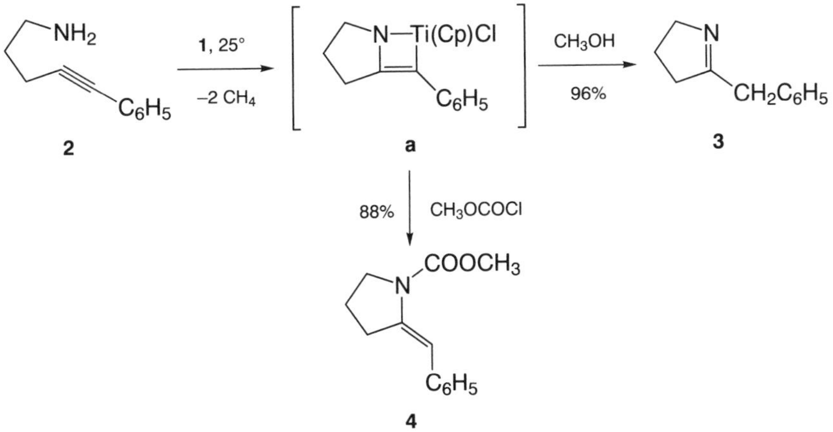

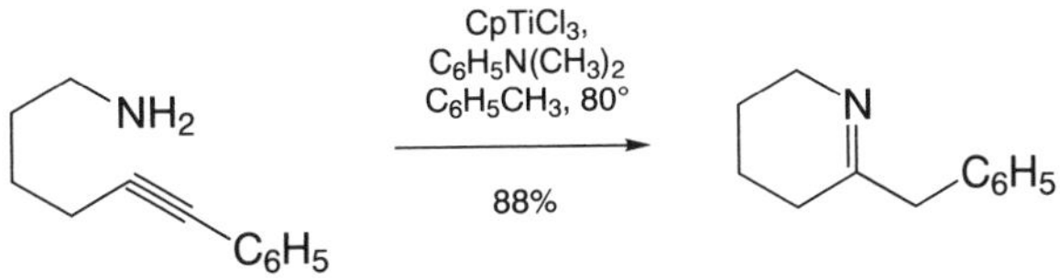

[1] P. L. McGrane, M. Jensen, and T. Livinghouse, *Am. Soc.*, **114**, 5459 (1992).

(−)-**B-Chlorodiisocampheylborane,** (−)-Ipc$_2$BCl (**1**), **13**, 72; **14**, 82.

$$O$$
$$\parallel$$

Reduction of acetylenic ketones, RC≡CCR.[1]   Alpine-Borane can effect this reduction in high enantioselectivity, but this borane is very sensitive to steric effects. Thus it fails to reduce a carbonyl group adjacent to a *t*-butyl group. In contrast, asymmetric reduction of alkyl aryl ketones with **1** is increased by a hindered alkyl group. The same effect obtains in reduction of actylenic ketones (**2**). Thus as the steric bulk of R^1 increases, the enantioselectivity also increases.

R^1 = *i*-Pr	86%	26% ee
R^1 = C(CH$_3$)$_2$C$_2$H$_5$	76%	>99% ee
R^1 = *t*-Bu	72%	>99% ee

[1] P. V. Ramachandran, A. V. Teodorovic, M. V. Rangaishenvi, and H. C. Brown, *J. Org.*, **57**, 2379 (1992).

4-Chloro-2,3-disubstituted-2-cyclobutenones,

(1)

Regiospecific synthesis of phenols.[1]   These substrates, obtained by reaction of 4-hydroxycyclobutenones with P(C$_6$H$_5$)$_3$/CCl$_4$, undergo Pd-catalyzed coupling with various vinyl- and arylstannanes and vinylzirconiums to form products that are transformed on thermolysis (100°) to tri- and tetrasubstituted phenols.

1) $Bu_3Sn(EtO)C=CH_2$, Pd cat.
2) 100°

54%

+

$Pd[P(C_6H_5)_3]_4$, $P(C_6H_5)_3$

(E)

100°

41%

[1] D. K. Krysan, A. Gurski, and L. S. Liebeskind, *Am. Soc.*, **114**, 1412 (1992).

(Chloromethyl)dimethylsilyl chloride, $ClCH_2Si(CH_3)_2Cl$ (1).

Stereocontrolled Diels–Alder reaction.[1] The Diels–Alder reaction of a 2-substituted 1,3-diene with acrylates gives mainly the cyclohexene in which the substituents are *para*. This intrinsic *para* effect can be reversed by joining the dienophile and the diene with a disposable tether. Thus the chloro group of chloroprene is replaced by a (β-hydroxyethyl)dimethylsilyl group by reaction of the Grignard reagent with (chloromethyl)dimethylsilyl chloride (**1**) followed by a second Grignard reaction with formaldehyde. The product (**2**) is then coupled with the acrylic acid (**3**) to form the precursor (**4**) for an intramolecular Diels–Alder reaction. The product (**5**) is then oxidized under the conditions of Tamao (**12**, 243–245) to the cyclohexenone (**6**). This methodology results in a single adduct that is formed by "*meta*" addition.

65% | 180°

1) cleavage
2) oxid

50%

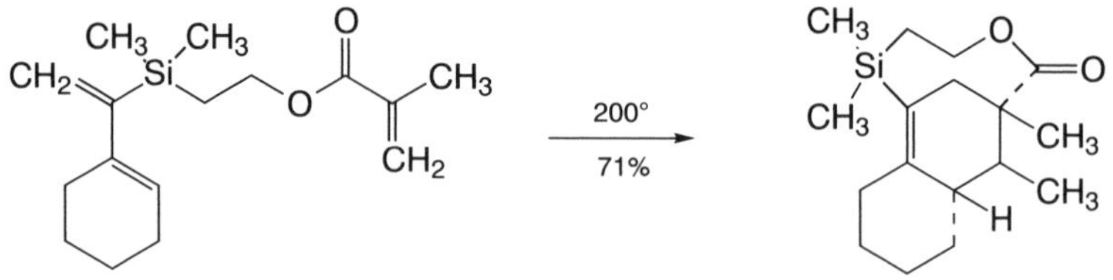

[1] J. Shea, A. J. Staab, and K. S. Zandi, *Tetrahedron Letters,* 2715 (1991).

m-Chloroperbenzoic acid.

Stereoselective epoxidation of enoates.[1]   The final step in the synthesis of (+)-aphidicolin (**4**) requires a stereoselective conversion of the cyclic norketone (**1**) to a *vic*-1,2-diol, >C(OH)-CH$_2$OH. Methylenation of the ketone followed by a Sharpless asymmetric dihydroxylation provides a 1:1 mixture of epimeric 1,2-diols. Reaction with a chiral oxaziridine also provides a 1:1 mixture of epimeric epoxides. The transformation is effected successfully by conversion of the ketone to the enol triflate, which is converted to the enoate (**2**) by Pd-catalyzed carbonylation in methanol (13,234). Epoxidation of **2** with *m*-CPBA in buffered CH$_2$Cl$_2$ with a radical scavenger (4,85–86) results in a single epoxy ester (**3**) in 90% yield. This product is reduced with lithium aluminum hydride (excess) to aphidicolin (**4**) in 67% overall yield from the ketone **1**.

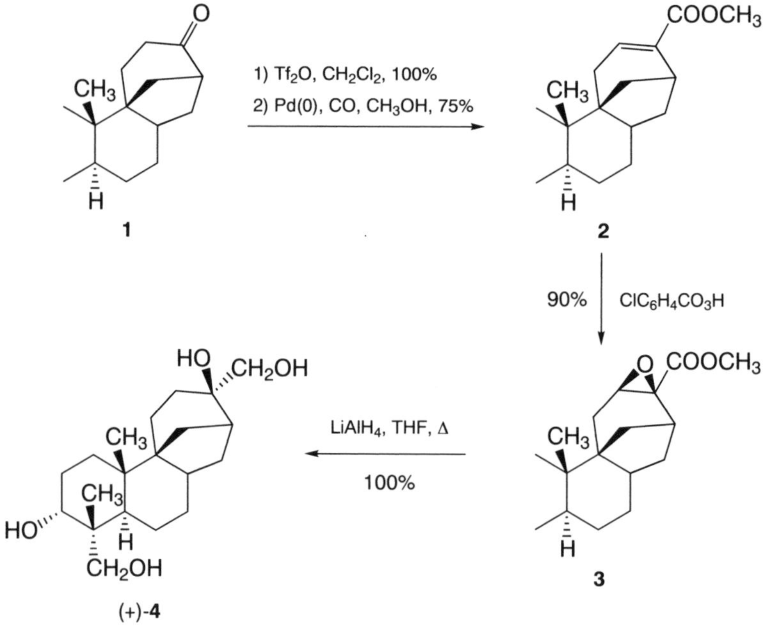

[1] C. J. Rizzo and A. B. Smith, III, *J. C. S. Perkin I,* 969 (1991).

Chloro(phenylethynyl)dimethylsilane, $Cl(CH_3)_2SiC \equiv CC_6H_5$ **(1).**

Stereoselective C-styryl glycosidation. Stork[1] has developed a stereoselective route to styryl C-glycosides by a radical intramolecular cyclization of an ethynyl group linked as a silyl ether to an α- or β-hydroxyl group of a phenylselenoglycoside. Thus reaction of the 3,4,6-tribenzyl ether of the 2-hydroxyphenylselenoglucoside **2** with **1** provides the tethered silyloxy precursor **3**. Reaction of **3** with Bu_3SnH and AIBN followed by desilylation (Bu_4NF) provides the α-C-styryl glucoside **4** in 83% yield.

In contrast cyclization and desilylation of **5**, in which the phenylethynylsilyl group is linked to the 3 β-hydroxyl group of a phenylselenoglucoside, provides the β-C-glucoside **6**. The styryl group of **4** and **6** is useful as such and as a precursor of an aldehyde or a carbinol. Note that a free hydroxy group is liberated in this cyclization and can be useful for subsequent transformation.

This overall process is also an efficient route to C-furanoside in the ribose series (equations I and II).

(I)

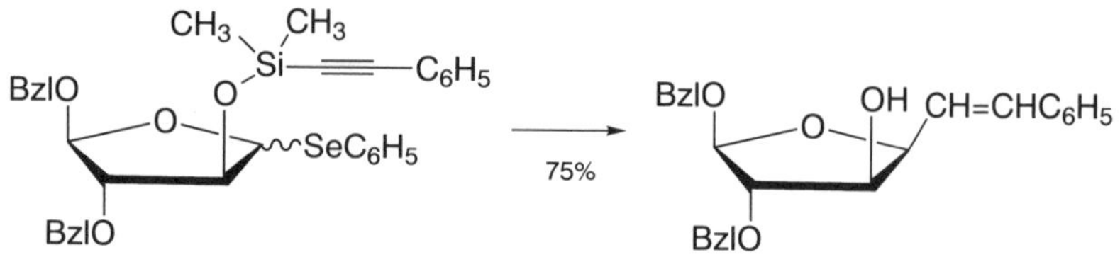

75%

(II)

80%

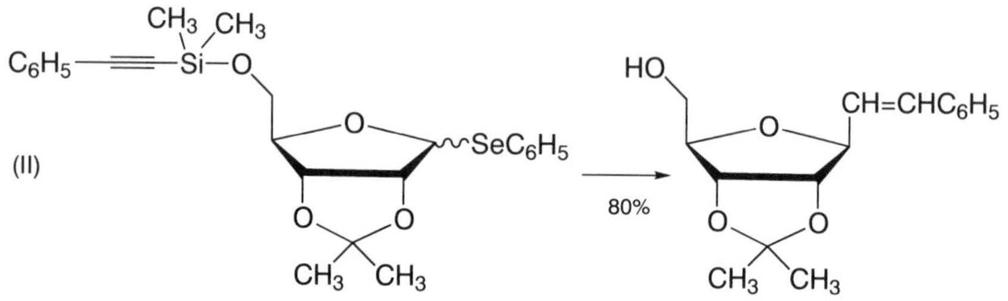

[1] G. Stork, H. S. Suh, and G. Kim, *Am. Soc.*, **113**, 7054 (1991).

1-Chloro-1-phenylsilacyclobutane, **(1)**

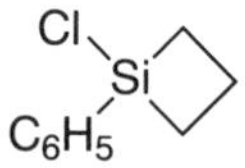

Preparation.[1]

Acceleration of aldol reactions.[2] In the absence of a catalyst, silyl enol ethers ordinarily do not react with aldehydes even at high temperatures. Thus the dimethylphenylsilyl ether **2a** does not react with C_6H_5CHO at 150°. In contrast, the silacyclobutane analog (**2b**), prepared from **1**, reacts with C_6H_5CHO at 27° to form *syn*- and *anti*-aldols in 12:1 ratio. This stereoselectivity is the opposite of that obtained by the $TiCl_4$-catalyzed reaction. 1-Chloro-1-methylsilacyclobutane, $CH_3(Cl)Si(CH_2)_3$, is even more effective than **1** for this aldol reaction. It is prepared from (3-chloropropyl)dichloromethylsilane.[3] It is particularly useful for aldol reactions with acyclic silyl ketene acetals (equation II).

(I)   + C_6H_5CHO ⟶

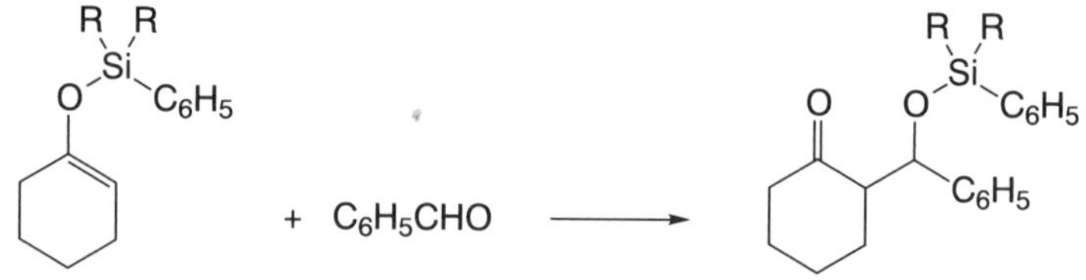

2a, R = CH_3 0%   **3b** (12 :1)
2b, R = $(CH_2)_3$, 100°   84%

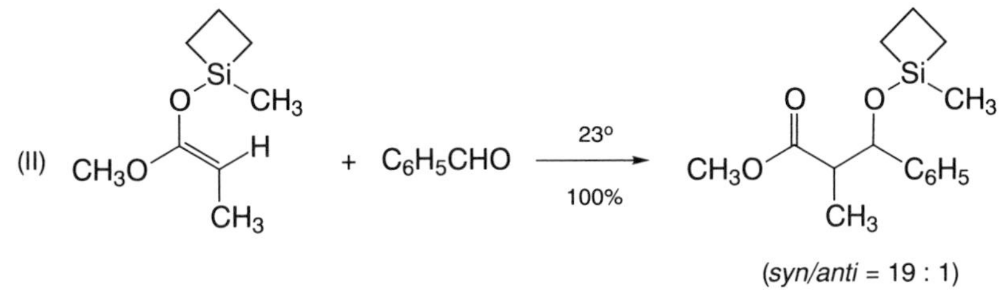

(II)

+ C_6H_5CHO

$\xrightarrow[100\%]{23°}$

(*syn/anti* = 19 : 1)

[1] N. Auner and J. Grobe, *J. Organomet. Chem.*, **188**, 25 (1980).
[2] A. G. Myers, S. E. Kephart, and H. Chen, *Am. Soc.*, **114**, 7922 (1992).
[3] K. V. Brown *et al.*, *J. Org.*, **56**, 698 (1991).

Chlorotrimethylsilane.

1,2-Addition of R_2CuLi to γ-alkoxy enals.[1] $ClSi(CH_3)_3$ is known to facilitate 1,4-addition of cuprates to enones and enals, but reactions of cuprates with γ-alkoxy enals in the presence of $ClSi(CH_3)_3$ results in exclusive 1,2-addition. Remarkably, this reaction shows significant 1,4-asymmetric induction.

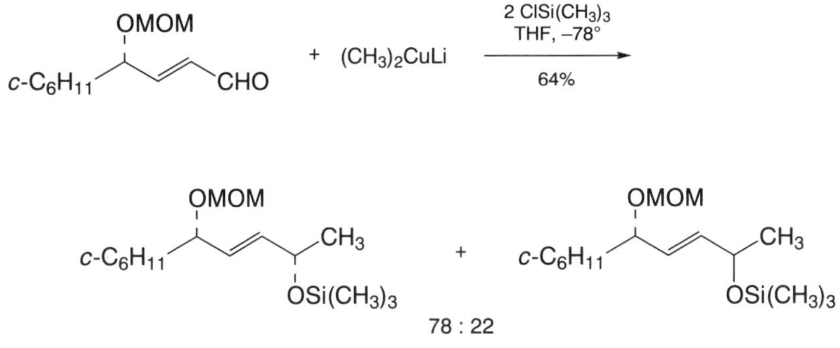

+ $(CH_3)_2CuLi$

$\xrightarrow[64\%]{\substack{2\ ClSi(CH_3)_3 \\ THF,\ -78°}}$

78 : 22

[1] M. Arai, T. Nemoto, Y. Ohashi, and E. Nakamura, *Synlett*, 309 (1992).

Chloro(triphenylphosphine)copper tetramer, $[CuCl(C_6H_5)_3P]_4$ (1).

Preparation.[1]

Stereoselective reactions with oxaziridines.[2]   This reagent can effect N—O cleavage to provide either pyrrolines or aziridines with marked selectivity (equations I and II).

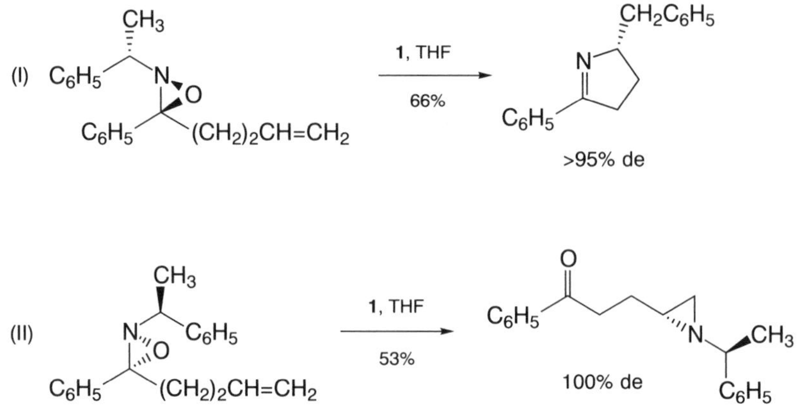

(I)

1, THF
66%

>95% de

(II)

1, THF
53%

100% de

[1] M. R. Churchill, S. A. Bezman, and J. A. Osborn, *J. Inorg. Chem.*, **11**, 1818 (1972).
[2] J. Aubé, X. Peng, Y. Wang, and F. Takusagawa, *Am. Soc.*, **114**, 5466 (1992).

Chromium aminocarbenes

Asymmetric synethesis of dipeptides.[1]   The chromium amino-carbene **1**, pentacarbonyl [(R-1–aza-2,2-dimethyl-3-oxa-5-phenylcyclopentyl) (methyl)carbene]chromium(0), derived from (R)-alanine, undergoes carbonylative coupling with amino esters under irradiation (visible light) at 0° to form (S,S)-dipeptides with high diastereoselectivity.

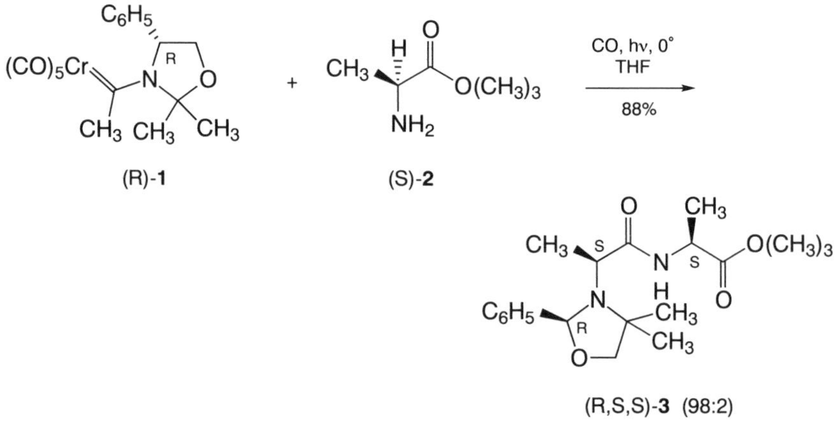

(R)-**1**   (S)-**2**

CO, hv, 0°
THF
88%

(R,S,S)-**3** (98:2)

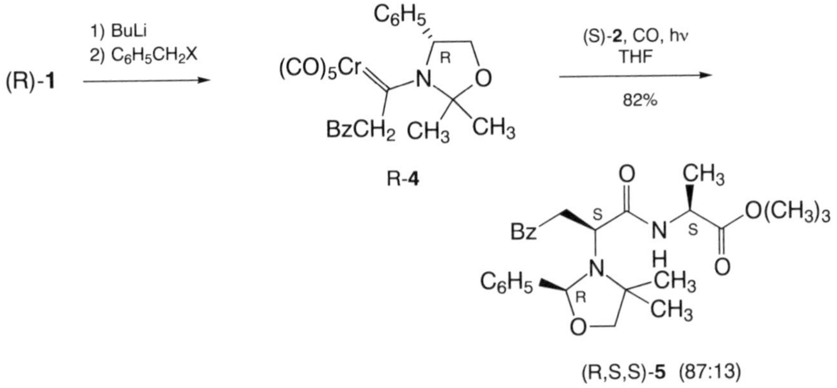

$$R\text{-}4$$

$$(R,S,S)\text{-}5 \ (87{:}13)$$

(R)-1 can also be converted into the (R)-chromium **4**, which couples with the *t*-butyl ester of (S)-alanine (**2**) to form the protected dipeptide **5**.

Cyclopropanation.[1] Chromium and molybdenum carbene complexes can react with unactivated 1,3-dienes as well as electron-rich or -poor dienes (**14**,91) to form vinylcyclopropanes with high regio- and diastereoselectivity. In fact, reactions with alkylmethoxycarbenes result in a single diastereomer. In addition, cyclopropanation occurs selectively with the least hindered double bond of the diene and more readily with 1,3-dienes that can adopt the s-*cis* conformation.

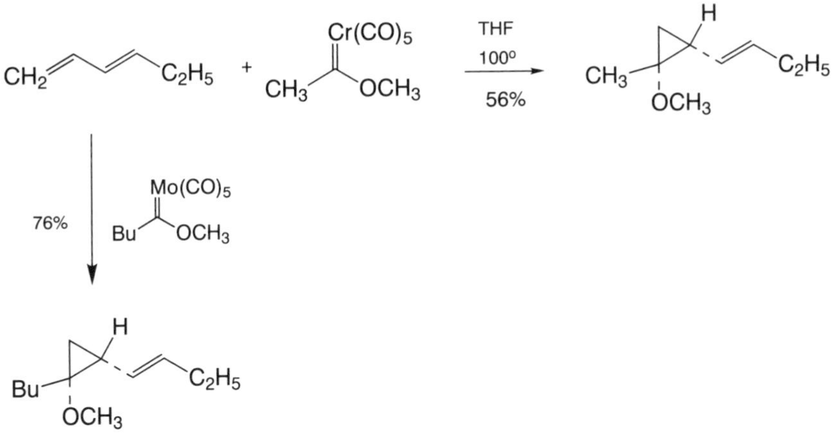

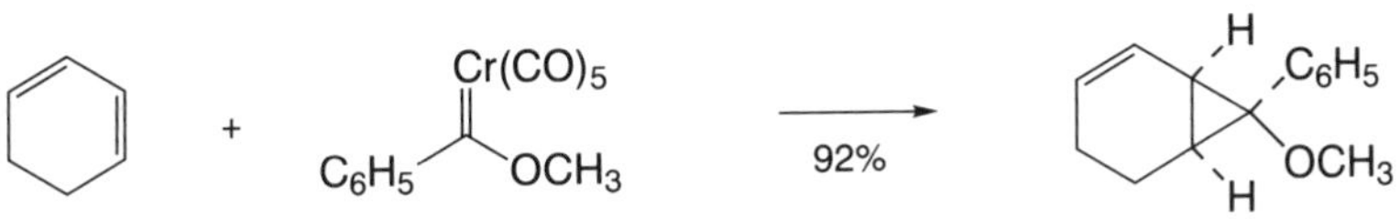

o-Alkoxyphenols.[2]   Chromium carbenes have been used for some time for synthesis of *p*-alkoxyphenols. A novel route to *o*-alkoxyphenols employs chromium dienylcarbenes. Thus the chromium carbene **1**, prepared as shown, when irradiated under carbon monoxide is converted into a complexed ketene (**a**) that cyclizes to **2** in 90% yield.

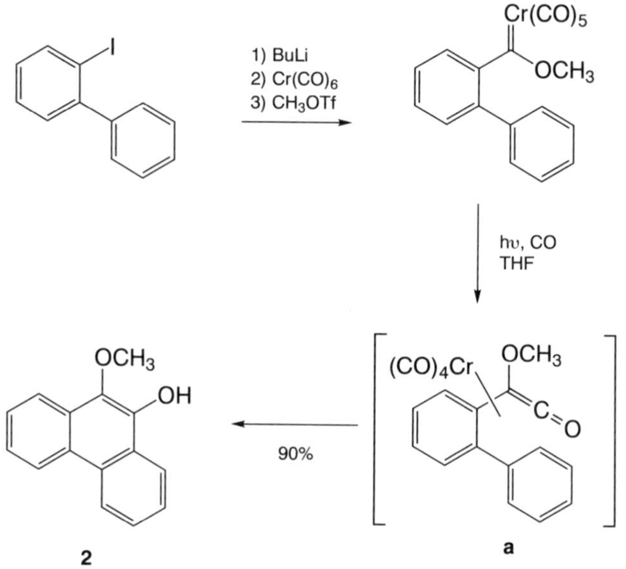

2   **a**

Optically active arylglycines.[3]   The optically active chromium–carbene complex **1**, prepared from (1S,2R)-2-amino-1,2-diphenylethanol, can be used for preparation of optically active arylglycines. Thus irradiation of **1** in CH_3CN containing DMAP

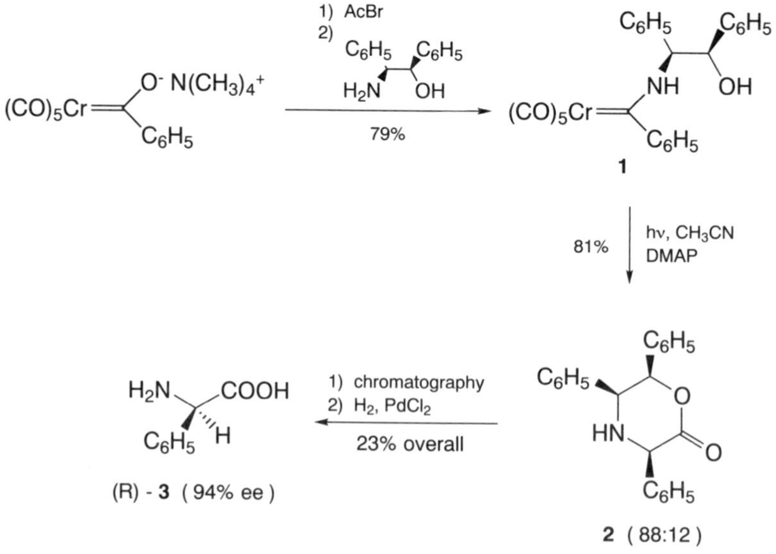

(R) - **3** (94% ee)

2 (88:12)

provides the oxazinone **2** in good yield, but modest diastereoselectivity (76% ee). Optically pure **2** can be obtained by chromatography. The final step involves removal of the chiral auxiliary, effected by hydrogenation.

Reaction with ketene ketals.[4]   The chromium carbene complex **1** reacts with the acyclic ketene ketal **2** to give two orthoesters, **3** and **4**, which are converted by aqueous acid to the butyrolactones **5** and **6**. In the example cited the carbene reacts with high selectivity with an ethyl group rather than a methyl group.

1   2

cis-**3** (76%)   **4** (<1%)

5   **6**

The reaction was used for a regioselective synthesis of eldanolide (**6**).

6 (*trans/cis* = 24:1)

(CO)$_5$Cr=C(OR)R complexes; allenes.[5]   Photolysis of a mixture of the chromium carbene **1** and the stabilized phosphorus ylide **2** under CO gives the allene **3** in 60% yield. Hydrolysis of **3** with H$_3$O$^+$ in ether provides the β-substituted (Z)-enone **4**. The allene **3** rearranges to a (Z)-1,3-diene (**5**). Pyridinium p-toluenesulfonamide is an effective catalyst for this rearrangement.

[1] J. R. Miller, S. R. Pulley, L. S. Hegedus, and S. DeLombaert, *Am. Soc.*, **114**, 5602 (1992).

[2] D. F. Harvey and K. P. Lund, *Am. Soc.*, **113**, 8916 (1991).

[3] C. A. Merlic and D. Xu, *ibid.*, **113**, 7418 (1991).

[4] J.-M. Vernier, L. S. Hegedus, and D. B. Miller, *J. Org.*, **57**, 6914 (1992).

[5] S. L. B. Wang, J. Su, W. D. Wulff, and K. Hoogsteen, *Am. Soc.*, **114**, 10665 (1992).

[6] M. R. Sestrick, M. Miller, and L. S. Hagedus, *ibid.*, **114**, 4079 (1992).

Chromium(II) chloride, $CrCl_2$.

Asymmetric addition of allylic bromides to aldehydes.[1]   The 1,2-addition fo crotyl bromides to aldehydes promoted by $CrCl_2$ (Hiyama reaction) provides homoallylic alcohols with *anti*-selectivity (**8**,111–112). A model study of the reaction of benzaldehyde with allylic bromides with a stereogenic group at the δ position indicates that this center controls the configuration at the generated centers (γ and β^1) to give as the major product the diastereomer (**2**) with the all-*syn* arrangement at three chiral centers. An additional stereocenter at the ε position of the allylic bromide merely increases the all-*syn* selectivity. Reactions of chiral allylic bromides with chiral aldehydes indicate that the configuration of the bromide can override that of the aldehyde. This reaction therefore can provide access to complex chiral adducts such as **4**, which is converted by standard

reactions into nephromopsinic acid, $(-)$-**5**.

[1] J. Mulzer, L. Kattner, A. R. Strecker, C. Schröder, J. Buschmann, C. Lehmann, and P. Luger, *Am. Soc.*, **113**, 4218 (1991).

Chromium(II) chloride–Lithium iodide (1:1).

Allenic alcohols. Propargylic halides react with carbonyl compounds in the presence of $CrCl_2 \cdot LiI$ (2 equiv.) in DMA to form allenic alcohols as the major products.[1] (Use of other solvents increases the formation of homopropargylic alcohols as minor products.) Ester, cyano, or chloro groups have no effect on the regioselectivity, but an α-substituent can favor formation of propargylic alcohols.

Addition of γ-disubstituted allylic phosphates to RCHO.[2]   Both (E)- and (Z)-allylic halides add to RCHO when catalyzed by $CrCl_2$ to form *anti*-homoallylic alcohols. Similar stereoselectivity obtains in additions of (E)- and (Z)-allylic phosphates. In this case addition of LiI and use of DMPU as solvent are usually required for satisfactory yields. In contrast, addition of γ-disubstituted phosphates to aldehydes is both stereoselective and stereodivergent (equations I and II).

(I) HexCHO + →(CrCl₂, LiI, 25°, 90%)→

99:1

(II) HexCHO + →(64%)→

98:2

[1] K. Belyk, M. J. Rozema, and P. Knochel, *J. Org.*, **57**, 4070 (1992).
[2] C. Jubert, S. Nowatny, D. Kornemann, I. Antes, C. E. Tucker, and P. Knochel, *J. Org.*, **57**, 6384 (1992).

Chromium(II) chloride–Nickel(II) chloride.

Cycloenediynols.[1]   The ω-iodo enediynals **1a** and **1b**, when treated with $CrCl_2$ (5–8 equiv.) and a catalytic amount of $NiCl_2$ (**14**,98), undergo an intramolecular addition of the alkynyl iodide to the aldehyde group to produce 10- and 11-membered cycloenediynols, isolated as the acetates **2a** and **2b**. These products are unstable, but **2b** was shown to undergo cycloaromatization to **3** when treated with 1,4-cyclohexadiene. This reaction is typical of some antitumor antibiotics and is believed to be responsible for the biological activity.

| **1a**, n = 1 | 34% | **2a** |
| **1b**, n = 2 | 76% | **2b** |

3

[1] C. Crévisy and J.-M. Beau, *Tetrahedron Letters*, **32**, 3171 (1991).

Chromium(VI) oxide (CrO₃).

$RCH=CH_2 \rightarrow RCH_2COOH$.[1]   This conversion can be effected in >80% overall yield by hydroboration of 1-alkenes with $HBBr_2 \cdot S(CH_3)_3$, $H_2BBr \cdot S(CH_3)_2$, thexyl-borane (H_2BThx), or dicyclohexylborane ($HBChx_2$), followed by hydrolysis to form $RCH_2B(OH)_2$. The final step is oxidation of $RCH_2B(OH)_2$ to $RCOOH$, which can be effected with CrO_3 in 90% aqueous acetic acid, pyridinium dichromate (PDC), or sodium dichromate (Na_2CrO_3) in aqueous H_2SO_4. In general the highest overall yields (80−92%) are obtained with CrO_3 in CH_3COOH/H_2O.

[1] H. C. Brown, S. V. Kulkarni, V. V. Khanna, V. D. Patil, and U. S. Racherla, *J. Org.*, **57**, 6173 (1992).

Cobalt(II) acetylacetonate, $Co(acac)_2$

Intramolecular homo Diels−Alder reaction of dienynes.[1]   Norbornadienes bearing a tethered alkyne group at C_2 undergo an intramolecular homo Diels−Alder reaction when treated with 8 mol% of $Co(acac)_2$, $(C_2H_5)_2AlCl$ (4 equiv.), and dppe at 25°. This reaction is particularly facile when a 5- or 6-membered ring is formed. No reaction is observed on thermal treatment of **1** at 140−170°.

R = H, n = 1   76%
R = H, n = 2   64%
R = CH_3, n = 1   69%

[1] M. Lautens, W. Tam, and L. G. Edwards, *J. Org.*, **57**, 8 (1992).

Copper(I) iodide−Tetrabutylammonium chloride.

Skipped diynes and triynes.   A new route to skipped diynes involves coupling of propargylic halides or tosylates with 1-alkynes catalyzed by CuI, $NaCO_3$, and Bu_4NCl in DMF or CH_3CN.[1] This reaction can be extended to skipped triynes (equation II).

(I)   Et—≡—CH$_2$I   +

CuI, Bu$_4$NCl
NaCO$_3$, DMF
────────────→
82%

(II)

H—≡—CH$_2$OH

+

C$_8$H$_{17}$—≡—CH$_2$Br

──────→
87%

CBr$_4$ / P(C$_6$H$_5$)$_3$

←────── H—≡—(CH$_2$)$_3$CO$_2$CH$_3$ ←──────
51%

The skipped di- or triynes are reduced to skipped (Z,Z)-dienes or (Z,Z,Z)-trienes by partial reduction with NaBH$_4$/Ni(OAc)$_2$ · 4H$_2$O (equation III).

(III)   C$_8$H$_{17}$—≡—≡—CH$_2$OH   ──────→   90%

C$_8$H$_{17}$ CH$_2$OH

[1] T. Jeffery, S. Gueugnot, and G. Linstrumelle, *Tetrahedron Letters,* **33**, 5757 (1992).

Copper(I) iodide–Triphenylphosphine.

Coupling of 1-alkynes and ArX.[1]   This coupling has been effected with Pd catalysts, but can also be effected with CuI/P(C$_6$H$_5$)$_3$ (1:2) in the presence of K$_2$CO$_3$ at 80–120°.

C$_6$H$_5$I   +   HC≡CC$_6$H$_5$

CuI/P(C$_6$H$_5$)$_3$, K$_2$CO$_3$
DMF , 120°
──────────────→
98%

C$_6$H$_5$C≡CC$_6$H$_5$

C$_6$H$_5$CH=CHBr   +   HC≡CC$_5$H$_{11}$-*n*   ──────→   C$_6$H$_5$CH=CHC≡CC$_5$H$_{11}$-*n*
(E/Z = 99:1)   85%   (E/Z = 99:1)

[1] K. Okuro, M. Furuune, M. Miura, and M. Nomura, *Tetrahedron Letters,* **33**, 5363 (1992).

Cyanotrimethylsilane.

Cyanosilylation of aldehydes.[1]   This reaction has been effected by use of a Lewis acid catalyst such as ZnI_2 or $AlCl_3$ (**4**,542–543). Amines such as triethylamine or ethyldiisopropylamine are also highly effective catalysts, providing cyanohydrin trimethylsilyl ethers in 90–100% yield. Presumably, the reactive intermediate is a pentavalent silicate, $(CH_3)_3Si(CN)N(C_2H_5)_5$. Tributylphosphine is as effective as a tertiary amine. The reaction can proceed in high enantioselectivity (90% ee) in the presence of the chiral tin(II) Lewis acid **1**, prepared by reaction of (+)-cinchonine with triflic acid and 1,1'-dimethylstannocene (**12**,201–202).

1

(90% ee)

[1] S. Kobayashi, Y. Tsuchiya, and T. Mukaiyama, *Chem. Letters,* **537**, 541 (1991).

Cyclam (1,4,8,11-tetraazacyclotetradecane), **1**.

(1)

Epoxidation.[1]   An iron complex prepared by reaction of $Fe(CF_3SO_3)_2$ with 1 equiv. of **1** is an effective catalyst for epoxidation of alkenes with H_2O_2 (30%) in CH_3OH or CH_3CN. Thus reaction with cyclohexene provides the epoxide as the main product

40%   17%

26%   <2%

with only traces of the allylic alcohol. *cis*-Stilbene is oxidized with high stereoselectivity to the *cis*-epoxide. Iodosylbenzene, C_6H_6IO, can also be used as the oxidant for this reaction.

[1] W. Nam, R. Ho, and J. S. Valentine, *Am. Soc.*, **113**, 7052 (1991).

Cyclic sulfates, sulfites, 15, 105–107; 165.

Review.[1]   These products have become of special interest since they are now available in optically pure form by asymmetric dihydroxylation of alkenes. In addition, the regioselectivity in reactions with nucleophiles can differ from that of epoxides and the yields can be superior. The reference covers literature from 1863 through 1991 (104 references).

[1] B. B. Lohray, *Synthesis*, 1035 (1992).

Cyclobutenediones.

Naphthoquinones from cyclobutenediones (**13**, 97–98; 209–120; **14**, 253–254; **16**, 102*)*.   Liebeskind *et al.*[1] have developed a stereocontrolled route to 2,3-disubstituted quinones from 2-alkyl-3-alkoxycyclobutenediones (**1**). Thus addition of ArLi to **1** followed by methylation provides **2** in 50–75% yield. Reaction of **2** with RLi followed by a quench with trifluoroacetic anhydride and NH_4Cl provides the cyclobutenone **3**. These products are converted into quinones **4** by pyrolysis at 140° followed by oxidation (CAN).

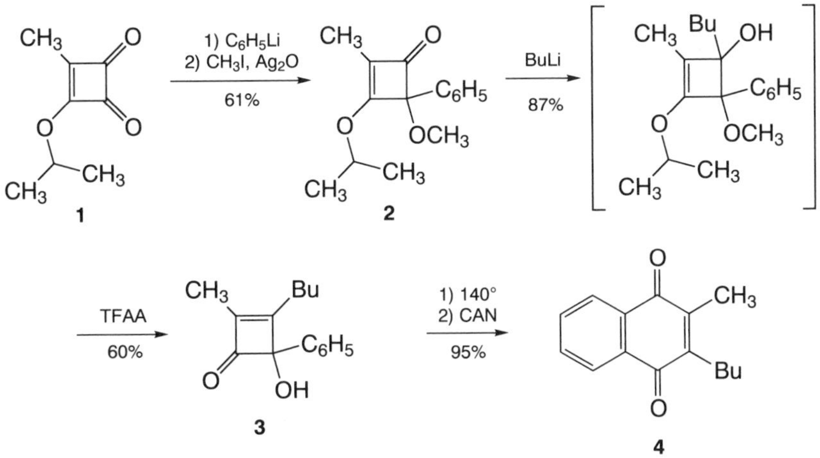

[1] L. S. Liebeskind, K. L. Granberg, and J. Zhang, *J. Org.*, **57**, 4345 (1992).

1S,2S-Cyclohexanediol,

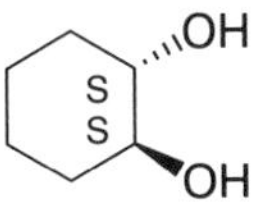

Asymmetric alkylation of β-keto esters.[1]   The α-carbomethoxycyclopentanone, when protected as the acetal **2** with (1S,2S)-cyclohexanadiol, undergoes highly stereoselective alkylation (equation I).

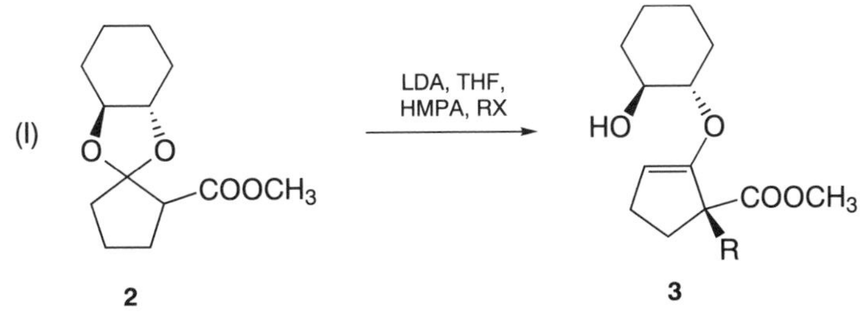

RX	= CH₃I	57%	92% de
	= C₁₉H₁₉Br	66%	>99% de

[1] K. Kato, H. Suemune, and K. Sakai, *Tetrahedron Letters*, **33**, 247 (1992).

(R,R)- and (S,S)-1,2-Cyclohexanedi(2,2–dimethylpropyl)amine.

Asymmetric olefination of methylcyclohexanones **(12, 396).**   The asymmetric ethylidenation of 4-methylcyclohexanone with the chiral phosphonamide R,R-**1**, derived from (R,R-1,2-cyclohexanedi(2,2-dimethylpropylamine), provides the (S)-ethylidene **2** in high enantioselectivity. This product undergoes an ene reaction with 2,3-di-O-benzyl-D-glyceraldehyde (**3**) to form the triol **4** (>90:10 selectivity), with two new chiral centers. Ozonization of **4** provides the optically active acyclic **5** with 1,5-*anti*-selectivity.

(R,R) - **1**

(S) - **2**, α_D + 23°

5, α_D - 2.6°

4 (>90 : 10)

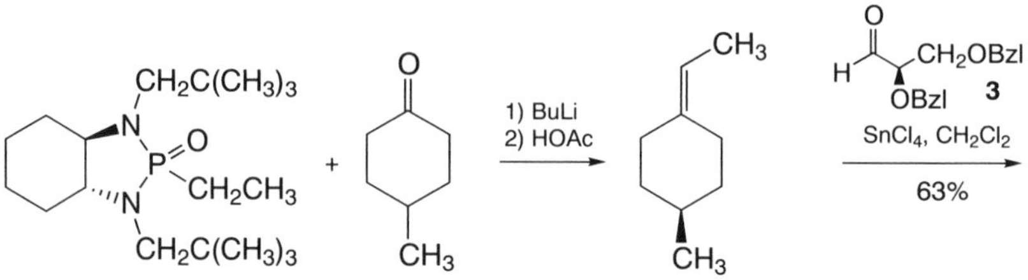

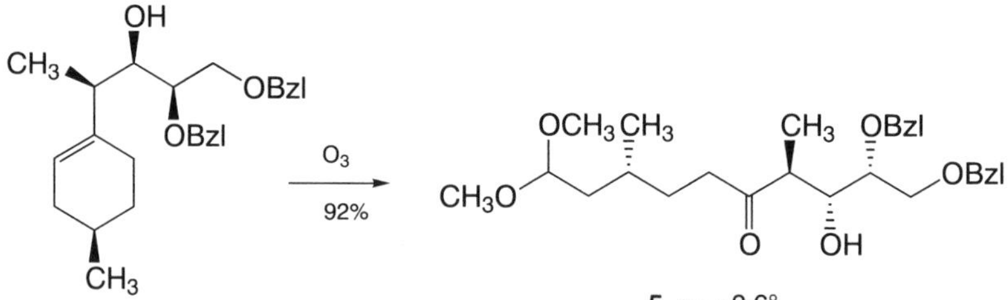

A similar reaction with (R)-**6** and 2,3-O-dibenzyl-L-glyceraldehyde followed by oxidative cleavage provides, 1,6-*anti*-**7** in 60% yield.[1]

(R) - **6**

7

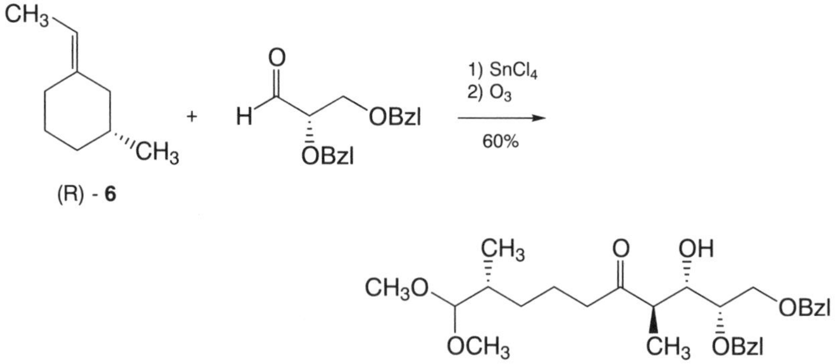

[1] S. Hanessian and S. Beaudoin, *Tetrahedron Letters,* **33**, 7659 (1992).

(−)-*trans*-Cyclohexane-(1R,2R)-disulfonamides, [chemical structure: cyclohexane with NHSO$_2$R substituents]   **16**, 102–103.

Enantioselective cyclopropanation of allylic alcohols.[1]   Enantioselective cyclopropanation of allylic alcohols is possible with the Simmons–Smith reagent catalyzed by this C$_2$-symmetric disulfonamide **1**, R = C$_6$H$_4$NO$_2$-*p*.

$$C_6H_5 \diagup\!\!\!\diagdown OH \xrightarrow[\substack{82\%}]{\substack{(C_2H_5)_2Zn/CH_2I_2 \\ \textbf{1}, CH_2Cl_2, -23°}} C_6H_5 \triangleright\!\!\diagdown OH$$

76% ee

[1] H. Tokahashi, M. Yoshioka, M. Ohno, and S. Kobayashi, *Tetrahedron Letters*, **33**, 2575 (1992).

(Cyclooctadiene)(cyclooctatriene)ruthenium, Ru(COD)(COT).

Coupling of 1,3-dienes with acrylates or acrylamides.[1]   Ru(COD)(COT) is the most efficient catalyst for this coupling to provide derivatives of 3,5-dienoic acids. In the absence of the Ru catalyst, Diels–Alder adducts are obtained.

$$CH_2\!=\!\diagup\!\!\diagdown\!CH_2 + CH_2\!=\!\diagdown CON(CH_3)_2 \xrightarrow[\substack{51\%}]{\substack{Ru(COD)(COT) \\ 80°}}$$

[chemical structures] ...CON(CH$_3$)$_2$ with CH$_3$ +   ...CON(CH$_3$)$_2$ with CH$_3$

80 : 20

[2 + 2]Cycloadditions.[2]   In the presence of this catalyst and a base, allylamines and acrylates undergo cycloaddition to give cyclobutane-β-aminocarboxylic acid derivatives. The reaction involves prior isomerization of the allylamine to an enamine, which then undergoes thermal cycloaddition with the acrylate.

$$CH_2\!=\!CHCH_2N(C_2H_5)_2 \xrightarrow{Ru} \left[CH_3CH\!=\!CHN(C_2H_5)_2\right] \xrightarrow[\substack{85\%}]{\substack{CH_2=CHCOOCH_3 \\ \text{[piperidine-NCH}_3\text{]}, 70°}}$$

[chemical structure: cyclobutane with COOCH$_3$, CH$_3$, N(C$_2$H$_5$)$_2$]

[1] T. Mitsudo, S.-W. Zhang, T. Kondo, and Y. Watanabe, *Tetrahedron Letters*, **33**, 341 (1992).
[2] T. Mitsudo, S.-W. Zhang, and N. Satake, T. Kondo, and Y. Watanabe, *ibid.*, **33**, 5533 (1992).

(Cyclopentadienone)(cyclopentadienyl)dicarbonylmolybdenum hexafluorophosphate (1).

Preparation:

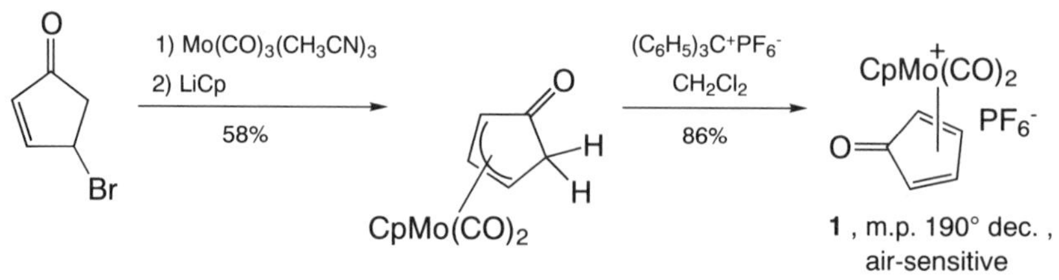

cis-4,5-Disubstituted-2-cyclopentenones.[1]   Although nucleophiles react with un-complexed cyclopentadienone β to the carbonyl group, **1** reacts with nucleophiles α to the carbonyl group of the terminus of a diene group. Thus reaction of **1** with a Grignard reagent provides **2**, which undergoes decomplexation by protonation with TFA to **3**. Oxidative decomplexation of **2** with $C_6H_5I(OCOCF_3)_2$ provides the *cis* 4,5-disubstituted cyclopentenone **4**.

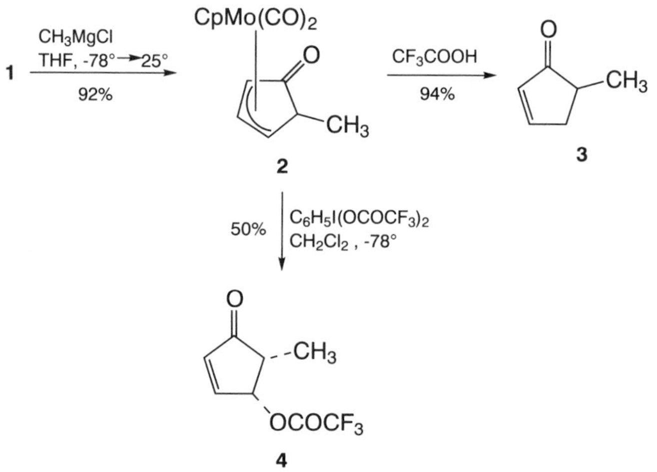

[1] L. S. Liebeskind and A. Bombrun, *Am. Soc.*, **113**, 8736 (1991).

Cyclopentadienyltantalum(V) carboxylates, $CpTaCl_3(OCOR)$.

These chemicals are prepared by reaction of a carboxylic acid with $CpTaCl_4$.

Acylation.[1]   These complexes react readily with amines or amino acids to form amides or dipeptides in high yield. The reaction with amino acid complexes provides

$$\text{CpTaCl}_3[\text{OCOCH(CH}_3)_2] \quad + \quad \text{C}_6\text{H}_5\text{CH}_2\text{NH}_2 \quad \longrightarrow$$

$$\text{CpTa(O)Cl}_2 \quad + \quad \text{C}_6\text{H}_5\text{CH}_2\text{NHC(O)CH(CH}_3)_2$$

90%

amides in 75–81% yield with little racemization.

[1] K. Joshi, J. Bao, A. S. Goldman, and J. Kohn, *Am. Soc.,* **114**, 6649 (1992).

D

Dialkylbis(cyclopentadienyl)zirconium, Cp_2ZrR_2.

Hydrosilylation of 1-alkenes.[1] These reagents are available by reaction of Cp_2ZrCl_2 with excess BuMgBr or C_2H_5MgBr in THF at $-78°$. Reaction of 1-octene with $H_2Si(C_6H_5)_2$ catalyzed by either $Cp_2Zr(C_6H_5)_2$ or Cp_2ZrBu_2 results in n-OctSiH$(C_6H_5)_2$ in 73–75% yield with >99% regioselectivity (equation I).

$$(I) \quad n\text{-Hex-CH=CH}_2 + H_2Si(C_6H_5)_2 \xrightarrow[\text{73 - 75\%}]{Cp_2ZrR_2} n\text{-Hex-CH}_2CH_2SiH(C_6H_5)_2$$

[1] T. Takahashi, M. Hasegawa, N. Suzuki, M. Saburi, C. J. Rousset, P. E. Fanwick, and E. Negishi, *Am. Soc.*, **113**, 8564 (1991).

Dialkylzinc, R_2Zn.

These compounds are relatively stable in sealed tubes, but ignite when exposed to air; they are very sensitive to moisture. Only diethylzinc is readily available in solutions packaged under nitrogen. These hazardous compounds can be generated relatively safely *in situ* from Grignard reagents in ether by transmetallation with $ZnCl_2$ (equation I).[1] This reaction generates R_2Zn and MgX_2, which is precipitated as a complex with

$$(I) \quad 2R_2MgX + ZnCl_2 \rightleftharpoons[\text{ether}] R_2Zn + MgX_2 + MgCl_2$$

$$\downarrow \text{dioxane}$$

$$X_2Mg\cdot\text{dioxane}$$

dioxane. Filtration under an inert atmosphere gives a solution of R_2Zn for further reactions. This transmetalation is slow in the case of hindered Grignard reagents.

[1] D. Seebach, L. Behrendt, and D. Felix, *Angew. Chem. Int. Ed.*, **30**, 1008 (1991).

Di-μ-allylpalladuim chloride), $[(C_3H_5)ClPd]_2$.

α-Cyclopropyl esters.[1] In the presence of TMEDA (large excess) and carbon monoxide, this π-allylpalladium complex reacts with enolates of carboxylic esters to form 2-cyclopropyl esters as the major product. The minor products of allylic alkylation

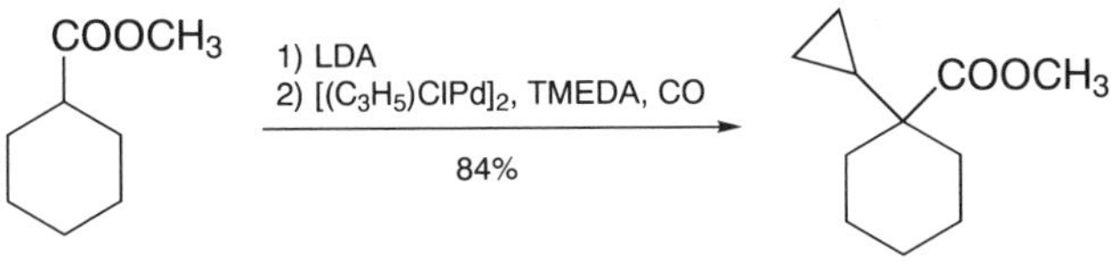

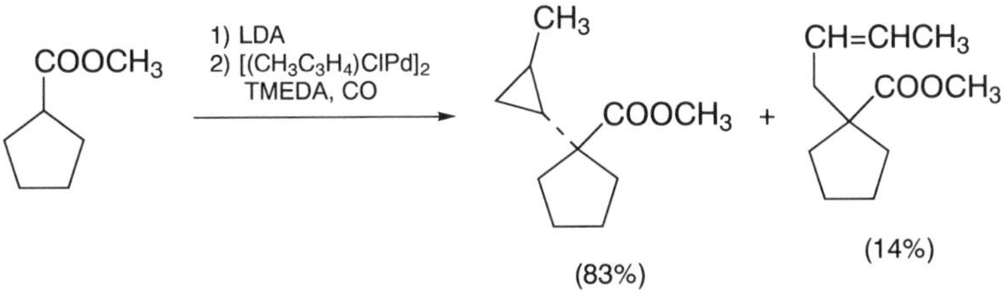

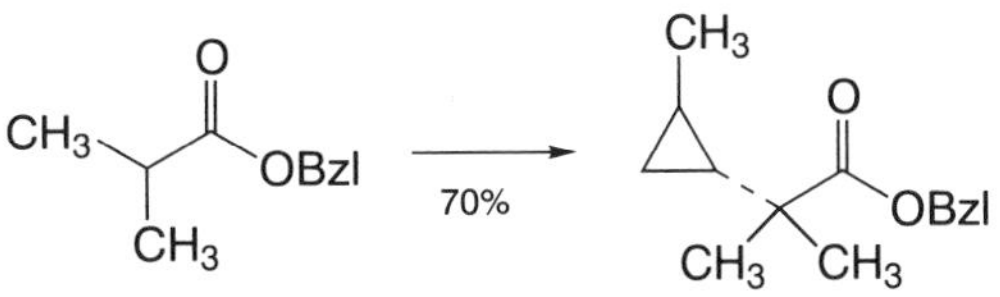

can be isolated after osmylation to the 1,2-diols.

[1] H. M. R. Hoffmann, A. R. Otte, and A. Wilde, *Angew. Chem. Int. Ed.,* **31**, 234 (1992).

(1R,2R)- or (1S,2S)-*trans*-1,2-Diaminocyclohexane.

Asymmetric epoxidation.[1]   Jacobsen's salen-based catalyst **1**, derived from (R,R)- or (S,S)-diphenyl-1,2-diaminoethane (**16**,157) can effect asymmetric epoxidation of alkenes with NaOCl, but the enantioselectivity is generally only moderate (~70% ee) in the case of *cis*-alkenes. Subsequently, this group has examined salen-based catalysts

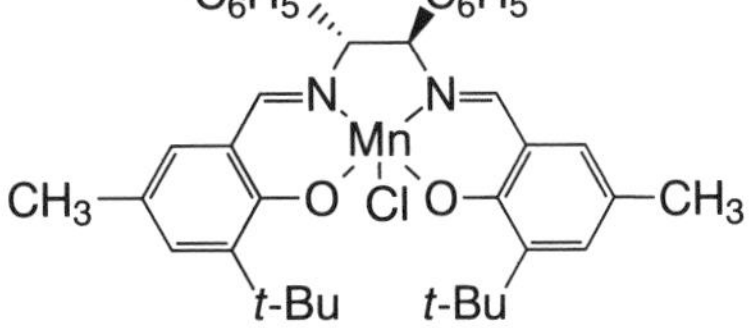

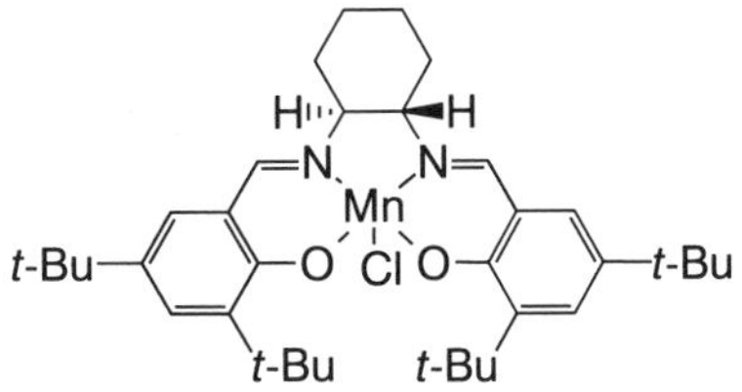

derived from the readily available *trans*-1,2-diaminocyclohexane, which can be resolved by crystallization with tartaric acid. Of these new catalysts, **2** proved to effect epoxidation of *cis*-alkenes with >90% ee. Thus it effects epoxidation of *cis*-β-methylstyrene in 92% ee and of a ketal in 94% ee. Even an α, β-unsaturated ester can be epoxidized in 89% ee, but in this case the reaction was conducted in the presence of 4-phenylpyridine N-oxide.

84%, 92% ee   63%, 94% ee   65%, 89% ee

Asymmetric epoxidation of 1,3-dienes and eynes. Jacobsen has extended his method for effecting asymmetric epoxidation of alkenes with a chiral (salen)Mn (II) complex (**16**, 157–158) to conjugated dienes and enynes by use of the chiral (salen)Mn(III) complex **2**.[2]

Epoxidation of 1,3-dienes with NaOCl catalyzed by S,S-**2** provides only mono-epoxides with only moderate enantioselectivity (~45% ee). In the case of (Z,E)-dienes, epoxidation occurs with high selectivity (10:1) at the (Z)-alkene (equation I). In contrast, high enantioselectivity can be obtained with *cis*-enynes (equation II).

(*trans/cis* = 7 : 1), 66%ee

(*trans/cis* = 2.5 : 1), 90%ee

Asymmetric oxidation of ArSR with H_2O_2.[3]   The chiral (salen)MnCl complexes derived from this diamine can effect asymmetric oxidation of alkyl aryl sulfides. The highest enantioselectivity is obtained with (R,R)-**3**.

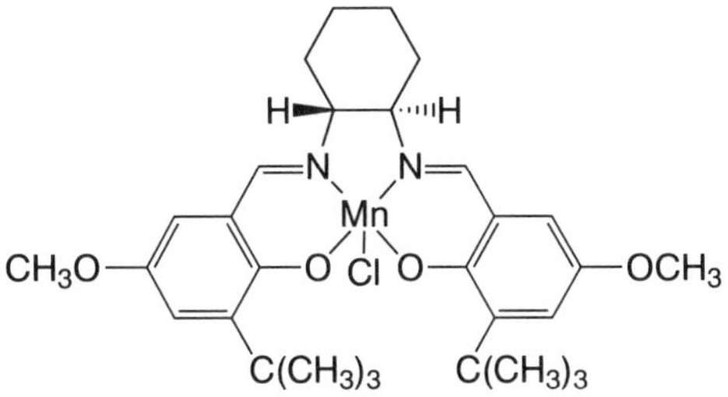

(R,R)-**3**

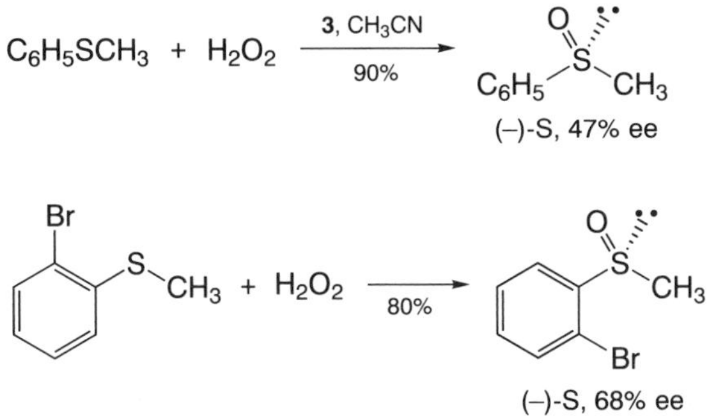

$C_6H_5SCH_3$ + H_2O_2 $\xrightarrow[90\%]{\textbf{3, }CH_3CN}$

(–)-S, 47% ee

(–)-S, 68% ee

[1] E. N. Jacobsen, W. Zhang, A. R. Muci, J. R. Ecker, and L. Deng, *Am. Soc.*, **113**, 7063 (1991).
[2] N. H. Lee and E. N. Jacobsen, *Tetrahedron Letters*, **32**, 6533 (1991).
[3] M. Palucki, P. Hanson, and E. N. Jacobsen, *ibid.*, **33**, 7111 (1992).

1,8-Diazabicyclo[5.4.0]undecene-7 (DBU).

α-Alkyl α, β-enones.[1]   These products can be prepared by addition of an enone to a Michael acceptor with DBU (0.20 equiv.) in 1,3-dimethyl-2-imidazolidinone (DMEU).

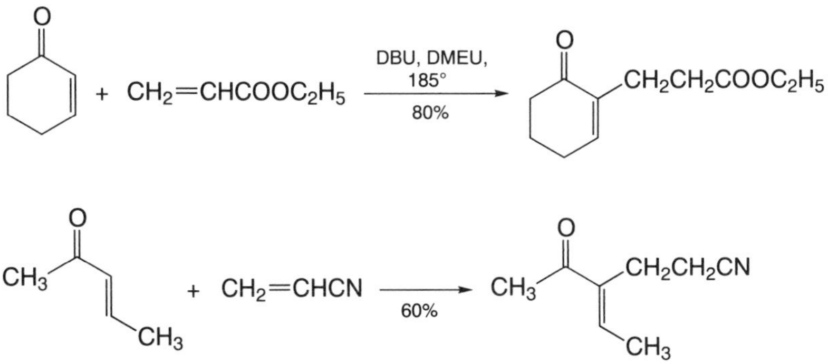

+ CH_2=$CHCOOC_2H_5$ $\xrightarrow[80\%]{\substack{DBU, DMEU, \\ 185°}}$

+ CH_2=$CHCN$ $\xrightarrow{60\%}$

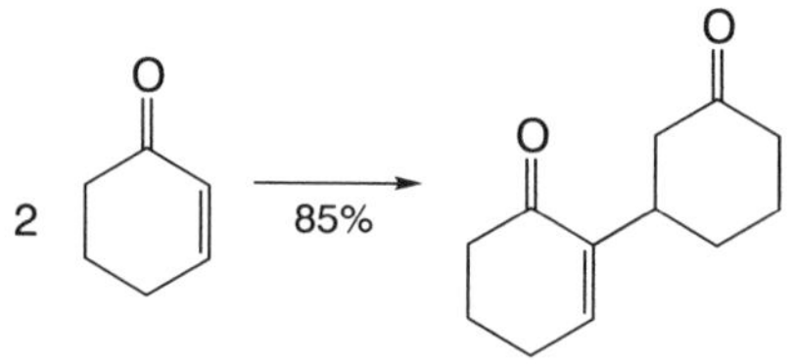

1,8-Diazabicyclo[5.4.0]-7-undecene hydrobromide, 1.

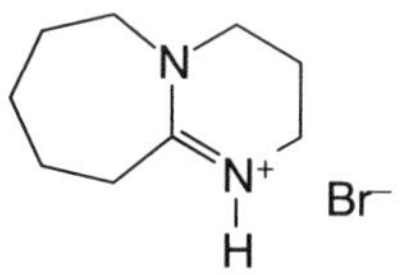

1, m.p. 120 - 122°

Prepared by reaction of DBU with Br$_2$ in HOAc in the presence of HBr.

Bromination of arenes.[2] This reagent brominates activated arenes at room temperature in DMF/H$_2$O (phenol → tribromophenol, 89% yield). Less reactive aromatics are brominated by **1** in the presence of HgCl$_2$ as catalyst (mesitylene → 2-bromomesitylene, 76% yield). Polycyclic arenes are brominated by **1** in refluxing acetic acid (naphthalene → 1-bromonaphthalene, 65% yield).

[1] J. R. Hwu, G. H. Hakimelahi, and C.-T. Chou, *Tetrahedron Letters,* **33**, 6469 (1992).
[2] H. A. Muathen, *J. Org.,* **57**, 2740 (1992).

Diazomethane.

Cyclopropanation.[1] This reaction can be conducted with diazomethane generated *in situ* from N-methyl-N-nitrosourea (excess) with KOH in CH$_2$Cl$_2$/O(C$_2$H$_5$)$_2$. A number of Pd compounds catalyze the cyclopropanation and are equally effective for *ex situ* and *in situ* reactions.

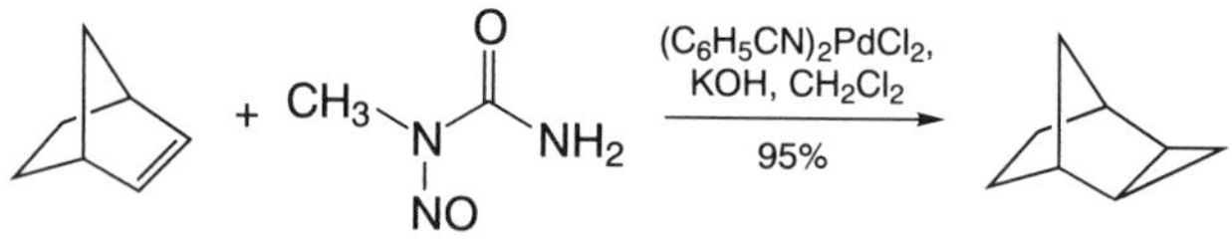

[1] O. M. Nefedov, Y. V. Tomilov, A. B. Kustitsyn, U. M. Dzhemilev, and V. A. Dokitchev, *Mendeleev Comm.,* 13 (1992).

Diborane.

Diastereoselective reduction of 4-keto-2-alkylborate esters (1).   This reaction can be effected with several reductants, but the highest diastereoselectivity (*anti*) is obtained with $BH_3 \cdot THF$ at $-78°$ in THF.[1]

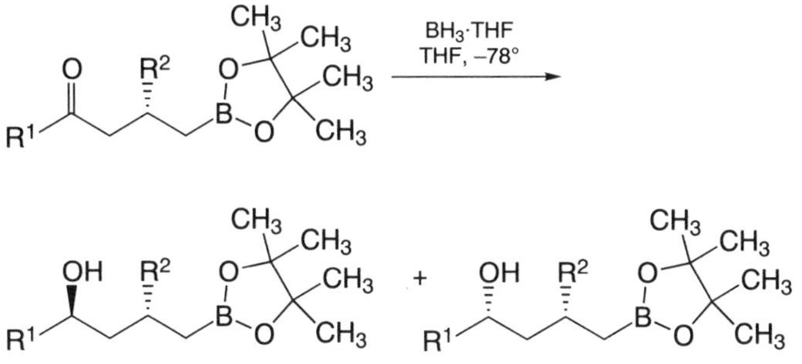

R¹ = CH₃, R² = CH₃	71%	34:1
$R^1 = CH_3$, $R^2 = i\text{-Pr}$	89%	62:1
$R^1 = i\text{-Pr}$, $R^2 = CH_3$	92%	>50:1

[1] G. A. Molander, K. L. Bobbitt, and C. K. Murray, *Am. Soc.*, **114**, 2759 (1992).

Dicarbonylbis(triphenylphosphine)palladium, $Pd(CO)_2[P(C_6H_5)_3]_2$ **(1)**

1,4-Silylstannation of 1,3-dienes.[1]   This is the most effective catalyst for this reaction, which can be highly regio- and (E)-selective. Thus the reaction of (trimethylsilyl)tributylstannane with butadiene provides a single (E)-alkene (equation I). Other organosilylstannanes with more bulky groups are less reactive and the products are obtained in lower yields.

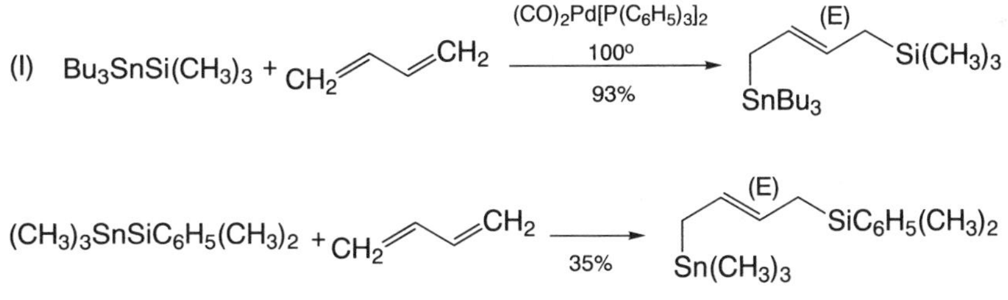

[1] Y. Tsuji and Y. Obora, *Am. Soc.*, **113**, 9368 (1991).

Dicarbonylcyclopentadienylcobalt, CpCO(Co)₂.

[2+2+2]Cycloaddition (cf. **12**,163).[1]   In the presence of CpCo(CO)₂, the mono-cyclic enyne (Z)-**1** cyclizes to the tetracyclic diene (**2**) as a 2:1 mixture of epimers. The product has the carbon skeleton present in the diterpene stemodin (**3**), and can be converted in three steps to an intermediate in a total synthesis of the diterpene. This cycloaddition fails with (E)-**1**.

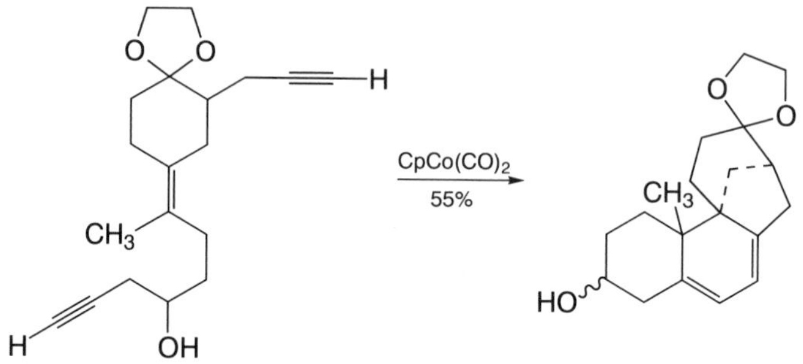

[1] J. Germanas, C. Aubert, and K. P. C. Vollhardt, *Am. Soc.,* **113**, 4006 (1991).

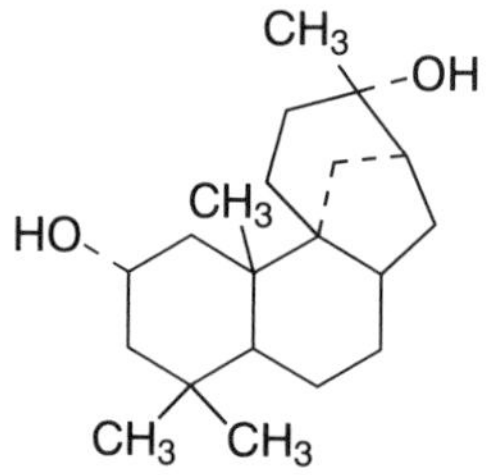

Di-μ-carbonylhexacarbonyldicobalt, Co₂(CO)₈.

Ring expansion-carbonylation.[1]   The ring expansion of aziridines to β-lactams by a Rh(I) catalyst (**15**,82–83) has been extended to expansion of pyrrolidines to piperidones by cobalt carbonyl-catalyzed carbonylation (equation I).

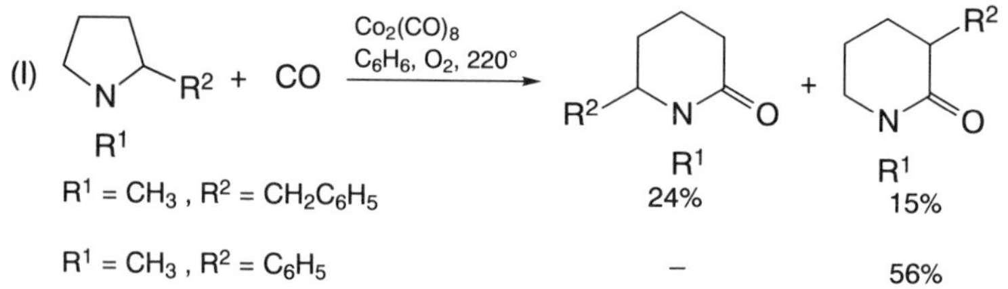

The yield of the carbonylation−expansion reaction is generally improved when catalyzed by both $Co_2(CO)_8$ and $Ru_3(CO)_{12}$, dodecacarbonyl-tri-*triangulo*-ruthenium (equation III).

(II)   + CO $\xrightarrow[46\%]{Co_2(CO)_8}$

+ $Ru_3(CO)_{12}$ 79%

This combined catalyst can also effect cyclization of 2,6-dimethylpiperidinyl ke-tones to tetrahydroindolizines (equation III), and this cyclization can predominate over rearrangement (equation IV).

(III) $\xrightarrow[86\%]{\substack{CO \\ Co_2(CO)_8,\ Ru_3(CO)_{12}}}$

(IV) $\xrightarrow[77\%]{\substack{CO \\ Co_2(CO)_8,\ Ru_3(CO)_{12}}}$ 10:1

Cyclic enediynes.   The realization that a number of antitumor antibiotics contain an enediyne core which is essential for their activity has led to extensive methods for synthesis of this ring system. One approach involves an aldol reaction with an acetylenic aldehyde protected as the $Co_2(CO)_6$ adduct. Thus the keto aldehyde **1**, prepared in five conventional steps from cyclopentene-1,3-dione, on treatment with dibutylboron triflate and $N(C_2H_5)_3$ cyclizes to **2** in 69% yield. Attempted decomplexation with N-methyl-

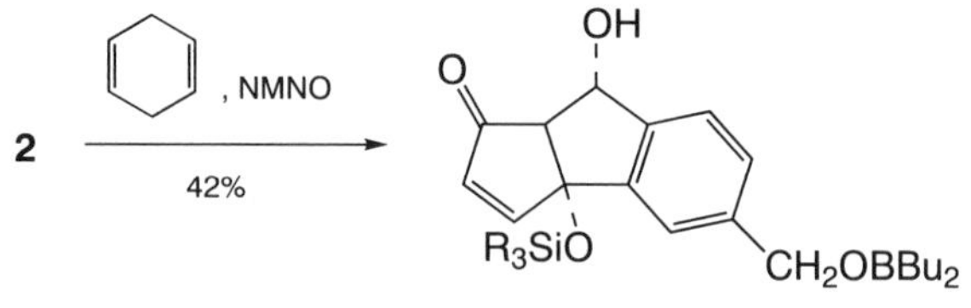

3

morpholine N-oxide (NMNO) in cyclohexene results in aromatization to **3**. Such aromatization is characteristic of compounds of this group of antitumor agents.[2]

A synthesis of the tetrahydroquinoline enediyne structure (**6**) of the antibiotic dynemicin[2] involves condensation of a Co(CO)$_6$-protected propargylic alcohol with an enol.[3] Thus treatment of **4** with triflic anhydride and 2,6–di-t-butyl-4-methylpyridine in CH$_2$Cl$_2$/CH$_3$NO$_2$ results in **5** in 52% yield. (Use of nitromethane is crucial for satisfactory results.) This product on decomplexation with iodine provides **6**, which undergoes aromatization when heated to form **7**.

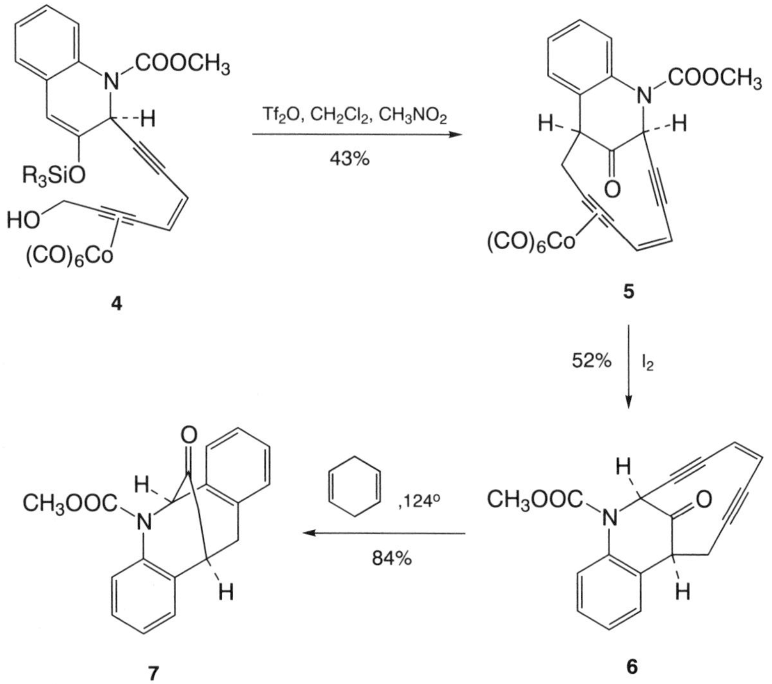

Diastereoselective aldol reactions of propynals.[4]   The cobalt complex **1** of propynal reacts with O-silyl ketene O,S-acetate (**2**) to form the *syn* aldol almost exclusively.

1   **2** (E/Z = >98 : <2)   **3** (syn/anti = >98 : <2)

The uncomplexed propynal undergoes the same reaction with high *anti*-selectivity (95:5).

Intramolecular Pauson–Khand reaction. Addition of $Co_2(CO)_8$ (1 equiv.) to alkynyl allylamino chromium or tungstrum carbene complexes (**1**) at 25° results in an intramolecular Pauson–Khand reaction to provide **2** in 70–75% yield.

1, M = Cr, W   **2**

[1] M. D. Wang and H. Alper, *Am. Soc.,* **114**, 7018 (1992).
[2] P. Magnus and T. Pitterna, *J. C. S. Chem. Comm.,* 541 (1991).
[3] P. Magnus and S. M. Fortt, *ibid.,* 544 (1991).
[4] C. Mukai, O. Kataoka, and M. Hanaoka, *Tetrahedron Letters,* **32**, 7553 (1991).
[5] F. Camps, J. M. Moretó, S. Ricant, and J. M. Viñas, *Angew. Chem. Int. Ed.,* **30**, 1470 (1991).

Dichlorobis(cyclopentadienyl)hafnium–Silver perchlorate,
Cp_2HfCl_2-$AgClO_4$, **15,**119–120; **16,**120–121.

Aryl C-glycosides.[1] This C-glycosidation was used for synthesis of the antibiotic (+)-gilvocarcin (**4**). An unusual feature of this synthesis is that the first step is the C-glycosidation of the A ring, followed by addition of the remaining three rings. Thus reaction of the L-acetylfucose derivative **1** with the iodophenol **2** promoted by Cp_2HfCl_2/$AgClO_4$ (2 equiv. each) in CH_2Cl_2 at $-78 \rightarrow -20°$ provides the C-glycoside **3** in 87% yield. This product was converted in five steps to **4** in 38% overall yield.

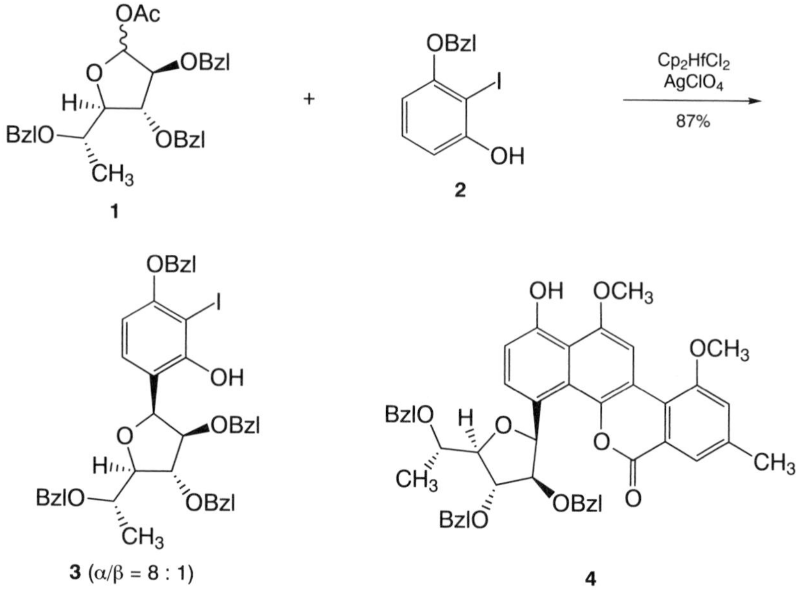

1   +   **2** $\xrightarrow[\text{87\%}]{\text{Cp}_2\text{HfCl}_2 \; \text{AgClO}_4}$

3 ($\alpha/\beta = 8 : 1$)   **4**

[1] T. Matsumoto, T. Hosoya, and K. Suzuki, *Am. Soc.,* **114**, 3568 (1992).

Dichlorobis(cyclopentadienyl)zirconium.

Coupling of 1-alkenes with Grignard reagents.   Two laboratories[1,2] have reported that Cp$_2$ZrCl$_2$ can catalyze coupling of 1-alkenes with C$_2$H$_5$MgBr or (C$_2$H$_5$)$_2$Mg. Thus 1-decene reacts with C$_2$H$_5$MgBr in the presence of Cp$_2$ZrCl$_2$ to form the bisGrignard reagent (**a**) that on quenching with H$_3$O$^+$ provides 3-methylundecane (**1**) The high

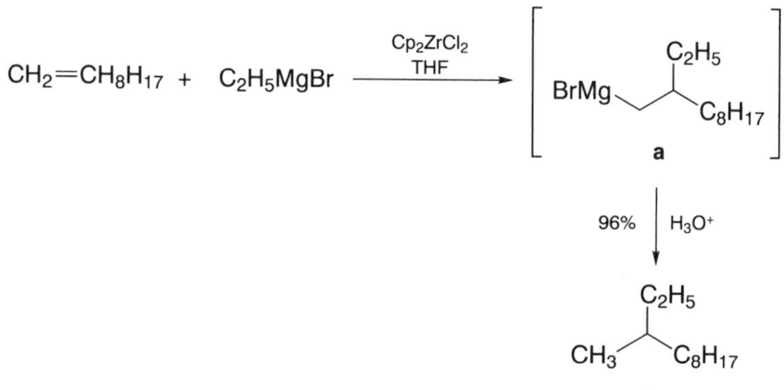

$$\text{CH}_2\text{=CH}_8\text{H}_{17} + \text{C}_2\text{H}_5\text{MgBr} \xrightarrow[\text{THF}]{\text{Cp}_2\text{ZrCl}_2} \quad$$

regioselectivity is notable. Another advantage is that the intermediate is a Grignard reagent subject to further functionalization.

Ethylmagnesiation of alkenes.[3]   In the presence of Cp_2ZrCl_2 (5 mol. %), ethylmagnesium chloride adds to alkenes to give an intermediate (**a**) that can react with various electrophiles (equation I).

(I) $n\text{-}C_8H_{17}CH=CH_2$ + C_2H_5MgCl → [intermediate **a**]

This carbomagnesiation can also be applied to allylic alcohols and ethers. In the case of the free alcohol, the reaction affords the *syn*-diol with 95:5 diastereoselectivity.

R = H	70%
R = CH$_3$	80%

95:5
11:89

Protection of the alcohol with a *t*-butyldimethylsilyl group completely blocks carbomagnesiation. The actual catalyst in these reactions may be "Cp_2Zr."

Ethylmagnesiation of homoallylic alcohols (ethers).[4]   Ethylmagnesium chloride can add to the double bond of homoallylic alcohols on catalysts with Cp_2ZrCl_2 with high stereoselectivity. In the case of the *anti*-homoallylic alcohol **1** (R=H) the reaction proceeds with high *anti*-selectivity in the case of both the alcohol and the methyl ether. In contrast to the reaction of the *syn*-homoallylic alcohol **3**, the reaction proceeds with

1, R = H	75%
R = CH₃	65%

anti-2

>99:1
97:3

3, R = H	55%	85:15
R = CH₃	38%	65:35

less *anti*-selectivity, and the *anti*-selectivity is markedly less in the case of the methyl ether (**3**, R=CH₃).

Although carbomagnesiations of disubstituted alkenes proceeds sluggishly, the reaction of *endo*-5-norbornen-2-ol (**5**) proceeds in high yield and regioselectivity. However, the reactivity and regioselectivity are decreased in the reaction of the methyl ether of **5**.

5, R = H	97%	>99:1
R = MEM	75%	97:3

[1] T. Takahashi, T. Seki, Y. Nitto, M. Saburi, C. J. Rousset, and E. Negishi, *Am. Soc.*, **113**, 6266 (1991).

[2] K. S. Knight and R. M. Waymouth, *ibid.*, **113**, 6268 (1991).

[3] A. H. Hoveyda and Z. Xu, *Am. Soc.*, **113**, 5079 (1991).

[4] A. H. Hoveyda, Z. Xu, J. P. Morken, and A. F. Houri, *ibid.*, **113**, 8950 (1991).

Dichlorobiscyclopentadienylzirconium – Butyllithium, "ZrCp₂," 14, 122–123.

Reductive coupling of enynes or dienes; perhydroindoles.[1]   Reaction of "ZrCp₂" with the cyclohexenylamine **1** in THF at 25° results in a zirconacycle (**a**) that is hydrolyzed by HCl (10%) to the perhydroindole **2** (75% yield). The Zr—C bonds can be cleaved by various other electrophiles to give substituted perhydroindoles. Treatment of **a** with CO gives the tricyclic ketone **3**.

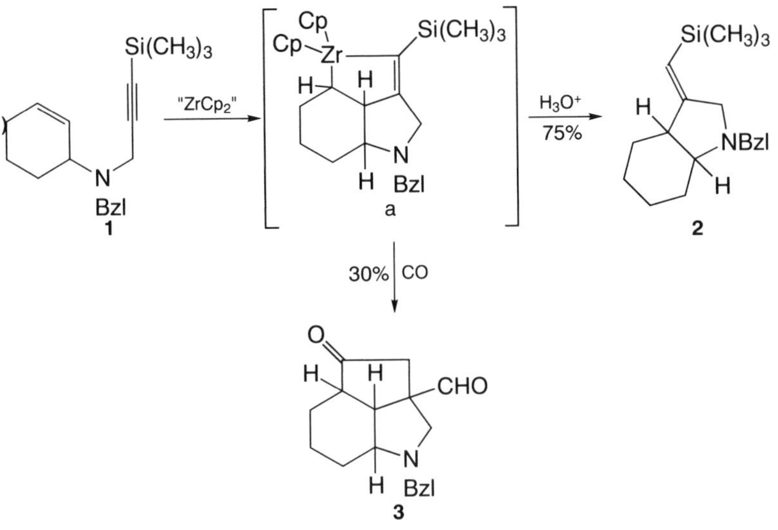

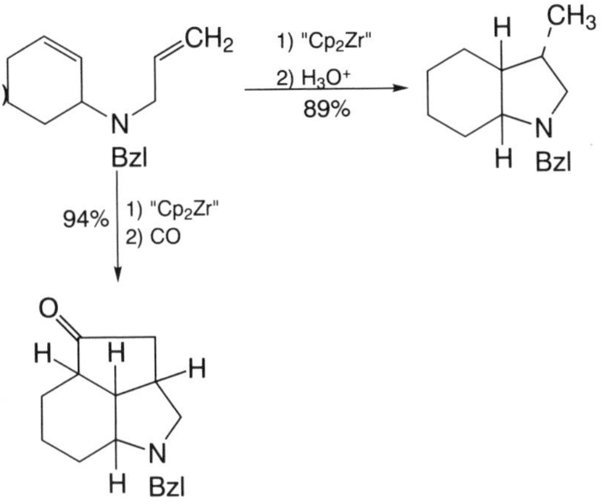

Triquinane synthesis.[2]   A highly stereocontrolled synthesis of the triquinane pentalenic acid (**6**) employs the zirconium promoted bicyclization–carbonylation of an enyne **1** with an allylic hydroxyl group to provide the bicyclic enone **2**, corresponding to the A-B rings. Addition of the C ring was effected by a novel Michael addition of an allylphosphonate group. This approach is useful because hydroboration of a vinylphosphonate followed by oxidation provides an (α-hydroxyalkyl)phosphonate, which is hydrolyzed by a mild base to an aldehyde (equation I). Thus addition of the anion (BuLi) of diethyl (Z)-crotylphosphononate to **2** provides **3**, which, after protection of the carbonyl group, is converted to the aldehyde **3**. Reaction of **3** with pyridinium *p*-toluenesulfonate effects deprotection of the carbonyl group followed by aldolization. Dehydration of the aldol followed by hydrogenation provides the tricyclic ketone **5**. This product is converted into (+)-**6** by standard reactions. This cyclopentannelation to provide the C ring proceeds with high stereo- and regiocontrol. Since allylic alcohols can be resolved by Sharpless kinetic asymmetric epoxidation, it should be possible to prepare optically active triquinanes by this route.

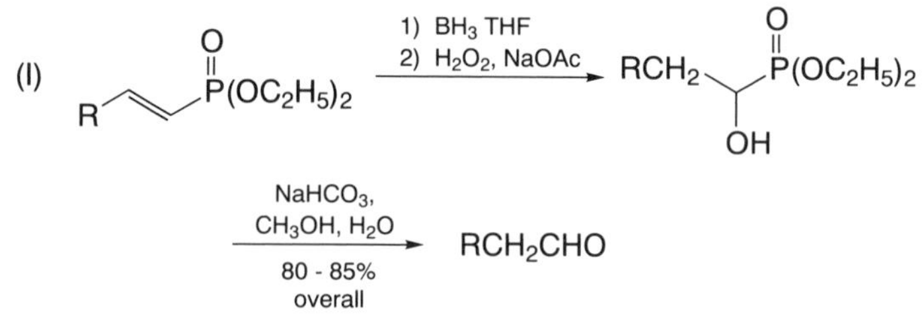

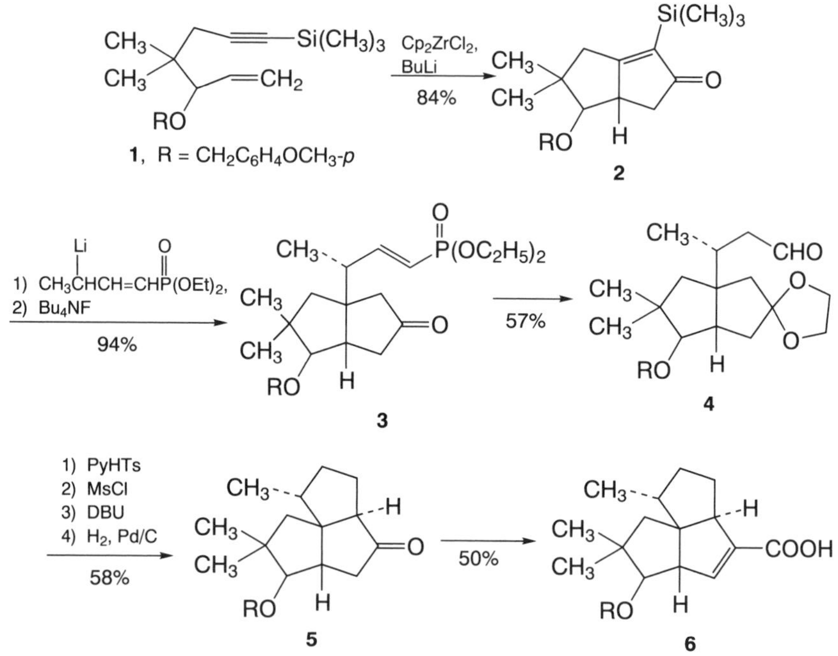

Bicyclization of 1,6-heptenynes.[3]   Treatment of the chiral 3-methyl-1,6-octenyne **1**, prepared from (+)-citronellene, with the reagent formed from Cp_2ZrCl_2/BuLi (1:2), followed by protonolysis and carbonylation, provides the bicyclic ketone **2**. This product was used for a synthesis of (+)-iridomyrmecin (**3**).

Allenylzirconium reagents.[4]   "Cp$_2$Zr" reacts with propargylic ethers to form an allenylzirconium reagent (**a**) that reacts *in situ* with aldehydes in the presence of BF$_3$ etherate to form *anti-β*-acetylenic alcohols as the major product. In addition to the *syn*- and *anti-β*-acetylenic alcohols, a significant amount of an α-allenic alcohol is usually formed.

[1] M. Mori, N. Uesaka, and M. Shibasaki, *J. Org.,* **57**, 3519 (1992).
[2] G. Agnel and E. Negishi, *Am. Soc.,* **113**, 7424 (1991).
[3] G. Agnel, Z. Owezarczyk, and E. Negishi, *Tetrahedron Letters,* **33**, 1543 (1992).
[4] H. Ito, T. Nakamura, T. Taguchi, and Y. Hanzawa, *ibid.,* **33**, 3769 (1992).

Dichlorobis(triphenylphospine)palladium(II).

Carbonylation of o-allylbenzyl halides.[1]   Carbonylation (600 psi) of *o*-allylbenzyl chloride (**1**) in the presence of triethylamine (2 equiv.) and in the presence of this palladium catalyst (5 mol %) provides the benzoannelated enol lactone (**2**) in 78% yield.

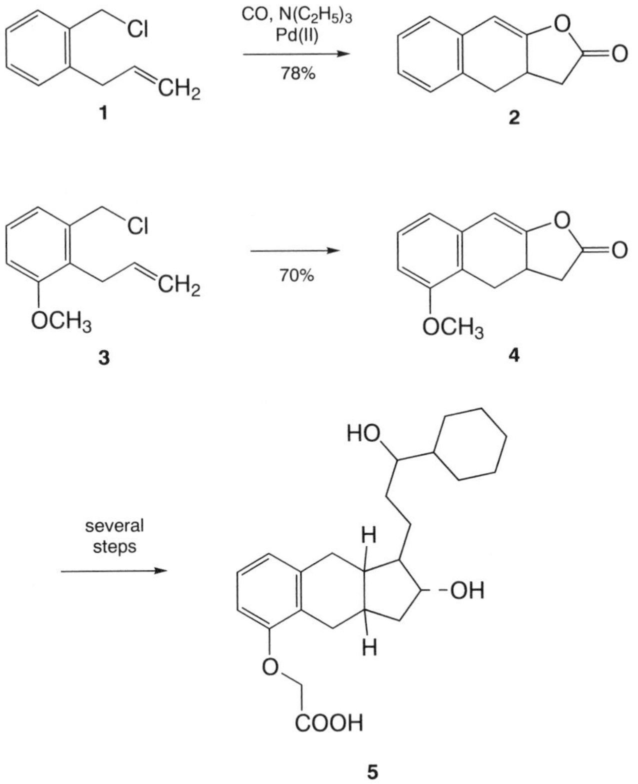

A similar reaction with **3** gives **4**, which is an intermediate in a synthesis of the antiulcer agent U-68,215 (**5**).

[1] G. Wu, I. Shimoyama, and E. Negishi, *J. Org.,* **56**, 6506 (1991).

Dichlorobis(triphenylphosphine)palladium(II) – Copper(I) iodide.

Coupling of dienol ditriflates and propargylic alcohols.[1]   This coupling can be used to obtain open-chain (E)- and (Z)-dienediyne diols, analagous to the cyclic di-enediyne system of the neocarzinostatin chromophore, responsible for the cytotoxity of the antitumor antibiotic. Both (E)- and (Z)-dienol ditriflates (**1**) can be prepared from 2-formylcyclopentanone as shown in equation (I). The dienol ditriflates couple with propargylic alcohols in the presence of 5 mol % of the Pd catalyst, 10% of CuI,

(E) - 1   (Z) - 1

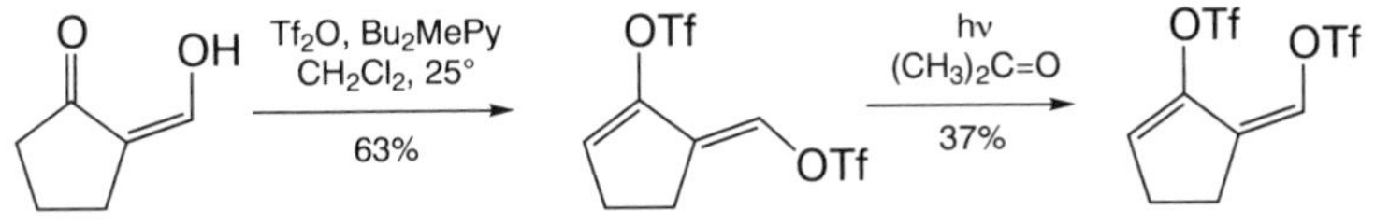

(E) - 2

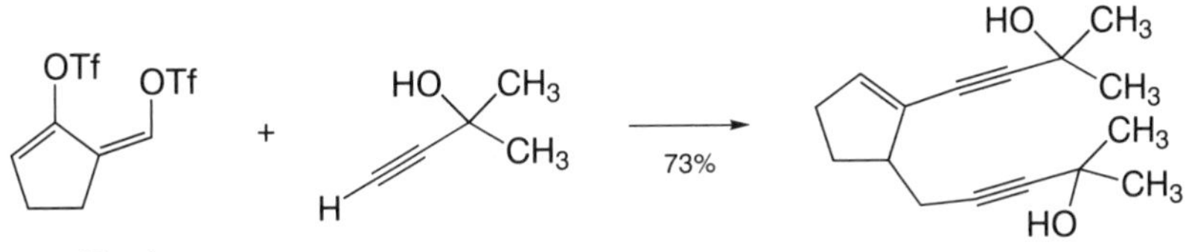

(Z) - 1

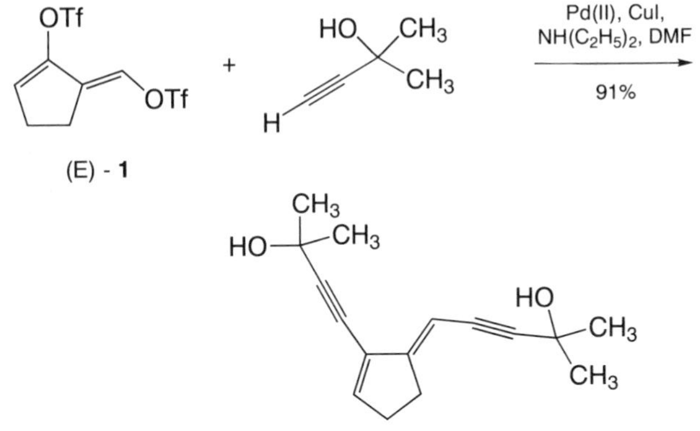

and 2 equiv. of diethylamine to give the (E)- and (Z)- dienediyne system **2**. Both (E)- and (Z)-2 exhibit comparable cytotoxity.

[1] K. Nakatani, K. Arai, K. Yamada, and S. Terashime, *Tetrahedron Letters*, **32**, 3405 (1991).

Dichlorodimethylsilane, $(CH_3)_2SiCl_2$.

Reductive coupling of RCHO and allylic alcohols[1]   A synthesis of tunicaminyl-uracil (**5**) involves a novel coupling of an aldehyde with an allylic alcohol, effected with dichlorodimethylsilane. Thus treatment of the alcohol **1** and the aldehyde **2** with benzeneselenol and $(CH_3)_2SiCl_2$ (excess) in pyridine at 23° provides the product **3** in 92% yield as a mixture of epimers (1:1). Radical cyclization of **3** in acetonitrile with Bu_3SnH and $(C_2H_5)_3B$ followed by siloxane hydrolysis provides the protected derivate **4** of tunicaminyluracil (**5**).

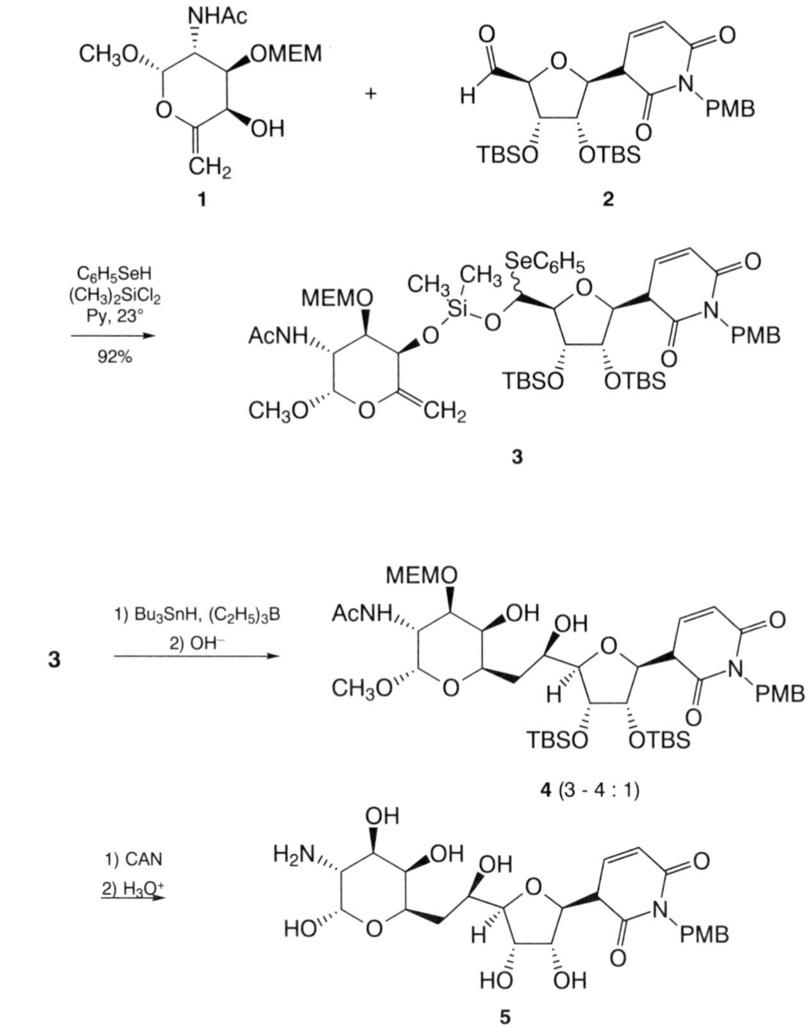

[1] A. G. Myers, D. Y. Gin, and K. L. Widdowson, *Am. Soc.*, **113**, 9661 (1991).

Dichloro(ethylene)platinum dimer, $[Pt(C_2H_4)Cl_2]_2$, Zeise's dimer (**1**).

Isomerization of silyloxycyclopropanes to allyl silyl ethers.[1]   This Pd(II) complex effects isomerization of silyloxycyclopropanes to allyl silyl ethers. The last example indicates that rearrangement involves inversion of the silyloxy group.

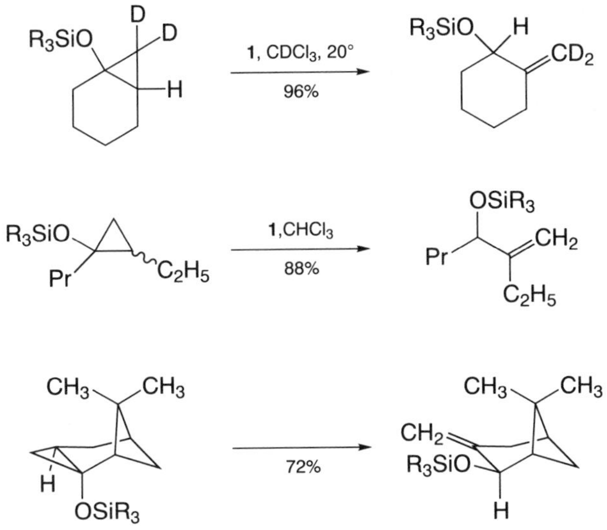

[1] K. Ikura, I. Ryu, N. Kambe, and N. Sonoda, *Am. Soc.*, **114**, 1520 (1992).

Diethylzinc–Chloroiodomethane, $(C_2H_5)_2Zn–ClCH_2I$ (1).

Cyclopropanation of alkenes.[1]   The Simmons–Smith reagent is generally prepared from diethylzinc and CH_2I_2, and presumably is (iodomethyl)zinc, ICH_2ZnX. The related reagent prepared from $(C_2H_5)_2Zn$ and $ClCH_2I$ in the ratio 1:2 is more reactive than the usual Simmons–Smith reagent, particularly when used with dichloroethane as solvent. Moreover, cyclopropanation of allylic alcohols can show high diastereoselectivity (equation I).

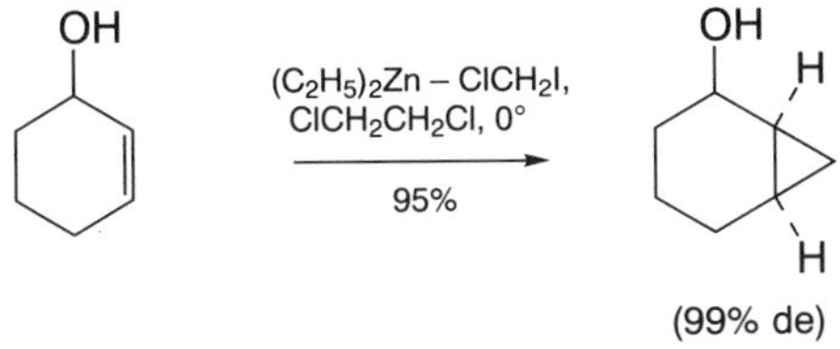

[1] S. E. Denmark and J. P. Edwards, *J. Org.*, **56**, 6974 (1991).

Diethylzinc–Methylene iodide (Simmons–Smith reagent).

Asymmetric cyclopropanation.[1] Cyclopropanation of substituted allylic alcohols (**1**) attached to a carbohydrate derived from D-glucose with $(C_2H_5)_2Zn$ and CH_2I_2 (excess) can proceed with high diastereoselectivity. The triflate of the product (**2**) when heated in

aqueous DMF undergoes contraction to the aldehyde **3** with release of the cyclopropyl-methanol **4**. Note that the presence of the free hydroxyl group at C_2 of the glucopyranoside **1** is essential for the high diastereoselectivity of the product **2**.

[1] A. B. Charett, B. Côté, and J.-F. Marcoux, *Am. Soc.*, **113**, 8166 (1991).

2,2′-Dihydroxy-1,1′-binaphthyl (BINOL).

3- and 3,3′-Substitution. The directed ortho lithiation route can be used to effect 3- and 3,3′-functionalization of BINOL by use of the MOM or SEM ethers or carbamates. Generation of the monoanion is best effected with 2.2 equiv. of *t*-BuLi (THF, −78°). Generation of the dianion is best effected with 3.0 equiv. of BuLi (ether, 25°). This method when applied to (S)-BINOL proceeds with retention of configuration.

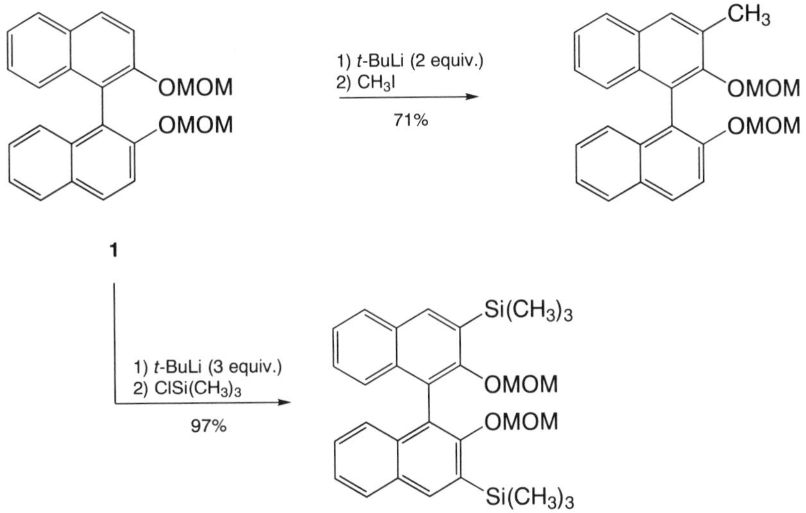

1

3,3′-Diaryl BINOLs can be obtained by conversion to the dibromo derivative of BINOL followed by Suzuki coupling with arylboronic acids mediated by $Pd[P(C_6H_5)_3]_4$ (equation I).

(I)   **1**

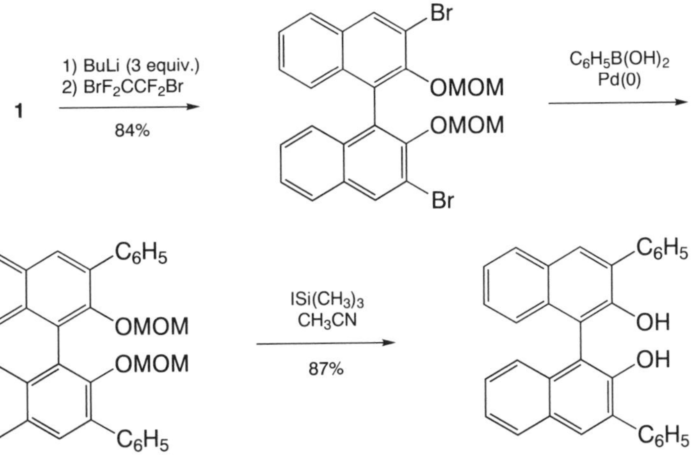

4,4-Disubstituted butan-4-olides.[2]   Products of this type can be obtained in high enantioselectivity by addition of Grignard reagents to γ-keto esters, (R)-**1**, of the chiral auxiliary BINOL. The reaction can proceed with high 1,7-asymmetric induction.

(R) - 1

CH₃MgBr
MgBr₂·O(C₂H₅)₂

72%

(S) - **2**, 95% ee

Review.[3]   This review is a general introduction to the use of 1,1'-binaphthyl derivatives as chiral auxiliaries; 86 references from 1974 to 1992.

[1] P. J. Cox, W. Wang, and V. Snieckus, *Tetrahedron Letters*, **33**, 2253 (1992).
[2] Y. Tamai, M. Akiyama, A. Okamura, and S. Miyano, *J. C. S. Chem. Comm.*, 687 (1992).
[3] C. Rossini, L. Franzini, A. Raffaelli, and P. Salvadori, *Synthesis*, 503 (1992).

2,2'-Dihydroxy-1,1'-binaphthyl – Trimethylaluminum.

cis-4-Alkoxycarbonyl-2-oxazolines.   In the presence of this organoaluminum complex (**1**) 5-alkoxyoxazoles (**2**) undergo a [3 + 2] cycloaddition with aryl aldehydes to form *cis*-4-alkoxy-2-oxazolines with high *cis*-selectivity.

2   +   C₆H₅CHO

1, CH₂Cl₂, CH₃CN
0°

77.7%

cis - **3**   98 : 2   *trans* - **3**

+

[1] H. Suga, S. Shi, H. Fujieda, and T. Ibata, *Tetrahedron Letters*, **32**, 6911 (1991).

2,2'Dihydroxy-1,1'-binaphthyl – Triphenyl borate

Asymmetric aza-Diels–Alder reactions.[1]   This reaction can be effected in 72–90% ee by catalysis with the chiral boron reagent **1**, prepared by reaction of (R)-BINOL with triphenyl borate in CH₂Cl₂ at 25°. Thus reaction of an aldimine and the Danishefsky

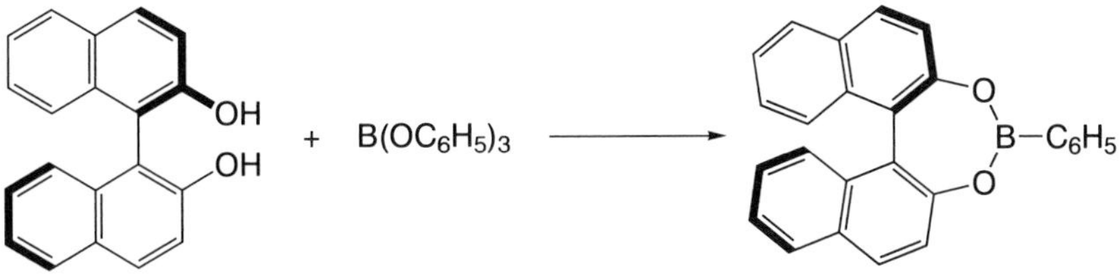

(R)   +   B(OC$_6$H$_5$)$_3$ ⟶   (R) - **1**

(I)   C$_6$H$_5$CH=NBzl   +

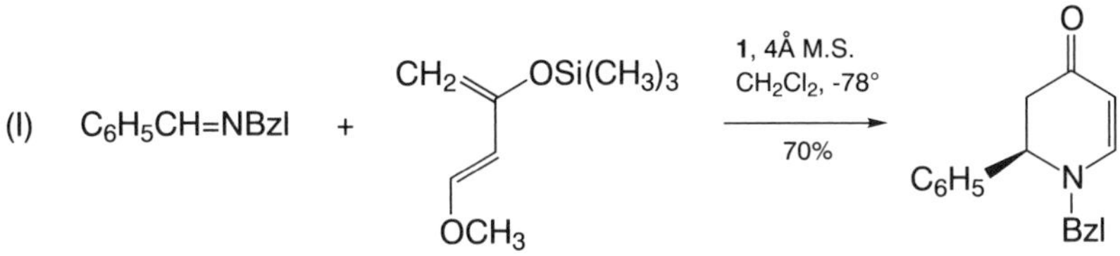

2, 85% ee

diene promoted by (R)-**1**, generated *in situ*, provides the adduct **2** in 85% ee (equation I). This asymmetric Diels–Alder reaction can be used to prepare (−)-anabasine (**5**), an

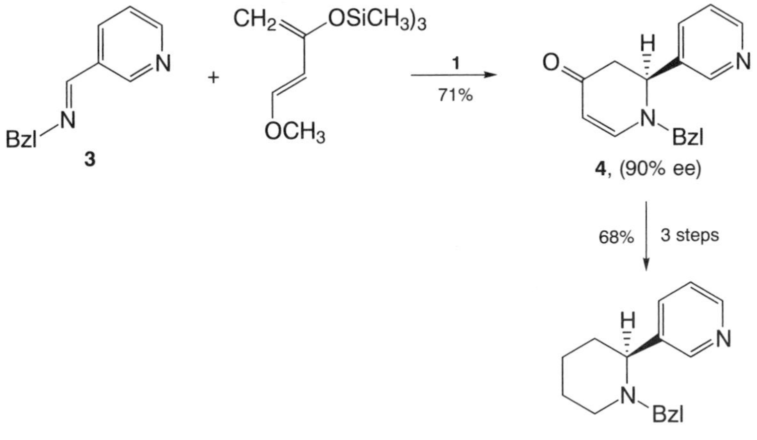

3   +   CH$_2$=C(OSiCH$_3$)$_3$ / OCH$_3$ $\xrightarrow[71\%]{\textbf{1}}$ **4**, (90% ee)

68% | 3 steps

5, α$_D$ = -79.2°

alkaloid derived from nicotinic acid.

[1] K. Hatltori and H. Yamamoto, *J. Org.*, **57**, 3264 (1992).

Dichloroketene.

Indolizidines; quinolizidines.[1]   Generation of dichloroketene in the presence of N-benzyl-2-vinylpiperidine (**1**) generates an intermediate (**a**) that cyclizes on warming to the 10-membered lactam **2**. A similar reaction with the N-benzyl-2-vinylpyrrolidine (**3**) provides the 9-membered lactam **4**.

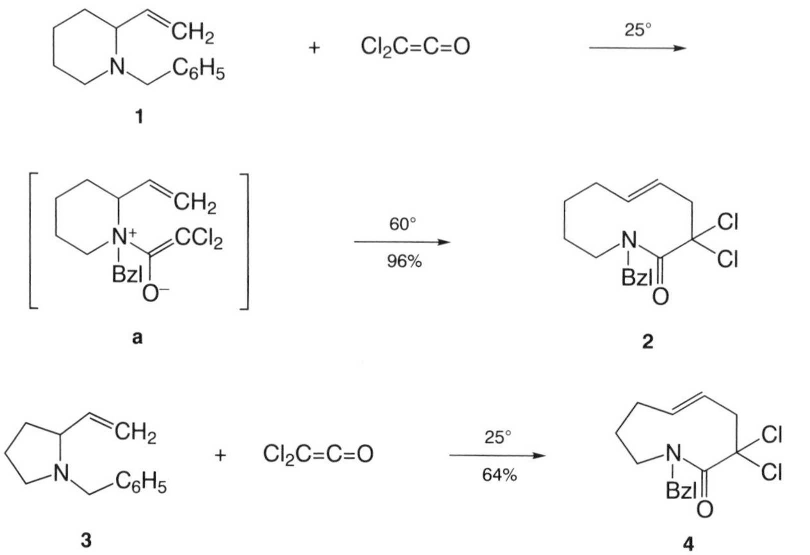

Exposure of either **2** or **4** to I_2 or C_6H_5SeBr triggers a transannular cyclization and debenzylation to provide iodo- or phenylseleno-substituted quinolizidines (**5**) from **2** or the analogous indolizidines from **4**.

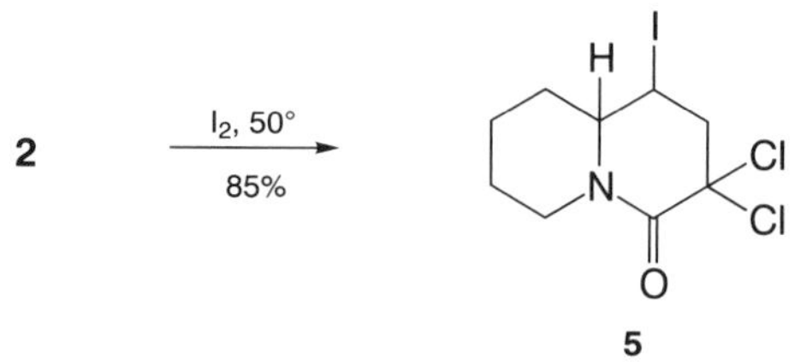

[1] E. D. Edstrom, *Am. Soc.*, **113**, 6690 (1991).

2,6-Dichloropyridine N-oxide,

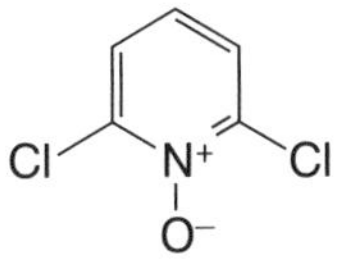

(1)

Oxidation.[1]   In the presence of ruthenium porphyrin oxo complexes, such as RuTMPO$_2$, and a trace of HBr or HCl, pyridine N-oxide **1** can oxidize alkanes to secondary alcohols at 25° in high yield. The reactive intermediate is not known. Adamantane is oxidized mainly to the 1-hydroxy derivative (68%); adamantane-1,3-diol (15–25%) and adamantanone-2 (1–2%) are minor products.

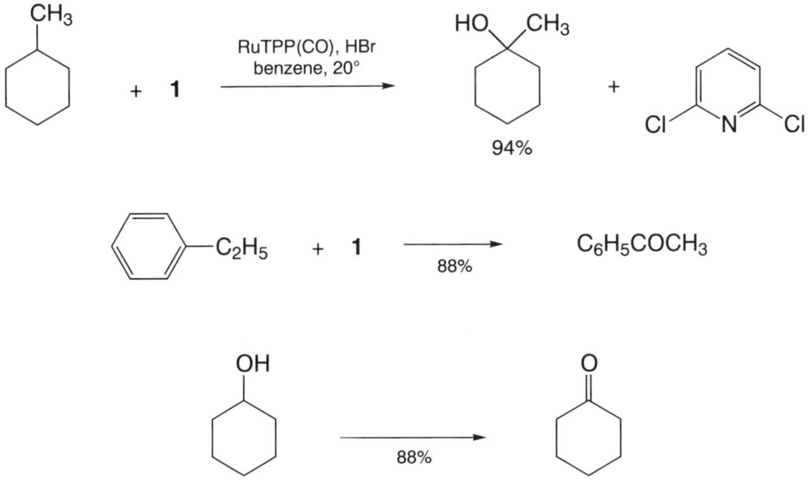

[1] H. Ohtake, T. Higuchi, and M. Hirobe, *Am. Soc.,* **114**, 10660 (1992).

1,3-Dichloro-1,1,3,3-tetraisopropyldisiloxane (TIPDS-Cl$_2$), **1**, [(CH$_3$)$_3$CH]$_2$Si(Cl)OSi(Cl)CH(CH$_3$)$_2$]$_2$ (**16**,125–126).
   Regio- and stereoselective dihydroxylation.[1]   This reagent adds to the oxepin **2** exclusively with the isolated cycloalkene double bond to provide **3** (84% yield). Whereas direct osmylation of **2** provides a mixture of diols in low yield, osmyla-

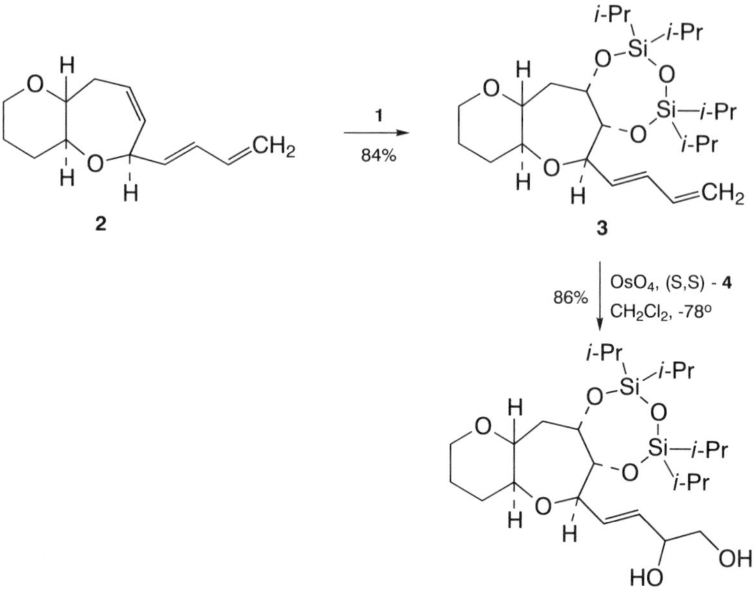

(2S) - **5** (major product)

tion of **3** results in almost exclusive attack of the terminal double bond. When the osmylation is carried out in the presence of (S,S)-N,N'-dineohexyl-2,2'-bipyrrolidine (**4**) (**16**,150–151), the major product (86%) is (2S)-**5**. However, the same reaction, but catalyzed by (R,R)-**4**, also provides (2S)-**5** as the major product, but in lower yield (71%). These results show that the diastereoselectivity and the regioselectivity are controlled by the sterically demanding TIPDS group.

[1] O. Sato and M. Hirama, *Synlett*, 705 (1992).

(Diethylamino)(phenyl)oxosulfonium methylide,

$[(C_2H_5)_2N(C_6H_5)]^+SO(CH_3)BF_4^-$ (**1**), 3, 105–106; **5**, 231.

1-Aminocyclopropane-1-carboxylic acids.[1]   These cyclopropyl amino acids can be obtained in high chemical and optical yield by reaction of this ylide with the α,β-dehydro lactones (**3**) prepared from (5S,6R)-4-*t*-butoxycarbonyl-5,6-diphenyl-2,3,5,6-tetrahydro-4H-1,4-oxazin-2-one (**2**, **14**,58–59). The reaction of **3** with the ylide derived from

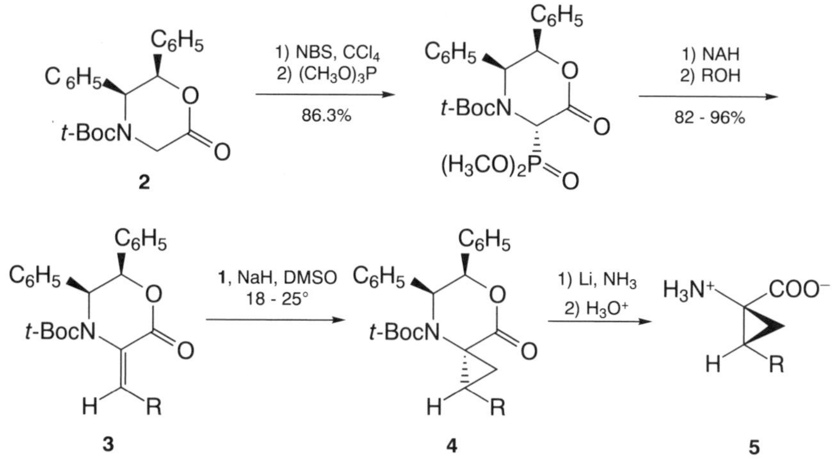

R = CH₃   82%   1 : 0
R = C₆H₅   96%   1 : 0

1 usually gives a single product **4**, regardless of R. Cyclopropanation of **3** with dimethyloxosulfonium methylide results in two diastereomeric products in the ratio 2–3:1. Use of diazomethane also results in a diastereomeric mixture. The products are converted into the desired amino acids by reduction with Li in NH₃ followed by deprotection of the amino group.

[1] R. M. Williams and G. J. Fegley, *Am. Soc.,* **113**, 8796 (1991).

Diisobutylaluminum hydride.

α-Trimethylsiloxy aldehydes.[1]   The α-trimethylsiloxy nitriles **1**, obtained in high yield by addition of cyanotrimethylsilane to ketones catalyzed by ZnI₂ (4,542–543), are reduced by DIBAH (1.5–2 equiv.) in hexane at 0° to imines (**a**), which can be hydrolyzed by dilute solutions of 1–1.6N H₂SO₄ at 15° to α-silyloxy aldehydes (**2**). These aldehydes can be hydrolyzed by dilute HCl to α-hydroxy aldehydes (**3**).

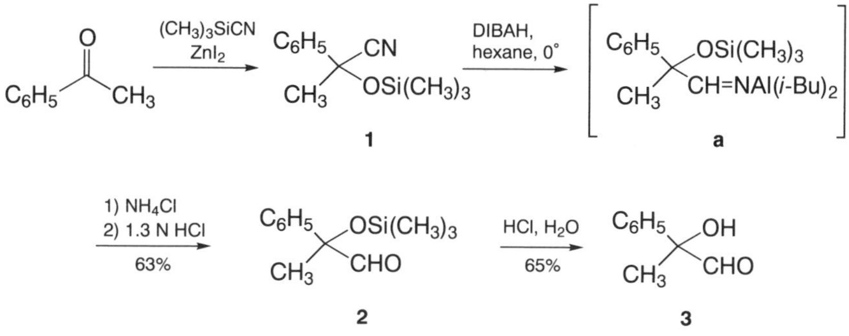

Regioselective reduction of terminal epoxides.[2]   By appropriate choice of i-Bu$_2$AlH or i-Bu$_3$Al, terminal epoxides can be reduced with marked regioselectivity to the primary or the secondary alcohol. Alkyl-substituted epoxides are reduced by i-Bu$_2$AlH mainly to a secondary alcohol, whereas use of i-Bu$_3$Al favors reduction to the primary alcohol. This regioselectivity is reversed in the case silyl-substituted epoxides. Phenyl-substituted epoxides are reduced by i-Bu$_3$Al mainly to primary alcohols. Reduction of these epoxides to secondary alcohols is best effected with LiAlH$_4$ in THF.

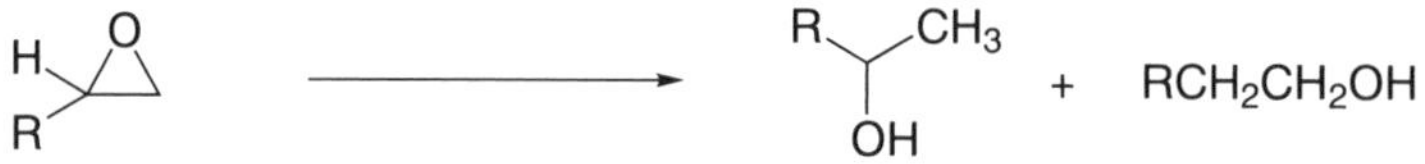

R = n-C$_8$H$_{17}$	i-Bu$_2$AlH	95%	100 : 1
	i-Bu$_3$Al	93%	0 : 100
R = (C$_6$H$_5$)$_3$Si	i-Bu$_2$AlH	90%	0 : 100
	i-Bu$_3$Al	90%	65 : 35
R = C$_6$H$_5$	LiAlH$_4$	95%	97 : 3
	i-Bu$_3$Al	90%	18 : 82

Asymmetric synthesis of 1,3-diols[3]   The (R)-β, δ-diketo sulfoxides (1), prepared by reaction of the dianion of β-diketones with (S)-menthyl p-toluenesulfinate, are reduced by

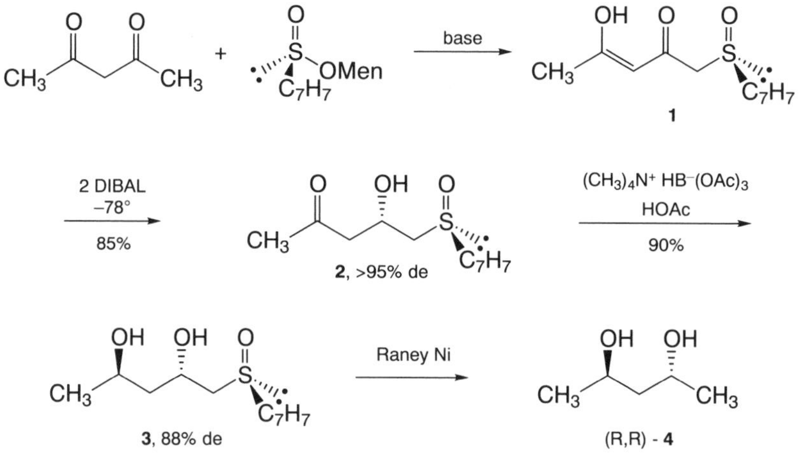

DIBAH (2 equiv.) to the (2S)-hydroxy ketone **2** with high diastereoselectivity (>95% de). The products can be converted to the *anti*-1,3-diol **4** by reduction with tetramethyl-ammonium triancetoxyborohydride (Evans reagent) followed by desulfurization.

Regioselective reduction of 2,3-epoxy alcohols.[4]   These epoxy alcohols can be converted into enantiomerically pure 2-alkanols by reduction of their tosylates with DIBAH (3 equiv.) in CH_2Cl_2 at $-15 \rightarrow 25°$.

Reductive cleavage of t-butyldimethylsilyl ethers.[5]   *t*-Butyldimethylsilyl ethers are generally cleaved by fluoride ion, but they can also be reductively cleaved by DIBAH in CH_2Cl_2 at 23° in 1–2 hours to yield the corresponding alcohols in 84–91% yield.

[1] M. Hayashi, T. Yoshiga, and N. Oguni, *Synlett*, 479 (1991).
[2] J. J. Eisch, Z.-R. Liu, and M. Singh, *J. Org.*, **57**, 1618 (1992).
[3] G. Solladié and N. Ghiatou, *Tetrahedron Letters*, **33**, 1605 (1992).
[4] J. M. Chong, *ibid.*, **33**, 33 (1992).
[5] E. J. Corey and G. B. Jones, *J. Org.*, **37**, 1028 (1992).

(1S,2R,4S,5R)-2,5-Diisopropylcyclohexane-1,4-diol (1).
   Preparation.[1]

Chiral titanium catalyst (2) for enantioselective isomerization of alkenes.[2]   The chiral titanium catalyst, an *ansa*-bis(indenyl)titanium dichloride (**2**), is prepared by the reaction of the dimesylate of **1** with indenyllithium to form a product that is con-

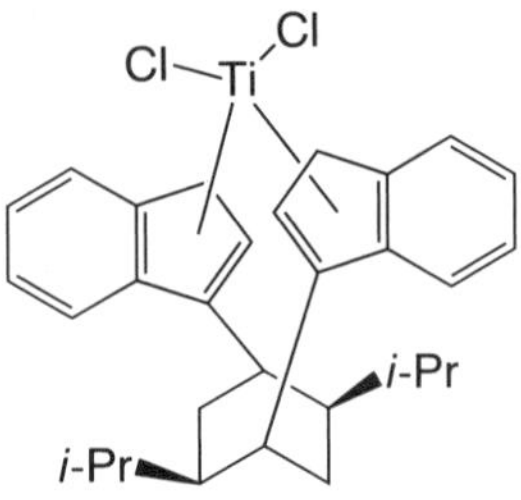

2

verted into **2** by reaction with BuLi and then with TiCl₃ in THF at −78 → 65°. A single product is formed in which the indenyl groups are placed away from the isopropyl group. This reagent, when reduced by LiAlH₄ (4 equiv.) can catalyze the isomerization of the *meso-trans*-4-butyl-1-vinylcyclohexane (**3**) to (S)-**4** as the only product in 76% ee.

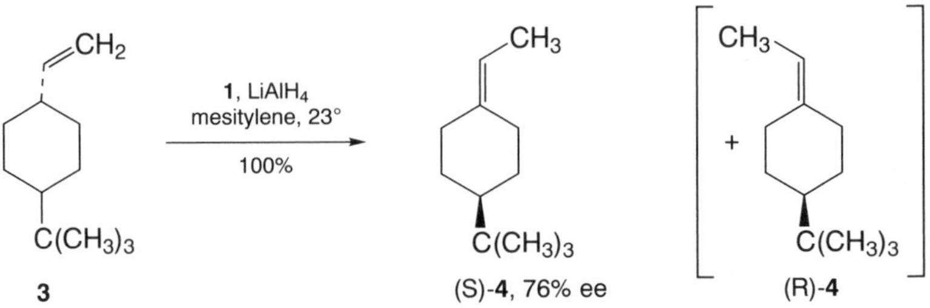

[1] Z. Chen and R. L. Halterman, *Synlett,* 142 (1990).
[2] *Idem, Am. Soc.,* **114**, 2276 (1992).

Dimethylsulfoxonium methylide.

Cyclopropanation.[1]   The key step in a biomimetic synthesis of (±)-colchicine (**4**) is a regio- and stereoselective reaction of the tricyclic **1**, prepared in several steps from 5-bromo-2-methoxyphenol, with dimethylsulfoxonium methylide to give a single product (**2**) in 75% yield. When treated with trifluoroacetic acid at 25°, this product rearranges in 89% yield to the α-tropolone O-methyl ether **3**. This product is converted into (±)-colchicine (**4**) in four known steps.

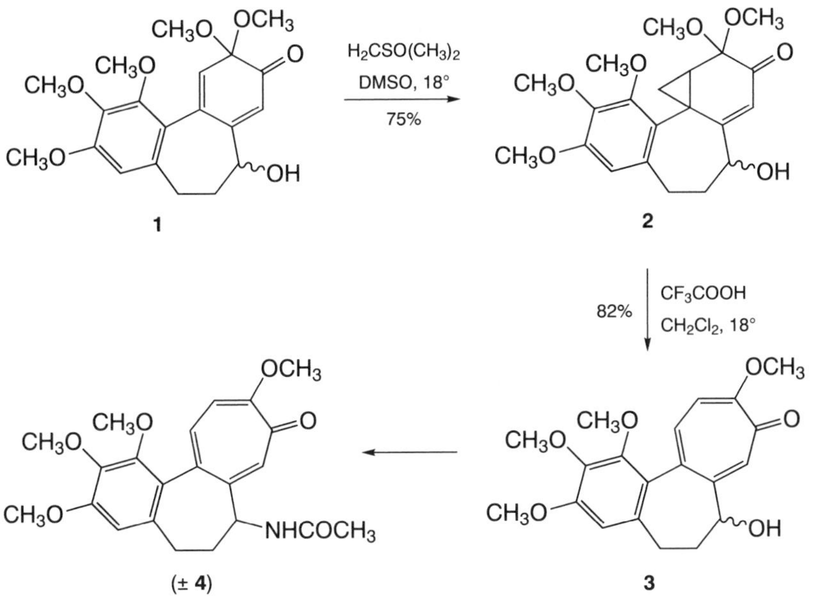

[1] M. G. Banwell, J. N. Lambert, M. F. Mackay, and R. J. Greenwood, *J. C. S. Chem. Comm.*, 974 (1992).

Dimethyl sulfoxide–Triphosgene.

Triphosgene $(CCl_3O)_2C=O$, is a readily available[1] substitute for phosgene.[2]

Oxidation.[3]   This combination (1) is comparable to DMSO-oxalyl chloride for both small- and large-scale Swern oxidations.

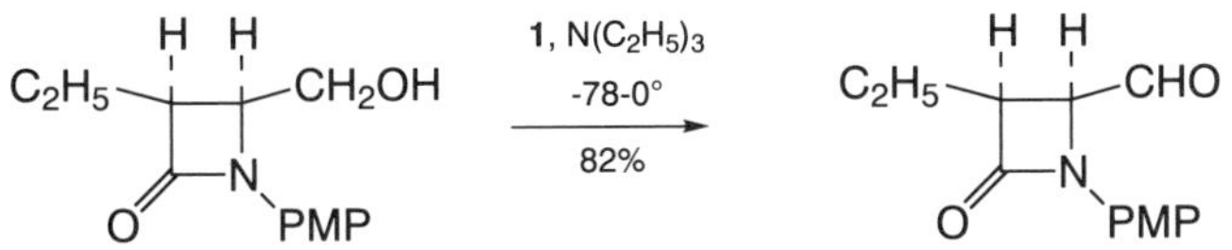

[1] H. Eckert and B. Forster, *Angew. Chem. Int. Ed.*, **26**, 894 (1987).
[2] *Aldrichimica Acta,* **21**, 47 (1988).
[3] C. Palomo, F. P. Cossio, J. M. Ontoria, and J. M. Odriozola, *J. Org.,* **56**, 5948 (1991).

1,2-Diphenyl-1,2-ethanediamine, 16, 153–158.

Asymmetric synthesis of 1,2-diamines.[1]   This synthesis involves as the first step reaction of (R,R)-1 with a 1,2-dione to form a dihydropyrazine (1,2-diimine, **a**), which can be reduced to a piperazine 2 by various hydride reagents. The highest diastereose-lectivity obtains with NaBH₃CN and an acid catalyst, pyridinium p-toluenesulfonate, in CH₃OH at −30°, which affords the diaxial product (2). Reduction of the C—N bonds can be effected by acetylation and cleavage with lithium/NH₃. Somewhat better results can be obtained by conversion of 2 to the biscarbamate 3 by reaction with isobutyl chloroformate and cleavage with Li/NH₃. This 1,2-diamine derivative is cleaved to the bishydrobromide salt 4 in 68% yield and 99% ee.

Asymmetric Ireland–Claisen rearrangement.[2]   The chiral diazaborolidine S,S-1 can effect highly enantio- and diastereoselective rearrangement of allylic esters. This reagent has previously been used to convert (E)-crotyl propionate (2) into the (E)- or (Z)-enolate (3) depending on the solvent and the base (16,155). Thus 2 is converted into (E,E)-3 by reaction with 1 and ethyldiisopropylamine in CH₂Cl₂ in CH₂Cl₂ at −78°. On standing at −20° for several days it rearranges to the *threo*-acid 4 in >97% ee. Reaction of 2 with 1 and triethylamine in toluene–hexane results in (E,Z)-3, which rearranges at −20° to the *erythro*-acid 4 in >96% ee. The by-product of rearrangement is the bis-sulfonimide precursor of 1. Lower enantioselectivity obtains in the case of esters of allyl alcohol itself, but rearrangement of various allylic propionate and butyrate esters proceeds with high enantioselectivity.

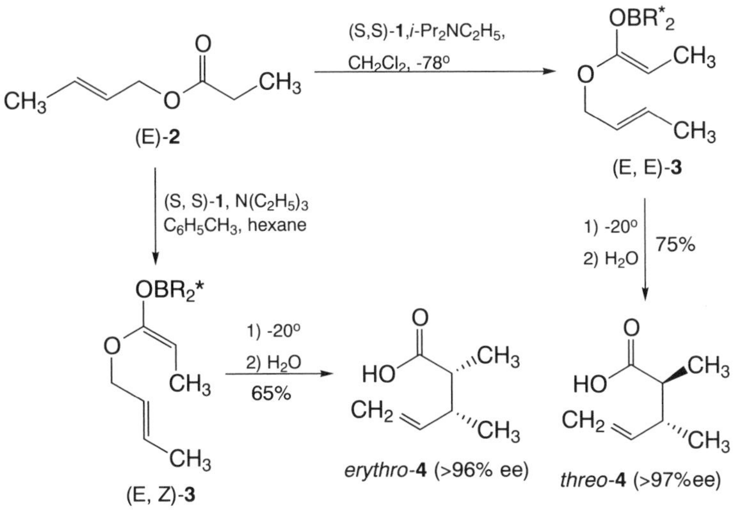

Enantioselective Darzens reaction.[3]   Reaction of *t*-butyl bromoacetate with aldehydes when promoted by (R,R)-**1** results in an *anti*-adduct (**2**) in >91% ee. The product can be debrominated by Bu_3SnH/AIBN or converted to a glycidic ester (**3**) by treatment

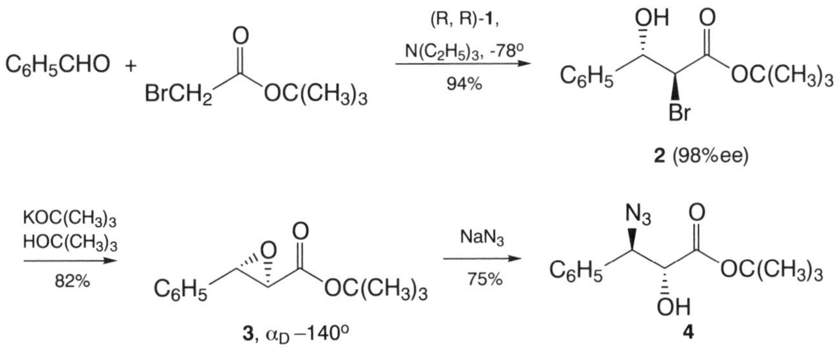

with base. These epoxides are easily converted to β-azido α-hydroxy esters (**4**), which are easily reduced to β-amino α-hydroxy esters.

Asymmetric synthesis of β-amino acid esters.[4]   The chiral diazaborolidine **1** (**16**,155) also effects diastereo- and enantioselective reactions of (S)-*t*-butyl thiopropionate (**2**) with N-benzyl or N-allyl aldimines **3** to form β-amino acid esters **4**, precursors to chiral *trans*-β-lactams (**5**) in 90–99% ee.

$$C_6H_5CH=NCH_2CH=CH_2 \quad + \quad CH_3CH_2\overset{\text{O}}{\underset{}{\,\,}}SC(CH_3)_3$$

3   **2**

$$\xrightarrow[\text{74%}]{\substack{(S,S)\text{-}1,\ N(C_2H_5)_3 \\ C_6H_5CH_3,\ \text{hexane},\ -78°}}$$

4 (*anti/syn* = >99:1)

$$\xrightarrow{t\text{-BuMgX}}$$

5, 90% ee

[1] M. H. Nantz, D. A. Lee, D. M. Bender, and A. H. Roohi, *J. Org.,* **57**, 6653 (1992).
[2] E. J. Corey and D.-H. Lee, *Am. Soc.,* **113**, 4026 (1991).
[3] E. J. Corey and S. Choi, *Tetrahedron Letters,* **32**, 2857 (1991).
[4] E. J. Corey, C. P. Decicco, and R. C. Newbold, *Tetrahedron Letters,* **32**, 5287 (1991).

(R,R)-1,2-Diphenylethane-1,2-diol, dimethyl ether (1)

(R,R) -**1** (**16** ,158-159).

Asymmetric aromatic alkylation; binaphthyls.[1]   The reaction of 1-naphthyllithium with a naphthylimine (**2**) catalyzed by **1** provides **3**, which is hydrolyzed (H_2O, CF_3COOH, Na_2SO_4), with recovery of **1**, to the aldehyde **4** in 85% ee. The stereochemistry can be improved by use of a bulky R group in **2**.

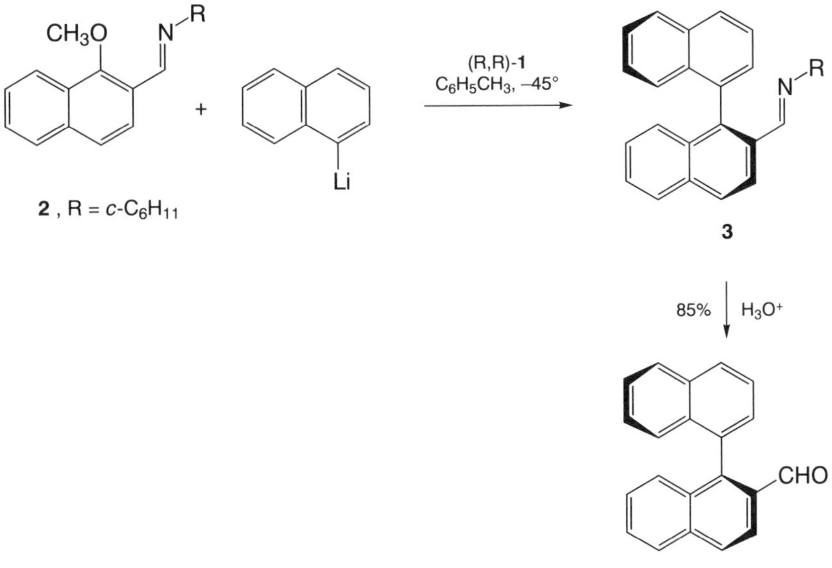

2 , R = c-C$_6$H$_{11}$

85% | H$_3$O$^+$

4 , 85%ee

[1] M. Shindo, K. Koga, and K. Tomioka, *Am. Soc.*, **114**, 8732 (1992).

(S)-(+)- or (R)-(−)-Diphenyl(1-methyl-2-pyrrolidinyl)methanol (1).

Enantioselective cyanomethylation.[1] Cyanomethylzinc bromide, prepared by re-action of BrCH$_2$CN with Zn/Cu, adds to aryl aldehydes in the presence of (S)-**1** (1 equiv.) to give β-hydroxy nitriles in 87–93% ee. If only 0.3 mol% of (S)-**1** is present the optical yield is markedly decreased (in the case of C$_6$H$_5$CHO ee falls from 93 to 78%). Therefore, **1** serves both as a ligand and a catalyst.

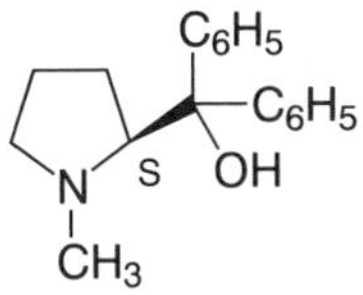

(S) - **1**

$$C_6H_5CHO \quad + \quad BrZnCH_2CN \quad \xrightarrow[76\%]{\substack{(S) - 1 \\ THF, -13°}} \quad R\overset{*}{C}H(OH)CH_2CN$$

(S), 93%ee

[1] K. Soai, Y. Hirose, and S. Sakata, *Tetrahedron Asymmetry,* **3**, 677 (1992).

Diphenylsilane/AIBN.

Alkenes from vic-diols.[1]   Bis(dithiocarbonates) can be converted into alkenes by reaction with tributyltin hydride/AIBN (**8**,499). Diphenylsilane and an initiator are equally effective and avoid use of toxic tin compounds. This radical reaction gives particularly high yields in reactions of nucleoside derivatives.

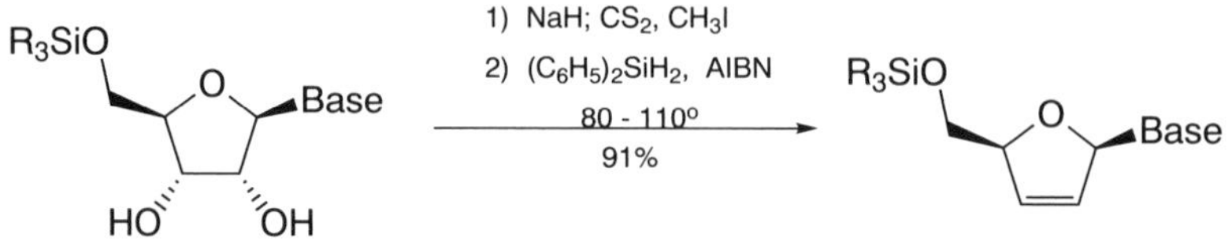

$$
\begin{array}{c}
\text{1) NaH; CS}_2\text{, CH}_3\text{I} \\
\text{2) (C}_6\text{H}_5)_2\text{SiH}_2\text{, AIBN} \\
\hline
\underline{\text{80 - 110}^\circ} \\
\text{91\%}
\end{array}
$$

[1] D. H. R. Barton, D. O. Jang, and J. Cs. Jaszberenyi, *Tetrahedron Letters,* **32**, 2569 (1991).

E

Enzymes.

The earliest use of enzymes as reagents for enantioselective reactions employed the common bakers' yeast (*Saccharomyces cerevisiae*), which had been used for centuries in the wine industry. Although it was known in 1966 that benzaldehyde-1-*d* is reduced by fermenting yeast to benzyl-1-*d* alcohol in 100% ee, this example of an asymmetric reaction was viewed as a challenge to chemists to develop chemical systems as efficient as the biochemical one. Indeed only two examples[1] of enzymatic reactions have been reported in *Organic Syntheses*. However, yeast reductions have proved useful in numerous ways as shown by two exhaustive reports listing nearly 700 references.[2] Since these reports, more examples from 1988 through 1991 have been listed if the reported enantiomeric excess is at least >95%.[3] This last review covers the general use of enzymes for preparation of enantiomerically pure compounds (880 references).

The recent decisions of the Food and Drug Administration to give preference to single enantiomeric drugs rather than to the racemates will undoubtedly provide further impetus to use of enzymes or other asymmetric reactions in drug design.[4]

Bakers' yeast is used almost exclusively for reduction, principally of ketones, by a dehydrogenase which usually follows the Prelog–Cram rule. As with chemical reductions, highest enantioselectivity obtains with aromatic aldehydes and aryl methyl ketones. β-Keto esters are also reduced with high eanantioselectivity by yeast. Some β-keto acids can also be reduced efficiently to (R)-β-hydroxy acids.

Yeast contains a variety of enzymes, and in some cases use of a single purified enzyme is preferable. These are divided into oxidoreductases, transferases, hydrolases, lyases, isomerases, and lipases. Many of these are commercially available (but expensive). Purified reductases usually require expensive cofactors. In addition individual microbes can be used as biocatalysts. A general review of microbial asymmetric reductions is available.[5] These reductions can be the opposite of those of yeast.

The most widely used enzymes are the hydrolytic enzymes: lipases, proteases, and nitrilases, probably because these enzymes do not require cofactors and are available commercially. They are particularly useful for resolution of esters,[6,7] and for organic synthesis.[8,9] Esterases can also catalyze esterification if the water concentration is low. Enzyme-catalyzed transesterification can be used for resolution of secondary alcohols and diols.[10]

A promising new area of enzymatic reactions is that of anaerobes. These organisms use redox systems that are not pyridine nucleotide dependent, but can use methylvirologen or benzylviologen (commercially available) as mediators. Electron donors can be hydrogen gas, formate, or carbon monoxide rather than glucose.[11] The first new enzyme from this source is a 2-enoate reductase, effected with stereospecific *trans*-addition of hydrogen. An additional double or triple bond in conjunction is not reduced. 2-Enals are also reduced, but reversibly. A 2-hydroxycarboxylate viologen oxidoreductam has also been identified. Carboxylic acid reductases from anaerobes can reduce acids to aldehydes in the presence of methylviologen.

Another new development in enzymatic reactions is the use of monoclonal antibodies as catalysts. These antibodies show regio- and stereoselectivity, substrate specificity, and rate acceleration. Hilvert[12] has discussed three different reactions that are catalyzed by these antibodies: decarboxylation, Diels–Alder reactions, and Claisen rearrangements. This catalysis is a new field with tremendous possibilities.

[1] D. Seebach et al., Organic Synthesis, 63, 1 (1985); K. Mori and H. Mori, ibid., 68, 56 (1990).
[2] S. Servin, Synthesis, 1 (1990); R. Csuk and B. I. Glanzer, Chem. Rev., 91, 49 (1991).
[3] E. Santaniello, P. Ferraboschi, P. Grisenti, and A. Manzocchi, Chem. Rev., 92, 1071 (1992).
[4] "Chiral Drugs," C&EN, Sept. 28, 46 (1992).
[5] H. Yamada and S. Shimizu, Angew. Chem. Int. Ed., 27, 622 (1988).
[6] C.-S. Chen and C. J. Seh, Angew. Chem. Int. Ed., 28, 695–707 (1989).
[7] W. Baland, C. Frescsel, and M. Lorenz, Synthesis, 1047 (1991).
[8] A. I. M. Janssen, A. J. H. Khunder, and B. Zwanenburg, Tetrahedron, 47, 4513, 7409, 7645 (1991).
[9] U. Ader, P. Andersch, M. Bergu, U. Goesgens, R. Sumayev, and M. Schneider, Pure Appl. Chem., 64, 1164 (1992).
[10] H. G. Leuenberger, Pure Appl. Chem., 62, 753 (1990).
[11] H. Simon, Pure Appl. Chem., 64, 1181 (1992).
[12] D. Hilvert, Pure Appl. Chem., 64, 1103 (1992).

Ephedrine.

Enantioselective addition of Bu$_2$CuLi to enones.[1]   Preliminary results indicate that the phosphorus derivative 1 of ephedrine can serve as a chiral ligand for dibutyl-cuprates. The ligand (+) is obtained by reaction of (−)-ephedrine with HMPA. Highest enantioselectivity is obtained by use of two equiv. of 1 for one equiv. of the cuprate

$$
\begin{array}{c}
C_6H_5 \\
CH_3 \!-\!\!\!-\!\!\!-O \\
\diagdown_{N}\!\!-\!\!PN(CH_3)_2 \\
| \\
CH_3
\end{array}
$$

1

and by addition of 4 equiv. of LiBr. However the enantioselectivity can vary from 0 to 75% ee depending on the source of CuI. Several related chiral phosphorus ligands have also been known to show moderate enantioselectivity.

[1] A. Alexakis, S. Mutti, and J. F. Normant, Am. Soc., 113, 6332 (1991).

(2S,4R)-2-Ethoxycarbonyl-4-phenyl-1,3-oxazolidine (1).
Preparation:

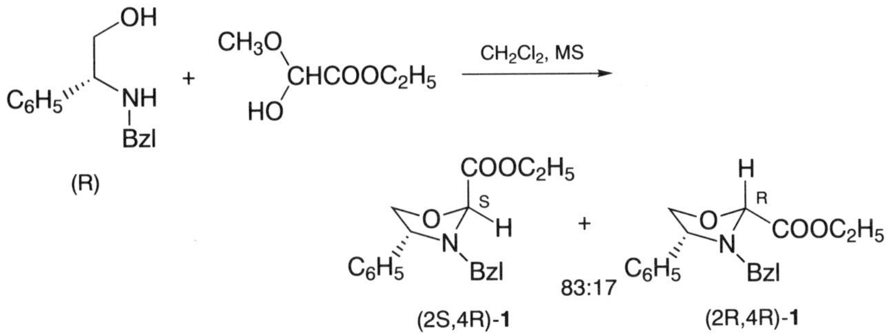

α-Amino esters.[1]   (2S,4R)-**1** is cleaved by dialkylzinc reagents in ether at 0° with high diastereoselectivity. The diastereomers can be obtained as optically pure products by flash chromatography. Bisdebenzylation affords pure ethyl (R)-α-aminocarboxylates (**3**) in >98% ee.

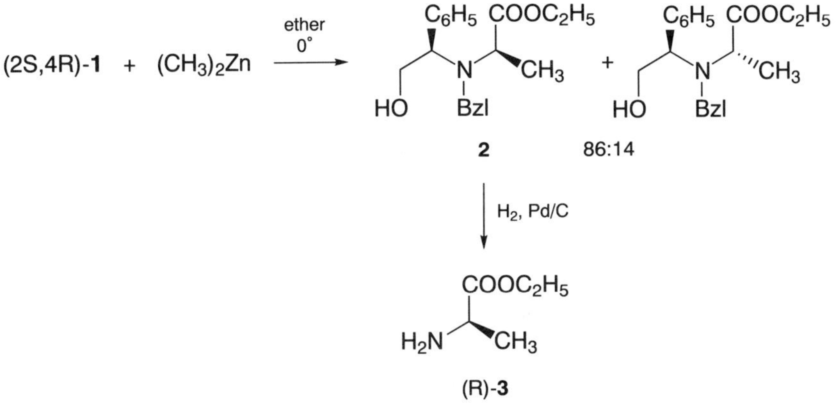

[1] C. Andrès, A. González, R. Pedrosa, A. Pérez-Encabo, S. Garcia-Granda, M. A. Salvado, and F. Gómez-Beltrán, *Tetrahedron Letters,* **33**, 4743 (1992).

Ethyl (S)-lactate.
Chiral sulfoxides.[1]   These have been prepared by a Sharpless-type asymmetric oxidation of sulfides (**13**,53), but this route is only efficient for preparation of aryl methyl sulfoxides (ee up to 92–96%). A more general route involves reaction of a pure chiral

cyclic sulfite (2) with organometallic reagents. It is particularly useful in the case of the sulfite 2, prepared as shown from ethyl (S)-lactate. The reaction of 2 with an alkyllithium

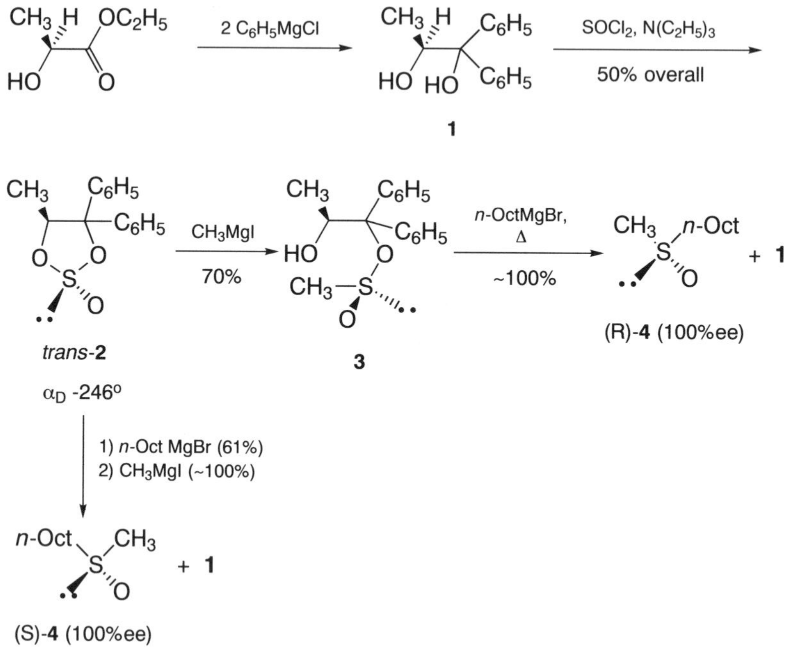

or alkyl Grignard reagent proceeds with marked regioslectivity and complete inversion of sulfur to provide a sulfinate (3), which on reaction with a second organometallic gives a chiral sulfoxide (4) in quantitative yield together with the chiral auxiliary diol 1. The method is particularly useful in preparation of alkyl t-butyl sulfoxides since t-BuMgCl reacts with trans-2 with a selectivity of 95:5 in contrast to that of CH_3MgCl (80:20).

[1] F. Rebiere, O. Samuel, L. Ricard, and H. B. Kagan, J. Org., 56, 5991 (1991).

(E)-Ethylidenecyclopropanone ketal (1).

[3+2]Cycloaddition to alkenes.[1]   The ketal 1 when heated (60–100°) can undergo regio- and *endo*-selective [3 + 2]cycloaddition to electron-deficient alkenes. Since hydrolysis of the adduct proceeds with >90% stereoselectivity, this cycloaddition can

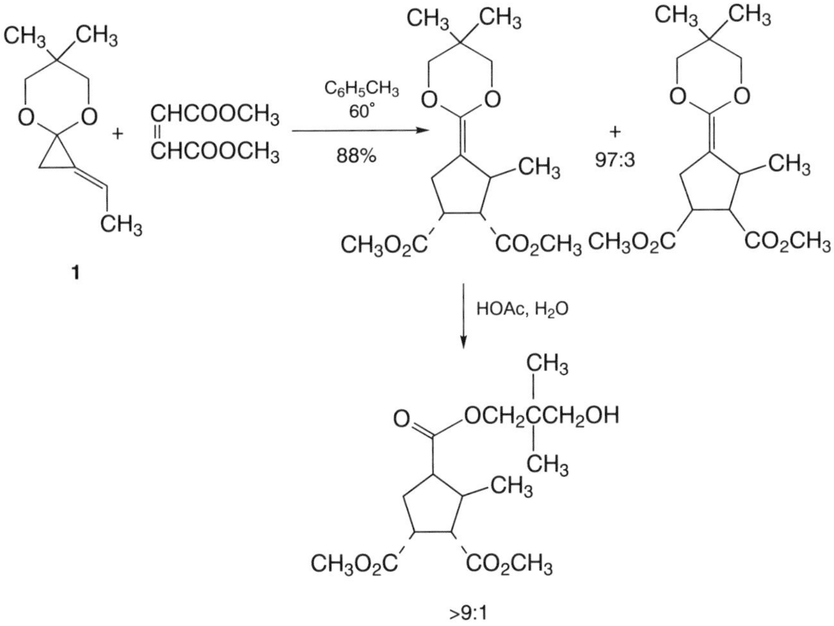

result in products with four chiral centers with high stereocontrol.

[1] S. Ejeri, S. Yamago, and E. Nakamura, *Am. Soc.*, **114**, 8707 (1992).

F

Ferric chloride.

Perylenequinone (2).[1]   A novel route to this quinone (**2**) involves double coupling of the 5-bromo-1,2-napthoquinone (**1**), prepared as shown, by treatment with FeCl$_3$

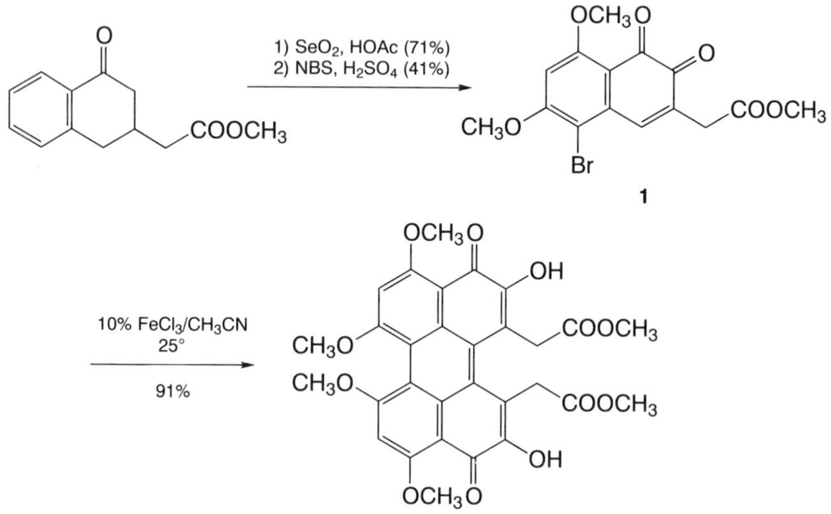

in anhydrous acetonitrile at 25°.

Anomerization catalyst.[2]   In the course of debenzylation/4-methoxycinnamoylation of the α-glycopranoside **1** promoted by FeCl$_3$, Nakanishi *et al.* observed that the anomeric β-methoxyl group is anomerized to give α-glycoside **2** (α/β = 95:5).

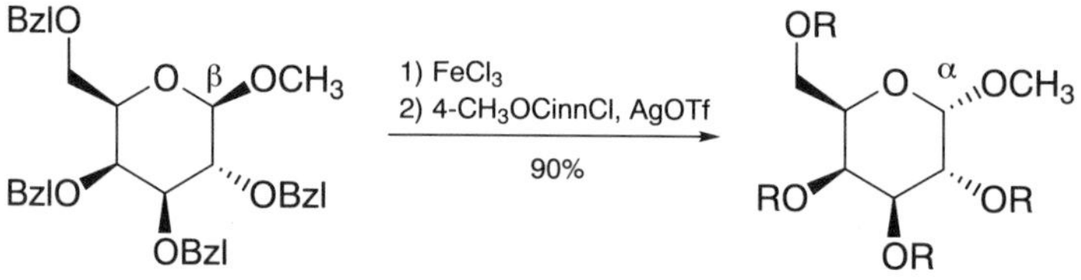

This is a fairly general anomerization. tetracetyl, Acetonides and benzylidine acetals are labile to $FeCl_3$, but reactions are faster with acetyl derivatives. The method fails with furanosides. The reaction of the glycoside **1** is carried out in CH_2Cl_2 at room temperature and is complete in 15 minutes to 5 hours.

Oxidative coupling of ArOAlCl₂.[3] Oxidative coupling of phenols results in complicated mixtures, but oxidative coupling of dichloroaluminum phenolates is a useful route to hydroxylated bi- and tetraaryls. The starting material is obtained by reaction of phenols with $AlCl_3$ in CH_3NO_2 and need not be isolated. Of several oxidants, $FeCl_3$ (1 equiv.) is most satisfactory, and CH_3NO_2 is the preferred solvent.

[1] Z. Diwu and J. W. Lown, *Tetrahedron*, **48**, 45 (1992).
[2] N. Ikemoto, O. K. Kim, L.-C. Lo, V. Satyanarayana, M. Chang, and K. Nakanishi, *Tetrahedron Letters*, **33**, 4295 (1992).
[3] G. Sartori, R. Maggi, F. Bigi, A. Arienti, G. Casnati, G. Bocelli, and G. Mori, *Tetrahedron*, **48**, 9483 (1992).

Fluorosilicic acid, H_2SiF_6.

Desilylation.[1] This acid is less acidic than HF, and can effect desilylation without effect on acid-labile protecting groups. A further advantage is that it can effect desilylation of *t*-butyldimethylsilyl ethers in the presence of *t*-butyldiphenylsilyl or triisopropylsilyl ethers. The selectivity of H_2SiF_6 can be enhanced by addition of triethylamine (more effective than DMAP). The cleavage is effected in CH_3CN at 0° and is usually complete after one hour.

[1] A. S. Pilcher, D. K. Hill, S. J. Shimshock, R. E. Waltermire, and P. DeShong, *J. Org.*, **57**, 2492 (1992).

G

Gallium(II) chloride, $GaCl \cdot GaCl_3$ (Ga_2Cl_4).

Reductive Friedel–Crafts reaction.[1]   Reaction of anisole with Ga_2Cl_4 (2 equiv.) and an aromatic aldehyde in CS_2 at room temperature results in diphenylmethanes. Similar results obtain with ketones and aliphatic aldehydes. This reaction involves the reducing ability of GaCl and the Lewis acid activity of $GaCl_3$.

[1] Y. Hashimoto, K. Hirata, N. Kihara, M. Hasegawa, and K. Saigo, *Tetrahedron Letters,* **33,** 6351 (1992).

Gallium(III) chloride–Silver perchlorate.

Friedel-Crafts acylation.   This reaction is generally effected with $AlCl_3$, but a stoichiometric amount is required. Since a combination of Lewis acids is sometimes more effective than a single Lewis acid, Mukaiyama *et al.*[1] have examined the reactivity of several Lewis acids in combination with $AgClO_4$ and report that the combination of $AgClO_4$ with $GaCl_3$, $AlCl_3$, or BCl_3 results in a catalyst that can effect Friedel-Crafts acylation in good yield. Highest yields are obtained when $AgClO_4$ and $GaCl_3$ are present in the ratio of 2:1. The actual catalytic species is presumably $GaCl(ClO_4)_2$, and high yields are obtained by use of 10 mol % of this catalyst.[2]

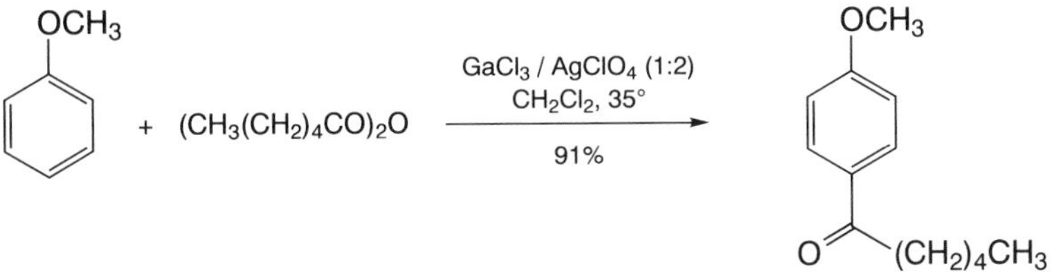

[1] T. Mukaiyama, T. Ohno, T. Nishimura, S. Suda, and S. Kobayashi, *Chem. Letters,* 1059 (1991).
[2] T. Harada, T. Ohno, S. Kobayashi, and T. Mukaiyama, *Synthesis,* 1216 (1991).

Grignard reagents.

Modified Grignard reagents, RMgL.[1]   These Grignard reagents can be prepared by reaction of RLi and MgL_2 [L=OSO_2CF_3, OSO_2CH_3, $OCOCH_3$, $OCOC(CH_3)_3$]. The ligand can exert a marked effect in addition to carbonyls. Thus $CH_3MgOCOC(CH_3)_3$ reacts exclusively with an aldehyde in the presence of ketone to form the adduct of the aldehyde in high selectivity (equation I).

(I)

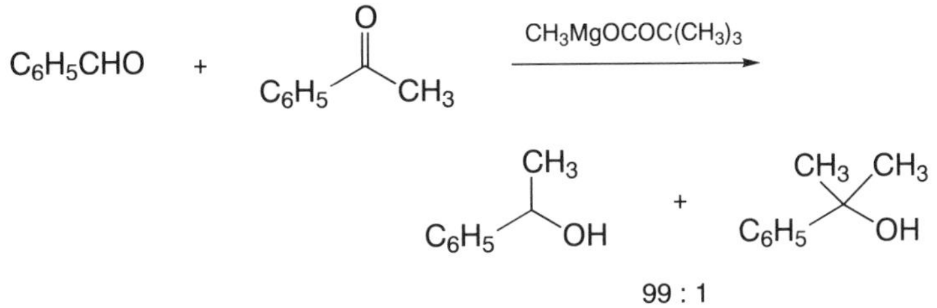

(II)

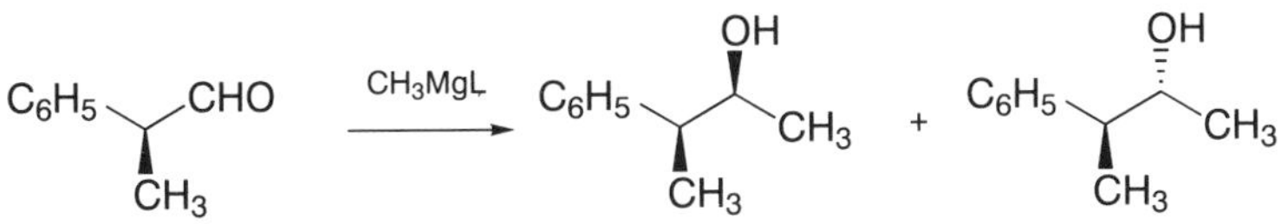

L = Cl, Br, I	80%	70 : 30
L = OTs	76%	92 : 8
L = OAc	83%	91 : 9

The ligand has a marked effect on the reaction of Grignard reagents with α-chiral aldehydes (equation II). It can also affect the addition of reagents to substituted cyclohexanones (equation III).

(III)    (CH₃)₃C~~~=O

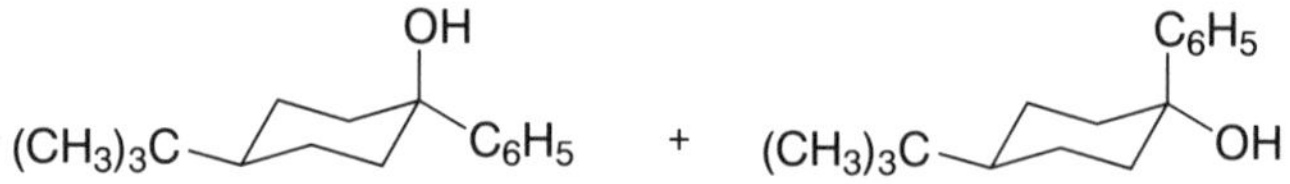

→

OH

(CH₃)₃C~~~C₆H₅   +   (CH₃)₃C~~~

C₆H₅MgBr	94%	49 : 51
C₆H₅MgOTf	87%	73 : 27
C₆H₅MgOTs	90%	85 : 15

[1] M. T. Reetz, N. Harmat, and R. Mahrwald, *Angew. Chem. Int. Ed.,* **31**, 342 (1992).

H

$$\text{CH}_3$$
$$|$$
1,1,1,2,3,3,3-Heptamethyltrisilane, $(CH)_3)_3Si\text{-}SiH\text{-}Si(CH_3)_3$

Preparation.[1]

Radical reductions.[2]   This silane is less reactive than Bu_3SnH, but is useful for reduction of substrates when the product is formed by slow hydrogen transfer. In the general procedure it is used in combination with AIBN (10–20%) in toluene or benzene at 75–90°. It is effective for reduction of halides, selenides, thionoesters, and isocyanides.

[1] M. Kumada, M. Ishikawa, and S. Maeda, *J. Organomet. Chem.,* **2**, 478 (1964).
[2] C. Chatgilialoglu, A. Guerrini, and M. Lucarini, *J. Org.,* **57**, 3405 (1992).

Hexamethylditin.

[4+1]Radical annelation.[1]   Irradiation of a mixture of the N-propargyl bromopyridone **1**, phenyl isocyanide (5 equiv.), and $(CH_3)_3SnSn(CH_3)_3$ (1.5 equiv.) in benzene at 80° results in the tetracycle **2** in 40% yield.

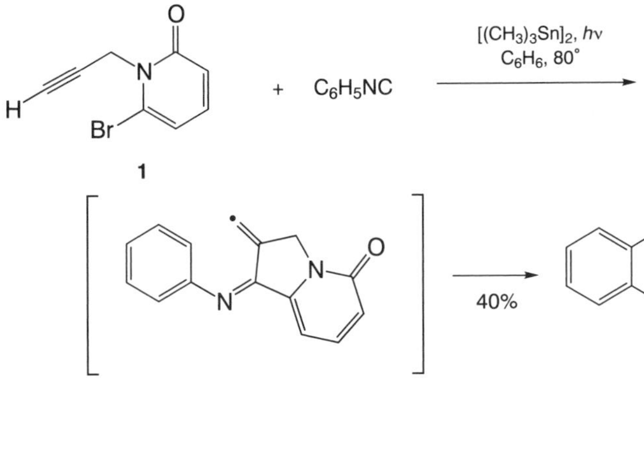

A modification of this radical annelation provides a synthesis of (±)-campothecin (**5**) by radical annelation of the pyridone **3** with phenyl isocyanide to provide **4**, a known precursor to **5**.

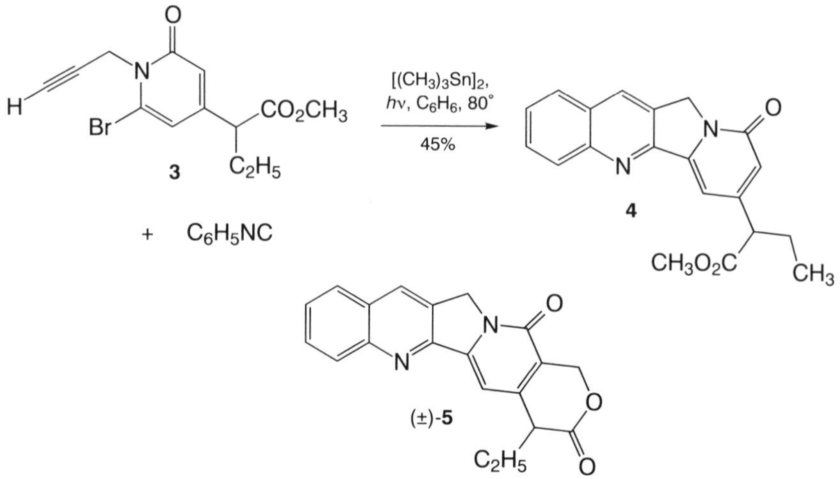

Dimerization of 1,3-dienes.[2]   In the presence of a Pd(0) catalyst, particularly bis(dibenzylideneacetone)palladium [Pd(dba)$_2$], 1,3-dienes react with **1** to afford adducts (**2**) of dimerization–double stannation.

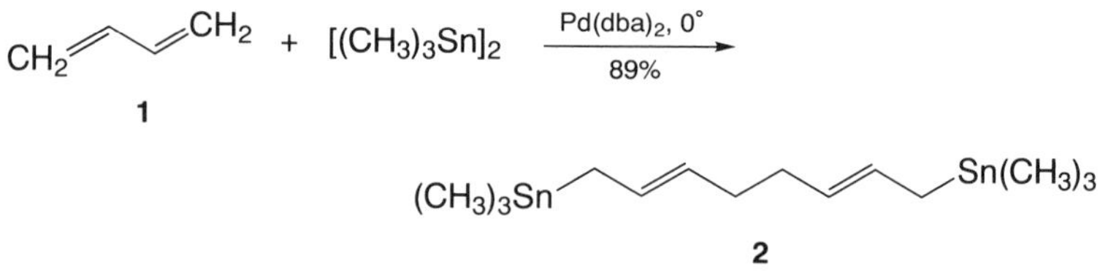

---

[1] D. P. Curran and H. Liu, *Am. Soc.*, **114**, 5863 (1992).
[2] Y. Tsuji and T. Kakehi, *J. C. S. Chem. Comm.*, 1000 (1992).

5-Hexen-1-yllithiums.

Cyclization to (cyclopentyl)methyllithiums.[1]   5-Hexen-1-yllithiums, prepared by iodine–lithium exchange at −78°, on warming to 25° cyclize to (cyclopentyl)-methyllithiums. Cyclization can be facilitated by addition of THF, TMEDA, or PMDTA (pentamethyldiethylenetriamine). Substituted substrates show considerable stereoselectivity.

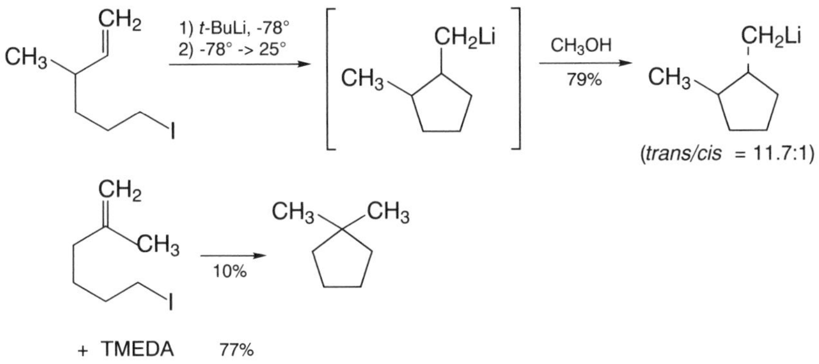

[1] W. F. Bailey, A. D. Khanolkar, K. Gavaskar, T. V. Ovaska, K. Rossi, Y. Thiel, and K. B. Wiberg, *Am. Soc.,* **113**, 5720 (1991).

Hydrogen peroxide.

Hydroxylation of naphthalene.[1]   Reaction of naphthalene with H_2O_2 (90%) in HF or in HF (70%)/pyridine (30%) at $-10°$ to 20° gives mixtures of naphthols in 26–43% yield in which 1-naphthol is the major product. If the reaction is conducted in a superacid (HF/SbF_5, HF/BF_3), 2-naphthol is obtained almost exclusively. The difference is

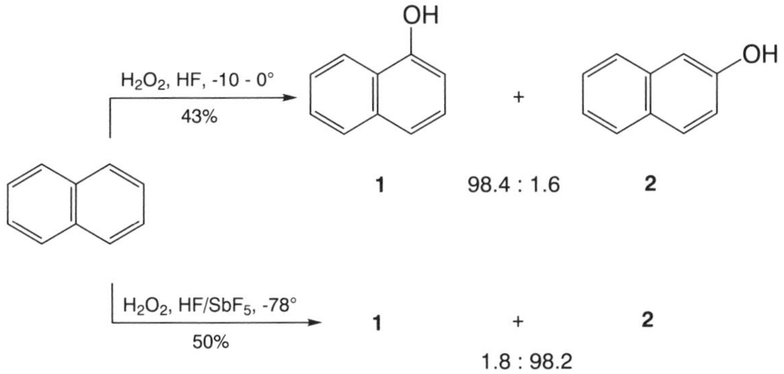

not a result of rearrangement of the hydroxyl group. The reaction in the presence of a superacid may involve a naphthalenium ion.

Alkyl hydroperoxides.[2]   Direct alkylation of hydrogen peroxide as a route to ROOH is usually unsatisfactory, but alkyl peroxymercuriation of alkenes (**1**) to provide hydroxymercurials **2** proceeds readily. After protection of the peroxide group, selective demercuration ($NaBH_4$) followed by deprotection (HOAc, H_2O) provides alkyl hydroperoxide **3** in 30–55% overall yield.

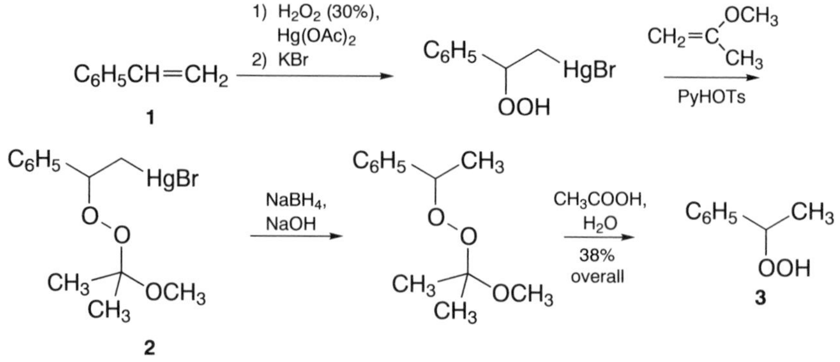

[1] G. A. Olah, T. Keumi, J. C. Lecoq, A. P. Fung, and J. A. Olah, *J. Org.,* **56**, 6148 (1991).
[2] A. J. Bloodworth, C. J. Cooksey, and D. Korkodilos, *J. C. S. Chem. Comm.,* 926 (1992).

Hydrogen peroxide–Peroxotungstophosphates.

The metal catalyst is prepared by reaction of 12-tungstophosphoric acid with H_2O_2 and cetylpyridinium chloride.

Oxidations with H_2O_2.[1]   In the presence of this catalyst, internal alkynes are oxidized by H_2O_2 (35%) to epoxides and α, β_2-enones.

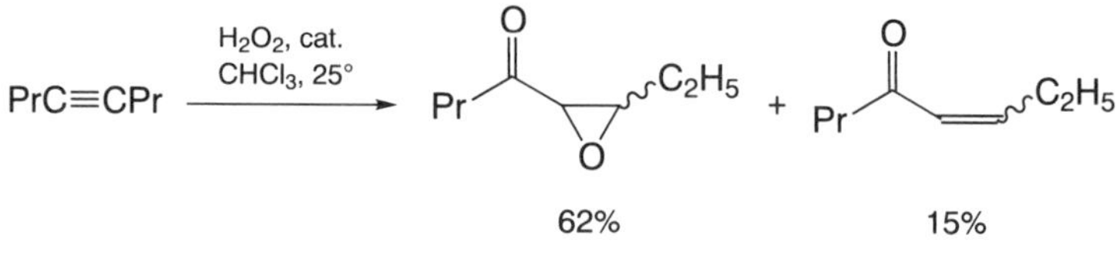

Under the same conditions, a primary–secondary vic-diol is oxidized to a 1-hydroxy-2-alkanone in 45–90% yield, whereas a primary–tertiary vic-diol is not oxidized. Secondary–secondary vic-diols are oxidized to α-ketols in modest yield or overoxidized to carboxylic acids.

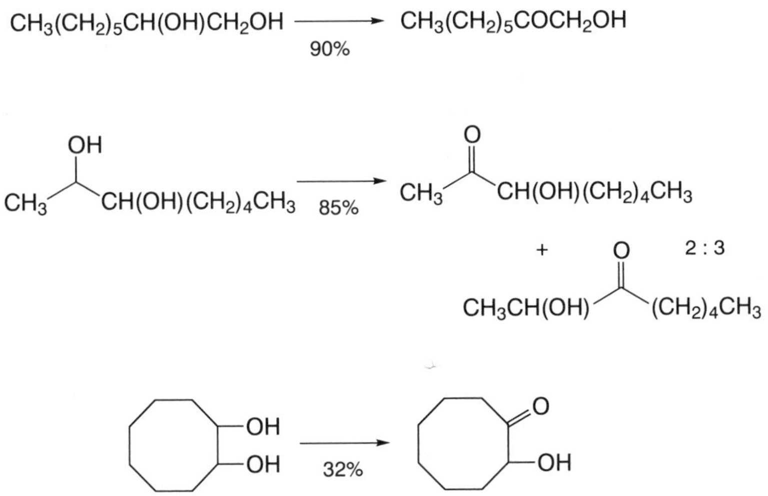

$$CH_3(CH_2)_5CH(OH)CH_2OH \xrightarrow[90\%]{} CH_3(CH_2)_5COCH_2OH$$

$$
\underset{CH_3}{\overset{OH}{\underset{}{\big|}}}CH(OH)(CH_2)_4CH_3 \xrightarrow{85\%}
$$

CH₃—CO—CH(OH)(CH₂)₄CH₃

+   2 : 3

CH₃CH(OH)—CO—(CH₂)₄CH₃

32%

[1] Y. Ishii and Y. Sakata, *J. Org.,* **55**, 5549 (1990); *idem, ibid.,* **56**, 6233 (1991).

Hydrogen peroxide–Rhenium(VII) oxide, H_2O_2–Re_2O_7.

Dihydroxylation of alkenes.[1]   Oxidation of terminal, internal, or cyclic alkenes by H_2O_2 at 90° catalyzed by Re_2O_7 results in *vic*-diols in 60–80% yield.

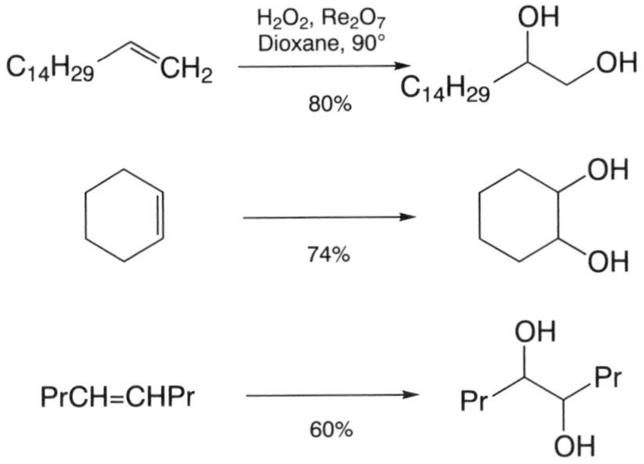

$$C_{14}H_{29}\text{—CH=CH}_2 \xrightarrow[\substack{80\%}]{\substack{H_2O_2,\ Re_2O_7 \\ Dioxane,\ 90°}}$$

$$C_{14}H_{29}\text{—CH(OH)CH}_2\text{OH}$$

74%

PrCH=CHPr $\xrightarrow[60\%]{}$ Pr—CH(OH)—CH(OH)—Pr

[1] S. Warwel, M. Rüschgen Klaas, and M. Sojka, *J. C. S. Chem. Comm.,* 1578 (1991).

Hydrogen peroxide–Tungstic acid, H_2O_2–$WO_3 \cdot nH_2O$.

Oxidation of cycloalkenes to dialdehydes.[1]   This reaction is usually effected with ozone, but can also be effected with H_2O_2 catalyzed by tungstic acid in *t*-butyl alcohol at 35°. The dialdehyde is accompanied by the cycloalkane-1,2-diol. Since a cycloalkene oxide is also oxidized to a dialdehyde under these conditions, the overall process probably involves oxidation to the epoxide, which is hydrolyzed to the 1,2-diol or oxidized further to the dialdehyde. Glutaraldehyde is obtained from cyclopentene by this reaction in 80% yield; adipaldehyde is obtained from cyclohexane in 47% yield.

[1] D. Jingfa, X. Xinhua, C. Haiying, and J. Anren, *Tetrahedron*, **48**, 3503 (1992).

Hydrogen peroxide/Urea–Phthalic anhydride.

$$R^1SR^2 \rightarrow R\overset{\overset{\displaystyle O}{\|}}{S}R^2.$$[1]   One problem with this conversion is overoxidation to a sulfone, commonly observed with most oxidants. The use of the complex of H_2O_2 with urea in combination with phthalic anhydride[2] overcomes this problem and provides sulfoxides in yields of 89–95%. Methanol is generally the solvent of choice.

N-Oxidation of amines.[3]   N-Oxides of tertiary amines or of N-heteroaromatic compounds can be obtained in high yield by N-oxidation with the H_2O_2/urea adduct in combination with phthalic anhydride (1:1), which is more effective than acetic or benzoic anhydride. Yields are generally >80%.

[1] R. Balicki, L. Kaczmarek, and P. Nantka-Namirski, *Ann.*, 883 (1992).
[2] C. Lu, E. W. Hughes, and P. A. Giguere, *Am. Soc.*, **63**, 1507 (1941).
[3] L. Kaczmareki, R. Balicki, and P. Nantka-Namirski, *Ber.*, **125**, 1965 (1992).

(R,R) or (S,S)-6,6'-(1-Hydroxy-2,2-dimethylpropyl)-2,2'-bipyridine (1).

Preparation[1]

Asymmetric addition of diethylzinc to RCHO.[2]   In the presence of 5 mol % of this C_2-symmetric 2,2'-bipyridine, (R,R)-**1**, diethylzinc adds to a wide variety of aldehydes with high enantioselectivity (equation I).

Surprisingly, the simple pyridine (R)-**2** is almost as effective as (R,R)-**1** as the chiral catalyst. Moreover it can show high asymmetric amplification. Thus use of R-**2** with an ee of 14% can provide an adduct with 87% ee.

Enantioselective addition of R₂Zn to enones.[3] This conjugate addition of R_2Zn to chalcones can be effected by catalysis with Ni(acac)₂–(S,S)-**1** in the ratio 1:20. In this case use of aceto- or propionitrile as solvent is essential. In this reaction, the pyrid-

ine (R)-**2** is more effective than (S,S)-**1** as the chiral ligand. Thus use of Ni(acac)₂– (R)-**2** produces the same adduct in 82% ee. Asymmetric amplification with (R)-**2** was also observed in this reaction.

[1] C. Bolm, M. Ewald, M. Felder, and G. Schlingloff, *Ber.,* **125,** 1169 (1992).
[2] C. Bolm, G. Schlingloff, and K. Harms, *Ber.,* **125,** 1191 (1992).
[3] C. Bolm, M. Ewald, and M. Felder, *Ber.,* **125,** 1205 (1992).

1-Hydroxy-3-isothiocyanatotetrabutyldistannoxane, HOSn(Bu)₂OSn(Bu)₂NCS (**1**).

Acetalization.[1] Distannoxanes of this type are useful catalysts for esterification (**15,89**) because they can activate both alcohols and carbonyl groups on the same template. Of a number of stannoxanes, **1** is found to be the most efficient catalyst for acetalization of aldehydes and ketones with ethylene glycol. In particular it can promote acetalization of cyclic α, β-enones, which usually proceeds in low yield.

[1] J. Otera, N. Dan-oh, and H. Nozaki, *Tetrahedron*, **48**, 1449 (1992).

[Hydroxy(tosyloxy)iodo]benzene, 14, 179–180; **16,** 179.

α-Tosyloxylation of ketones.[1]   This reaction is usually conducted in refluxing acetonitrile. The rate of the reaction is markedly increased by sonication. Thus it can be conducted under sonication at 55° in 10–30 minutes. This version can be used to tosyloxylate even alicyclic ketones.

[1] A. Tuncay, J. A. Dustman, G. Fisher, C. I. Tuncay, and K. S. Suslick, *Tetrahedron Letters*, **33**, 7647 (1992).

N-Hydroxyurethane, $HONHCOOC_2H_5$ **(1).** Available from Aldrich.

N—O Linked glycosides.[1]   The trisaccharide group of several potent antitumor antibiotics contains an unusual N—O linkage. A general method for stereoselective synthesis of such saccharides employs N-hydroxyurethane as the key reagent. The carboethoxy group deactivates the nitrogen group so that glycosylation occurs selectively with the hydroxyl group, but it facilitates deprotonation of the nitrogen to permit subsequent S_N2 displacement with inversion. The requisite glycosyl urethanes such as **2** can be prepared directly by reaction of a glycosyl sulfoxide with **1.** Deprotonation of **2** (NaH) provides an anion on nitrogen that reacts with a triflate (**3**) with inversion in high yield. The coupled product is deprotected by NaOH (solid) in CH_3OH at 0° → 25° to provide the N—O disaccharide **4.**

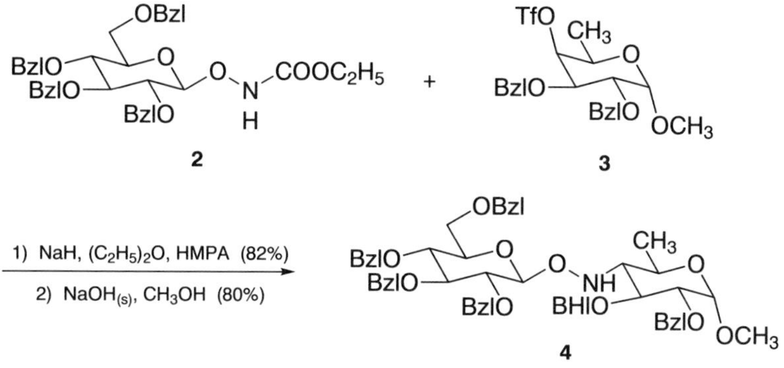

$$\text{2} + \text{3}$$

1) NaH, $(C_2H_5)_2O$, HMPA (82%)

2) NaOH$_{(s)}$, CH$_3$OH (80%)

4

[1] D. Yang, S.-H. Kim, and D. Kahne, *Am. Soc.*, **113**, 4715 (1991); *Idem,* in press.

I

Indium.

Allylation in water.[1] Unlike most metals, indium is stable to boiling water or alcohol. Indium can effect allylation of aldehydes in water at 25° with no need for a promotor.

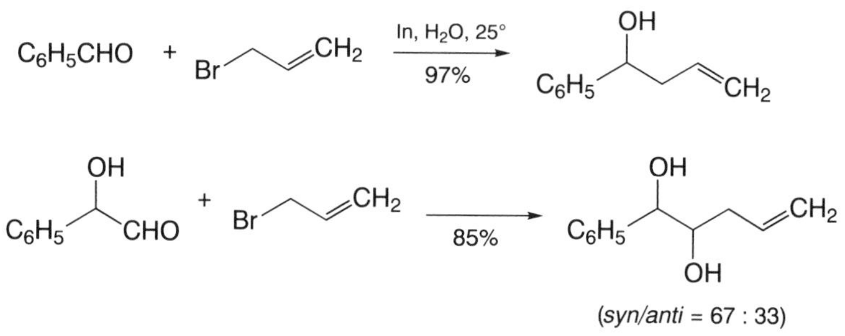

$$C_6H_5CHO + Br\diagup\hspace{-0.3em}=CH_2 \xrightarrow[97\%]{In, H_2O, 25°} C_6H_5\diagup\overset{OH}{|}\diagup\hspace{-0.3em}=CH_2$$

(*syn/anti* = 67 : 33)

The reaction when extended to acrylates provides hydroxy acrylic esters, which are precursors to methylene γ-lactones.

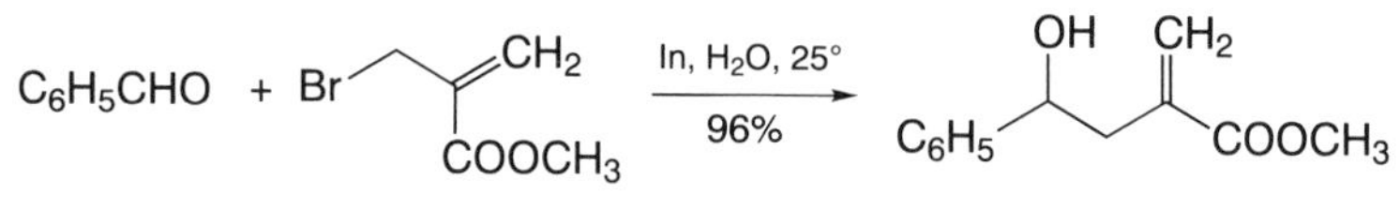

$$C_6H_5CHO + Br\diagup\overset{=CH_2}{\underset{COOCH_3}{|}} \xrightarrow[96\%]{In, H_2O, 25°} C_6H_5\diagup\overset{OH}{|}\diagdown\overset{CH_2}{\underset{COOCH_3}{||}}$$

[1] C. J. Li and T. H. Chan, *Tetrahedron Letters,* **32,** 7017 (1991).

Iodine monobromide, IBr.

Diastereoselective cyclization of homoallyl carbonates.[1] The conversion of the carbonate **1** to **2** has traditionally been effected with I_2 in CH_3CN at −20°. It can be carried out with higher selectivity with IBr in toluene, particularly at temperatures of −80–85°.

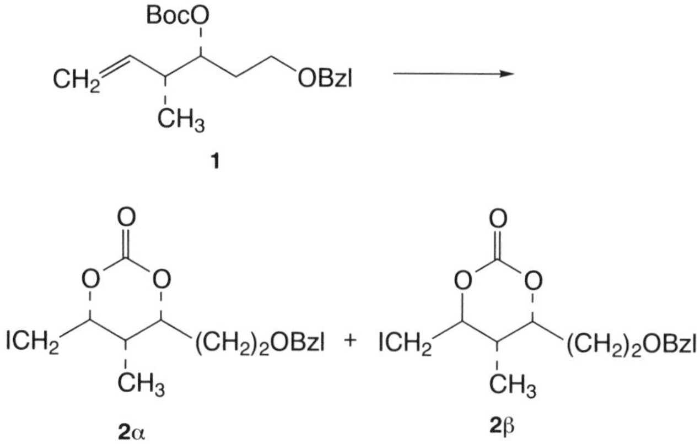

$$I_2, CH_3CN, -20°$$

I_2, CH_3CN, -20°	79%	5.7:1
IBr, $C_6H_5CH_3$, -80-85°	80%	14.9:1

The generally higher selectivity of IBr was observed with a number of other homoallylic carbonates.

[1] J. J.-W. Duan, P. A. Sprenger, and A. B. Smith, III, *Tetrahedron Letters,* **33**, 6439 (1992).

Iodine/Pyridine.

α-Iodination of cycloalkenones.[1] This reaction can be effected with I_2 (1–4 equiv.) dissolved in pyridine/CCl_4 (1:1) at $0 \rightarrow 25°$.

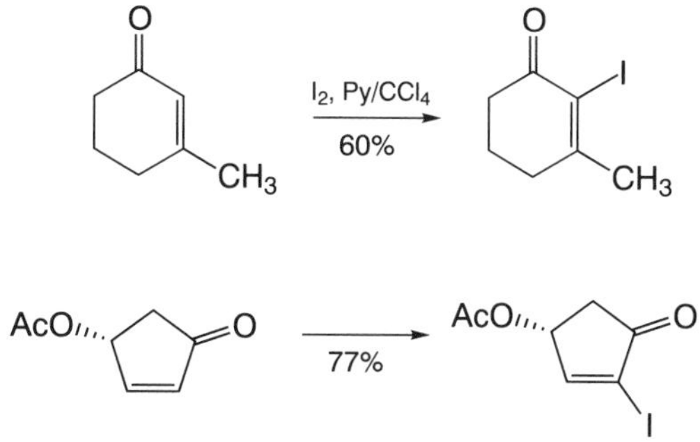

[1] C. R. Johnson, J. P. Adams, M. P. Braun, C. B. W. Senanayaka, P. M. Wovkulich, and M. R. Ukoković, *Tetrahedron Letters,* **33**, 917 (1992).

Iodine–Silver trifluoroacetate, I_2–AgO_2CCF_3.

Cyclization of homoallylic alcohols to tetrahydrofurans. Lipshutz[1] has prepared the four isomers of the homoallylic alcohol **1**, and has shown that by a proper choice

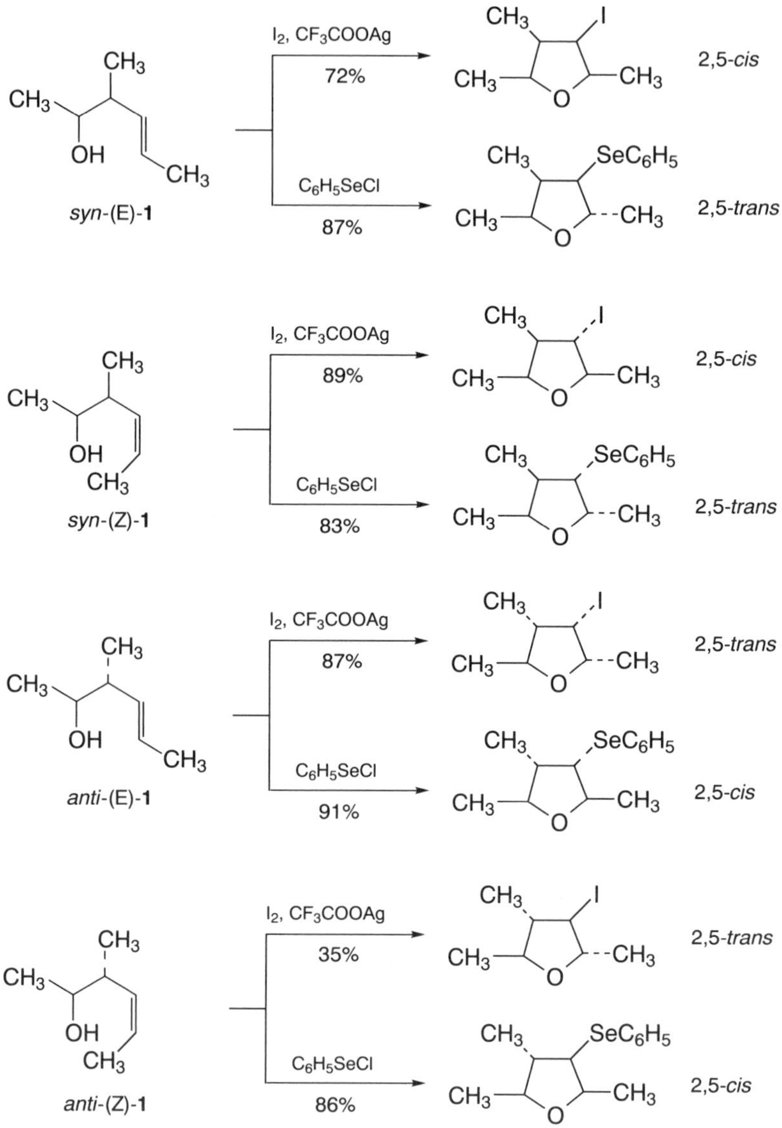

of the electrophile, either iodine or phenylselenyl chloride, the 2,5-*cis*- or 2,5-*trans*-tetrasubstituted furan can be prepared selectively.

[1] B. H. Lipshutz and J. C. Barton, *Am. Soc.*, **114**, 1084 (1992).

(Iodomethyl)zinc iodide, ICH_2ZnI, (1).

α-Methylene-γ-butyrolactones.[1] A new one-pot preparation of these lactones involves stereoselective addition of a copper/zinc reagent to an acetylenic ester to form an alkenylcopper reagent **2** that on homolgation with **1** reacts with an aldehyde (or ketone) to form an α-methylene-γ-butyrolactone (**3**) with *cis*-selectivity.

$$HC\equiv CCOOC_2H_5 \ + \ C_6H_5CH_2Cu(CN)ZnX \ \longrightarrow \ \left[\begin{array}{c} COOC_2H_5 \\ Bzl \quad Cu(CN)ZnX \end{array} \right]$$

2

↓ **1**

$$\begin{array}{c} C_6H_5CH_2 \quad CH_2 \\ C_6H_5 \quad O \quad O \end{array} \xleftarrow[78\%]{C_6H_5CHO} \left[\begin{array}{c} COOC_2H_5 \\ Bzl \quad CH_2Cu(CN)ZnX \end{array} \right]$$

3 (*cis/trans* = 92:8)

An intramolecular version is also possible (equation I).

(I)

$$\begin{array}{c} O \\ C_6H_5 \quad (CH_2)_3Cu(CN)ZnI \end{array} \xrightarrow[76\%]{\begin{array}{c} 1)\ HC\equiv CCOOC_2H_5 \\ 2)\ C_6H_5CHO;\ Zn(CH_3)_2 \end{array}}$$

[structure product with C_6H_5, O, CH_2, H]

[1] A. Sidduri and P. Knochel, *Am. Soc.*, **114**, 7579 (1992).

Iodonium di-*syn*-collidine perchlorate, (1), 10, 212; 11, 269.

Enantioselective iodolactonization of a bis-γ,δ-unsaturated chiral amide.[1] This reaction can be effected with high enantioselectivity using a sultam derived from D-camphorsulfonic acid as the chiral auxiliary. The cyclization of **2** to (−)-**3** can also be effected with KI/I_2 in >98% ee, but the chemical yield is only 39%.

[structure of reactant **2** with CH_3, CH_2, S–N, O_2, O, CH_2 groups]

$$\xrightarrow[87\%]{\begin{array}{c} 1,\ -40° \\ CH_2Cl_2,\ CH_3OH \end{array}}$$

[structure of product with CH_2, O, O, CH_2I groups]

(−) **3** (>98% ee)

2

[1] T. Yokomatsu, H. Iwasawa and S. Shibuya, *J. C. S. Chem. Comm.*, 728 (1992).

Iodosylbenzene–Triflic anhydride (C_6H_5IO/Tf_2O), 1.

Glycosidation of thioglucosides. Reaction of these two reagents in the presence of 4Å MS in CH_2Cl_2 provides a reagent (1) that promotes glycosidation of tetrabenzyl-protected methylthioglucosides.[1] The reaction of tetraacetyl-protected methylthiogluco-sides with 1 proceeds with marked β-selectivity, particularly when SiO_2 is also present.

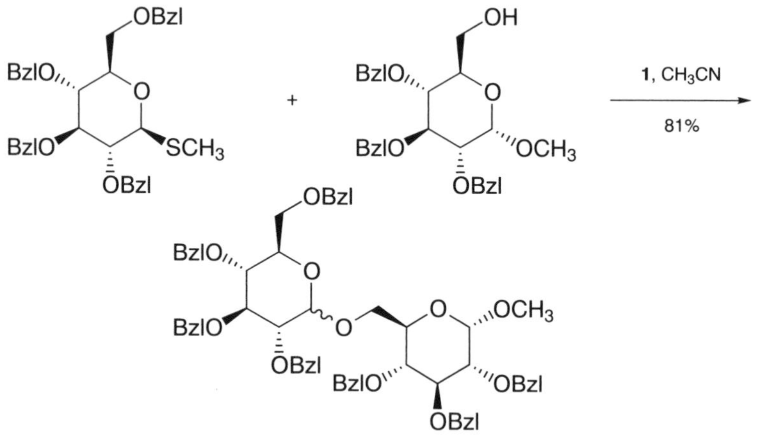

$(\alpha/\beta = 1.12)$

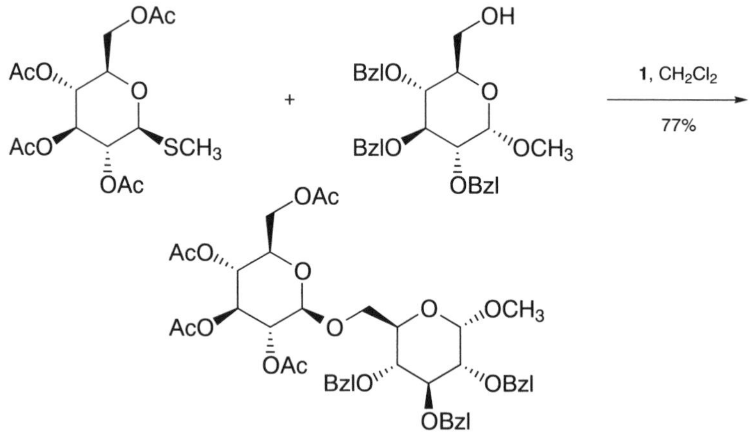

$(\beta, 100\%)$

[1] K. Fukase, A. Hasuoka, I. Kinoshita, and S. Kusumoto, *Tetrahedron Letters*, **33**, 7165 (1992).

Iodosylbenzene–Trimethylsilyl azide, $C_6H_5IO/N_3Si(CH_3)_3$.

Cyclic α,β-enones. Reaction of triisopropyl (TIPS) enol ethers of cyclic ketones with the combination of C_6H_5IO and trimethylsilyl azide in CH_2Cl_2 at $-15°$ provides a β-azide, which on treatment with Bu_4NF at $0°$ is converted into the α,β-enone. Examples:

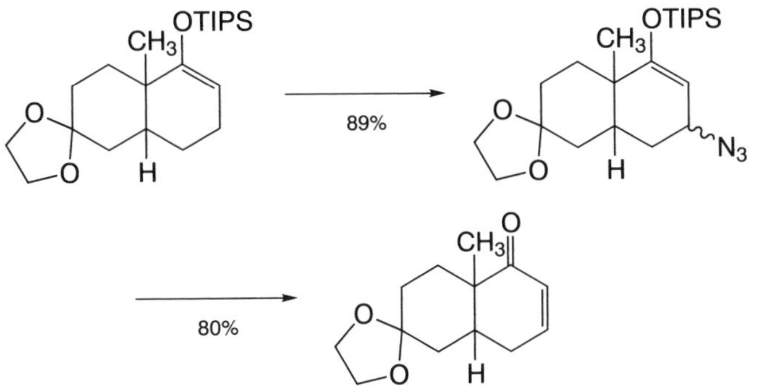

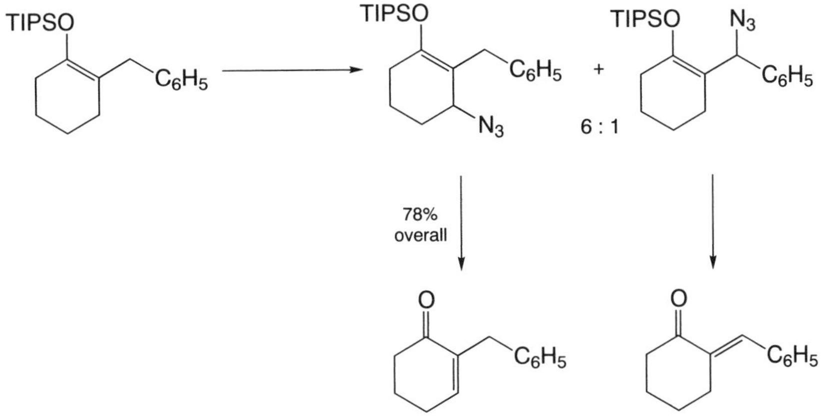

[1] P. Magnus, A. Evans, and J. Lacour, *Tetrahedron Letters,* **33**, 2933 (1992).

Iron powder.

Aerobic oxidation of alkanes.[1] Various metal complexes are known to catalyze air oxidation of unactivated C—H bonds. Murahashi *et al.* have found that both ruthenium and iron complexes are useful catalysts for aerobic oxidation in combination with an aldehyde and an acid. Iron powder is the most effective catalyst, but $FeCl_3 \cdot 6H_2O$, $RuCl_3 \cdot H_2O$, and $RuCl_2[P(C_6H_5)_3]_3$ can be used. Useful aldehydes are heptanal, 2-methylpropanal, and even acetaldehyde. A weak acid is suitable; thus acetic acid is preferred to chloroacetic acid. By using the most satisfactory conditions, cyclohexane is converted to cyclohexanone and cyclohexanol (2.3:1 ratio), and alkylarenes are oxidized selectively at the benzylic position. This system is clearly different from the Gif iron system (14,184–185); the tertiary/secondary C—H bond selectivity differs markedly in the two systems.

[1] S.-I. Murahashi, Y. Oda, and T. Naota, *Am. Soc.,* **114**, 7913 (1992).

Iron(III) oxide (Fe$_2$O$_3$).

Baeyer–Villiger oxidation of ketones.[1]   This oxidation can be effected in high yield by oxygenation in the presence of an aldehyde (3 equiv.) and Fe$_2$O$_3$. Benzaldehyde is the aldehyde of choice but heptanal is also useful. Benzene is the only suitable solvent.

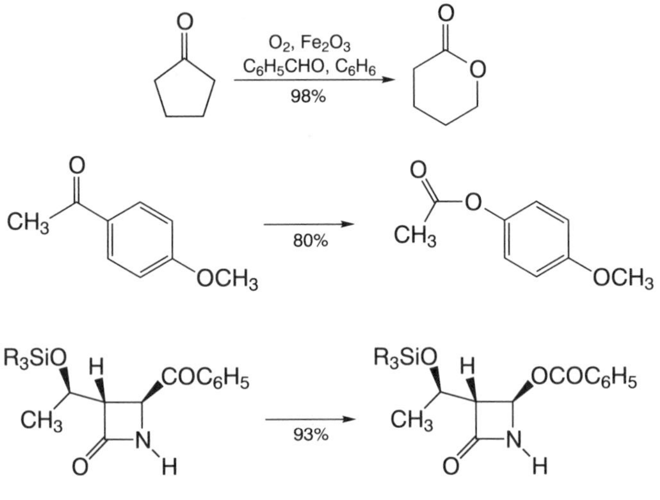

[1] S.-I. Murahashi, Y. Oda, and T. Nasta, *Tetrahedron Letters,* **33**, 7557 (1992).

N-Isopropyl-(2S,6S)-1,3,2-oxazaphosphorinane,

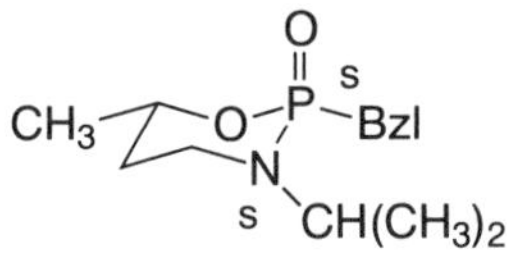

(1)

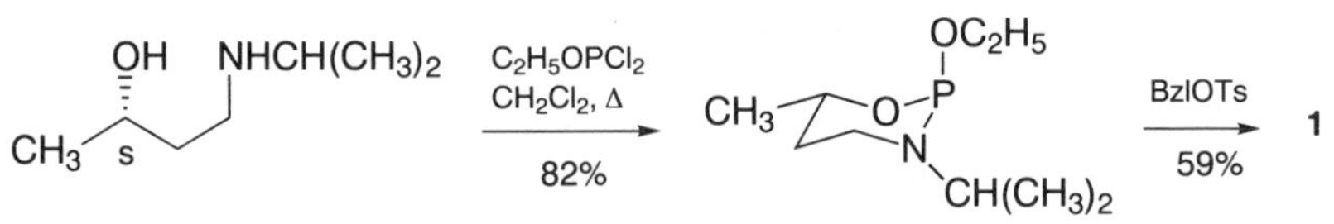

Asymmetric Wittig–Horner reactions.[1]   Wittig–Horner reactions have usually involved phosphine oxides, phosphonates, or phosphonamides. Highly enantioselective olefinations have now been achieved by use of this chiral phosphonamidate **1**. Thus the anion (*t*-BuLi) reacts with 4-*t*-butylcyclohexane to form **2** in >98% de. Elimination with trityl triflate/2,6-lutidine provides the alkylidene **3** with essentially complete stereospecificity. The reaction was shown to be applicable to a variety of 4-alkylcyclohexanones.

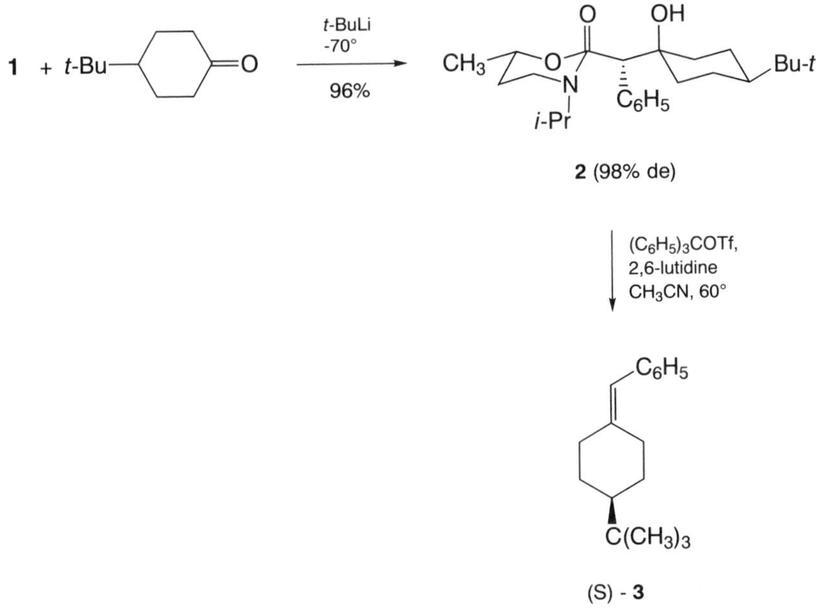

2 (98% de)

$(C_6H_5)_3COTf$,
2,6-lutidine
CH_3CN, 60°

(S) - **3**

[1] S. C. Denmark and C.-T. Chen, *Am. Soc.*, **114**, 10674 (1992).

L

Lanthanide(III) alkoxides.[1]

Lanthanide(III) isopropoxides can be prepared in 75–85% yield from hydrated lanthanide(III) chlorides by reaction with methyl orthoformate to give $MCl_3 \cdot 4CH_3OH$, which is converted to $MCl_3 \cdot 3i\text{-}P_2OH$ by reaction with isopropanol. After removal of the liberated CH_3OH, $MCl_3 \cdot 3$ i-PrOH is treated with 3 equiv. of BuLi to give $M(O\text{-}i\text{-}Pr)_3 \cdot LiCl$. This procedure has been used to prepare the isopropoxides of La, Ce, Sm, and Yb. These isopropoxides as obtained without removal of LiCl, can replace $Al(O\text{-}i\text{-}Pr)_3$ as the catalyst in Oppenauer oxidations. The most effective for oxidation of 1-phenyl-1-ethanol with 2-butanone is $Yb(O\text{-}i\text{-}Pr)_3$, followed by $La(O\text{-}i\text{-}Pr)_3$, $Sm(O\text{-}i\text{-}Pr)_3$, and $Ce(O\text{-}i\text{-}Pr)_3$. They can also be used for Meerwein–Ponndorf reductions. In this reaction, the most effective catalyst is $La(O\text{-}i\text{-}Pr)_3$, followed by $Sm(O\text{-}i\text{-}Pr)_3$, $Ce(O\text{-}i\text{-}Pr)_3$, and $Yb(O\text{-}i\text{-}Pr)_3$. The lanthanide alkoxides are also effective catalysts for epoxidation of geraniol with t-butyl hydroperoxide. Highest yields (96%) are obtained with $Yb(O\text{-}i\text{-}Pr)_3$ as catalyst. Addition of 4 Å Ms to remove water usually improves yields in reactions with lanthanide alkoxides, which are very sensitive to water.

[1] A. Lebrun, J.-L. Namy, and H. B. Kagan, *Tetrahedron Letters,* **32**, 2355 (1991).

Lanthanum(III) t-butoxide, $La_3(O\text{-}t\text{-}Bu)_9$ (1).

Catalytic asymmetric nitroaldol reaction.[1] In the presence of this alkoxide, α-chloro ketones or nitro alkanes undergo aldol reactions (equation I and II).

Asymmetric nitroaldol reactions are possible by use of the optically active lanthanum oxide **2**, prepared from (S)-(−)-2, 2′ dihydroxy-1,1′- (S)-(−)-binapthyl (BINOL) with $La_3(O\text{-}t\text{-}Bu)_9$.

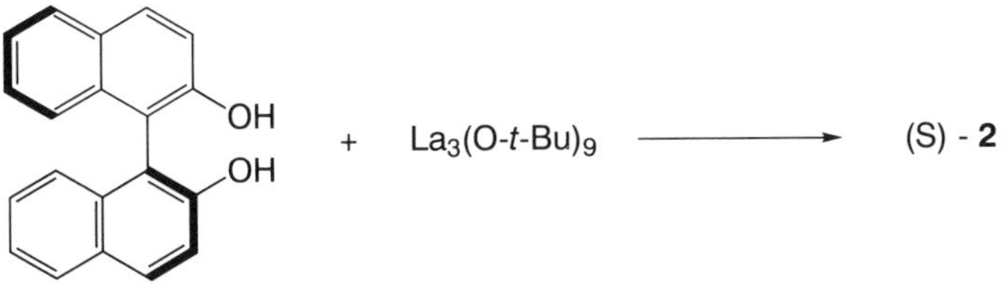

(S) - (–)

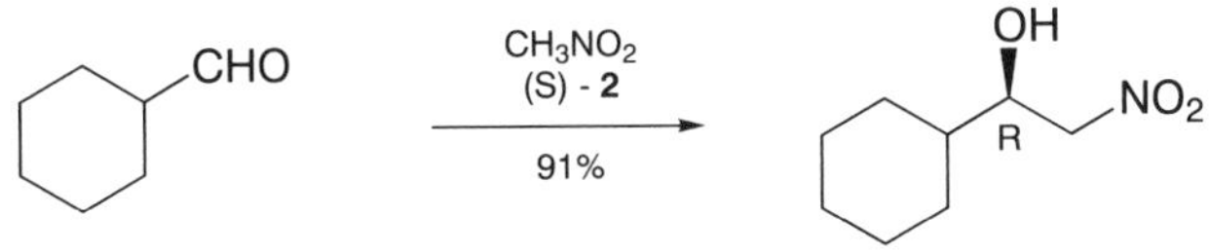

90% ee

[1] H. Sasai, T. Suzuki, S. Arai, T. Arai, and M. Shibasaki, *Am. Soc.*, **114**, 4418 (1992).

Lithium/Ammonia.

Reductive elimination of Bu₃Sn from carbocycles.[1]   The Bu_3Sn group of carbocy-cles can be reduced efficiently by treatment with lithium (excess) and *t*-BuOH in THF and NH_3. This reduction permits use of this group as a directing group in annelation re-actions as shown in Scheme (I). Thus reaction of the enone **1** with lithium (phenylthio)-(trimethylstannyl)cuprate followed by methylation provides **2** as a single product. This product was converted into an enone and alkylated to give **3**. Cyclization of **3**

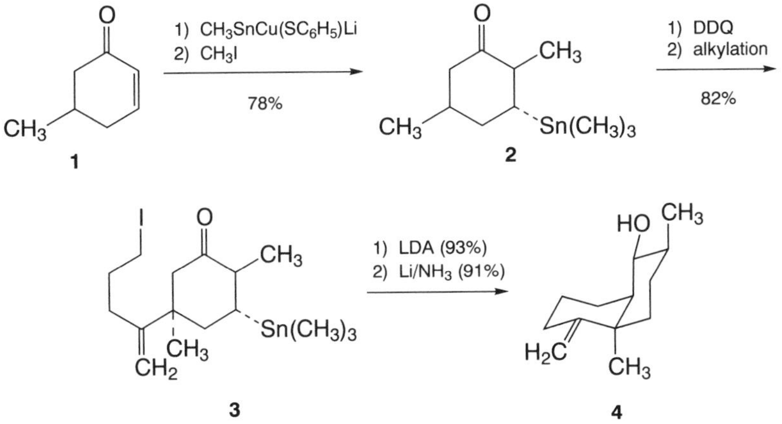

followed by addition of Li/NH_3 reduced the keto group and removed the tributyltin group to give **4**. The trimethyltin group was essential for control of the chiral centers of **4**.

[1] E. Piers and J. Y. Roberge, *Tetrahedron Letters,* **32**, 5219 (1991).

Lithium aluminum amides, LiAl(NHR)$_4$ (1).

These reagents can be prepared by addition (dropwise) of the amine (5 equiv.) to a suspension of LiAlH$_4$ in anhydrous ether or THF with stirring until precipitation is complete. R can be Pr, *i*-Pr, *t*-Bu, or CH$_2$C$_6$H$_5$.

$R^1COOR^2 \rightarrow R^1CONHR + R^2OH$.[1]   Lithium aluminum amides (**1**) can convert esters into amides in essentially quantiative yield.

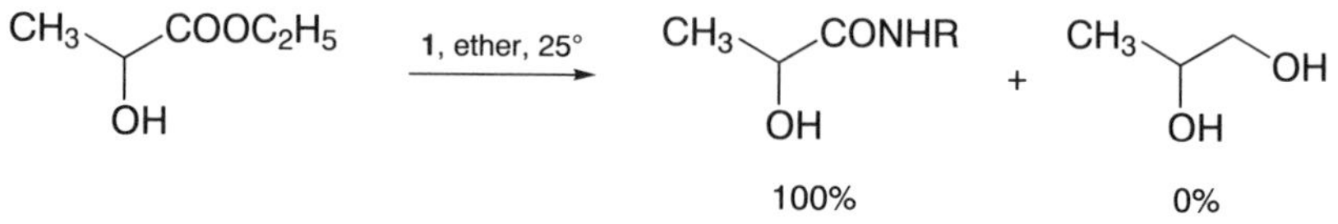

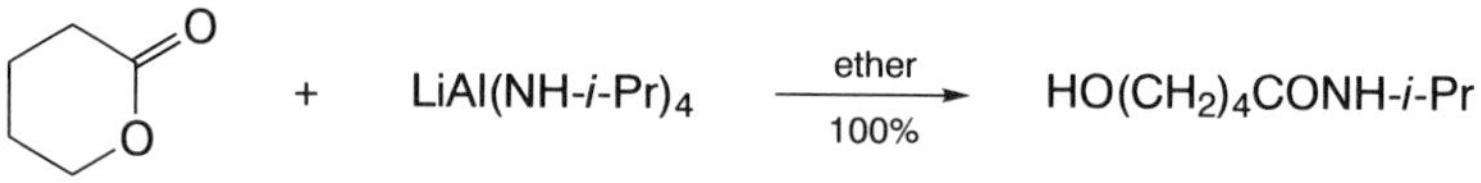

These reagents can also effect regioselective opening of aryl epoxides.

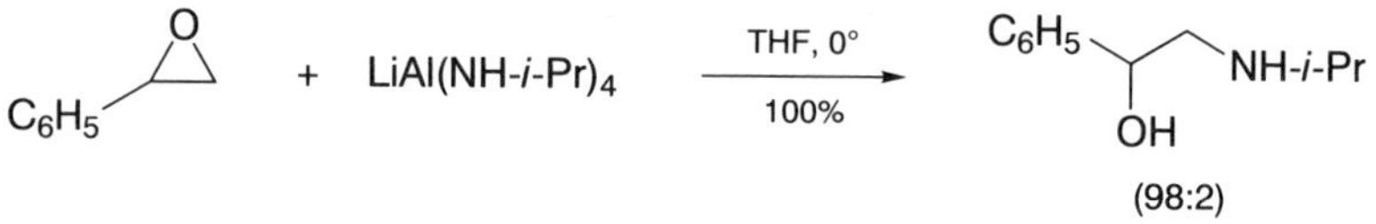

[1] A. Solladié-Cavallo and M. Bencheqroun, *J. Org.,* **57**, 5831 (1992).

Lithium aluminum hydride–(S)-2,2'-Dihydroxy-4,5,6,4',5',6'-hexamethoxybiphenyl (1).

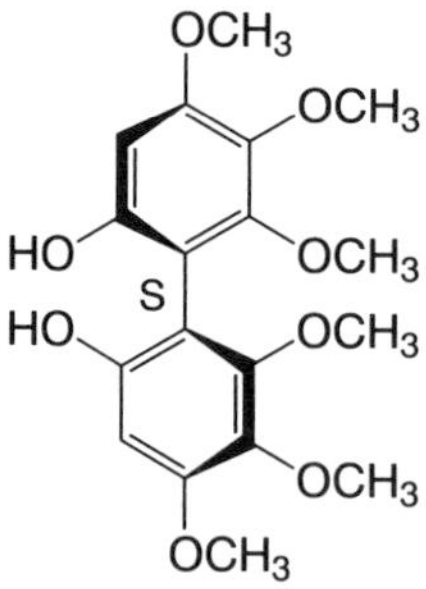

Asymmetric reduction of ketones.[1]   A reagent **2**, prepared by reaction of LiAlH$_4$ with **1** and C$_2$H$_5$OH (1 equiv. each) in THF at 20°, effects asymmetric reduction of dialkyl ketones or alkyl aryl ketones in 53–93% yield and 60–97% ee. The enantioselectivity is generally greater than that obtained with Noyori's reagent BINAL-H (**9**,169–170), particularly in reduction of dialkyl ketones in which the alkyl groups have similar steric effects.

[1] D. Rawson and A. I. Meyers, *J. C. S. Chem. Comm.*, 494 (1992).

Lithium (R,R)-bis(1-phenylethyl)amide,

1

Kinetic resolutoin of a β-lactam.[1]   Addition of this chiral base to the N-protected β-lactam **2** at −90° followed by quenching with excess ClSi(CH$_3$)$_3$ provides (R)-**2** and (3R,4S)-**3**, formed by silylation of (S)-**2**. This product (**3**) can be obtained in up to 72% ee

(±) - **2**, R$_3$ = *t*-BuMe$_2$ (3R,4S) - **3**, 72% ee   (R) - **2**, 100% ee

by controlling the amount of base. The recovered (R)-2 can be obtained in essentially 100% ee.

Similar results are obtained in an aldol reaction of (±)-2 with acetaldehyde, but the selectivity in this case is about 10:1, whereas it is about 7:1 in the silylation reaction.

(±) - **2**   **4** (10 : 1)   **5**

Regioselective deprotonation of a ketone; regioselective resolution.[2]   Treatment of the (R,R)-ketone **2** with (S)-**1** provides the Δ^6-enol silane **3**, whereas a similar reaction of the (S,S)-isomer of **2** provides the Δ^5-enol silane **4**. Since **3** and **4** are difficult to separate, they were identified from their corresponding enones. In contrast treatment

(R, R)-**2**   (S,S)-**2**

(S)-1, (CH$_3$)$_3$SiCl, -92°   (S)-1, (CH$_3$)$_3$SiCl, -92°

3 (94:6)   **4** (79:21)

(S)-1, (CH$_3$)$_3$SiCl, -70°

rac-**2** ⟶ **3** + **4**

60:40

of racemic-**2** shows only slight regioselectivity.

[1] P. Coggins and N.S. Simpkins, *Synlett,* 313 (1992).
[2] K. Bambridge, N.S. Simpkins, and B.P. Clark, *Tetrahedron Letters,* **33**, 8141 (1992).

Lithium 4,4'-di-*t*-butylbiphenylide (1, LDBB).

Reductive lithiation of tetrahydrofurans.[1]   Tetrahydrofurans can be reductively cleaved by this radical anion at -80° in the presence of 1 equiv. of BF$_3$ etherate to give a 4-lithiobutoxide best represented by **2**. This lithium reagent reacts with aldehydes and ketones to form 1,5-diols, which are readily cyclized to tetrahydropyrans. 2-Methyl-tetrahydrofuran is cleaved to the most substituted carbanion (4).

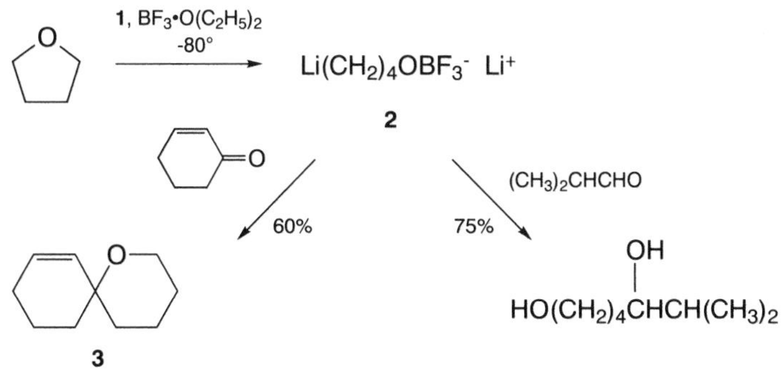

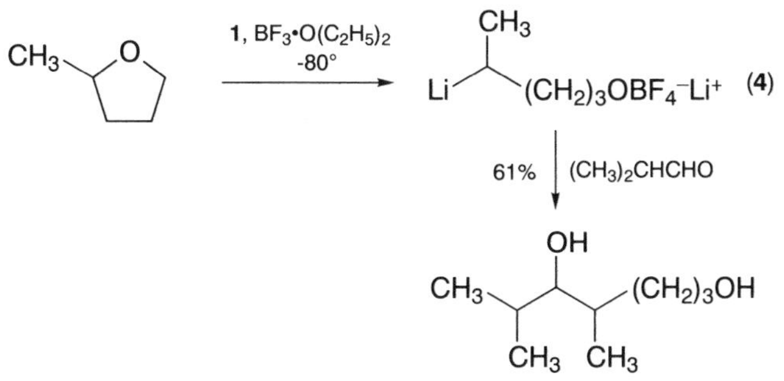

2-Vinyltetrahydrofuran (neat) is readily cleaved by 1 to a dianion, which on protonation provides the (Z)-alkenyl alcohol (5) as the major isolated product.

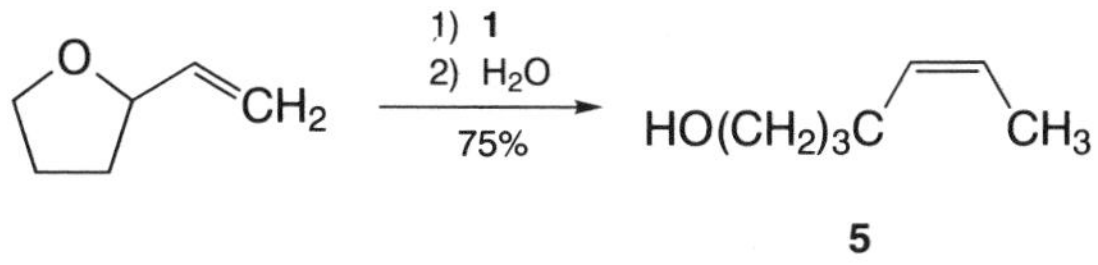

[1] B. Mudryk and T. Cohen, *Am. Soc.*, **113**, 1866 (1991).

Lithium diisopropylamide.

Anionic Fries carbamoyl transfer.[1]   Treatment of the biphenyl 2-O-carbamate **1** in which the *ortho*-position is protected by a methoxyl or triethylsilyl (TES) group with LDA results in transfer of the carbamoyl group to form the biphenyl **2**. Cyclization of

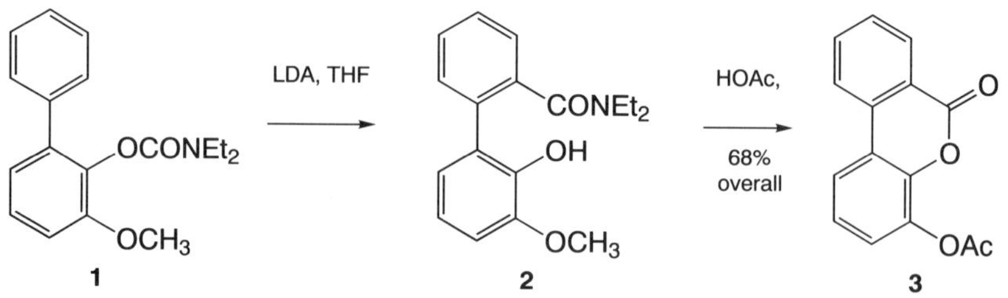

LDA, THF

OCONEt₂

1

CONEt₂
OH
OCH₃

2

HOAc,

68%
overall

O
O
OAc

3

2 results in the dibenzopyranone **3** in 68% overall yield.

This carbamoyl transfer was used to synthesize the fluorenone dengibsin (**6**) from the key intermediate **4**. An anionic rearrangement provides **5**. Methylation and desilylation provides an amide that cyclizes to a fluorenone that on dealkylation with BCl₃ provides **6**.

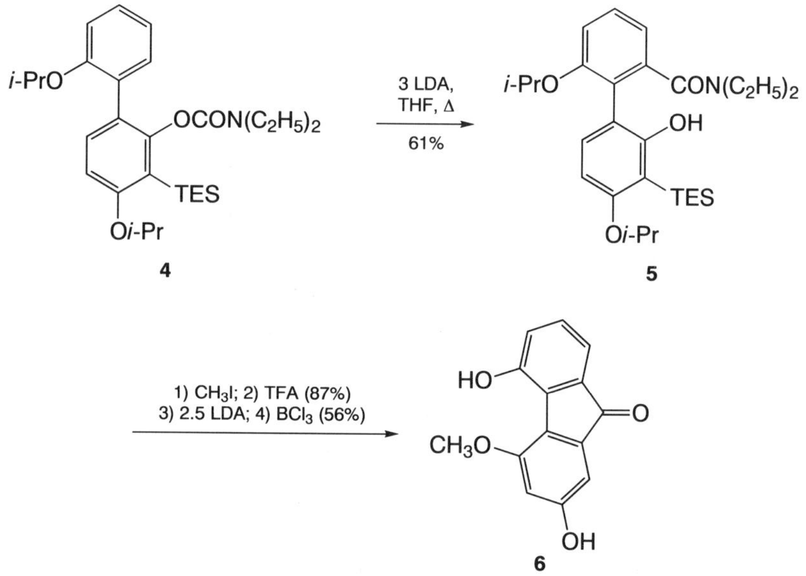

i-PrO

OCON(C₂H₅)₂

TES

O*i*-Pr

4

3 LDA,
THF, Δ

61%

i-PrO

CON(C₂H₅)₂
OH

TES

O*i*-Pr

5

1) CH₃I; 2) TFA (87%)
3) 2.5 LDA; 4) BCl₃ (56%)

HO

CH₃O

O

OH

6

Intramolecular aryne cycloaddition. [2]   A new approach to the basic skeleton of ergot alkaloids involves an intramolecular cyclization of the amide **1** to form **2**, effected with LDA at −30°.

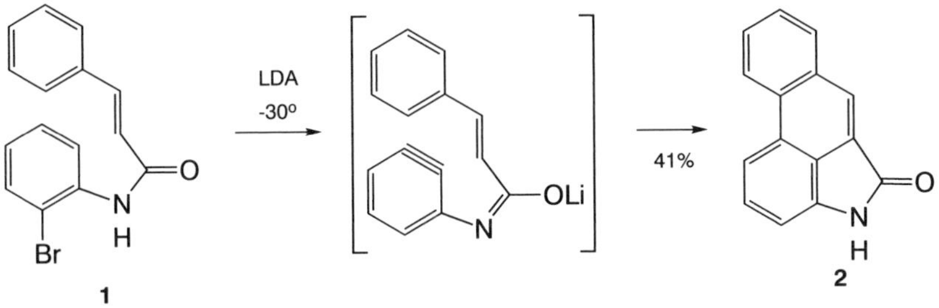

[1] W. Wang and V. Snieckus, *J. Org.*, **57**, 424 (1992).
[2] B. Gómez, E. Guitián, and L. Castedo, *Synlett.*, 903 (1992).

Lithium diisopropylamide/Butyllithium.

β-Lithio ketone enolates.[1] β-Stannyl ketones[2] such as **1** on deprotonation (LDA) to the enolate followed by Li/Sn exchange with BuLi at −78 to 0° are converted into the β-lithio ketone (Z)-enolate **a**. This species undergoes β-alkylation more readily than α-alkylation; β-alkylation followed by α-allylation is also possible.

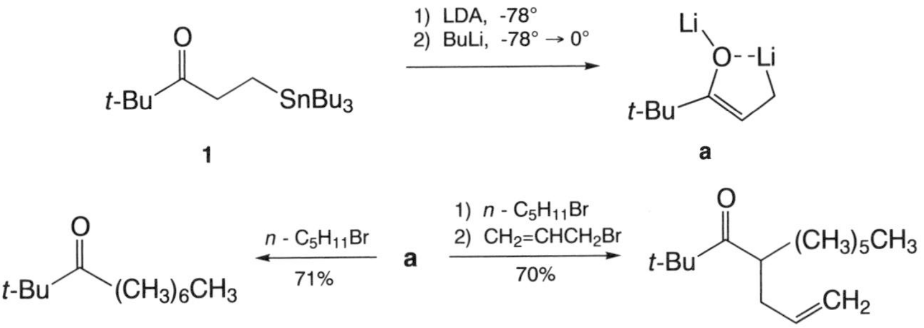

[1] H. Nakahira, I. Ryu, M. Ikebe, N. Kambe, and N. Sonoda, *Angew. Chem. Int. Ed.*, **30**, 177 (1991).
[2] I. Ryu, S. Murai, and N. Sonoda, *J. Org.*, **51**, 2389 (1986).

Lithium perchlorate.

[1,3]Sigmatropic rearrangement of allyl vinyl ethers.[1] Allylic vinyl ethers rearrange in 3M lithium perchlorate in diethyl ether at 25° to homoallylic aldehydes.

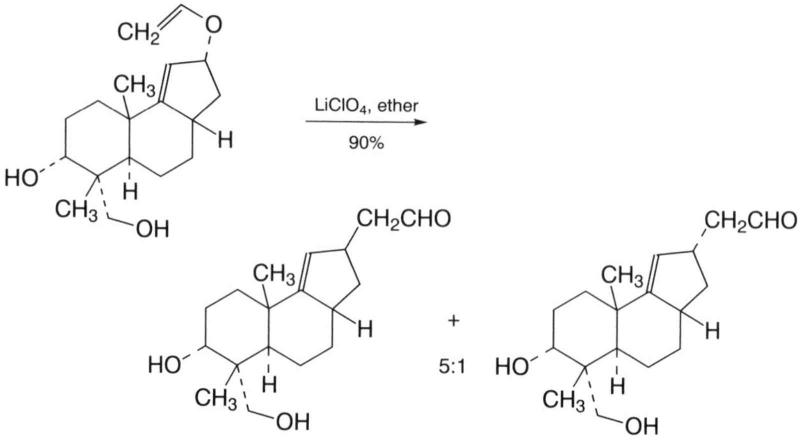

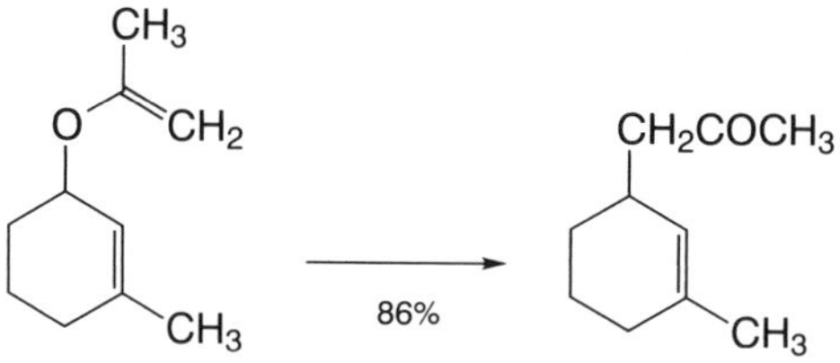

Conjugate addition of O-silyl ketene acetals to enones.[2]   Addition of 1-methoxy-1-(*t*-butyldimethylsilyloxy)ethylene to cyclohexenone proceeds in low yield when catalyzed by TiCl$_4$ or TiCl$_4$/Ti(O-*i*-Pr)$_4$, but is effected in 95% yield when catalyzed by 1.0 M LiClO$_4$ in diethyl ether.

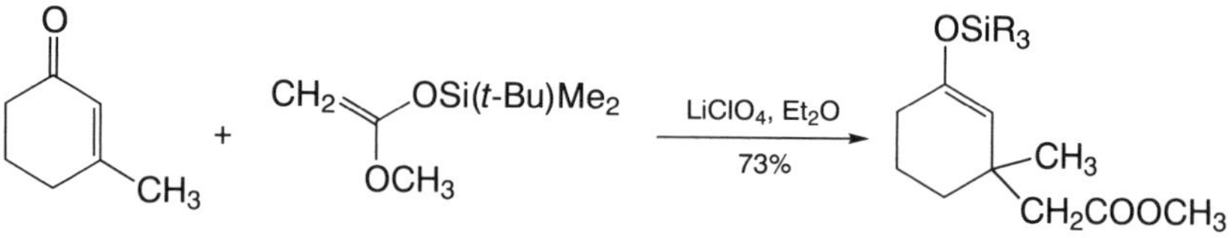

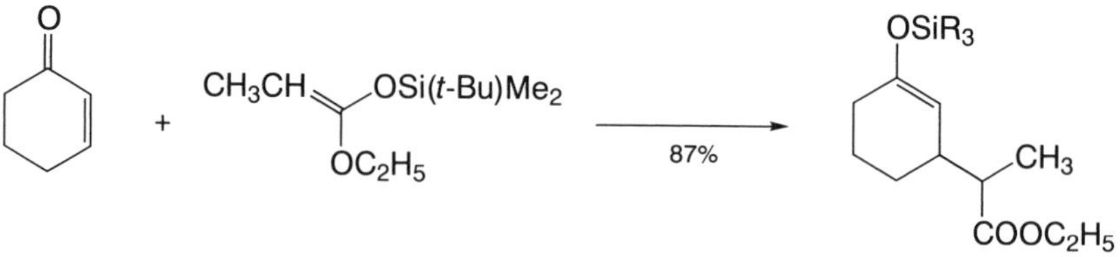

syn-Selective addition of allylstannanes to α-hydroxy aldehydes.[3]   This reaction is usually effected with Lewis acid catalysts (TiCl$_4$, MgBr$_2$). It can also be effected by use of 5 M LiCl$_4$O in ether.

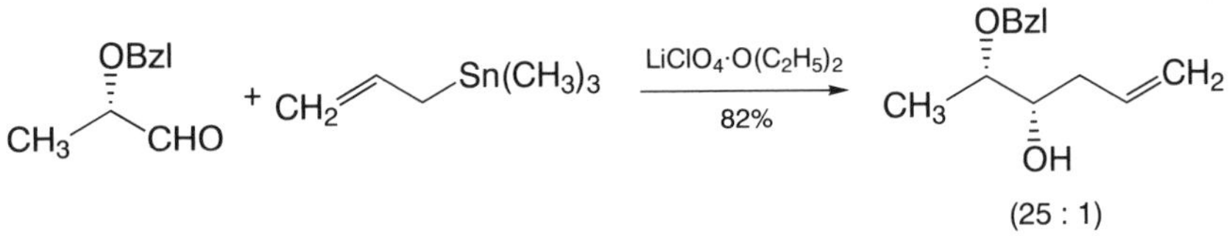

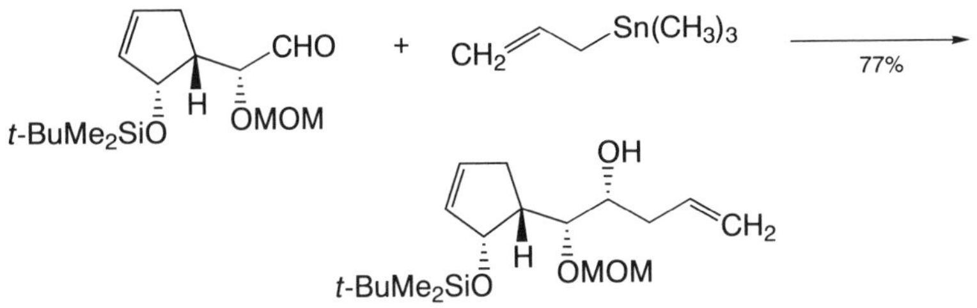

Aldol reaction of silyl enol ethers.[4]   Benzaldehyde reacts with the silyl ketene ketal **1** at 25° in the presence of LiClO₄ (3 mol %) to give the aldol **2** in 86% yield. Under these conditions a chiral α-alkoxy aldehyde reacts with **1** to give the aldol of

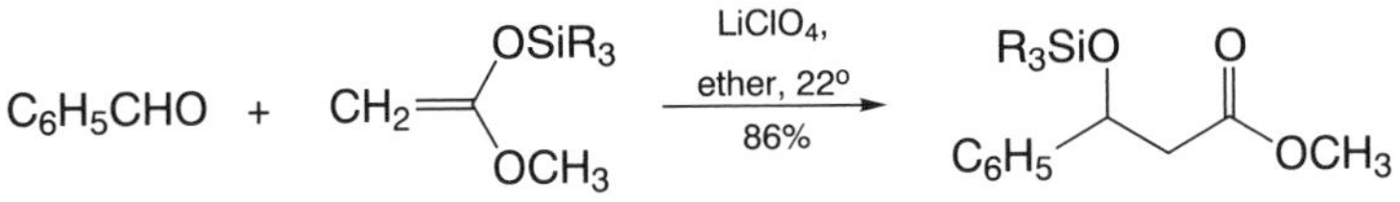

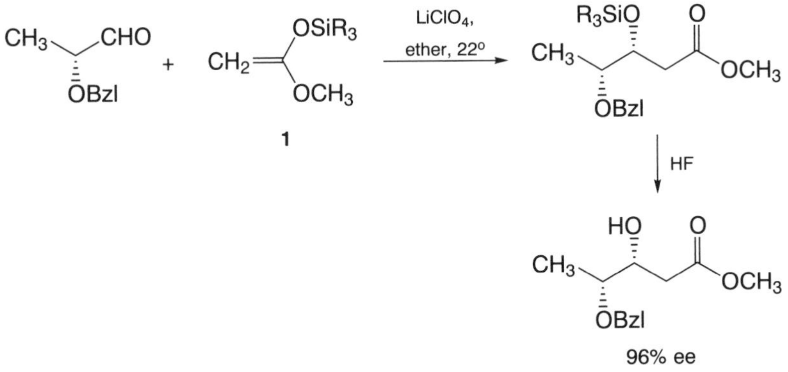

chelation control.

Substitution of allylic alcohols by silyl ketene acetals.[5]   Allylic alcohols undergo substitution with 1-methoxy-1-(*t*-butyldimethylsilyloxy)ethylene (**1**) in 3M LiClO₄ in diethyl ether.

Caution! A violent explosion has been reported from contact of cyclooctatetraene with LiClO$_4$ in refluxing ether.[6]

[1] P. A. Grieco, J. D. Clark, and C. T. Jagoe, *Am. Soc.*, **113**, 5488 (1991).
[2] P. A. Grieco, R. J. Cooke, K. J. Henry, and J. M. VanderRoest, *Tetrahedron Letters*, **32**, 4665 (1991).
[3] K. J. Henry, Jr., P. A. Grieco, and C. T. Jagoe, *ibid.*, **33**, 1817 (1992).
[4] M. T. Reetz, B. Raguse, C. F. Marth, H. M. Hügel, T. Bach, and D. N. A. Fox, *Tetrahedron*, **48**, 5731 (1992).
[5] P. A. Grieco, J. L. Collins, and K. J. Henry, Jr., *Tetrahedron Letters*, **33**, 4735 (1992).
[6] R. A. Silva, *Chem. Eng. News*, Dec. 2 (1992).

Lithium pyrrolindoborohydride. LiH$_3$B-N ⟨⟩   (1)

Lithium aminoborohydrides.[1]   The reagent **1** is a typical member of a number of lithium aminoborohydrides, prepared by reaction of BuLi with amine · boranes in quantitative yields. These reagents can be stored at 25° under N$_2$ for at least six months; they are not pyrophoric. They are comparable to LiAlH$_4$ as reductants. Thus **1** reduces carbonyl compounds (including esters) in high yield. Lactones and anhydrides are reduced, but carboxylic acids are not reduced. In addition **1** reduces amides, epoxides, oximes, nitriles, and even halides.

[1] G. B. Fisher, J. Harrison, J. C. Fuller, C. T. Goralski, and B. Singaram, *Tetrahedron Letters,* **33**, 4533 (1992).

Lithium 2,2,6,6-tetramethylpiperidide (LiTMP, 1).

(E)-Selective enolate formation.[1] Corey and Gross (**12,**285) have noted that (E)-lithium enolates are formed with high selectivity by treatment of ketones with a lithium dialkylamide followed by trapping with $ClSi(CH_3)_3$. The (E)-selectivity may be the effect of LiCl formed on trapping, since the addition of LiCl or LiBr to LiTMP also results in (E)-selective lithium enolates. The best experimental conditions involve metalation with crystalline 2,2,6,6-tetramethylpiperidinium bromide, which generates both LiTMP and LiBr under anhydrous conditions.

$$50 : 1$$

Vinylcyclopropanes.[2] Allylic bromides or chlorides, particularly prenyl halides, on deprotonation with LiTMP (THF,20°) are converted into a carbenoid that reacts with alkenes to provide vinylcyclopropanes.

$$(CH_3)_2C=CHCH_2Br + C_6H_5CH=CH_2 \xrightarrow[58\%]{\text{LiTMP, THF, 20°}}$$

(cis/trans = 74:26)

Anthracenes.[3] A convenient route to anthracenes involves the reactions in tetrahydropyran of LiTMP with benzocyclobutenols and halobenzenes (precursors to benzynes).

OCH₃ ... 1, THP, Δ ... 62%

OCH₃ ... OH ... Br ... 40% ... OCH₃ ... OCH₃ ... OCH₃

[1] P. L. Hall, J. H. Gilchrist, and D. B. Collum, *Am. Soc.*, **113**, 9571 (1991); P. L. Hall, J. H. Gilchrist, A. T. Harrison, D. J. Fuller, and D. B. Collum, *ibid.*, **113**, 9575 (1991).
[2] I. Jefferies, M. Julia, J.-N. Verpeaux, and T. Zahneisen, *Synlett.*, 647 (1991).
[3] J. J. Fitzgerald, N. E. Drysdale, and R. A. Olofson, *J. Org.*, **57**, 7122 (1992).

Lithium tetramethylthallate, Li(CH₃)₄Tl.

Conjugate addition to enones.[1] This ate complex reacts with cyclic enones at −40° to form the 1,4-adduct with high selectivity. Note that a mixed ate complex

Li(CH₃)₄Tl / ether, −40° / 70% ... CH₃ ... >95 : <5 ... HO CH₃

CH₃ ... CH₃ ... COOC₂H₅ ... 81% ... CH₃ O CH₃ ... COOC₂H₅

such as (CH₃)₂TlCl · 2BuLi delivers the butyl ligand more readily than the methyl ligand. Surprisingly, the mixed ate complex (CH₃)₃CTl · LiC≡CC₅H₁₁ delivers the acetylenic ligand exclusively.

[1] I. E. Markó and F. Rebière, *Tetrahedron Letters*, **33**, 1763 (1992).

Lithium triethylborohydride.

Hydroboration with dialkoxyboranes. Dialkoxyboranes, prepared by reaction of BH₃ in THF with an alcohol (2 equiv.), react very sluggishly with alkenes. However if

1% of $LiBH(C_2H_5)_3$ is present, hydroboration with $BH(OR)_2$ can proceed at a reasonable rate and in good yield. Lithium triethylborohydride can also promote the formation of $BH(OR)_2$.

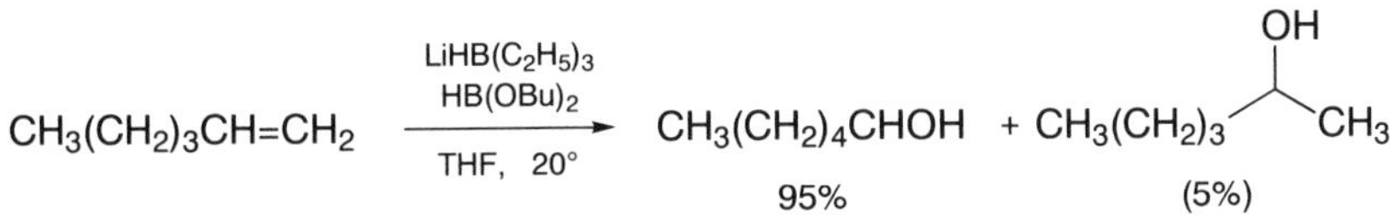

$$CH_3(CH_2)_3CH=CH_2 \quad \xrightarrow[\text{THF, 20°}]{\substack{LiHB(C_2H_5)_3 \\ HB(OBu)_2}} \quad \underset{95\%}{CH_3(CH_2)_4CHOH} \quad + \quad \underset{(5\%)}{CH_3(CH_2)_3\overset{OH}{\underset{}{\wedge}}CH_3}$$

[1] A. Arase, Y. Nunokawa, Y. Masuda, and M. Hoshi, *J.C.S. Chem. Comm.,* 51 (1992).

Lithium tris(diethylamino)aluminum hydride, $Li[N(C_2H_5)_2]_3AlH$.
Prepared by reaction of $LiAlH_4$ with $HN(C_2H_5)_2$ (3 equiv.) in THF.
RCONH$_2$ → *RCHO.*[1] Reduction of carboxamides to aldehydes with $LiAlH_4$ is useful only with N,N-disubstituted carboxamides. This new hydride (**1**) can reduce primary carboxamides to aldehydes at 25° in 12 hours in yields of 50–90%.

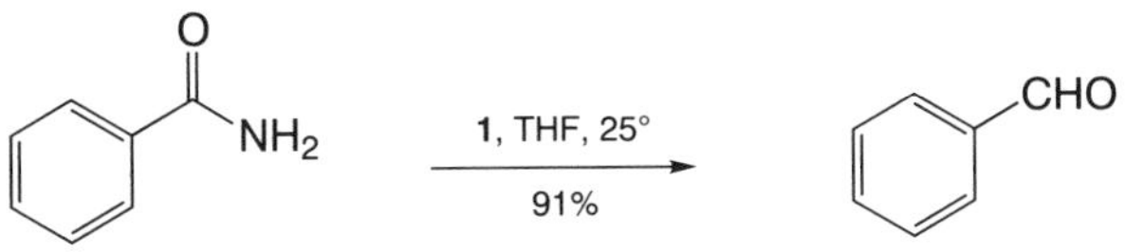

$$\text{(benzamide)} \xrightarrow[\text{91\%}]{\text{1, THF, 25°}} \text{(benzaldehyde)}$$

[1] J.S. Cha, J.C. Lee, H.S. Lee, S.E. Lee, J.M. Kim, O.O. Kwon, and S.J. Min, *Tetrahedron Letters,* **32**, 6903 (1991).

M

Magnesium(II) bromide etherate, $MgBr_2 \cdot O(C_2H_5)_2$.

***anti*-Aldol selectivity.**[1] *anti*-Selectivity can be increased by transmetalation of the lithium enolate of the aldehyde or ketone. Note that the *anti*-selectivity requires

$$C_6H_5COC_2H_5 \quad \xrightarrow[\text{73\%}]{\substack{\text{1) LDA; MgBr}_2\text{•O(C}_2\text{H}_5)_2 \\ \text{2) C}_6\text{H}_5\text{CHO}}}$$

(*anti/syn* = 98:2)

$$(CH_3)_3CCOC_2H_5 \quad \xrightarrow[\text{70\%}]{}$$

(*anti/syn* = 99:1)

1–16 hours of equilibration of the enolates, and this equilibrium is highly dependent on the substitution pattern of both the ketone and aldehyde. However, the order of addition of $MgBr_2 \cdot O(C_2H_5)_2$ (1.25 equiv.) is not important.

[1] K. A. Swiss, W.-B. Choi, D. C. Liotta, A. F. Abdel-Magid, and C. A. Maryanoff, *J. Org.,* **56,** 5978 (1991).

Magnesium iodide, MgI_2.

Iodohydrins; 1,2-diols.[1] The reaction of 2,3-epoxy alcohols with MgI_2 in toluene at $-60°$ results in 3-iodo-1,2-diols in 85–95% yield. These iodohydrins are reduced to 1,2-diols by Bu_3SnH in 50–85% yield. The method is applicable to derivatives of these alcohols (acetyl, benzyl). The high regioselectivity is attributed to magnesium chelation. The reaction can also be used to generate secondary diols (last example).

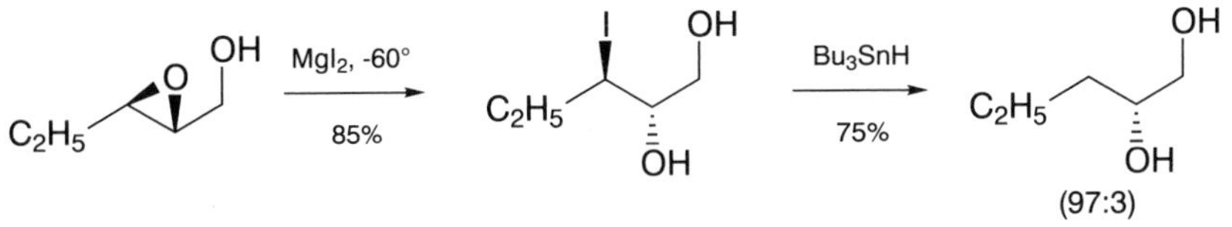

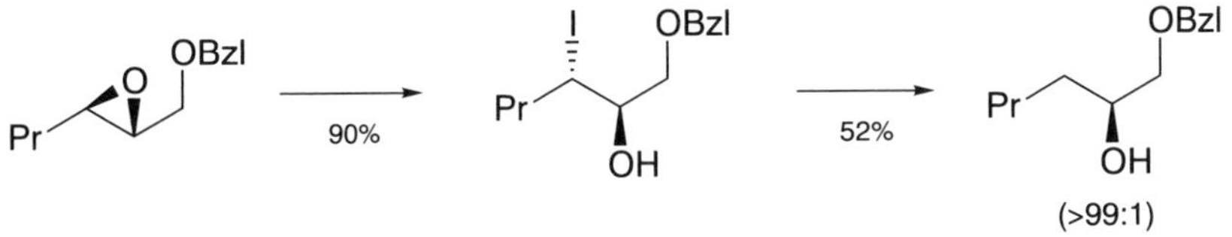

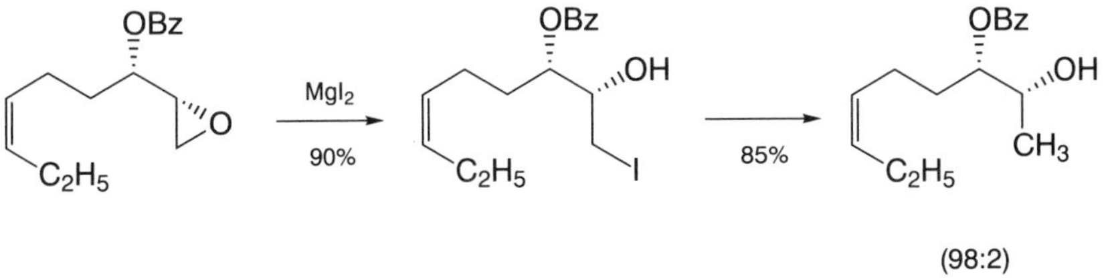

[1] C. Bonini, G. Righi, and G. Sotgiu, *J. Org.*, **56**, 6206 (1991).

Manganese(III) acetate.

Oxidative cyclization by addition to $C\equiv N$.[1]   The acetoacetate **1** undergoes cyclization in the presence of $Mn(OAc)_3 \cdot 2H_2O$ (0.5 equiv.) in CH_3COOH to give the ketone **2** (30–50%), which is evidently formed by hydrolysis of an intermediate imine.

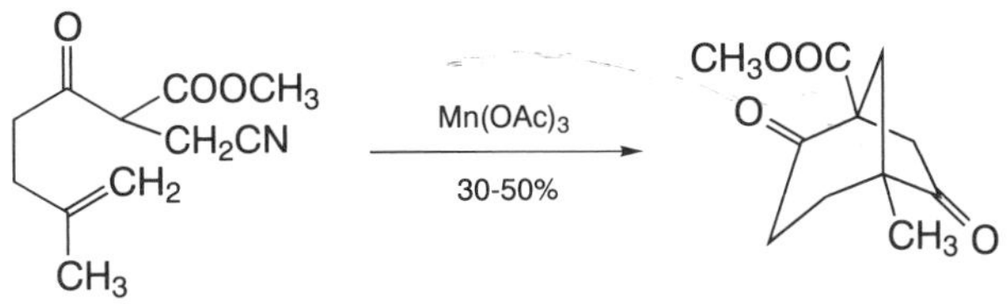

This reaction can be used for synthesis of a decalindione such as **4** from **3**, or for cyclization to cyclopentanones.

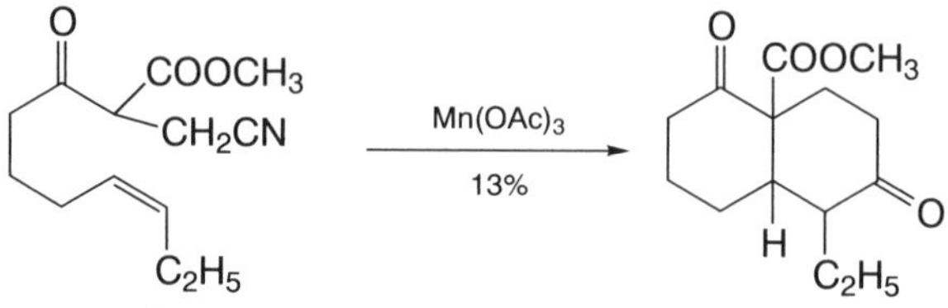

[1] B. B. Snider and B. O. Buckman, *J. Org.*, **57**, 322 (1992).

Manganese(III) acetate – Copper(II) acetate.

Radical cyclization of polyenes.[1]   Radical cyclization of the tetraunsaturated β-keto ester **1** with $MnO(OAc)_7/Cu(OAc)_2$ (2:1) affords the D-homo-5α-androstane **2** with seven chiral centers in 31% yield.

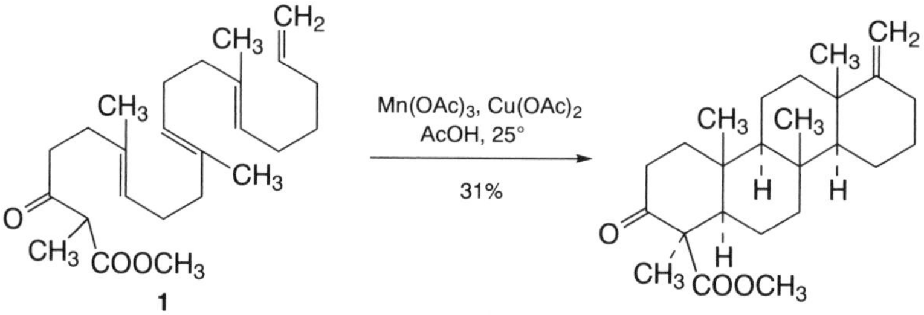

[1] P. A. Zoretic, X. Weng, M. L. Caspar, and D. G. Davis, *Tetrahedron Letters,* **32**, 4819 (1991).

Mercury(II) acetate

1,2,4-Trioxanes.   1,2,4-Trioxanes (**2**) can be prepared by reaction of hemiperoxy-acetals (**1**) with $Hg(OAc)_2$ (1 equiv.) catalyzed by $HClO_4$. This route can show high diastereoselectivity. Thus the trioxane **3** has the chair conformation with all three

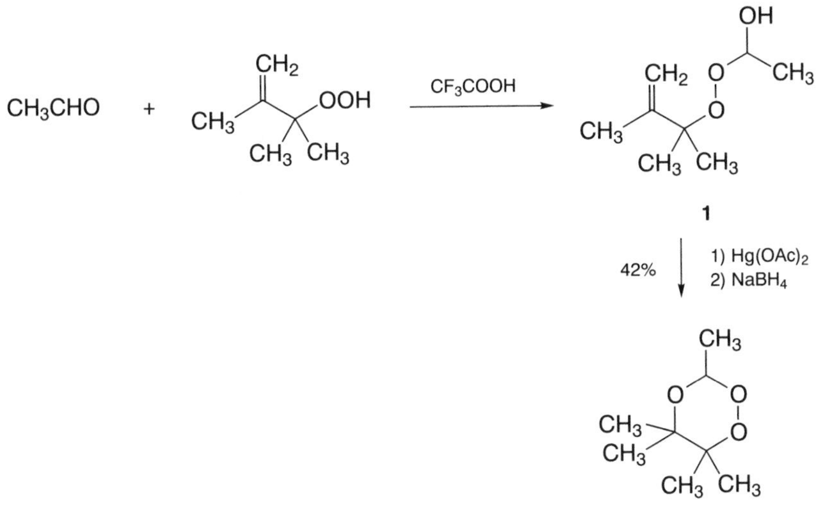

CH_3CHO + [structure: 2-methyl-3-methylenebutan-2-yl hydroperoxide] $\xrightarrow{CF_3COOH}$ [structure **1**]

1

42% ↓ 1) Hg(OAc)₂ 2) NaBH₄

[structure **2**]

2

CH_3CHO + [structure: 1-phenyl-2-propenyl hydroperoxide, $CH_2\!\!=\!\!$ ... OOH, C_6H_5] $\xrightarrow[\text{2) Hg(OAc)}_2]{\text{1) CF}_3\text{COOH}}$ [structure with $AcOHgCH_2$ ... CH_3 ... C_6H_5]

$50\text{-}75\%$ ↓ NaBH₄

[structure **3**]

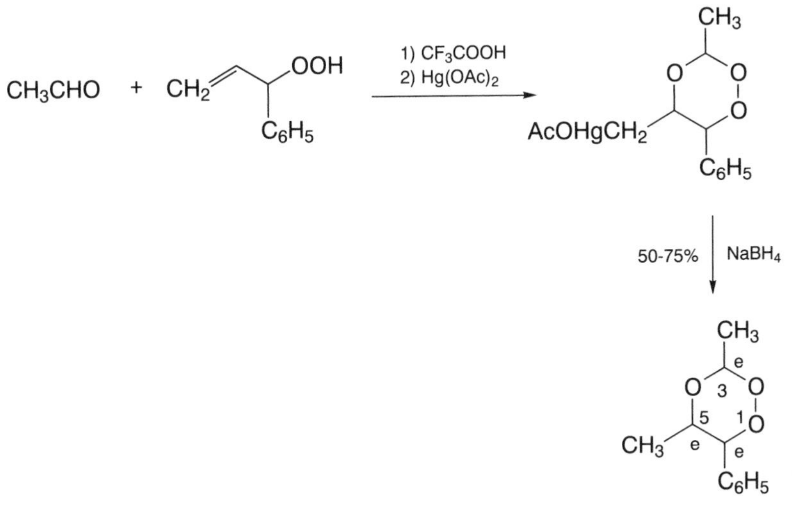

3 (e,e,e >97%)

substituents equatorial.

[1] A. J. Bloodworth and N. A. Tallant, *J. C. S. Chem. Comm.*, 428 (1992).

Mercury(II) triflouromethanesulfonate–N,N-Dimethylaniline,
$Hg(OTf)_2 \cdot C_6H_5N(CH_3)_2$ **(1), 12,** 307.

Polyene cyclizations.[1] This reagent **(1)** is particularly useful for cyclization of polyenes containing various oxygenated groups that can function as terminating groups. In this case cyclization occurs on oxygen rather than carbon, as observed with Lewis acid catalysts. Subsequent to cyclization, reduction with $NaBH_4/NaOH$ eliminates the mercury substituent.

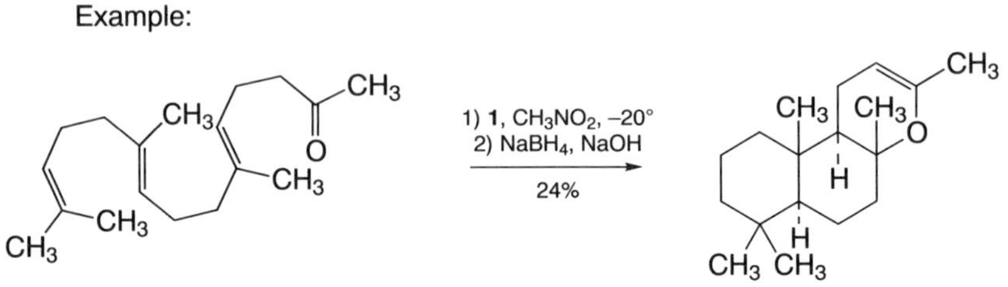

Example:

[1] A. S. Gopalan, R. Prieto, B. Mueller, and D. Peters, *Tetrahedron Letters,* **33,** 1679 (1992).

1-Mesityl-2,2,2-trifluoroethanol, 1.
  Preparation:[1]

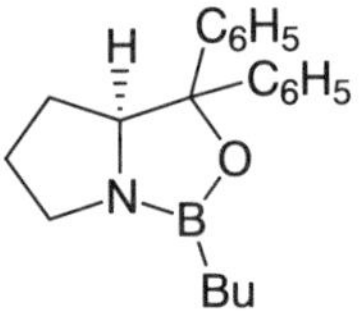

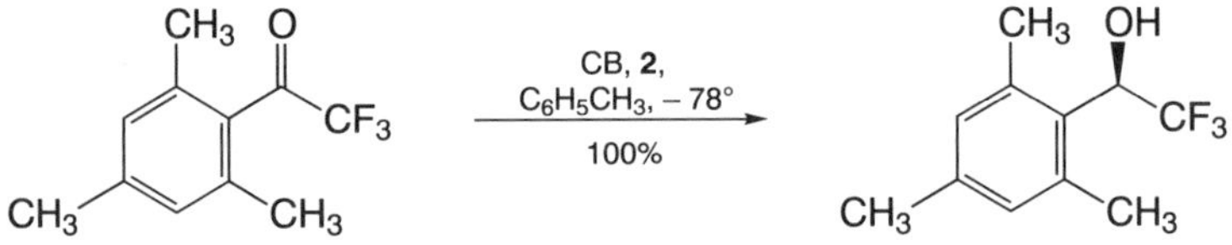

Asymmetric Diels-Alder reactions of α, β-unsaturated acids.[2] The acrylate ester **(2)** of this alcohol **(1)** undergoes Lewis acid-catalyzed reactions with cyclopentadiene in >97:3 diastereoselectivity.

2

3, > 97 : 3

4

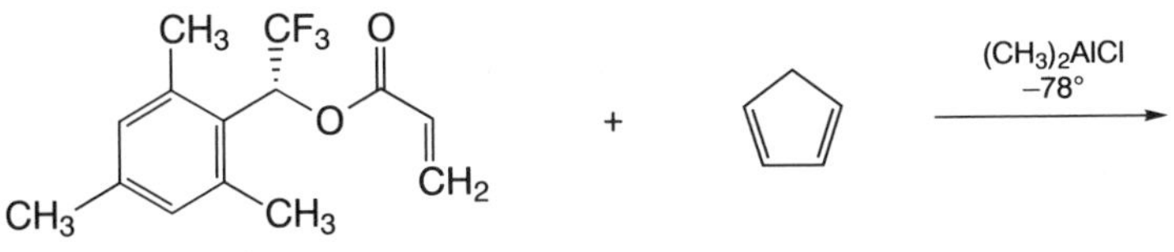

[1] E. J. Corey, X.-M. Cheng, K. A. Cimprich, and S. Sarshar, *Tetrahedron Letters,* **32**, 6835 (1991).
[2] E. J. Corey, X.-M. Cheng, and K. A. Cimprich, *ibid.,* **32**, 6839 (1991).

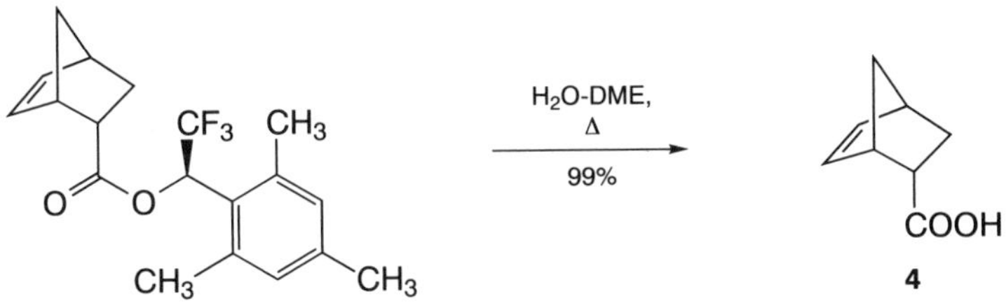

Metal halides.

Cleavage of terminal epoxides; halohydrins.[1] The cleavage of a terminal epoxide (**1**) can result in two halohydrins. By proper choice of a metal halide either one of the halohydrins can be obtained with about 95% regioselectivity.

$\longrightarrow$ $RCH(OH)CH_2X$ + $RCH(X)CH_2OH$

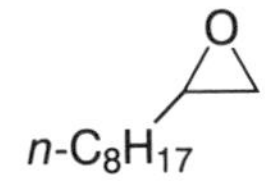

$TiCl_4$, C_7H_{16} , -25°	>95%	5:95
$TiBr_2(NR_2)_2$, C_7H_{16} , 0°	>95%	94:6
HCl , H_2O , 25°	>95%	89:11

[1] J. J. Eisch, Z.-R. Liu, X. Ma, and G.-X. Zheng, *J. Org.,* **57**, 5140 (1992).

Methanesulfonyl chloride/Sodium hydrogen carbonate.

β-Lactams.[1]   Treatment of the β-amino acid **1** with 1 equiv. of CH_3SO_2Cl at 45° in CH_3NO_2 containing suspended $NaHCO_3$ furnishes the β-lactam **2** in the highest yield reported to date for this cyclodehydration. The *cis*-isomer of **1** also undergoes this

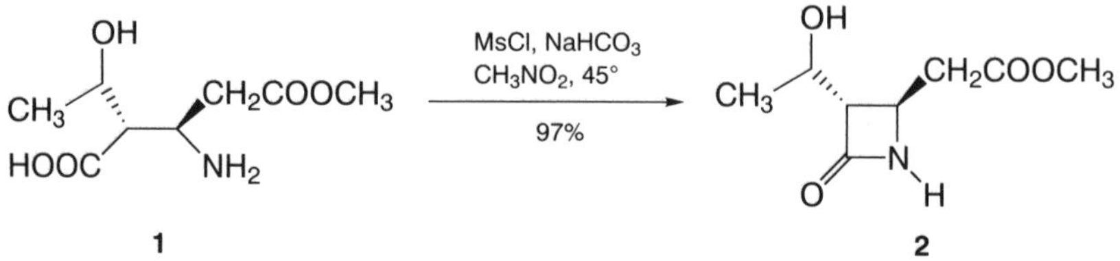

reaction to give *cis*-**2** (75% yield). Yields are lower when applied to simple β-amino acids, probably because of lower solubility in CH_3NO_2 and higher temperatures, 75–80°.

[1] M. F. Loewe, R. J. Cvetovich, and G. G. Hazen, *Tetrahedron Letters, 32*, 2299 (1991).

(S)- or (R)-2-Methoxy-2'-diphenylphosphino-1,1'-binaphthyl (1).

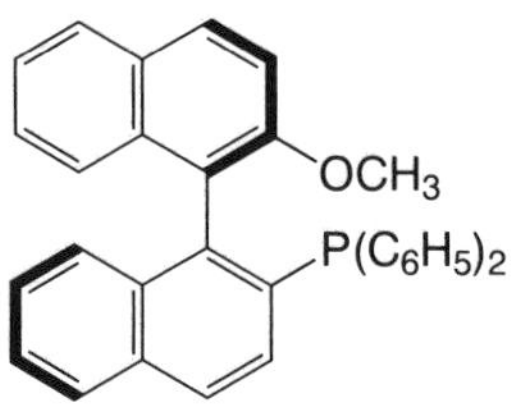

Asymmetric hydrosilylation of norbornene (**2**).[1]   This reaction can be effected in high regio- and enantioselectivity by reaction of **2** with Cl_3SiH catalyzed by (allyl)chloropalladium dimer complexed with (R)-**1**. The product can be converted to (1S,2S,4R)-norbornanol (**4**) in 96% ee.

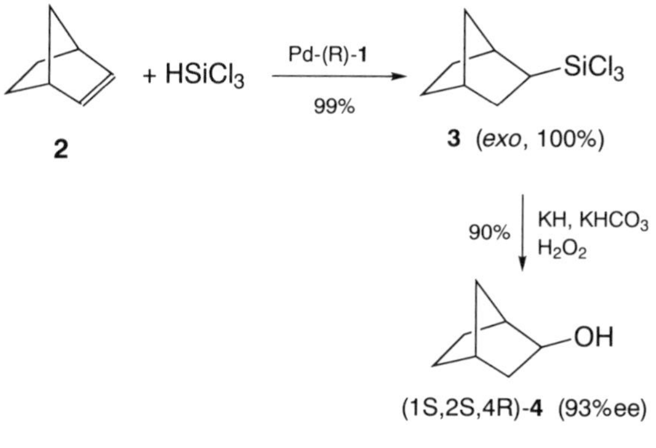
+ HSiCl3
Pd-(R)-1
99%
SiCl3
2
3 (exo, 100%)
90%
KH, KHCO3
H2O2
OH
(1S,2S,4R)-4 (93%ee)

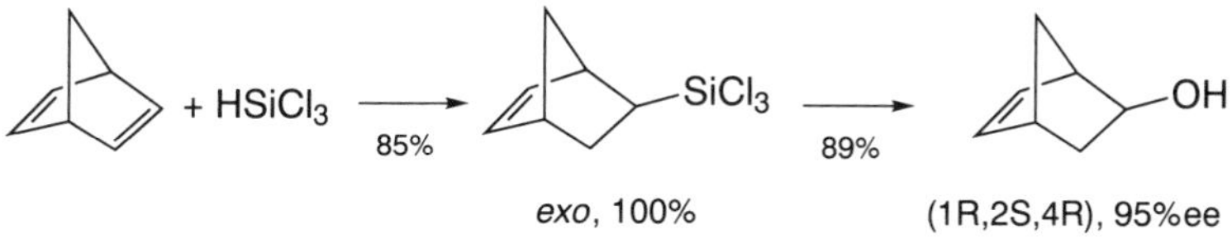
+ HSiCl3
85%
SiCl3
89%
OH
exo, 100%
(1R,2S,4R), 95%ee

[1] Y. Uozumi, S.-Y. Lee, and T. Hayashi, *Tetrahedron Letters,* **33**, 7185 (1992).

(1S,2R)-2-(*p*-Methoxyphenylsulfonyl)amino-1-phenylpropanol (1).

Enantioselective addition of $(C_2H_5)_2Zn$ to RCHO.[1] Of a variety of chiral N-sulfonylamino alcohols, **1** was found to be the most effective ligand for asymmetric addition of diethylzinc to aldehydes catalyzed by titanium(IV) isopropoxide in methylene chloride. Addition of calcium hydride or 4 Å molecular sieves does not affect the enantioselectivity but can increase the yield.

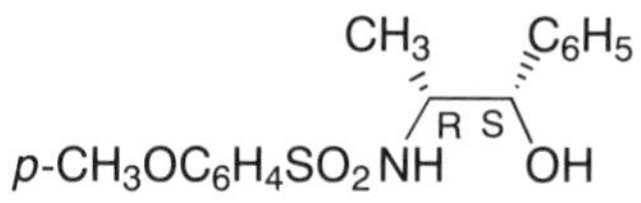
CH3
C6H5
R S
p-CH3OC6H4SO2NH
OH
(1)

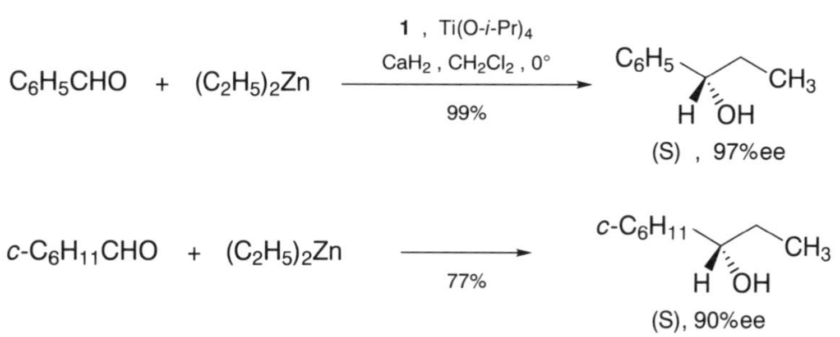
C6H5CHO + (C2H5)2Zn
1 , Ti(O-i-Pr)4
CaH2 , CH2Cl2 , 0°
99%
C6H5
CH3
H OH
(S) , 97%ee
c-C6H11CHO + (C2H5)2Zn
77%
c-C6H11
CH3
H OH
(S), 90%ee

[1] K. Ito, Y. Kimura, and H. Okamura, and T. Katsuki, *Synlett,* 573 (1992).

(S)-2-(Methoxymethyl)pyrrolidine, **(1)**

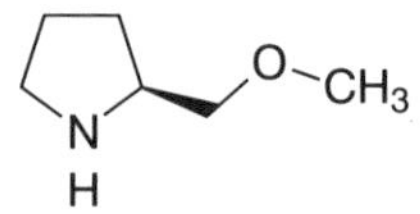

2-Methylenecyclohexenones.[1]   A diastereoselective synthesis of a 2-methylene-cyclohexenone (**5**) involves use of (S)-2-(methoxymethyl)pyrrolidine (**1**) as the chiral auxiliary and as the leaving group for generation of the *exo*-methylene group.

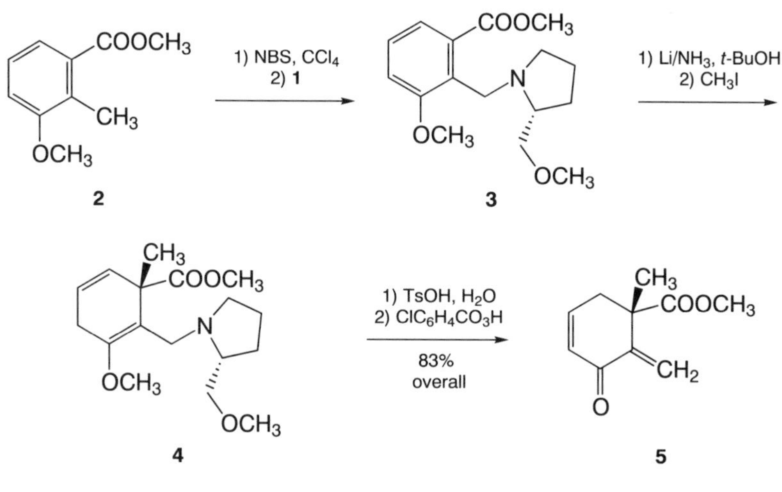

[1] A. G. Schultz and R. E. Taylor, *Am. Soc.,* **114**, 3937 (1992).

trans-1-Methoxy-3-trimethylsilyloxy-1,3-butadiene (Danishefsky's diene).
   Reaction with imines derived from α-amino esters.[1]   In the presence of ZnCl$_2$ (1 equiv.) the imine **1** derived from L-valine reacts with this diene to form 6-substituted 2,3-didehydro-4-piperidinones **2** and **3** with high diastereoselectivity (92:8). This reaction is believed to involve a Mannich reaction to form **a**, which undergoes Michael cyclization to form **2** and **3** in the ratio 92:8.

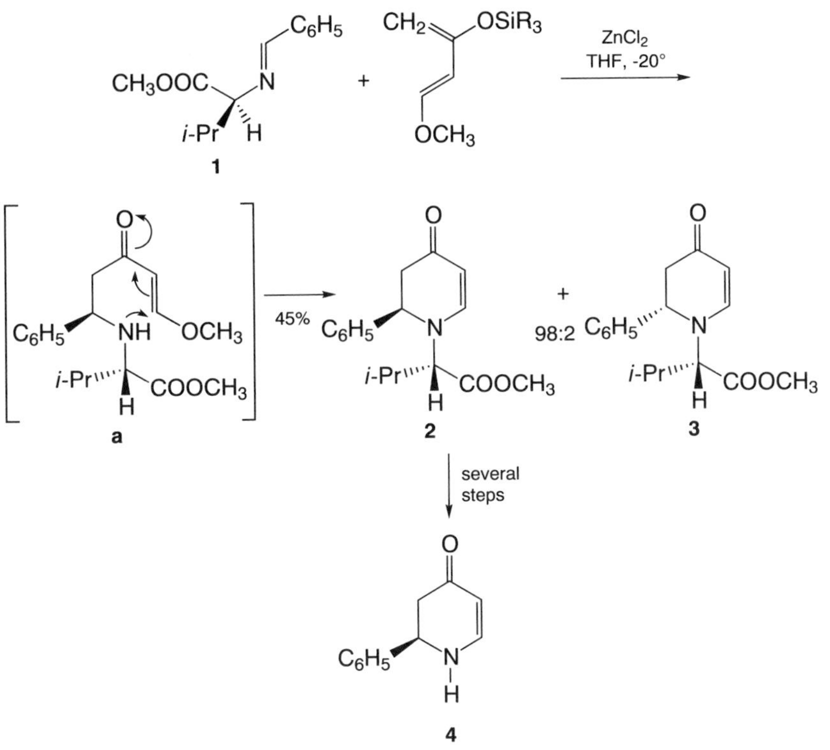

The chiral auxiliary can be recovered by a Curtius reaction.

[1] H. Waldmann and M. Braun, *J. Org.*, **57**, 4444 (1992).

Methyl acrylate.

[2+2+2]Cycloadditions.[1] The anion of the methyl cyclopentanecarboxylate **1** when treated with methyl acrylate undergoes sequential Michael–Michael–Dieckman cyclization to form the spirobicyclic cyclohexanone β-keto ester **2** as a mixture of two diastereomers (41% yield). Decarboxylation of **2** followed by addition of CH₃MgBr

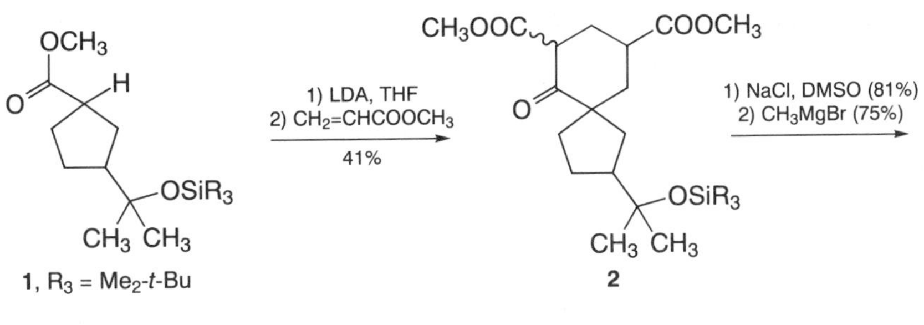

provides **3**. This product was converted by dehydration, oxidation, and dehydrogenation to the trienone **4** with only one chiral center at the spiro carbon atom. This hindered trienone was converted to the sesquiterpene β-vetivone **6** by conjugate methylation followed by decarboxylation.

[1] G. H. Posner and E. M. Shulman-Roskes, *Tetrahedron,* **48**, 4677 (1992).

Methylaluminum bis(4-bromo-2,6-di-*t*-butylphenoxide), 1, 15, 206; **16**, 209–212.

Asymmetric synthesis of β-hydroxy aldehydes.[1]   A novel route to β-hydroxy aldehydes is based on the ability of this reagent to effect stereoselective rearrangement of α,β-epoxy silyl ethers to β-silyloxy aldehydes. The stereoselectivity of this rearrangement depends on the silyl substituents, being highest with triphenylsilyl ethers and negligible with the more hindered triisopropylsilyl ethers. Thus the *syn*-epoxy silyl ether **2**, prepared by Sharpless asymmetric epoxidation of the allylic alcohol with (+)-DIPT followed by silylation, is rearranged by **1** to the *anti*-β-hydroxy aldehyde (40:1). The *anti*-epoxy silyl ether **2**, obtained by Mitsunobu inversion, rearranges to

(C$_6$H$_5$)$_3$SiO

Bu   S

CH$_3$

syn-2

73%   1

(C$_6$H$_5$)$_3$SiO

Bu   S   S   CHO   + *syn*-(R,S)-**3**

CH$_3$

anti-(S,S)-**3**   40:1

Mitsunobu inversion

(C$_6$H$_5$)$_3$SiO

Bu   R

CH$_3$

anti-2

81%   2

(C$_6$H$_5$)$_3$SiO

Bu   R   S   CHO   + *anti*-(R,R)-**3**

CH$_3$

syn-(S,R)-**3**   12:1

syn-(S,R)-**3**. Since the enantiomers of *syn*- and *anti*-**2** are available by Sharpless asymmetric epoxidation with $(-)$–DIPT, the four possible aldols of a β-hydroxy aldehyde are available from this rearrangement of α,β-epoxy triphenylsilyl ethers.

This rearrangement provides a practical route to a number of β-silyloxy aldehydes, even those possessing an asymmetric quaternary α-carbon.

(C$_6$H$_5$)$_3$SiO

O   CH$_3$

88%   1

(C$_6$H$_5$)$_3$Si O   CHO

CH$_3$

anti/syn = 100:1

(C$_6$H$_5$)$_3$SiO

Bu   O   CH$_3$

CH$_3$

64%

(C$_6$H$_5$)$_3$SiO   CHO

Bu

Et   CH$_3$

anti/syn = 200:1

OSi(C$_6$H$_5$)$_3$

O

85%

OSi(C$_6$H$_5$)$_3$

CHO

anti/syn = 100:1

Rearrangement of epoxides.[2]   This aluminum reagent (1 equiv.) effects rearrangement of trisubstituted epoxides to aldehydes with very high selectivity. Lewis acids are generally used for rearrangement of epoxides, but the ordinary Lewis acids rearrange trisubstituted epoxides to a mixture of aldehydes and ketones. Of various metal fluorides, antimony pentafluoride is the most effective for selective rearrangement of these epoxides to ketones.

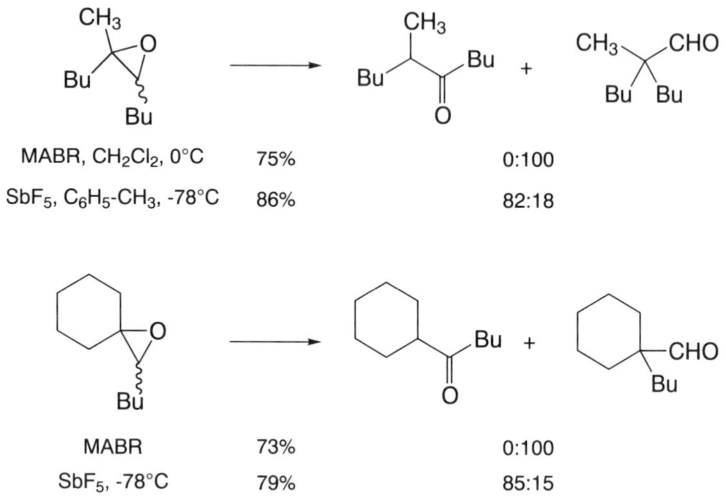

MABR, CH$_2$Cl$_2$, 0°C	75%	0:100
SbF$_5$, C$_6$H$_5$-CH$_3$, -78°C	86%	82:18

MABR	73%	0:100
SbF$_5$, -78°C	79%	85:15

The rearrangement of tri- and tetra-substituted epoxides by this reagent involves *anti*-migration of the alkyl groups.[3]

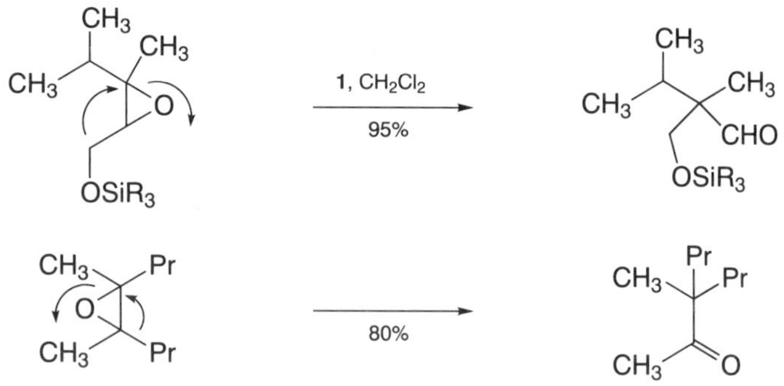

α-Alkylation of enol silyl ethers with RCH₂OTf.[4]   Alkylation of enol silyl ethers directly with primary alkyl halides is limited to allylic and benzylic halides. In contrast, in the presence of 1.1-2 equiv. of MABR, enol silyl ethers undergo α-alkylation with primary alkyl triflates (2 equiv.) in moderate yield. The reaction is applicable to enol silyl ethers of ketones, esters, and some aldehydes. *t*-Butyldimethylsilyl groups are preferred for reactions with ketene silyl acetals.

[1] K. Maruoka, J. Sato, and H. Yamamoto, *Am. Soc.,* **113**, 5449 (1991).
[2] K. Maruoka, R. Bureau, T. Ooi, and H. Yamamoto, *Synlett,* 491 (1991).
[3] K. Maruoka, T. Ooi, and H. Yamamoto, *Tetrahedron,* **48**, 3303 (1992).
[4] K. Maruoka, J. Sato, and H. Yamamoto, *Am. Soc.,* **114**, 4422 (1992).

Methylaluminum bis(2,6-di-*t*-butyl-4-methylphenoxide) (MAD).

Asymmetric Diels–Alder reactions of unsymmetrical fumarates.[1]   The complex formed from *t*-butyl methyl fumarate (1) with MAD reacts with cyclopentadiene at −78° to form the adduct **2** in 93% yield (99:1). Similar high discrimination is shown in reaction of the complex with 2-methylbutadiene (51% yield, 86:14). In contrast, Diels–Alder reaction of **1** catalyzed by $(C_2H_5)_2AlCl$ shows practically no selectivity. MAD can even distinguish between a methyl and an ethyl ester group.

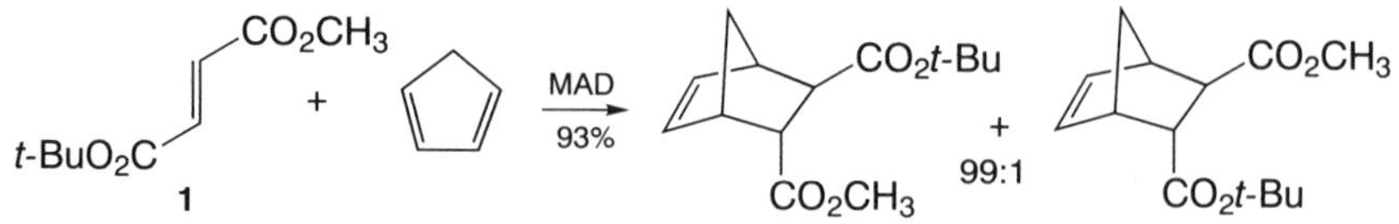

The MAD-mediated Diels–Alder reaction of *l*-menthyl methyl furmarate with cyclopentadiene (2 equiv., CH_2Cl_2, $-78°$) proceeds in 86% de with an *endo-exo* methoxycarbonyl ratio of 98:2.

[1] K. Maruoka, S. Saito, and H. Yamamoto, *Am. Soc.*, **114**, 1089 (1992).

Methyl (R)- or (S)-mandelate, $C_6H_5CHOHCOOCH_3$ (1).

Asymmetric synthesis of β-lactams.[1] The [2+2]cycloaddition of benzyloxyketenes and imines **2** derived from (R)-**1** results in 3,4-*cis*-disubstituted β-lactams (**3**), which can be converted to optically active 3-amino-2-hydroxybutyric acids (**5**).

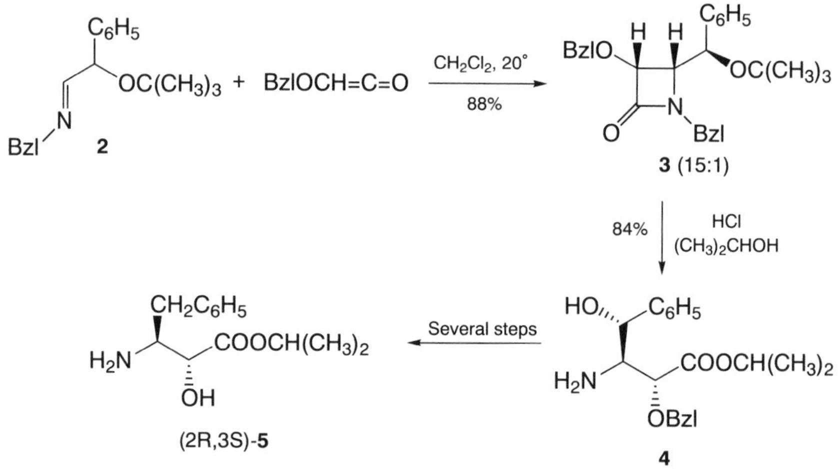

[1] Y. Kobayashi, Y. Takemoto, T. Kamijo, H. Harada, Y. Ito, and S. Terashima, *Tetrahedron*, **48**, 1853 (1992).

endo-**2-Methylbicyclo[2.2.1]hept-5-enyl ethyl ketone,**

Stereoselective aldol reactions of α-methyl aldehydes.[1] The lithium enolate of this ethyl ketone (**1**) reacts with various α-methyl aldehydes to form aldols, which after protection as silyl ethers and thermolysis (500°) were isolated as the α, β-unsaturated ketones **2** and **3**. The diastereoselectivity is highly dependent on the R group. Thus

R = C$_6$H$_5$-	98 : 2
R = H$_2$C=CH-	80 : 20
R = BzlOCH$_2$-	12 : 88
R = (*i*-Pr)$_3$SiOCH$_2$-	15 : 88

2-aryl- and 2-vinylpropionaldehydes react with *syn, syn*-selectivity, whereas the β-benzyloxy aldehydes and β-triisopropylsilyloxy aldehydes react with *syn, anti*-selectivity.

[1] M. Ahmar, R. Block, G. Mandville, and I. Romain, *Tetrahedron Letters*, **33**, 2501 (1992).

10-Methyl-9,10-dihydroacridene/Sodium borohydride,

Photocatalysis of radical cyclization. This system, when irradiated, can effect dehalogenation of aliphatic bromides and aryl halides via a radical intermediate.[1] It can also effect radical cyclization of 1-allyloxy-2-halobenzenes (equation I) in 76% isolated yield.[2]

[1] M. Ishikawa and S. Fukuzumi, *Am. Soc.*, **112**, 8864 (1990).
[2] G. Boisvert and R. Giasson, *Tetrahedron Letters*, **33**, 6587 (1992).

(2R,3R,5R)-4-Methyl-5-phenyl-3-phenylthio-2-morpholinol (1).

This chiral morpholine is obtained in 85% yield by reaction of glyoxal with (R)-N-methylphenylglycinol and thiophenol in water.[1]

Asymmetric synthesis of N-methyl-α-amino esters.[2]   This morpholine can be used as a chiral template for synthesis of N-methyl-α-amino esters. Thus reaction with an alkylcopper involves displacement of the phenylthio group by an alkyl group by the usual S_N2 process with inversion (about 90:10). In contrast, reaction with an alkylzinc iodide involves substitution with essentially complete retention, possibly via an iminium intermediate. The alkylated product (2) is then oxidized to an oxazinone (3), which on treatment with vinyl chloroformate followed by hydrolysis provides N-methyl-α-amino esters (4) in high optical purity. This approach to chiral amino acids is unusual in that either enantiomer can be formed from the same template depending on the choice of the organometallic reagent. Unfortunately, the chiral auxiliary (expensive) is not recovered for reuse.

[1] C. Agami, F. Couty, L. Hamon, B. Prince, and C. Puchot, *Tetrahedron*, **46**, 7003 (1990).
[2] C. Agami, F. Couty, B. Prince, and C. Puchot, *ibid.*, **47**, 4343 (1991).

Methyl(phenylseleno)malononitrile,   (1)

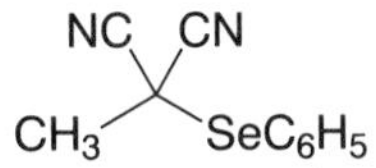

This reagent is prepared by reaction of the anion of methylmalononitrile with C_6H_5SeBr.

Radical additions to alkenes.[1]   Unlike iodomalonitriles which do not add to alkenes containing OR, SR, or NR_2 groups (**16**,183–184), this malononitrile in the presence of AIBN undergoes 1,2-addition to alkenes possessing these groups. The addition to 1,2-disubstituted alkenes proceeds in high yield and with high regioselectivity.

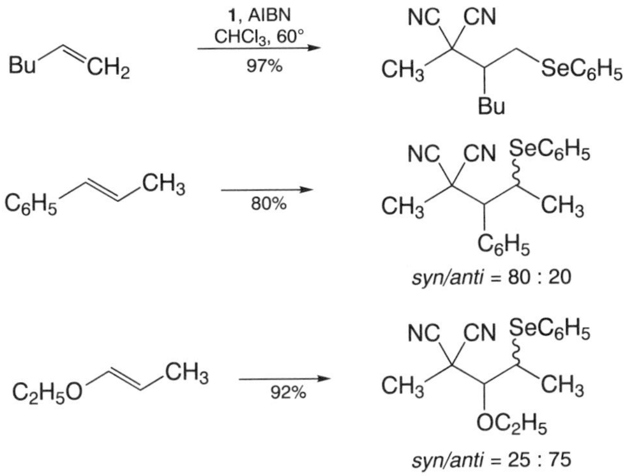

[1] D. P. Curran and G. Thoma, *Am. Soc.*, **114**, 4436 (1992).

Methyl trifluoromethanesulfonate, CH_3OTf.

[5+2]Cycloaddition.   A new route to complex seven-membered rings involves activation of an unsaturated pyrone such as **1** with CH_3OTf (2 equiv., 20°) to form a pyrylium salt (**a**), which undergoes [5+2]cycloaddition at 25° when exposed to CsF to give **2** in 84% yield.

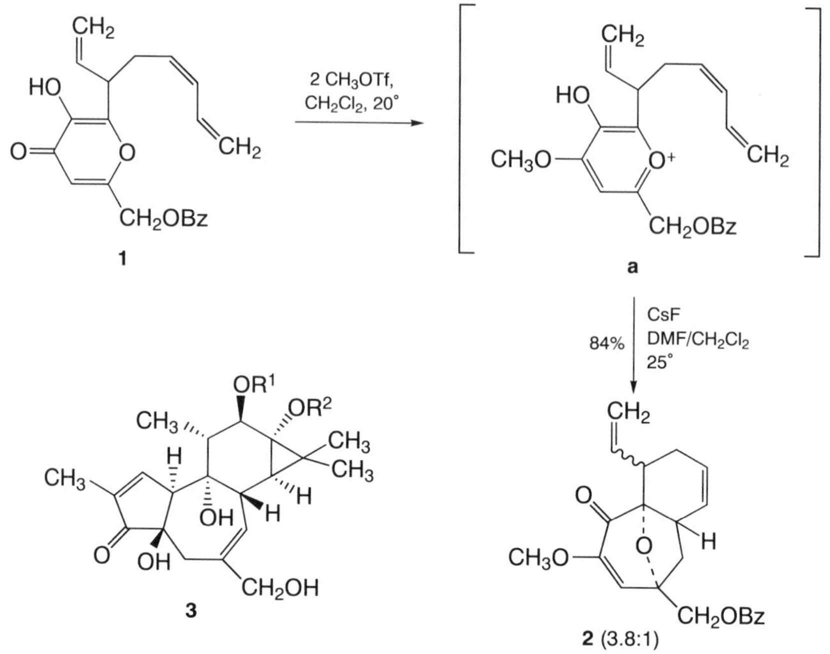

Note that **2** corresponds to the B/C ring of phoboids such as tigliane (**3**). Phoboids are of interest as highly potent tumor promoters.

[1] P. A. Wender and J. L. Mascareñas, *J. Org.*, **56**, 6267 (1991).

Methyltrioxorhenium, CH_3ReO_3 (1).

This oxide is prepared by reaction of Re_2O_7 with $HSn\,(CH_3)_3$ and purified by vacuum sublimation. It is soluble in organic solvents and also in water and is stable to air and acid.[1]

Olefin metathesis.[2] CH_3ReO_3 mixed with Al_2O_3/SiO_2 is an effective heterogeneous catalyst for metathesis of alkenes in CH_2Cl_2 or C_6H_5Cl at 25°.

$$CH_3CH=CHC_2H_5 \xrightarrow[100\%]{\overset{\textstyle 1}{Al_2O_3/SiO_2}} CH_3CH=CHCH_3 \quad + \quad C_2H_5CH=CHC_2H_5$$

$$C_4H_9CH=CH_2 \xrightarrow[100\%]{} C_4H_9CH=CHC_4H_9 \quad + \quad CH_2=CH_2$$

Epoxidation.[3] CH_3ReO_3 is an effective catalyst for epoxidation of alkenes with H_2O_2 in *t*-butyl alcohol, tetrahydrofuran, or water at $-10°$ to 80°. It also can catalyze hydrolysis of epoxides to *trans*-1,2-diols. This secondary reaction can be suppressed by addition of amines.

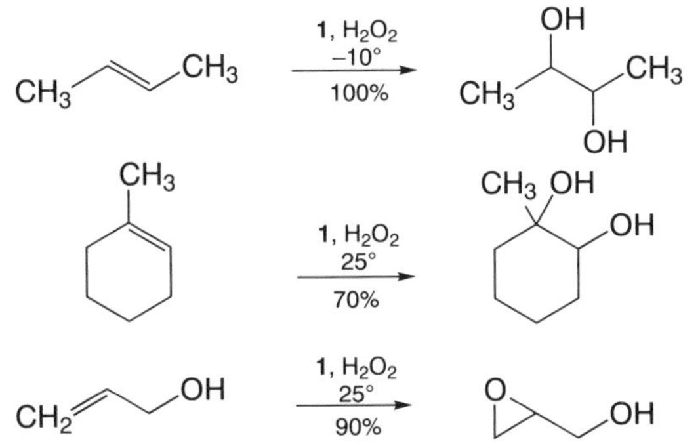

Aldehyde olefination.[4] CH_3ReO_3 (**1**) is an efficient catalyst for condensation of aldehydes with a diazoalkane and a phosphine to form an alkene and a phosphine oxide. The actual catalyst may be $CH_3ReO_2OPR_3$. The reaction is applicable to aliphatic and aromatic aldehydes and also to enals. Some cycloketones undergo this olefination but in only moderate yield.

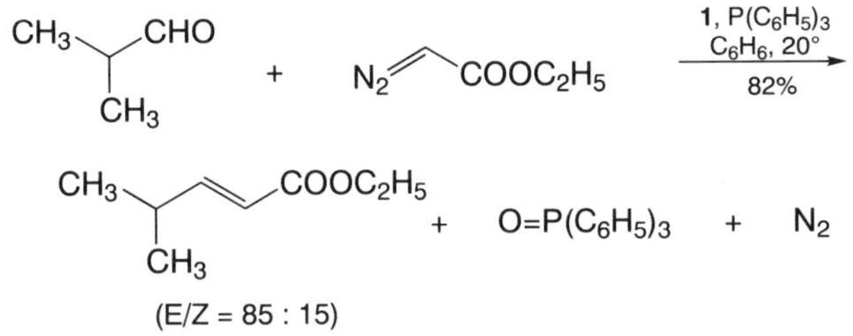

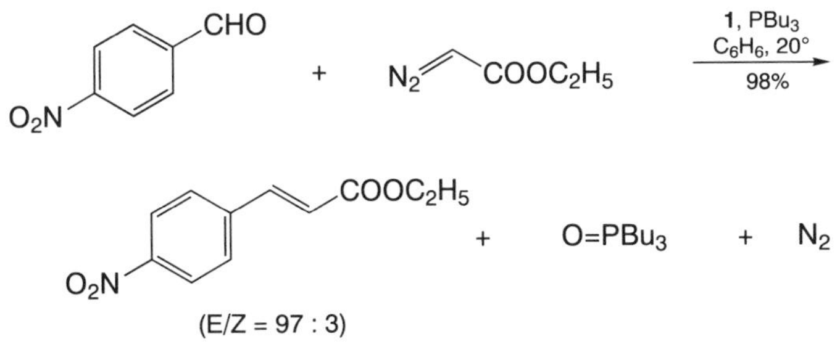

[1] W. A. Herrmann, J. G. Kuchler, G. Weichselbaumer, E. Herdtweck, and P. Kiprof, *J. Organomet. Chem.*, **372**, 351 (1989).
[2] W. A. Herrmann, W. Wagner, U. N. Flessner, U. Volkhardt, and H. Komber, *Angew. Chem. Int. Ed.*, **30**, 1636 (1991).
[3] W. A. Herrmann, R. W. Fischer, and D. W. Marz, *ibid.*, **30**, 1638 (1991).
[4] W. A. Herrmann and M. Wang, *ibid.*, **30**, 1641 (1991).

Molecular Sieves.

Stereoselectivity effects.[1]   There are several recent reports that molecular sieves can improve the stereoselectivity of various reactions such as the Sharpless epoxidation (e.g., **13**,51). The beneficial effect has been attributed to water-scavenging by the sieves. Molecular sieves also show a marked effect on the diastereoselectivity of the Pd-catalyzed cyclization of *cis*-1,2-divinylcyclohexane with chiral acids (equation I). In these reactions, additions of molecular sieves were found in some cases to increase the diastereo-

(I)

selectivity, in one case from 0 to 54% ee. In more cases, the addition of molecular sieves could even reverse the diastereoselectivity. The highest selectivity obtains with Lancaster 4 Å and 13X sieves, both of which have sodium as the counterion. The steric effects do not result from elimination of water, since sieves containing 20% of water show improved stereoselectivity. No explanation is available at present, but a surface effect is one possibility.

[1] L. Tottie, P. Baeckström, C. Moberg, J. Tegenfeldt, and A. Heumann, *J. Org.*, **57**, 6579 (1992).

Molybdenum carbene complexes.

These complexes can be prepared in about 55% yield by reaction of RLi with $Mo(CO)_6$ followed by methylation (equation I).

$$Mo(CO)_6 \xrightarrow[\substack{2) \ CH_3OSO_2F, \ 0° \\ 55\%}]{1) \ RLi, \ ether, \ 0°} (CO)_5Mo=\overset{OCH_3}{\underset{R}{C}}$$

Cyclopropanation.[1]   These carbenes are particularly useful for cyclopropanation of electron-poor olefins. The reaction occurs under milder conditions and at a faster rate with molybdenum carbenes than with chromium- or tungsten-derived complexes. This cyclopropanation has been used to trap a molybdenum vinylcarbene generated by intra-

$$(CO)_5Mo = \overset{OCH_3}{\underset{Bu}{<}} \quad \xrightarrow[25°]{CH_2=CHCN} \quad \overset{CH_3O}{\underset{Bu}{\triangle}}_{CN} \quad + \quad \overset{CH_3O}{\underset{Bu}{\triangle}}_{CN}$$

27%   35%

molecular cyclization of the carbene complex (1) of an alkyne. Thus reaction of the carbene 1 with methyl acrylate in THF at 65° provides a mixture of the vinylcyclopropanes
2 in 71% yield.[2]

$$I(CH_2)_3C \equiv CR \qquad \xrightarrow[\substack{3)\ CH_3OSO_2CF_3}]{\substack{1)\ BuLi \\ 2)\ Mo(CO)_6}} \qquad \overset{CH_3O}{\underset{Mo(CO)_5}{C}} {-}(CH_2)_3C \equiv CR$$

R = (CH_2)_2CH_3

1

1 $\xrightarrow{THF,\ 65°}$ $\left[\text{R=Mo(CO)}_4 \atop \text{OCH}_3 \right]$ $\xrightarrow[71\%]{CH_2=CHCOOCH_3}$ (structure) COOCH_3 OCH_3

2

The process has been extended to the dienyne 3. Reaction of 3 with butylmethoxy-
molybdenum carbene 4 at 60° results in a divinylcyclopropane (5), which undergoes
[3.3]sigmatropic rearrangement to hexahydroazulenes 6 in 87% yield.[3] Use of the
analogous chromium complex with 3 also results in a hexahydroazulene, but in much
lower yield. An electron-withdrawing group in the diene is essential for formation of 5.

(structure diagram)

3

$$\overset{Mo(CO)_5}{\underset{Bu}{C}} {OCH_3} \qquad \xrightarrow{60°}$$

$$\left[\text{(structure 5)} \right] \xrightarrow{87\%} \text{(structure 6)}$$

5

6   (1 : 4.8)

[1] D. F. Harvey and M. F. Brown, *Tetrahedron Letters*, **31**, 2529 (1990).
[2] *Idem*, *Am. Soc.*, **112**, 7806 (1990).
[3] D. F. Harvey and K. P. Lund, *ibid.*, **113**, 5066 (1991).

Molybdenum imido alkylidine complexes.

Olefin metathesis; oxygen and nitrogen heterocycles.   Schrock's group has developed a number of molybdenum catalysts of this type for olefin metathesis. A typical catalyst is **1**.[1]

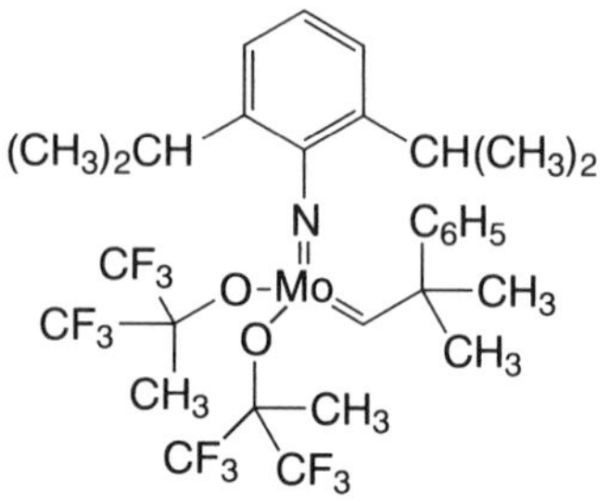

Reaction of a diallyl ether (**2**) with **1** at 20° results in cyclization to a dihydrofuran with loss of ethylene. This reaction is applicable to formation of tri- and tetrasubstituted alkenes and to synthesis of dihydropyrans and seven-numbered oxygen heterocycles.[2]

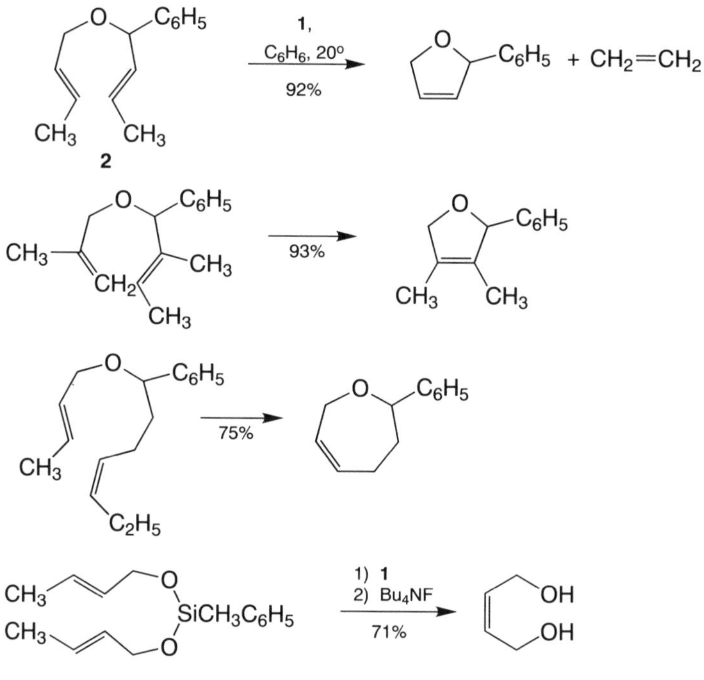

This cyclization also affords a route to nitrogen heterocycles from acyclic dienyl-amines and -amides. Pyrrolines can be obtained from diallylamines. This reaction can also form tetrahydropyridines and -azepines, as well as lactams.[3]

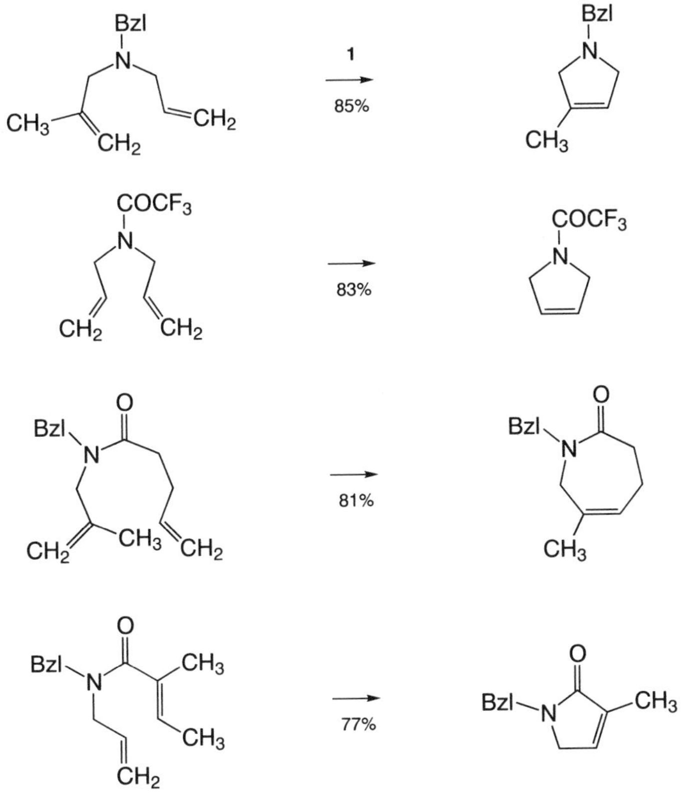

[1] R. R. Schrock et al., *Am. Soc.,* **112**, 3875 (1990).
[2] G. C. Fu and R. H. Grubbs, *ibid.,* **114**, 5426 (1992).
[3] G. C. Fu and R. H. Grubbs, *Am. Soc.,* **114**, 7324 (1992).

Monochloroalane, AlH$_2$Cl (1, 595–599).

Azetidines. (12, 333)[1] These cyclic trimethyleneamines are difficult to prepare directly, but can be obtained by reduction of β-lactams. Metal hydrides generally reduce β-lactams with cleavage of the 1,2-bond to afford γ-amino alcohols as the major product, but monochloroalane (AlH$_2$Cl) or dichloroalane (AlHCl$_2$) reduces β-lactams to azetidines in high yield (85–100%). Azido groups, if present, are converted to amine groups. *t*-Butyl ester groups are reduced to hydroxymethyl groups, and acetates are converted to hydroxy groups. Chiral β-lactams are reduced without loss of enantioselectivity. The 2-arylazetidines obtained in this way on hydrogenolysis catalyzed by Pd/C or Raney

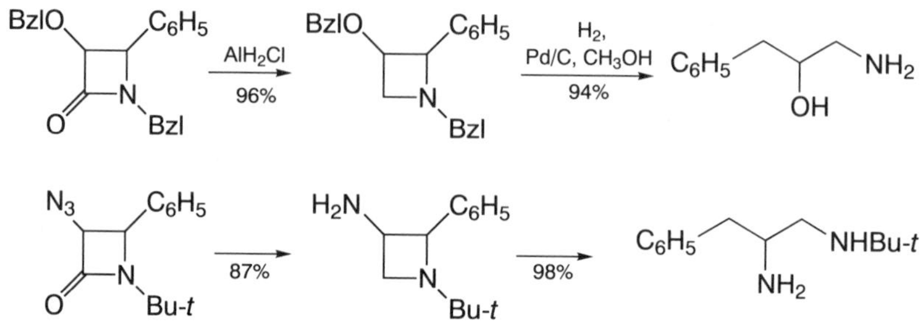

nickel undergo cleavage of the 1,2-bond to afford the corresponding acyclic amino alcohols or diamines.

This reduction has been extended to bis-β-lactams (2), prepared as shown from a benzylidine amino β-lactam (equation I). Two isomers are formed, which are separable by chromotography. These bis-β-lactams are reduced by ClAlH$_2$ to *anti-* and *syn*-bis-β-

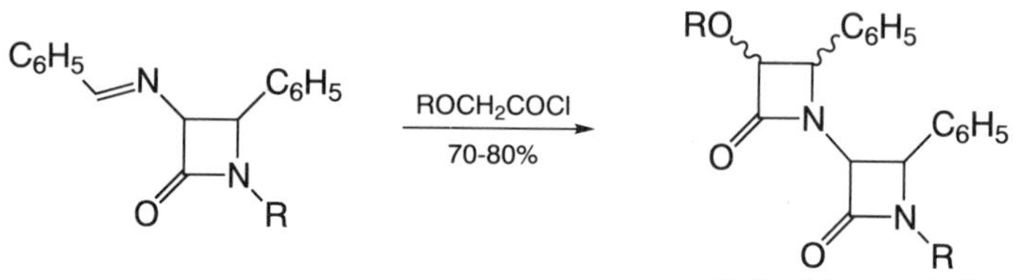

2 (*anti* /*syn* ≈ 2:1)

azetidines, which in turn undergo hydrogenolysis to open-chain polyamino alcohols and polyamino ethers.

[1] I. Ojima, M. Zhao, T. Yamato, K. Nakahashi, M. Yamashita, and R. Abe, *J. Org.*, **56**, 5263 (1991).

Monochloroborane–Dimethyl sulfide, ClBH$_2$ · S(CH$_3$)$_2$ (1).

Cleavage of epoxides, ethers, ketals.[1]   This borane cleaves epoxides to chlorohydrins with marked regioselectivity. It is also useful for cleavage of ethers and acetals.

$$ClBH_2 \cdot S(CH_3)_2$$
1

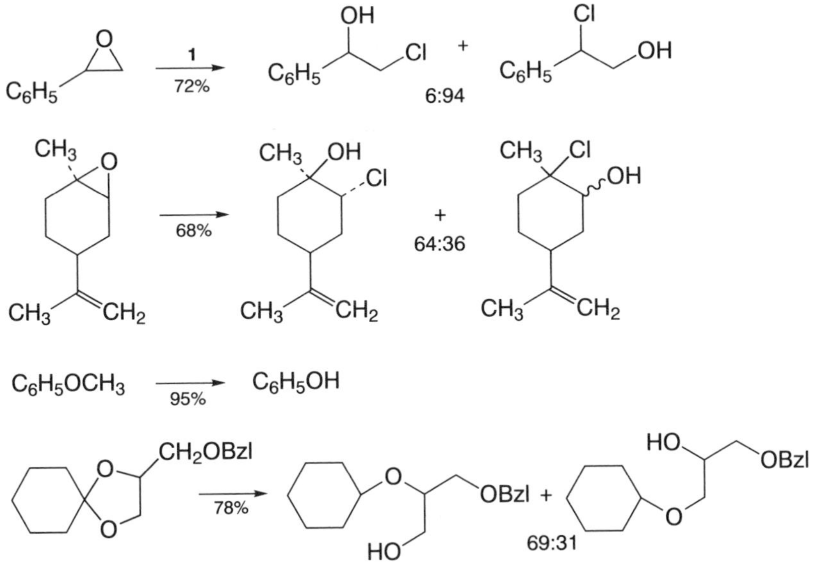

$$C_6H_5OCH_3 \xrightarrow[95\%]{} C_6H_5OH$$

Epoxides → chlorohydrins. [2]   This reagent converts epoxides into chlorohydrins. The same cleavage of α-hydroxy epoxides results in *anti*-chlorohydrins.

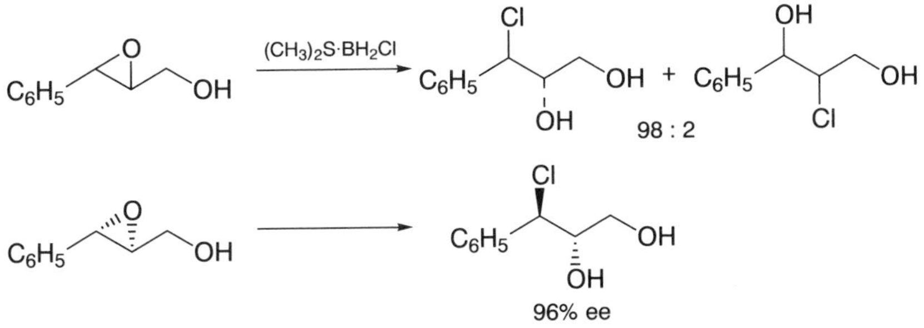

[1] P. Bovicelli, E. Mincione, and G. Ortaggi, *Tetrahedron Letters,* **32**, 3719 (1991).
[2] P. Bovicelli, P. Lupattelli, and M. T. Bersani, and E. Mincione, *ibid.,* **33**, 6181 (1992).

Monoperoxyphthalic acid (MPPA).

Epoxidation of allylic alcohols. [1]   Although *m*-chloroperbenzoic acid in aqueous $NaHCO_3$ can effect epoxidation of simple allylic alcohols in 80–95% yield, this reagent shows slight regioselectivity in the case of polyolefinic alcohols. Thus geraniol reacts with $ClC_6H_4CO_3H$ to give a mixture of the diepoxide and both monoepoxides.

In contrast epoxidation of geraniol with monoperoxyphthalic acid in the presence of cetyltrimethylammonium hydroxide (surfactant) effects almost exclusive epoxidation of the allylic double bond to give 2,3-epoxygeraniol in 90% yield. Epoxidation with MPPA in $NaHCO_3$ solution shows the opposite regioselectivity and provides 6,7-epoxygeraniol in 72% yield.

[1] F. Fringuelli, R. Germani, F. Pizzo, F. Santinelli, and G. Savelli, *J. Org.*, **57**, 1198 (1992).

N

1,4-Naphthalenedicarbonitrile (DCN, 1).

[3+2]Cycloaddition of azirines.[1] Irradiation of the azirine **2** in the presence of DCN (**1**) forms a radical cation (**3**), which is trapped by imines to form imidazoles (**4**).

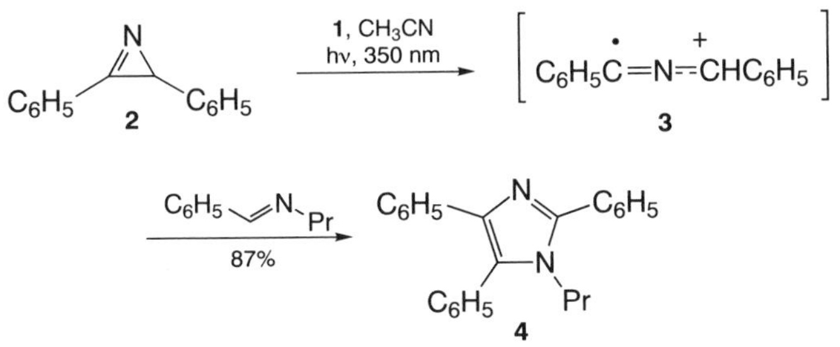

[1] F. Müller and J. Mattay, *Angew. Chem. Int. Ed.*, 1336 (1991).

Nickel(II) acetylacetonate, Ni(acac)$_2$.

Coupling of aryl O-carbamates and aryl triflates with RMgCl.[1] This coupling can be effected by catalysis with Ni(acac)$_2$. The particular value of this coupling is that O-carbamates can direct *ortho*-metalation, and thus can be converted into useful derivatives for coupling to complex substituted arenes.

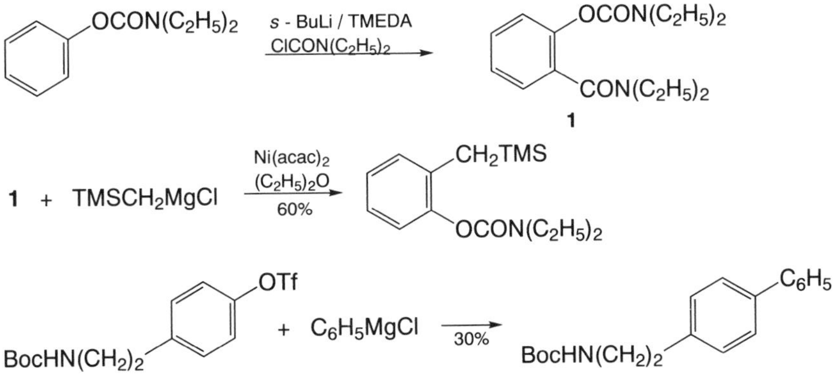

[1] S. Sengupta, M. Leite, D. S. Raslan, C. Quesnelle, and V. Snieckus, *J. Org.*, **57**, 4066 (1992).

Nickel acetylacetonate–Diisobutylaluminum hydride.

Cyclization of enynes, dienynes, and diynes with aryl isocyanides.[1]   This combination of reagents (1:2) generates a Ni(0) catalyst (**1**) which is easier to handle than the air-sensitive bis(1,5-cyclooctadiene)nickel(0), Ni(COD)$_2$. Reaction of 1,6-enynes (**2**) with an aryl isocyanide in the presence of this Ni(0) catalyst combined with Bu$_3$P (2 equiv.) results in bicyclic iminocyclopentenes (**3**) which can be hydrolyzed to the corresponding ketone (**4**). The overall reaction is an alternative to the Pauson–Khand reaction.

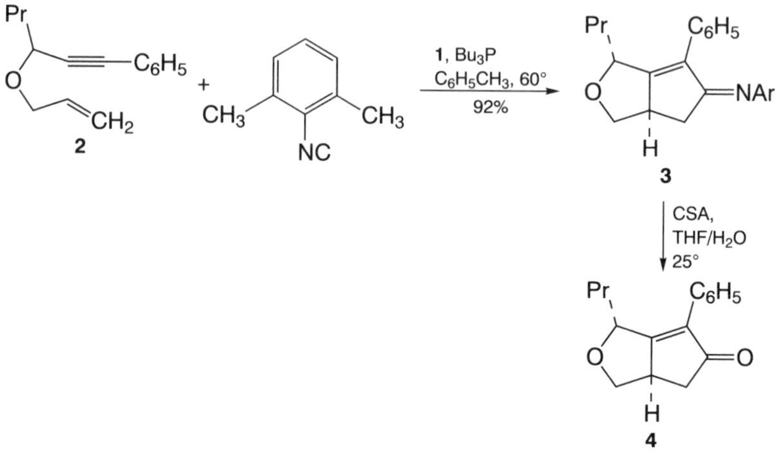

The same reaction but with a 1,*n*-diyne and in the absence of a phosphine can provide bicyclic iminocyclopentadienes. Highest yields obtain with 1,3-diynes. Yields of bicyclic products decrease as the number of intervening carbon atoms increases.

This nickel(0) catalyst can also effect cyclizations of 1,7-diynes with hydrosilanes. 1,6–Diynes can undergo a Ni(0)-catalyzed hydrosilylation to a bicyclic silacyclopentadiene (equation II).

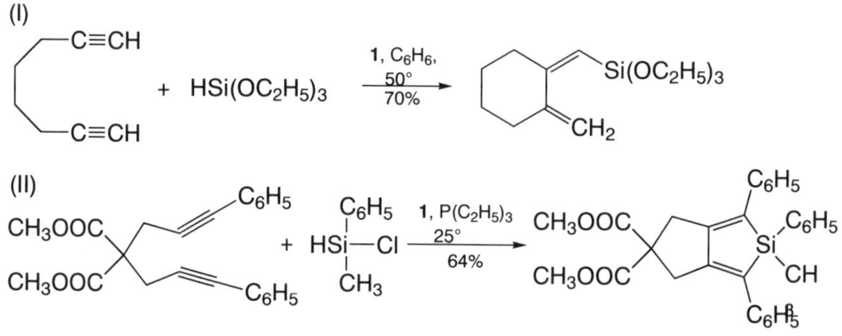

(I)

(II)

[1] K. Tamao, K. Kobayashi, and Y. Ito, *Synlett*, 539 (1992).

O

Organoaluminum reagents.

Stereoselective acetal cleavage.[1] Chiral acetals such as **1** are known to be cleaved by RLi, RMgX, R_2Zn, and a number of organometallic nucleophiles with inversion of configuration. This reaction is a useful route to various optically active alcohols (12,375–378). Surprisingly, reaction of **1** with $Al(CH_3)_3$ is not stereoselective, but highly stereoselective alkylation can be effected by reaction with an (aryloxy)dimethylaluminum prepared *in situ* by reaction of $Al(CH_3)_3$ with 1 equiv. of a phenol. The most reactive aluminum reagent is obtained from pentafluorophenol, but 2,4,6-tri-*t*-butylphenol is almost as effective. Of equal interest, alkylation with these organoaluminum reagents proceeds with high retention (97–99:3–1).

This cleavage usually also provides, as a minor product, the ketone **3**, formed by an intramolecular Meerwein–Ponndorf–Verley reduction and Oppenauer oxidation. β-Alkoxy ketones of this type (**6**) can be obtained as the major product by reduction of ketals (**5**) with diethylaluminum fluoride (1.2 equiv.) and pentafluorophenol (2.4 equiv.), (equation II). Note that the reduction is again effected with retention.

Stereoselective addition to 2-methyl-3-oxo amides (esters).[2]   Both R_2AlCl and R_3Al add to ketone groups of these substrates with high *anti*-selectivity.

[1] K. Ishihara, N. Hanaki, and H. Yamamoto, *Am. Soc.*, **113**, 7074 (1991).
[2] M. Taniguchi, H. Fujii, K. Oshima, and K. Utimoto, *Tetrahedron Letters*, **33**, 4353 (1992).

Organoantimony compounds.

Review.   Huang[1] has reviewed the synthetic use of organoantimony compounds. In addition to Wittig-type reactions., his laboratory has reported several novel uses for these compounds such as a selective acetalization of aldehydes (equation I). Acetalization can also be effected by the system $SbCl_3/Al$ and an alcohol in almost quantitative yield.

[1] Y.-Z. Huang, *Acc. Chem. Res.*, **25**, 182 (1992).

Organobarium reagents.

Allylbarium reagents; homoallylic alcohols.[1]   The reaction of anhydrous BaI_2 with lithium biphenylide (2 equiv.) in THF provides a reactive barium species that reacts with allylic chlorides at $-78°$ to form allylbarium chlorides. These reagents react with high α-selectivity with carbonyl compounds to form homoallylic alcohols with retention of configuration.

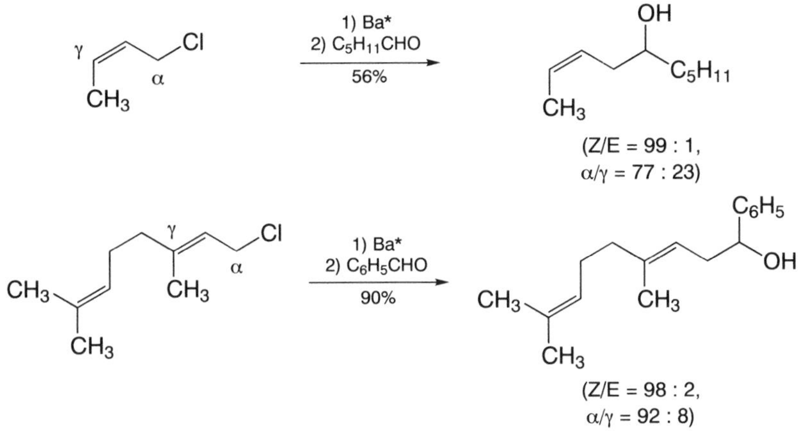

(Z/E = 99 : 1,
α/γ = 77 : 23)

(Z/E = 98 : 2,
α/γ = 92 : 8)

[1] A. Yanagisawa, S. Habaue, and H. Yamamoto, *Am. Soc.*, **113**, 8955 (1991).

Organocerium reagents.

Addition to RC≡N and >C=NH.[1]   Organocerium reagents (RCeCl₂) undergo double addition to nitriles to form *tert*-amines often in high yield. The rate of this double addition is about four times slower than that of addition to a carbonyl group.

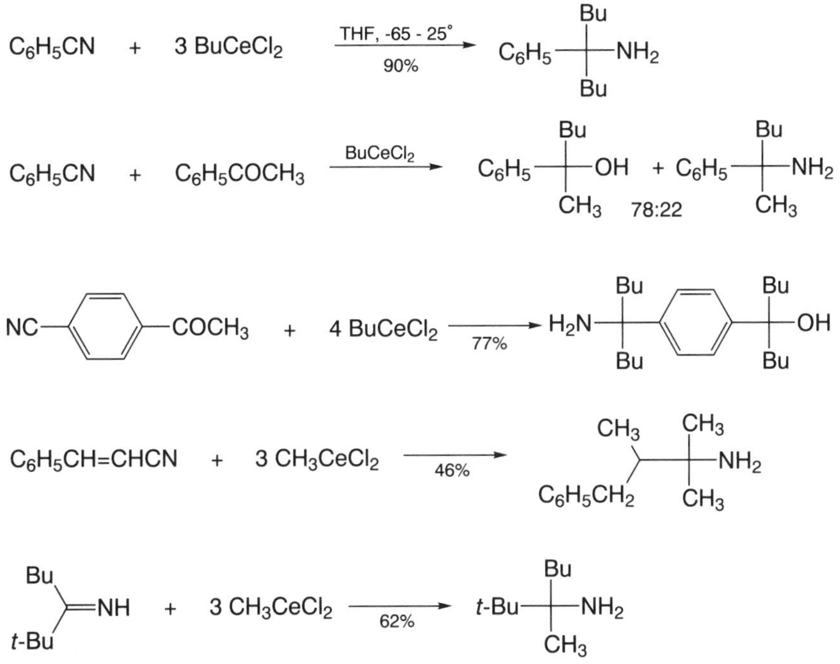

The last example shows that these organocerium reagents also react with ketimines to form tertiary amines.

α-Silyl ketones.[2] Organocerium reagents (**15**,221) add to the carbonyl group of trialkylsilylketenes to form β-silyl cerium enolates (**a**), which undergo alkylation on reaction with RI and HMPA to form α-silyl ketones. This tandem alkylation can be used to prepare either one of the two possible α-silyl ketones by the choice of RCeCl$_2$ and RX.

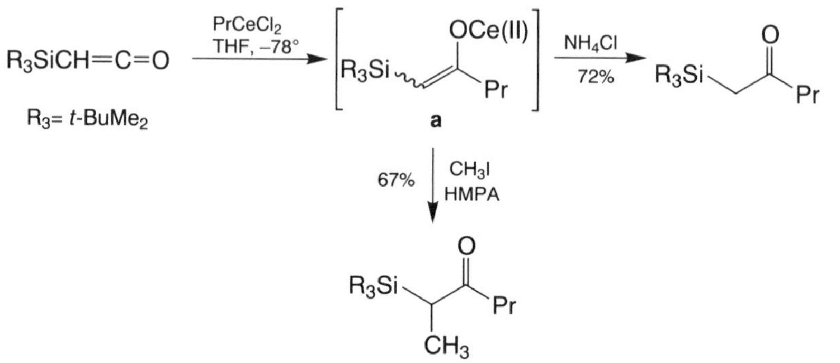

Asymmetric synthesis of dl-1,3-diphenyl-1,3-propanediamine (3).[3] The synthesis of **3** is an extension of a synthesis of chiral amines by addition of C$_6$H$_5$CeCl$_2$ (C$_6$H$_5$Li/CeCl$_3$) to hydrazones (**14**,217–218). Thus addition of C$_6$H$_5$CeCl$_2$ to the dihydropyrazole **1**, prepared as shown from cinnamaldehyde, provides **2**, which can be converted into **3**.

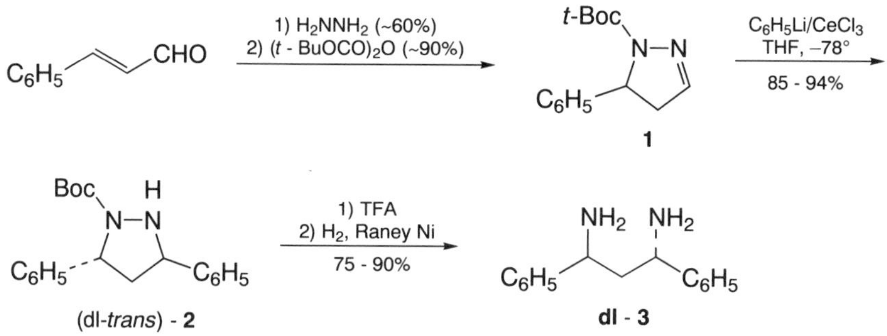

Ultrasound-assisted preparation.[4] CeCl$_3$ is not soluble in THF, but the heptahydrate when sonicated in THF is converted into the soluble CeCl$_3$ · THF. Reaction of CeCl$_3$ · THF with RLi provides RCeCl$_2$ in situ.

$$\text{CeCl}_3 \cdot 7\,\text{H}_2\text{O} \xrightarrow{\text{THF, ((((}} \text{CeCl}_3 \cdot \text{THF} \xrightarrow{\text{BuLi}} \left[\text{BuCeCl}_2\right]$$

77%

Bu   OH

[1] E. Ciganek, *J. Org.*, **57**, 4521 (1992).
[2] Y. Kita, S. Matsuda, S. Kitagaki, Y. Tsuzuki, and S. Akai, *Synlett*, 401 (1991).
[3] S. E. Denmark and J.-H. Kim, *Synthesis*, 229 (1992).
[4] N. Greeves and L. Lyford, *Tetrahedron Letters*, **33**, 4759 (1992).

Organochromium reagents.

$RCrCl_2$[1] Reagents of the type $RCrCl_2$ can be prepared *in situ* or in crystalline form by reaction of $CrCl_3(THF)_3$ in THF at $-20°$ with RLi, RMgX, or $(C_2H_5)_3Al$. These reagents alkylate aldehydes in yields of 40–90%, being highest with CH_3CrCl_2 and aromatic aldehydes. They generally do not alkylate simple ketones, but ketones bearing hydroxyl, methoxyl, or dimethylamino groups on the α- or β-position are alkylated by $CH_3CrCl_2(THF)_3$ in good yield with very high selectivity in the presence of ketones lacking such groups.

[1] T. Kauffmann, C. Beirich, A. Hamsen, T. Möller, C. Philipp, and D. Wingbermühle, *Ber.*, **125**, 157 (1992).

Organocopper reagents.

Stannylcupration of propargylamines.[1] Boc- or $Si(CH_3)_3$-protected propargyl-amines (**1**) react with $Bu_3Sn(Bu)Cu(CN)Li_2$ and then with an electrophile to form 2-substituted 3-(tributylstannyl)allylamines (**2**). These products can be converted into 1,2-disubstituted allylamines (**3**).

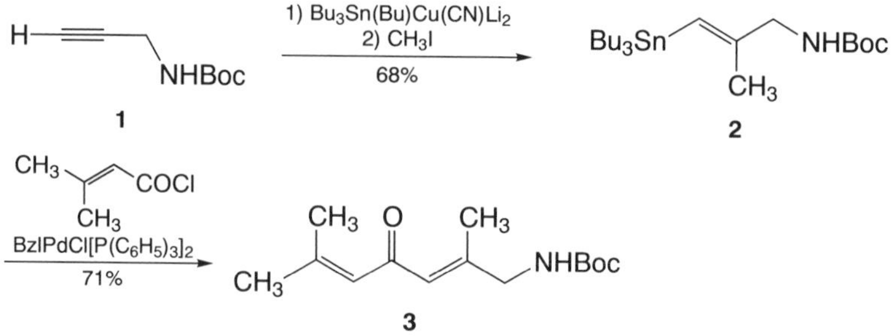

Cyclic α-alkoxyorganocuprates.[3]   Acyclic optically active α-alkoxyorganocopper reagents unfortunately undergo racemization during 1,4-addition reactions. In contrast, a cyclic α-alkoxyorganocopper reagent can be stable at −78°. Enantiomerically pure reagents of this type have been prepared from (R)-3-hydroxybutyric acid (**15,171**) via the aldehyde **1** to give 4-(tributylstannyl)-1,3-dioxanes (**3**), which are separable by

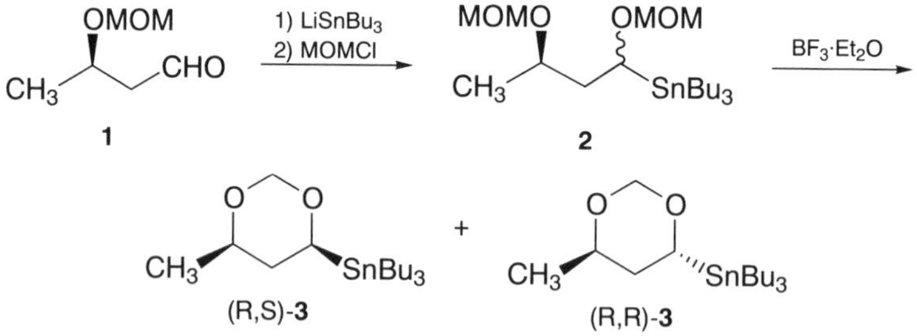

flash chromatography. The corresponding 4-lithio-1,3-dioxanes, formed by transmetalation with BuLi, differ markedly in conformational stability. The axial isomer, derived from the (R,R)-**3** is unstable above −78° and rearranges to the equatorial isomer. The α-alkoxycopper reagent **4**, formed from (R,S)-**3**, by treatment with BuLi, then CuI, and finally TMEDA at −78°, reacts with ethyl propiolate in the presence of ClSi(CH$_3$)$_3$ to form a single adduct (**5**) in 92% yield.

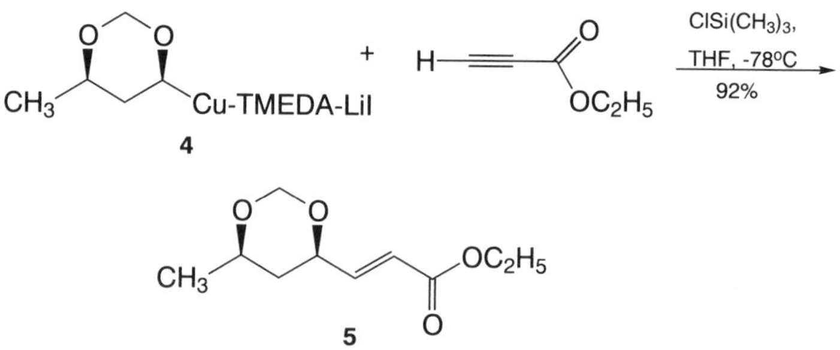

Allylic organocopper reagents.[4] The activated copper species (Cu*), produced by reduction of CuCN · LiX, reacts with allyl chlorides and acetates to afford allylic organocopper reagents, which couple with various electrophiles in good yields.

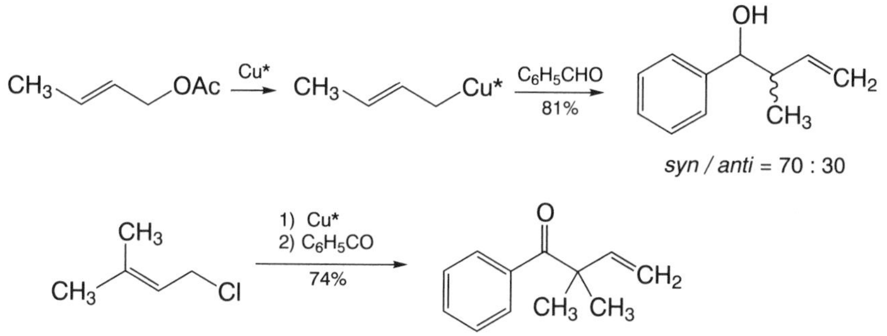

syn / anti = 70 : 30

The particular value of this route to organocopper reagents is that it tolerates a wide range of functional groups: enone, epoxide, carbamate, nitrile.

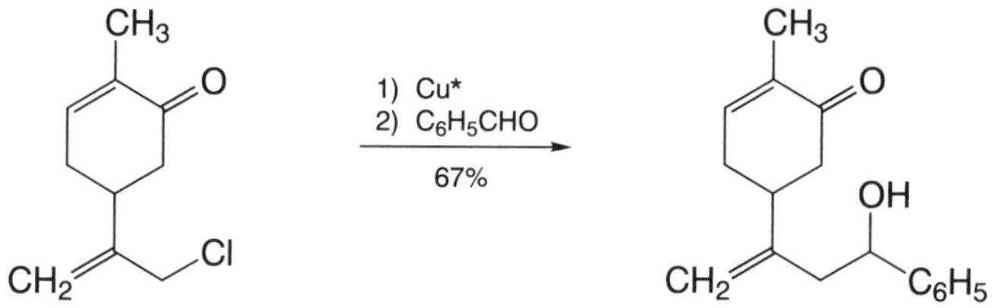

Amide cuprates; β-lactams.[5] Amide cuprates such as [BzlSi(CH₃)₃N]₂CuLi (1) react with 1,3-dienoates to give with high or exclusive regioselectivity the 1,4-adduct (equation I). This reaction can be extended to a three-component coupling. Thus reaction

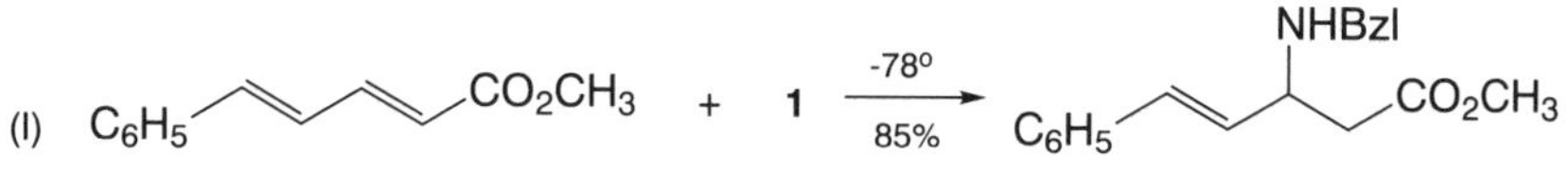

of **1** with the dienoate **2** followed by trapping with C₆H₅CHO gives **3** in 77% yield. Treatment of **3** with (C₆H₅)₂P/PyS)₂ gives the β-lactam **4**.

CH₃ ∿∿CO₂i-Pr   +   C₆H₅CHO   +   1   ———→ 77%

BzlNH   OH
CH₃∿∿...C₆H₅
        CO₂i-Pr
        3

———→

HO H H
C₆H₅ ∿∿CH₃
    N
  O Bzl
    4

This reaction has been extended to asymmetric three-component coupling using a (−)-bornane-sultam.

Chiral lithium amidocuprates.[6]   Lithium amidocuprates can add to enones in modest to high enantioselectivity if they bear a chiral ligand, LiCuL*R, such as (R)- or (S)-N-methyl-1-phenyl-2-(1-piperidinyl)ethanamine (L*). The addition of LiCuL*Bu with 2-cycloheptnone has been examined in detail. Although the enantioselectivity is satisfactory, the yield is only 54%. Use of 3 equiv. of the cuprate increases the yield to 82% with no appreciable effect on the enantioselectivity. Addition of ClSi(CH₃)₃ has little effect, but addition of HMPA lowers both the yield and the enantioselectivity significantly. Use of CuBr · S(CH₃)₂ and CuCN are unsatisfactory. Ether and dimethyl sulfide are the best solvents; low yields and complete loss of enantioselectivity obtain in THF. The report suggests that these cuprates react as dimers and that the mesodimeric complexes are unreactive.

C₆H₅
CH₃∿N ∿∿N⟨piperidine⟩   (S)-L*
      H
  LiCuL*Bu, ether, -78°
  ———————————→
        54%

(cycloheptenone) → (cycloheptanone with Bu)

96% ee

Vinylic cuprates.   Marek et al.[7] have prepared the versatile vinyl cuprate **1**, which can function as the dianion of acrolein.

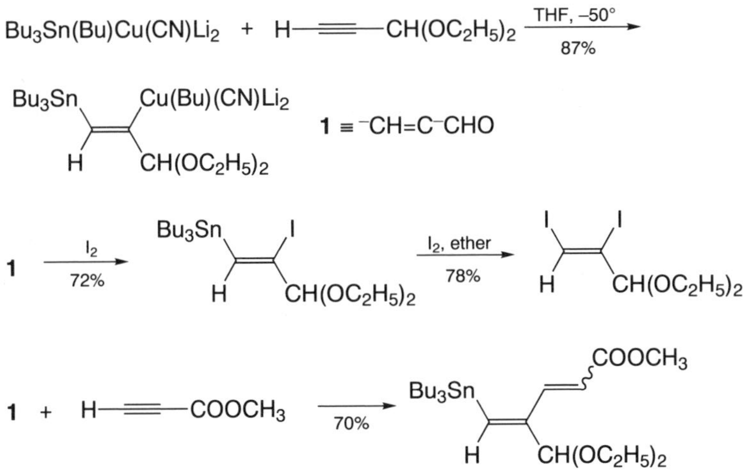

$$Bu_3Sn(Bu)Cu(CN)Li_2 \ + \ H\!\equiv\!\!\equiv\!\!-CH(OC_2H_5)_2 \xrightarrow[\ 87\%\ ]{THF,\ -50°}$$

Bu$_3$Sn, Cu(Bu)(CN)Li$_2$
H CH(OC$_2$H$_5$)$_2$

$1 \equiv \ ^-CH=C^-CHO$

$$1 \xrightarrow[72\%]{I_2}$$ Bu$_3$Sn, I / H CH(OC$_2$H$_5$)$_2$ $\xrightarrow[78\%]{I_2,\ ether}$ I I / H CH(OC$_2$H$_5$)$_2$

$$1 \ + \ H\!\equiv\!\!\equiv\!\!-COOCH_3 \xrightarrow{70\%}$$ Bu$_3$Sn, COOCH$_3$ / H CH(OC$_2$H$_5$)$_2$

(E/Z = 87 : 13)

***cis*-2-Tributylstannylvinyl(cyano)cuprate.**[8]   This cuprate (1) in combination with 2 eq. of LiCl undergoes conjugate additions to enones in high yield.

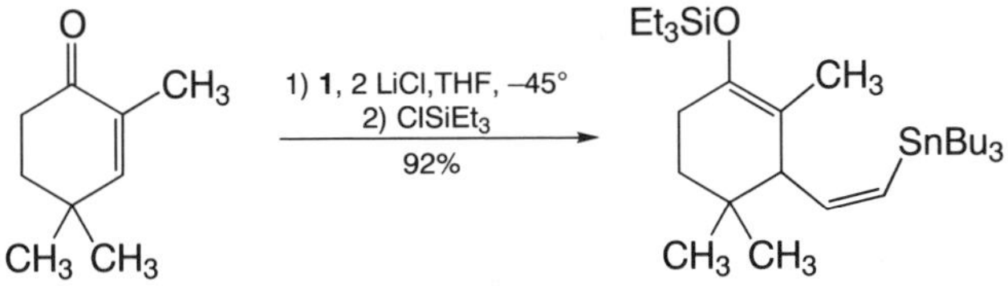

β- and γ-Amino acids.[9]   *t*-Butyl N-boc-aspartate (1) can be converted into a β-amino ester 2 without racemization by conversion of the free acid function α to the amino group to the CH$_2$OTs or CH$_2$I group. Coupling of these products with organocuprates affords N-protected β-amino acid esters. This procedure when extended to glutamic acid provides γ-amino acids.

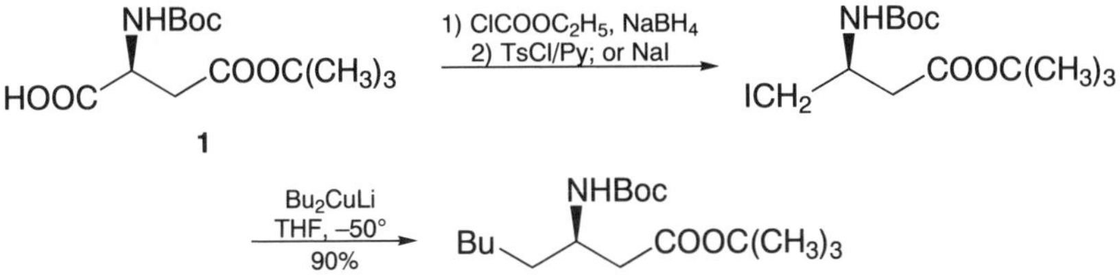

β-Amino alcohols.[10]   Addition of RLi or RMgX to trimethylsilyl imines (**1**) of α-hydroxy aldehydes provides syn-β-amino alcohols, probably because of chelation. In contrast, addition of organocopper reagents derived from Grignard reagents can give mixtures of *syn-* and *anti*-β-amino alcohols. The highest *anti*-diastereoselectivity obtains with $BuCuMgBrI \cdot I \cdot BF_3$.

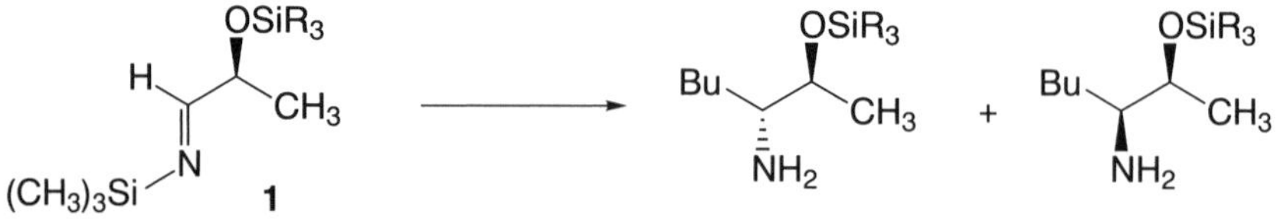

BuLi, THF	46%	2:98
BuCuMgBrl•BF$_3$ THF, S(CH$_3$)$_2$	54%	>99:1
Bu(CN)CuMgBr•BF$_3$ THF, 2 Li	37%	>99:1

Vinylcuprates.[11]   Vinylcuprates (**2**) that can also undergo substitution reactions can be prepared as shown in equation (II). These vinyl cuprates can convert a 1-alkyne

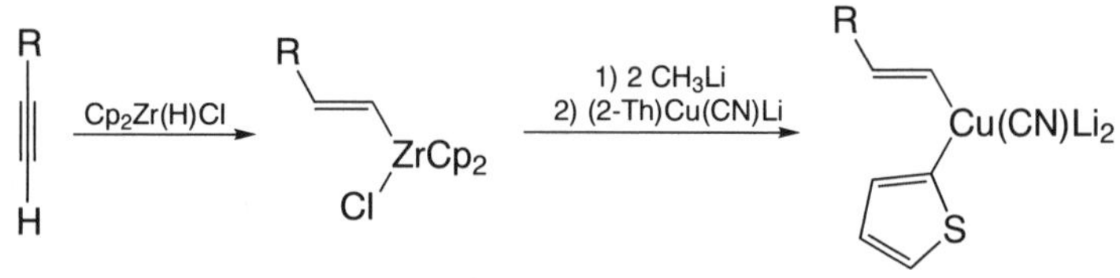

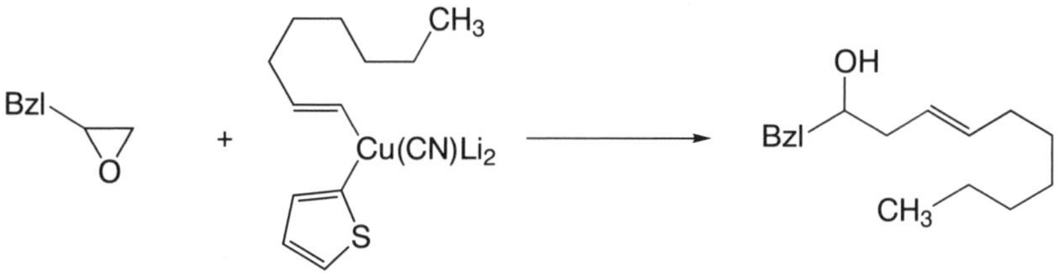

into an (E)-1,2-disubstituted alkene.

Vinylic lithiocyanocuprates.[12]   These cuprates (**1**) can be prepared by reaction of 1-alkynes with the Schwartz reagent and then with $(CH_3)_2Cu(CN)Li$ (equation I).

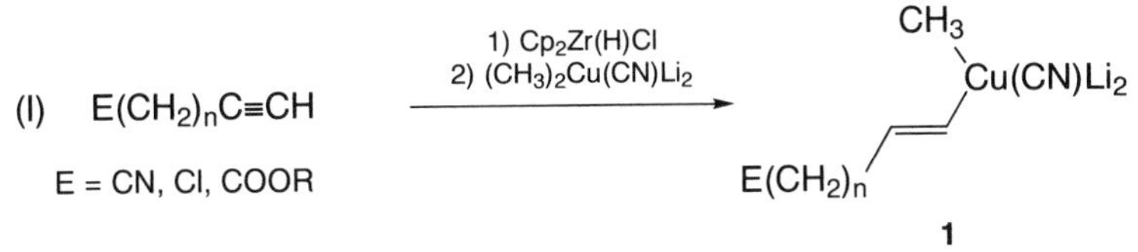

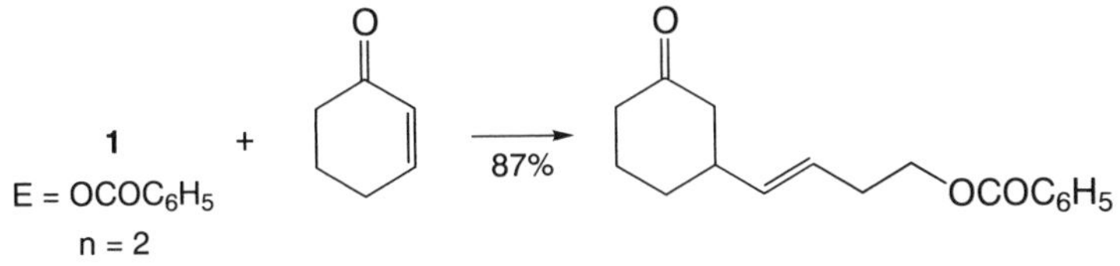

(I)   E(CH$_2$)$_n$C≡CH

    1) Cp$_2$Zr(H)Cl
    2) (CH$_3$)$_2$Cu(CN)Li$_2$

E = CN, Cl, COOR

CH$_3$

Cu(CN)Li$_2$

E(CH$_2$)$_n$

1

1   +

E = OCOC$_6$H$_5$
n = 2

87%

OCOC$_6$H$_5$

Biaryl synthesis.[13]   Diaryl higher-order cuprates can undergo oxidative coupling in
2-methyltetrahydrofuran to form unsymmetrical biaryls in good yields.

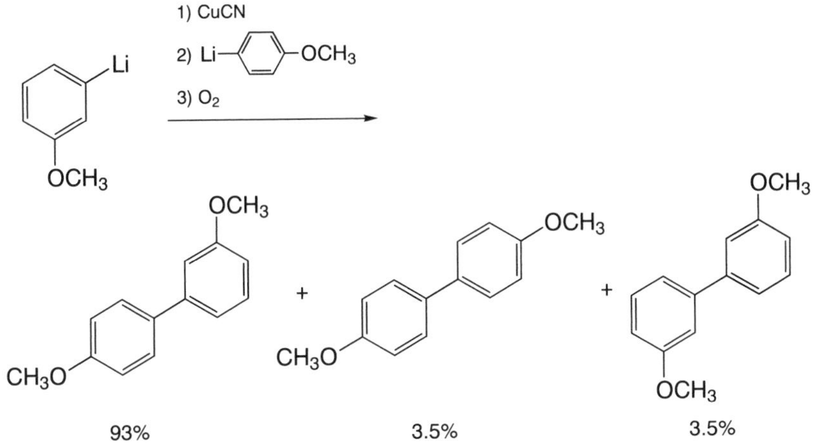

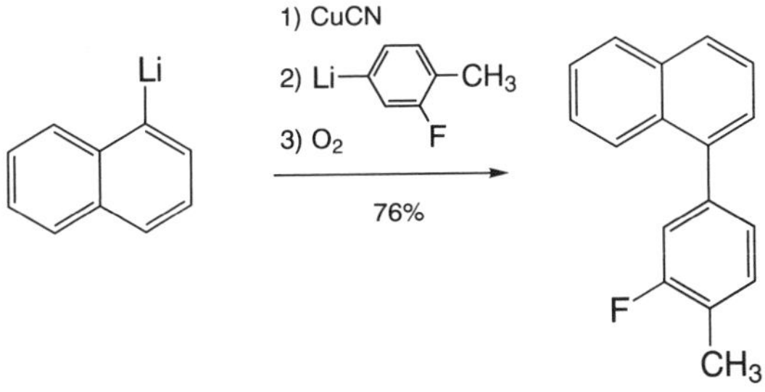

Li

    1) CuCN
    2) Li—⟨ ⟩—OCH$_3$
    3) O$_2$

OCH$_3$

OCH$_3$

CH$_3$O

93%

OCH$_3$

+

CH$_3$O

3.5%

OCH$_3$

+

OCH$_3$

OCH$_3$

3.5%

Li

    1) CuCN
    2) Li—⟨ ⟩—CH$_3$
          F
    3) O$_2$

76%

F

CH$_3$

α-Allenic alcohols.[14]   The reaction of RMgX/CuI catalyzed by BF_3 etherate with the cyclic carbonates or sulfites of benzyl ethers or propargylic alcohols proceeds in a S_N2' fashion to afford α-allenic alcohols in high *anti*-diastereoselectivity.

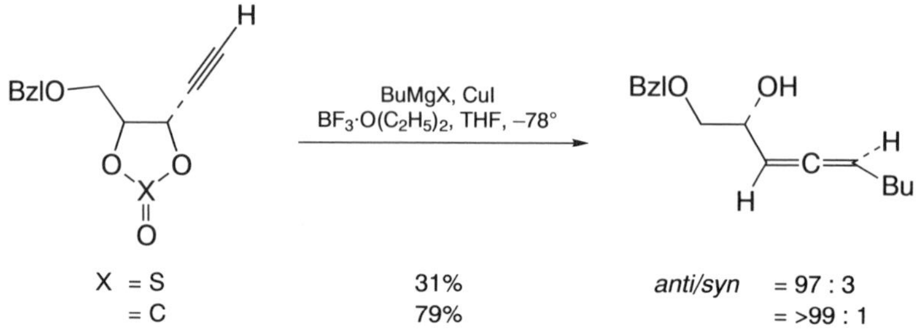

| X = S | 31% | anti/syn | = 97 : 3 |
| = C | 79% | | = >99 : 1 |

Conjugate addition to acyclic (E)- or (Z)-enones.[15]   Lithium dimethylcuprate reacts with both the (E)- and (Z)-steroidal 22-ene-24-ones (1) to form only the (22R)- methyl adduct (2). In contrast the same reaction but with added $ClSi(CH_3)_3$ and HMPA with (E)-1 provides the (22R)-methyl adduct, and with (Z)-1 provides the (22S)-methyl adduct.

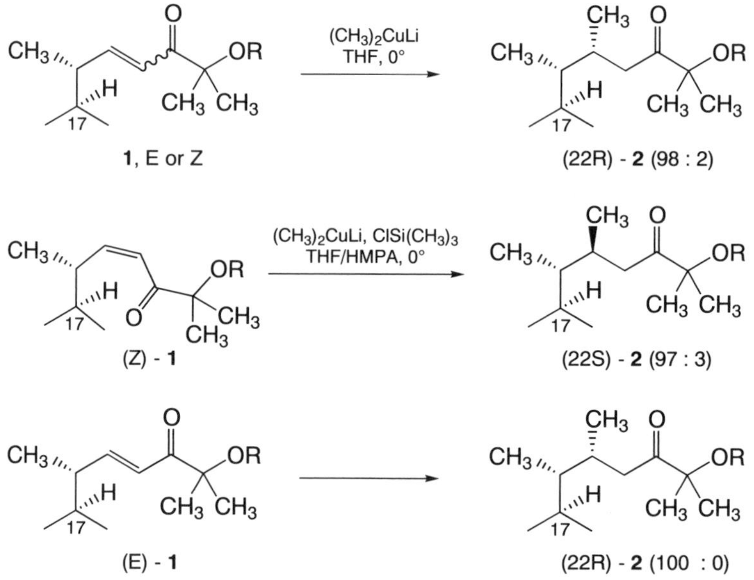

Conjugate addition of $(CH_3)_2CuLi$ to γ-alkoxy- and γ-ureido-α,β-enoates.[16] Lithium dimethylcuprate in combination with chlorotrimethylsilane adds to γ-alkoxy α,β-enoates with *anti*-selectivity. The highest *anti*-selectivity obtains with a benzyl-oxymethyl (BOM) ether such as 1. In contrast, when the γ-substituent is a ureido the cuprate addition proceeds with very high *syn* selectivity.

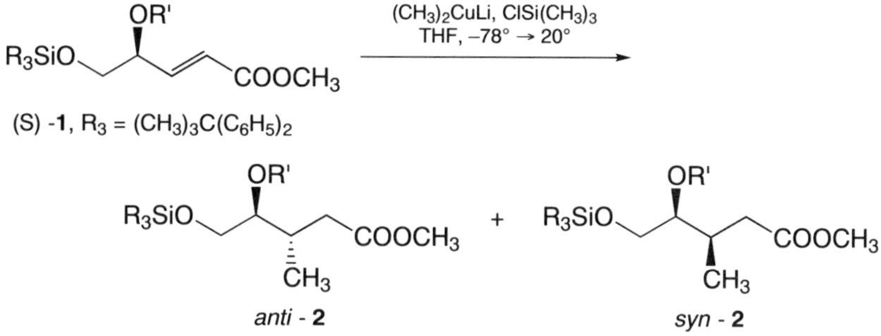

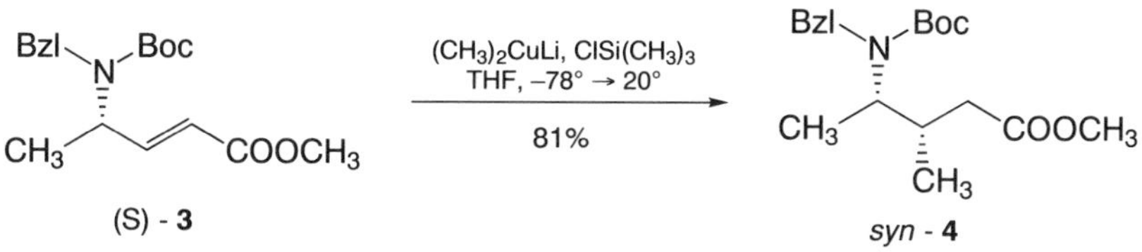

R' = BOM	87%	14 : 1
R = CH3	88%	9.9 : 1

Conjugate addition of BuCu.[17] BuCu activated by $ISi(CH_3)_3$ adds to the chiral crotonate **1** with excellent diastereoselectivity. In contrast, additions of lithium butylcuprates are highly dependent on the exact composition of the reagent and on the conditions, and

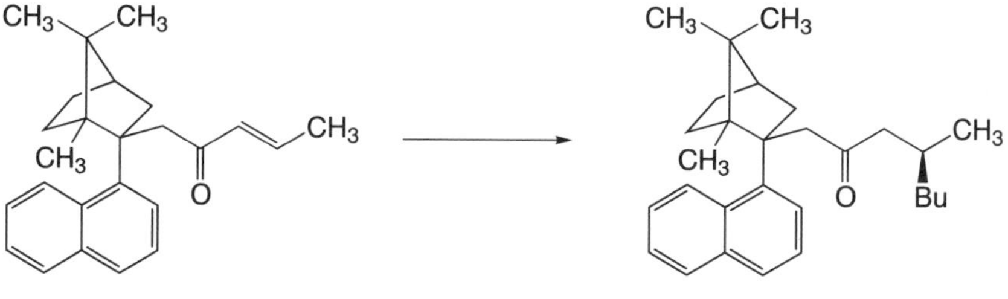

1 (S)-**2**

+ $BuCu{\cdot}ISi(CH_3)_3$	93%	98% de
$LiBu_2Cu$, 20°	95%	18% de

can result in (S)- or (R)-**2** as the major product.

Conjugate addition to γ-hydroxyenones.[18] A variety of cuprates undergo conjugate addition to γ-hydroxyenones such as **1** to give mainly the *anti*-adduct. The yields are highly dependent on the type of cuprate. In contrast, Grignard reagents generally

O

O

O

1

anti - 2   +   syn - 2

Li$_2$Bu$_2$CuCN	93%	93 : 7
Bu$_2$CuLi	22%	95 : 5
BuMgCl, DMPU	83%	0 : 100

O

O

O

2

syn   +   anti

2 Li$_2$(CH$_3$)$_2$CuCN	65%	10 : 90

react with the same γ-hydroxy enones to give 1,4-adduct in good yield and with almost complete *syn*-selectivity.

Conjugate addition of BzlCu to α,β-enoates.[19]   Although the usual benzylic copper reagents [BzlCu(CN)MgCl, Bzl$_2$CuMgCl] undergo ready conjugate addition to α,β-enones, they do not react with α,β-enoates, probably because of thermal instability. The preferred reagent for this reaction is BzlCu, prepared by reaction of BzlCl with Mg followed by transmetalation with CuI or CuBr · S(CH$_3$)$_2$. The best results are obtained by addition of ClSi(CH$_3$)$_3$ (5 equiv.) and TMEDA (1.1 equiv.) in THF at $-78°$ → $40°$.

CH$_3$O

C$_6$H$_5$

BzlMgCl, CuI
ClSi(CH$_3$)$_3$, TMEDA
―――――――――――→
86%

CH$_3$O

C$_6$H$_5$

Bzl

Enantioselective conjugate addition.[20]   The bidentate ligand **1**, prepared from L-proline, can effect enantioselective conjugate addition of lithium dimethylcuprate to chalcone at $-20°$ in 84% ee. When the acyl group of **1** is replaced by methyl, the adduct

$$\text{(pyrrolidine ring)}-\text{N}-\text{CH}_2-\text{P}(C_6H_5)_2 \quad \mathbf{1}$$
$$\overset{|}{\text{COC}(CH_3)_3}$$

$$C_6H_5\overset{\,}{\diagdown}\!\!\!\diagup\!\!\!\diagdown C_6H_5 \qquad \xrightarrow[\text{79\%}]{\substack{(CH_3)_2CuLi,\ \mathbf{1} \\ O(C_2H_5)_2,\ -20°}} \qquad C_6H_5\diagdown\overset{\text{S}}{\underset{CH_3}{\diagup}}\!\!\!\diagdown\overset{\,}{\underset{O}{\diagup}}C_6H_5$$

84% ee

is obtained in only 2% ee. The superiority of **1** is ascribed to the ability of the carbonyl group and the phosphorus atom to coordinate with the lithium and copper atoms of the cuprate.

Asymmetric conjugate addition of CH$_3$Li.[21]   The cuprate **2**, prepared by reaction of the chiral bornane **1**, (2R,2S,3S,4S)-3[(1-methylpyrrolyl)methylamino)-1,7,7-trimethylbicyclo[2.2.1]-heptanol-2, with CuI and then CH$_3$Li, undergoes 1,4-addition to (E)-cyclopentadecenone-2 (**3**) to form (R)-muscone **4** in 80% yield and 91% ee. If CH$_3$Li is replaced by CH$_3$MgCl, only the 1,2-adduct of **3** is formed.

$$\mathbf{2} \; + \; \text{(cyclopentadecenone)} \quad \xrightarrow[\text{80\%}]{\substack{CH_3Li \\ C_6H_5CH_3,\ -78°}} \quad \text{(R-muscone)}$$

3   (R) - (-) - **4**, 91% ee

[1] L. Capella, A. Degl'Innocenti, A. Mordini, G. Reginato, A. Ricci, and G. Seconi, *Synthesis,* 1201 (1991).

[2] B. H. Lipshutz, E. L. Ellsworth, S. H. Dimock, and D. C. Reuter, *Tetrahedron Letters,* **30,** 2065 (1989).

[3] R. J. Linderman and B. D. Griedel, *J. Org.,* **56,** 5491 (1991).

[4] D. E. Stack, B. T. Dawson, and R. D. Rieke, *Am. Soc.,* **114,** 5110 (1992).

[5] Y. Yamamoto, N. Asao, and T. Uyehara, *ibid.,* **114,** 5427 (1992).

[6] B. E. Rossiter, G. Miao, N. M. Swingle, M. Eguchi, A. E. Hernandez, and R. G. Patterson, *Tetrahedron Asymmetry,* **3,** 231 (1992).

[7] I. Marek, A. Alexakis, and J.-F. Normant, *Tetrahedron Letters,* **32,** 6337 (1991).

[8] J. P. Marino, M. V. M. Emonds, P. J. Stengel, A. R. M. Oliveira, F. Simonelli, and J. T. B. Ferriera, *ibid.,* **33,** 49 (1992).

[9] A. El Marini, M. L. Roumestant, P. Viallefont, D. Razafindrambora, M. Bonato, and M. Follet, *Synthesis,* 1104 (1992).

[10] G. Cainelli, D. Giacomini, M. Panunzio, and P. Zarantonello, *Tetrahedron Letters,* **33,** 7783 (1992).

[11] B. H. Lipshutz and K. Kato, *ibid.,* **32,** 5647 (1991).

[12] B. H. Lipshutz and R. Keil, *Am. Soc.,* **114,** 7919 (1992).

[13] B. H. Lipshutz, K. Siegmann, and E. Garcia, *ibid.,* **113,** 8161 (1991).

[14] S.-K. Kang, S.-G. Kim, and D.-G. Cho, *Tetrahedron Asymmetry,* **3,** 1509 (1992).

[15] K. Yamamoto, S. Yamada, and K. Yamaguchi, *Tetrahedron Letters,* **33,** 7521 (1992).

[16] S. Hanessian and K. Sumi, *Synthesis,* 1083 (1991).

[17] M. Bergdahl, M. Nilsson, T. Olsson, and K. Stern, *Tetrahedron,* **47,** 9691 (1991).

[18] K. A. Swiss, W. Hinkley, C. A. Maryanoff, and D. C. Liotta, *Synthesis,* 127 (1992).

[19] P. S. Van Heerden, B. C. B. Bezuidenhoudt, J. A. Steenkamp, and D. Ferreira, *Tetrahedron Letters,* **33,** 2383 (1992).

[20] M. Kanai, K. Koga, and K. Tomioka, *ibid.,* **33,** 7193 (1992).

[21] K. Tanaka, J. Matsui, H. Suzuki, and A. Watanabe, *J. C. S. Perkin I,* 1193 (1992).

Organocopper/magnesium reagents.

Reaction with chiral 1,3-dioxolan-4-ones.[1]   In contrast to RLi and R_2CuLi, which react with the carbonyl group of chiral 1,3-dioxolan-4-ones (**1**), prepared from (R)- or (S)-mandelic acid and an aldehyde, $RCu(MgBr_2)$ or $R_2CuMgBr$ react with the dioxolanone **1** to provide ethers of methyl mandelate (**2**) in 90–94% ee. Removal of the

chiral auxiliary is effected with $Pb(OAc)_4$ or O_2 + KO-*t*-Bu, to provide chiral secondary alcohols **3**.

[1] B. Heckmann, C. Mioskowski, J. Yu, and J. R. Falck, *Tetrahedron Letters,* **33,** 5201 (1992); B. Heckmann, C. Alayrac, C. Mioskowski, S. Chandrasckhar, and J. R. Falck, *ibid.,* **33,** 5205 (1992).

Organocopper/zinc reagents.

Heterobimetallic reagents.[1]   Reagents of this type can be prepared from primary 1,n-diiodoalkanes by reaction with zinc (2 equiv.) followed by transmetalation with $CuCN \cdot 2LiCl$ (1 equiv.). When separated by 4–6 CH_2 groups, these reagents undergo selective monocoupling with various electrophiles to give intermediates that can couple with a second electrophile to give polyfunctional products.

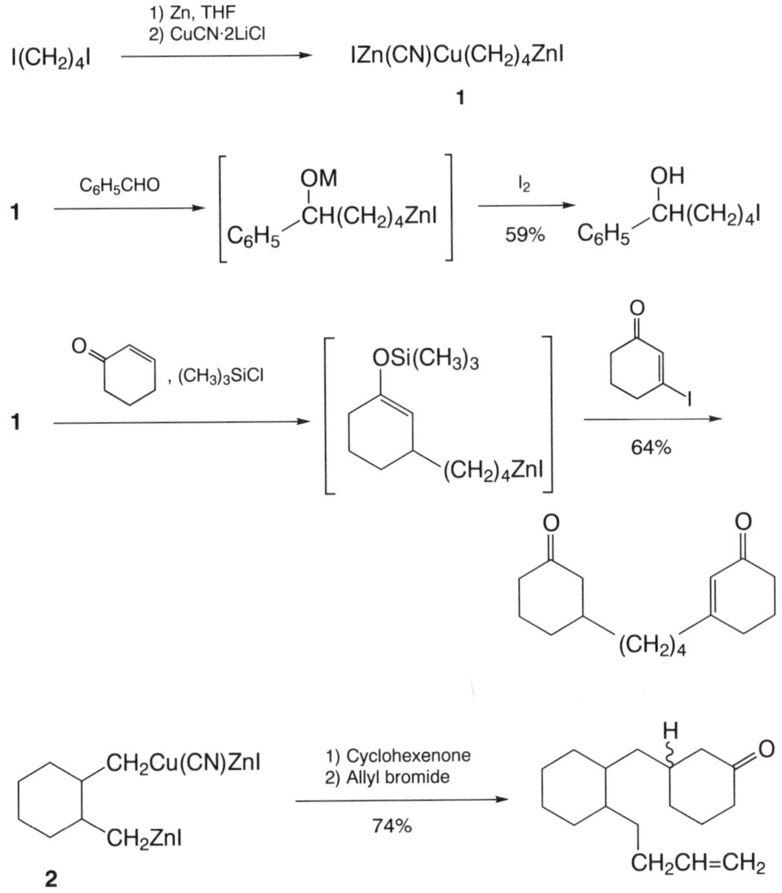

The unique selectivity of reagents **1** and **2** can be explained by the higher selectivity of the C—Cu bond compared to the C—Zn bond toward electrophiles. These mixed bimetallic reagents are useful for preparation of polyfunctional products.

(RCOCH=CH)Cu(CN)ZnI.[2]   Zinc reacts at 25° with β-iodo-α,β-enones or -enoates to form a moderately stable organozinc iodide reagent. These zinc reagents in the presence of a Pd(0) catalyst can couple with alkenyl and aryl iodides. They can also be converted to moderately stable alkenylcopper/zinc reagents by reaction with CuCN · 2LiCl in THF/S(CH$_3$)$_2$ at $-48°$. These copper/zinc reagents are less reactive than alkyl copper/zinc reagents. They do not react with aldehydes, acid chlorides, or alkyl halides, but they do react with (CH$_3$)$_3$SnCl, 1-iodoalkynes, and allylic halides.

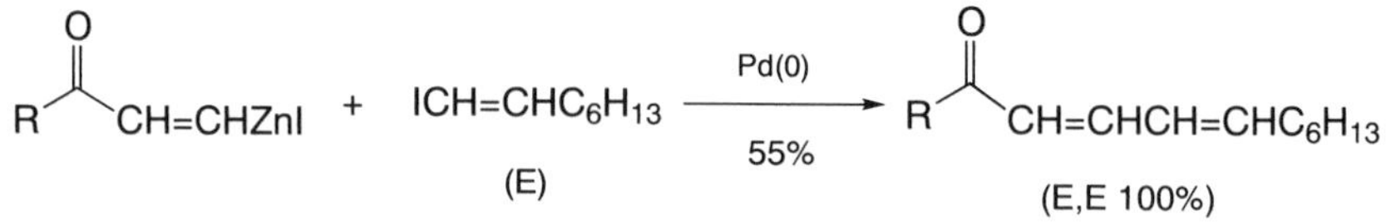

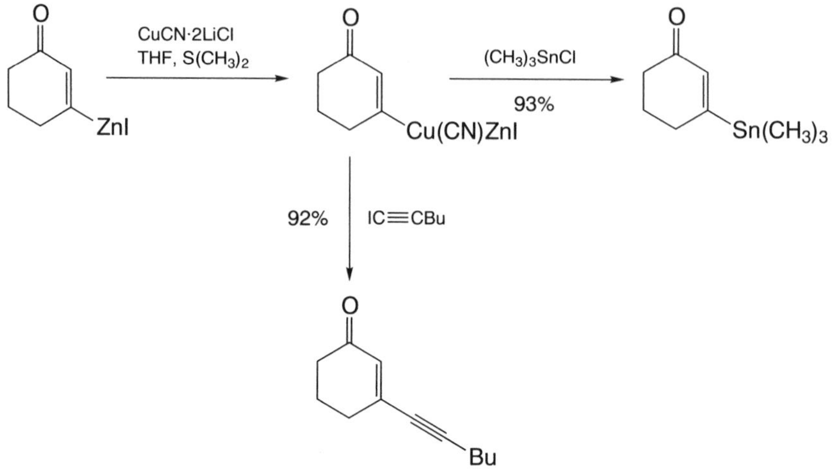

R CH=CHZnI + ICH=CHC6H13
(E)
Pd(0)
55%
R CH=CHCH=CHC6H13
(E,E 100%)

ZnI
CuCN·2LiCl
THF, S(CH3)2
Cu(CN)ZnI
(CH3)3SnCl
93%
Sn(CH3)3
92%
IC≡CBu
Bu

Propargylic copper/zinc reagents.[3] Reaction of alkynylcoppers with iodomethylzinc iodide (Simmons–Smith reagent) provides propargylic copper/zinc reagents, which react with aldehydes or ketones to form homopropargylic alcohols in 80–95% yield.

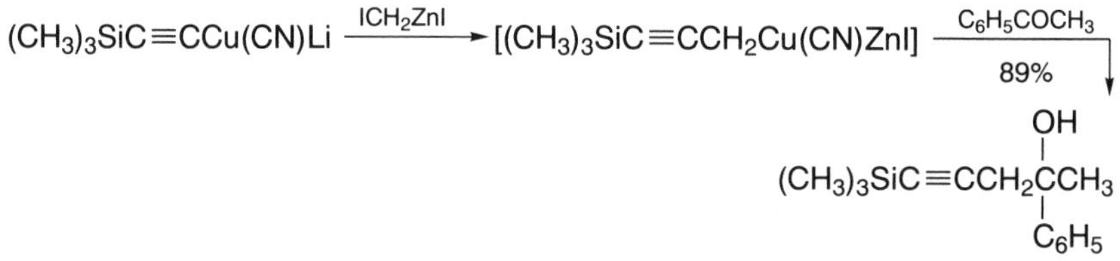

(CH3)3SiC≡CCu(CN)Li
ICH2ZnI
[(CH3)3SiC≡CCH2Cu(CN)ZnI]
C6H5COCH3
89%
OH
(CH3)3SiC≡CCH2CCH3
C6H5

Reaction of alkynylcoppers with 5 equiv. of iodomethylzinc iodide results in insertion of four methylene groups to provide organocopper/zinc reagents of type **2**, which can be trapped by allylic halides (equation I).

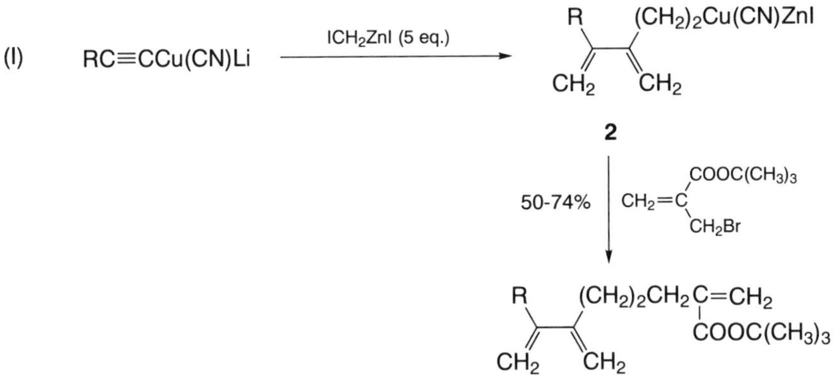
(I)
RC≡CCu(CN)Li
ICH2ZnI (5 eq.)
R
(CH2)2Cu(CN)ZnI
CH2
CH2
2
50-74%
COOC(CH3)3
CH2=C
CH2Br
R
(CH2)2CH2C=CH2
COOC(CH3)3
CH2
CH2

Reagents with amino or amido groups. Knochel *et al.*[4] report that reagents of this type can be prepared with only slight deprotonation of groups bearing acidic hydrogens such as primary or secondary amines or amides, or 1-alkynes (equation I).

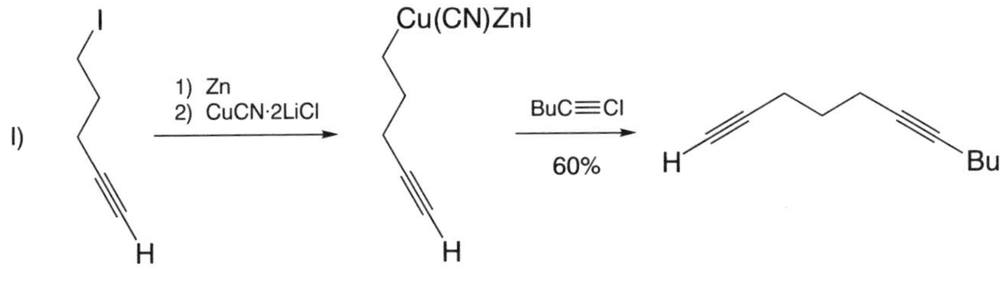
I)
I
H
1) Zn
2) CuCN·2LiCl
Cu(CN)ZnI
H
BuC≡CI
60%
H
Bu

Addition to alkynes.[5] The polyfunctional zinc/copper reagents such as **1**, prepared by reaction of organozinc halides with $(CH_3)_2Cu(CN)Li_2$, reacts slowly with activated alkynes to provide *syn*-adducts, which can react with various electrophiles to provide alkenes (equation I). An intramolecular version can result in alkylidenecyclopentanes (equation II). In contrast, attempts to use this intramolecular cyclization to obtain

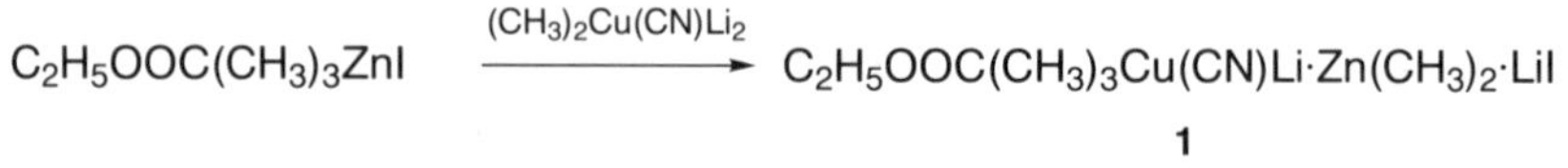

$$C_2H_5OOC(CH_3)_3ZnI \xrightarrow{(CH_3)_2Cu(CN)Li_2} C_2H_5OOC(CH_3)_3Cu(CN)Li\cdot Zn(CH_3)_2\cdot LiI$$

1

(I) 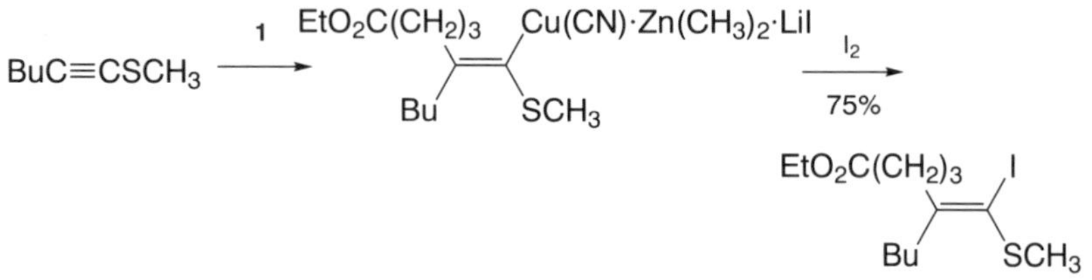 $BuC{\equiv}CSCH_3 \xrightarrow{\textbf{1}}$ $EtO_2C(CH_2)_3$ $Cu(CN)\cdot Zn(CH_3)_2\cdot LiI$ ... Bu ... SCH_3 $\xrightarrow[75\%]{I_2}$

$EtO_2C(CH_2)_3$ ... I ... Bu ... SCH_3

(II)

1) 25°C
2) $\overset{COOC_2H_5}{CH_2{=}C{-}CH_2Br}$

57%

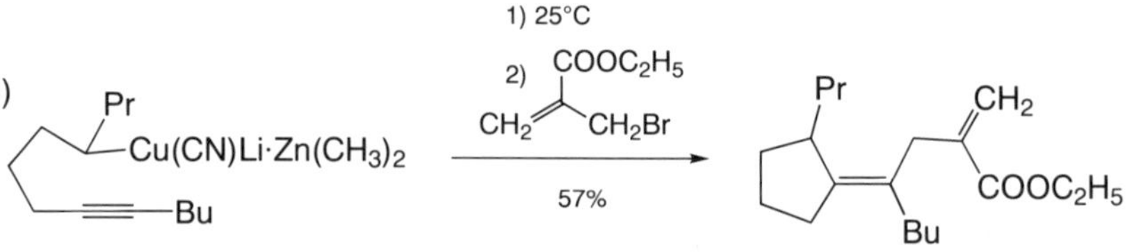

four- or six-membered rings were not successful.

3,4-Disubstituted cyclobutene-1,2-diones.[6] These products can be prepared by reaction of 3,4-dichlorocyclobutene-1,2-diones with functionalized organocopper/zinc reagents.

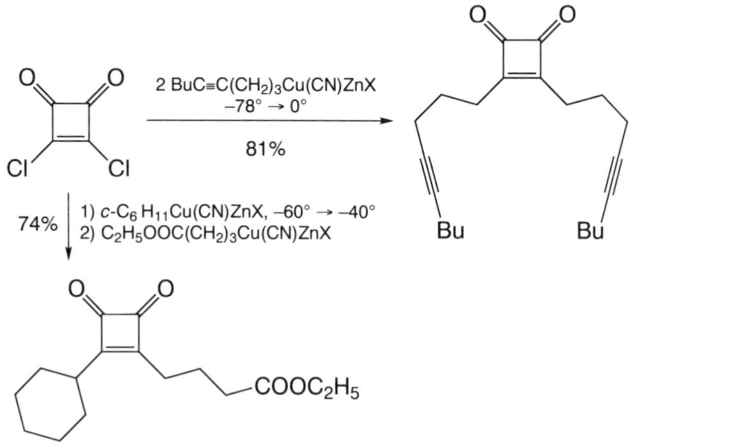

2 $BuC{\equiv}C(CH_2)_3Cu(CN)ZnX$
−78° → 0°

81%

74% │ 1) $c\text{-}C_6H_{11}Cu(CN)ZnX$, −60° → −40°
     │ 2) $C_2H_5OOC(CH_2)_3Cu(CN)ZnX$

[1] S. A. Rao and P. Knochel, *J. Org.*, **56**, 4591 (1991).
[2] C. J. Rao and P. Knochel, *ibid.*, **56**, 4593 (1991).
[3] M. J. Rozema and P. Knochel, *Tetrahedron Letters*, **32**, 1855 (1991).
[4] H. P. Knoess, M. T. Furlong, M. J. Rozema, and P. Knochel, *J. Org.*, **56**, 5974 (1991).

[5] S. A. Rao and P. Knochel, *Am. Soc.*, **113**, 5735 (1991).
[6] A. Sidduri, N. Budries, R. M. Laine, and P. Knochel, *Tetrahedron Letters*, **33**, 7515 (1992).

Organolead compounds.
Stereoselective aldol reactions with α-methoxy organolead reactions (cf. 16,242). The α-methoxy lead reagent **1**[1] reacts with 2-phenylpropanal in the presence of $TiCl_4$ with high diastereoselectivity at three contiguous centers (equation I). Only two of the four possible aldols are formed and these two are formed in the ratio 95:5. The tin reagent corresponding to **1** shows similar diastereoselectivity, but the total yield is only 19%.

2, *syn, syn*   **3**, *anti, syn*

The reaction of (S)-**1** with the steroidal aldehyde **2** catalyzed by $TiCl_4$ gives the *syn*-product **3** as a single steroisomer with control at three contiguous centers,

(C_{20}, C_{22}, and C_{23}).[2] The side chain of **3** is characteristic of brassinosterols (plant growth hormones).

[1] T. Furata and Y. Yamamoto, *J. C. S. Chem. Comm.*, 863 (1992).
[2] *Idem, J. Org.*, **57**, 2981 (1992).

Organomanganese reagents
Pentacarbonyl(trialkylsilyl)manganese (15,235). These reagents are useful for synthesis of [4,5]- and [5,5]spiroketals from tetrahydrofuran. Thus cleavage of THF with $TBDMS-Mn(CO)_5$ affords the manganese complex **1**, which undergoes insertion with methyl acrylate to form a manganacycle **2** in 71% overall yield. Photodemetalation followed by treatment with camphorsulfonic acid (CSA) results in a spirolactone (**3**). This

lactone was converted into the pheromones (**5**) of the common wasp by methylenation and hydrogenation.

1

2   **3**

4   **5** (4:1)
66% from **3**

The enol ether **4** was also converted into the pheromone (**7**) of the olive fruit fly by reaction with dimethyldioxirane to provide a very unstable epoxide that rearranges to the ketone **6** on treatment with CSA. Reduction of **6** with NaBH$_4$/CeCl$_3$ provides the pheromone **7** as a 13:1 mixture of epimers.

4   + epimer
78% from **3**   **6**   **7**   13:1

This route to spiroketals may be useful for synthesis for more complicated systems since preliminary results indicate that 2-methyltetrahydrofuran is cleaved by TBDMS–Mn(CO)$_5$ with about 10:1 regioselectivity and insertion of methyl methacrylate and methyl crotonate with complex **1** proceeds with complete regioselectivity.

Selective addition to carbonyls.[2]   The alkyl- or arylmanganese pivalates RMn-OCOC(CH$_3$)$_3$ (**1**), prepared by reaction of organolithiums with manganese pivalate, add to 4-*t*-butylcyclohexanone to give mainly the axial alcohol (equation I). The stereoselectivity is comparable to that obtained with $CH_3Ti(OC_6H_5)_3$, but higher than that shown by similar manganese, cerium, or magnesium reagents.

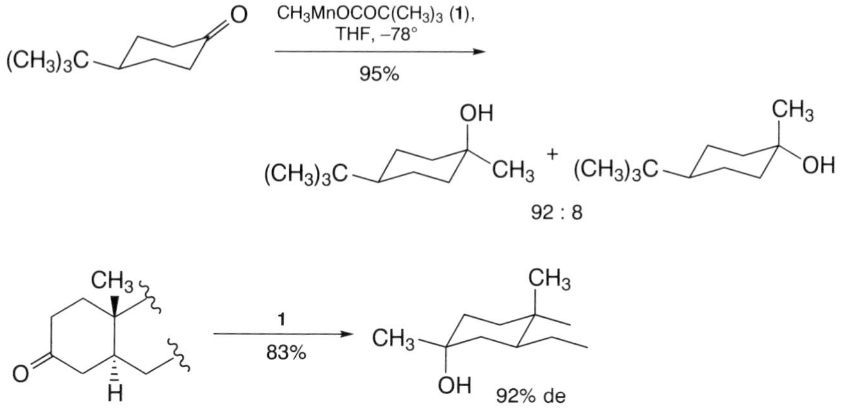

The organomanganese reagents are comparable to ceriummagnesium ate complexes in almost complete addition to aldehydes in the presence of ketones.

Acylfurans.[3]   2- and 3-Acylfurans can be prepared in excellent yield by reaction of furoyl chlorides with RMnCl catalyzed by CuCl. This reaction can provide the natural furanyl ketones naginata ketone (**1**) and perilla ketone (**2**).

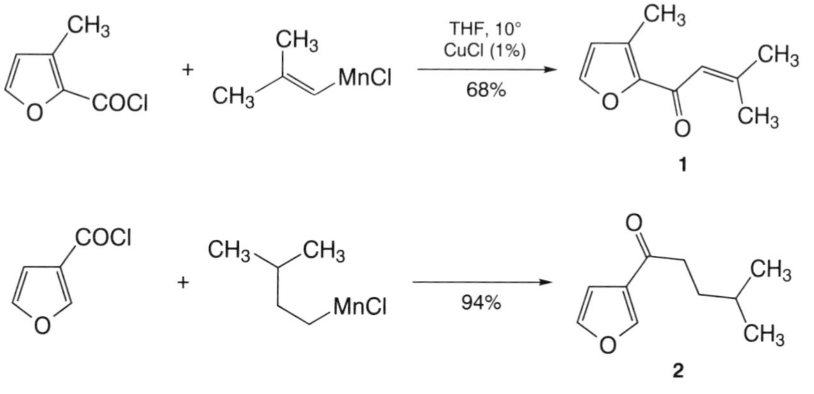

Another route to these furanyl ketones involves acylation of furylmanganese chloride catalyzed by iron(III) acetylacetonate (equation I).

MnCl₄Li₂ (1)   This soluble ate complex $MnCl_4Li_2$ (1) is prepared by mixing $MnCl_2$ and 2LiCl in THF at 20°.[4] In the presence of the complex, Grignard reagents react with acid chlorides to form ketones in high yield.

[1] P. DeShong and P. J. Rybcezynski, *J. Org.,* **56**, 3207 (1991).
[2] M. T. Reetz, H. Haning, and S. Stanchev, *Tetrahedron Letters,* **33**, 6963 (1992).
[3] G. Cahiez, P.-Y. Chavant, and E. Metais, *ibid.,* **33**, 5245 (1992).
[4] G. Cahiez and B. Laboue, *ibid.,* **33**, 4439 (1992).

Organosamarium(III) reagents.[1]

Reaction of a primary or secondary alkyl halide with SmI_2 results in an alkylsamarium(III) reagent, which is moderately stable and which reacts with various electrophiles. These reagents are undoubtedly intermediates in the Barbier reaction of alkyl halides with ketones initiated by SmI_2, and in fact higher yields are often possible if the Barbier reaction is conducted by reaction of the alkyl halide first with SmI_2 and then with the carbonyl compound.

[1] D. P. Curran and M. J. Totleben, *Am. Soc.,* **114**, 6050 (1992).

Organosilver reagents.

Cyano(methyl)argentates.[1]   A number of these silver complexes have been prepared by reaction of CH_3Li or CH_3MgBr with AgCN including: $CH_3Ag(CN)Li$, $(CH_3)_2Ag(CN)Li_2$, $(CH_3)_2Ag(CN)(MgBr)_2$. In comparison to similar reagents prepared from AgBr, these cyano(methyl)argentates are less sensitive to light. All these cyano(methyl)argentates react more readily with aldehydes than with ketones, with $(CH_3)_2Ag(CN)(MgBr)_2$ showing the highest selectivity. In addition, β-hydroxy ketones and β-amino ketones are methylated preferentially because of chelation. These reagents also react with complete selectivity with the epoxide group of epoxy ketones. Styrene oxide is methylated preferentially at the position α- to the phenyl group.

$$C_6H_5-\overset{O}{\triangle} \xrightarrow[\text{ether}]{(CH_3)_2Ag(CN)Li_2}$$

$$\underset{(65\%)}{C_6H_5\overset{CH_3}{\diagup}\text{OH}} + \underset{(13\%)}{C_6H_5\overset{OH}{\diagup}CH_3}$$

$$C_6H_5-\overset{O}{\underset{CH_3}{\triangle}} \xrightarrow{81\%} \underset{C_6H_5}{\overset{CH_3\ OH}{\diagup}CH_3}$$

[1] T. Kauffmann, C. Neiteler, and S. Robbe, *Ber.*, **125**, 2409 (1992).

Organothallium reagents.

Trialkylthallium reagents have received little attention from organic chemists, perhaps because they are sensitive to air, water, and light, and are toxic as all thallium compounds are. However, they are readily available by reaction of alkyllithiums or Grignard reagents with stable and unreactive dialkylthallium halides. These R_3Tl reagents react rapidly at room temperature with acid halides to form ketones with regeneration of the dialkylthallium halide. Even in the presence of excess R_3Tl, no products of addition

$$\underset{C_9H_{19}}{\overset{O}{\parallel}}Cl \;+\; (CH_3)_3Tl \xrightarrow{25°} \underset{C_9H_{19}\quad CH_3}{\overset{O}{\parallel}} \;+\; (CH_3)_2TlCl$$

$$\qquad\qquad\qquad\qquad\qquad\qquad 85\% \qquad\quad 95\%$$

to the carbonyl group are formed. The mixed $R_3Tl(III)$ reagent, dimethyl-(phenylacetenyl)thallium(III), prepared as shown, reacts with acid halides with selective transfer of the acetylenic group.[1]

$$C_6H_5C\equiv CH \xrightarrow{(CH_3)_3Tl} \left[(CH_3)_2Tl-C\equiv CC_6H_5\right] \xrightarrow[77\%]{C_9H_{19}COCl} \underset{C_9H_{19}\quad C\equiv CC_6H_5}{\overset{O}{\parallel}}$$

These R_3Tl reagents also transfer alkyl groups to reactive secondary or tertiary alkyl halides with formation of R_2TlX.[2]

$$(C_6H_5)_3CCl \;+\; (C_6H_5)_3Tl \xrightarrow{80\%} (C_6H_5)_4C \;+\; (C_6H_5)_2TlCl$$

The dialkylthallium chlorides can also be used as catalysts for condensation of alkyl halides with alkyllithiums as the stoichiometric reagent.[3]

$$(C_6H_5)_3CCl \ + \ C_6H_5C \equiv CLi \ \xrightarrow[82\%]{(CH_3)_2TlCl \ (cat.)} \ (C_6H_5)_3CC \equiv CC_6H_5$$

[1] I. E. Markó and J. M. Southern, *J. Org.*, **55**, 3368 (1990).
[2] I. E. Markó, J. M. Southern, and M. L. Kantam, *Synlett.*, 235 (1991).
[3] I. E. Markó and M. L. Kantam, *Tetrahedron Letters*, **32**, 2255 (1991).

Organozinc reagents.

Reactions catalyzed by transition metals.[1] Erdik has reviewed reactions of organozinc reagents that are catalyzed by various Ni(0), Ni(II), Pd(0), Pd(II), and Cu(I) complexes (144 references). Pd and Ni are particularly effective for coupling of allyl-, benzyl-, propargyl-, and arylzinc reagents. Nickel catalysts can promote conjugate addition of organozinc reagents to enones. The report includes a list of more than 35 examples of transition metal catalyzed reactions of Reformatsky reagents.

(RCH₂)Zn from RCH₂MgX.[2] Dialkylzincs are pyrophoric and moisture-sensitive. Only diethylzinc is readily available from commerical sources, but it is expensive. A new route to dialkylzincs involves reaction of RCH_2MgX with $ZnCl_2$ in ether to form $(RCH_2)_2Zn$ and MgX_2. The latter product is removed by precipitation with dioxane. The

$$2 \ RCH_2MgX \ + \ ZnCl_2 \ \underset{ether}{\overset{}{\rightleftharpoons}} \ (RCH_2)_2Zn \ + \ MgX_2$$

dialkylzincs obtained in this way add to C_6H_5CHO when catalyzed by Seebach's spirotitanate (**1**), derived from (R,R)-tartaric acid, to give alcohols in >95% ee.

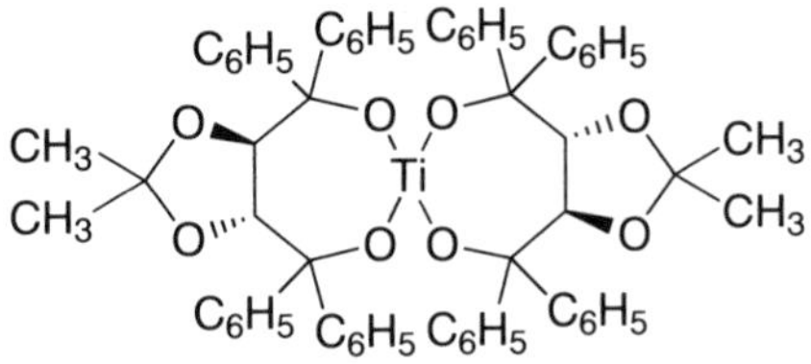

$$C_6H_5CHO \quad + \quad (RCH_2)_2Zn \quad \xrightarrow[\text{40 - 83\%}]{\substack{\textbf{1}, \text{Ti(O-}i\text{-Pr)}_4 \\ 30^\circ}} \quad \underset{>95\% \text{ ee}}{C_6H_5 \overset{OH}{\wedge} CH_2R}$$

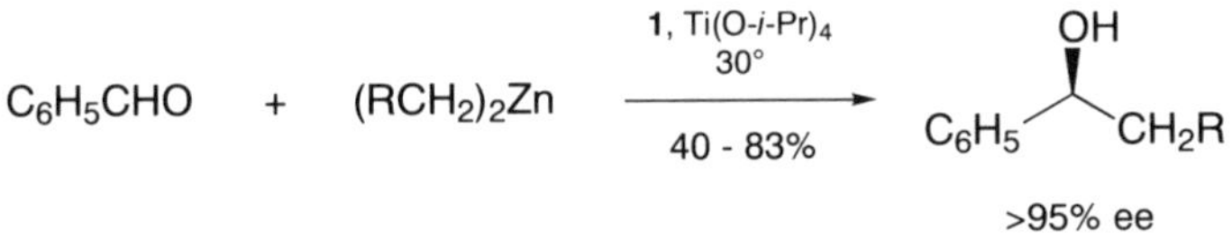

Allylethylzinc.[3]   A reagent (**2**) of this type is prepared by hydroboration of 1-hexyne with dicyclohexylborane (1 equiv.) to form an (E)-1-alkenylborane **1**, which undergoes transmetalation to form **2** on treatment with diethylzinc. This reagent adds to an aldehyde in the presence of the catalyst $(-)$-3-*exo*-(dimethylamino)isoborneol (DAIB, **3**) to form a secondary (E)-allylic alcohol in 70–95% yield with 79–98% ee.

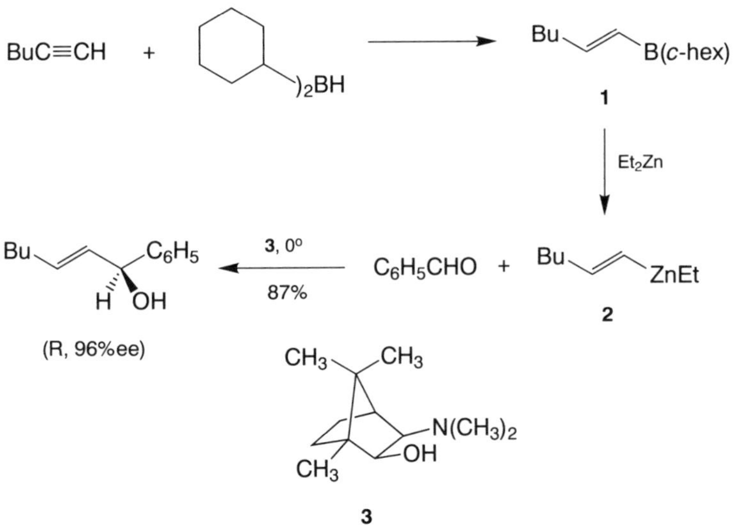

Dimethylzinc can replace diethylzinc. Aliphatic straight-chain or α- or β-branched aldehydes react with **2** in 80–91% ee, but pivaldehyde reacts in low yield and 73% ee.

Allylic dimetallic zinc reagents.[4]   A typical reagent (**1**) of this type can be prepared *in situ* as shown in equation (I). This reagent can undergo attack with two

(I) $(CH_3)_3Si-C\equiv C-CH_2OR$

$\xrightarrow[\text{2) } 2CH_2=CHCH_2ZnBr]{\text{1) BuLi}}$

(diagram of product)

1a, R = CH_3
1b, R = t-Bu

R = CH_3, t-Bu

electrophiles, but the first reaction always occurs on the carbon α to oxygen, while the second is on the carbon α to the silyl group. In all cases, the reaction is regio- and stereoselective. A wide variety of electrophiles can be used in the first step: carbonyl

1b $\xrightarrow{C_6H_5COCl}$ (intermediate **a**)

$\xleftarrow[78\%]{H_3O^+}$ $\xrightarrow[78\%]{I_2}$

2   **3** (97:3)

compounds, alkyl bromides. The second electrophile can be H_3O^+ or I_2 or even aryl or vinyl iodides with a Pd(0) catalyst. The t-butoxy group can be converted to an acetoxy group by treatment with $Ac_2O/FeCl_3$.

Dialkenylzinc reagents, $(RCH=CH_4)_2Zn$.[5]   In the presence of $Pd[P(C_6H_5)_3]_4$ (3 mol %), dialkenylzincs or an alkenylzinc bromide react with cyclic α-iodo- or α-triflyloxyenones to give α-alkenylenones. These products are reduced by $LiAlH(OCH_3)_3$ and CuBr to the corresponding α-alkenylketones. Similar results can be obtained with alkenylaluminum reagents.

(reaction diagram)

$+ \left(Bu\diagup\!\!\!\!\diagdown\right)_2 Zn$ $\xrightarrow[86\%]{\substack{Pd(0) \\ DMF}}$

Substituted dialkylzinc reagents.[6] Reagents of this type can be obtained by reaction of functional alkyl iodides with $(C_2H_5)_2Zn$. These zinc reagents react with various electrophiles in the presence of CuCN · 2LiCl.

Cl(CH2)4I + (C2H5)2Zn
45 – 55°
Cl(CH2)4ZnC2H5
Δ, 0.1 mm Hg
[Cl(CH2)4]2Zn
[Cl(CH2)4]2Zn
1
O
I
CuCN·2LiCl
86%
(CH2)4Cl

The reaction of these dialkylzincs with aldehydes in the presence of the catalyst **2** of Yoshioka and Ohno (**16**,103), derived from (1R,2R)-cyclohexanediamine, proceeds in 86–97% ee.

[Cl(CH2)4]2Zn
1
CHO
CH2
CH3
2, 50°
95%
HO
(CH2)4Cl
97% ee

SO2CF3
N
Ti[OCH(CH3)2]2
N
SO2CF3

2

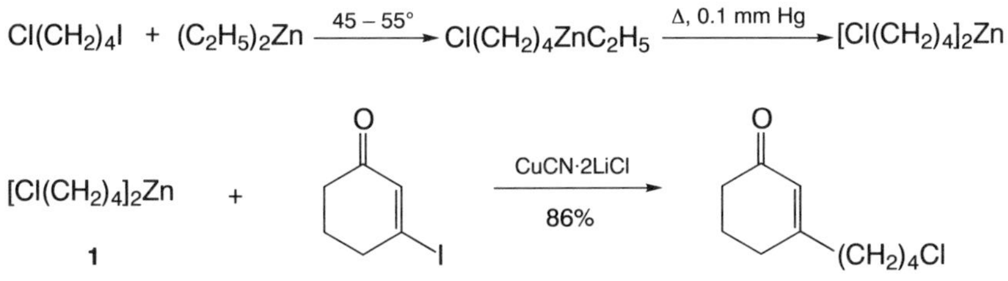

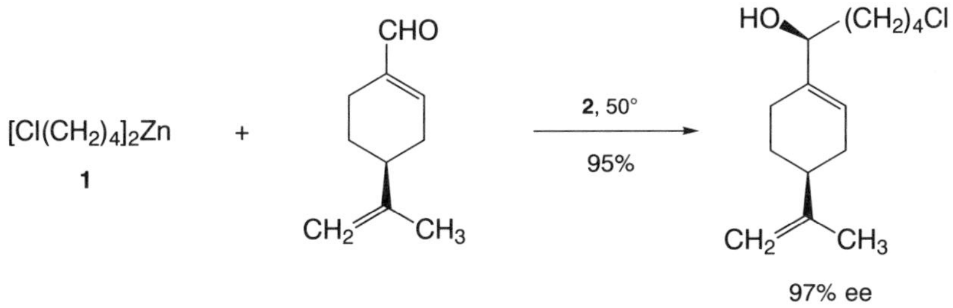

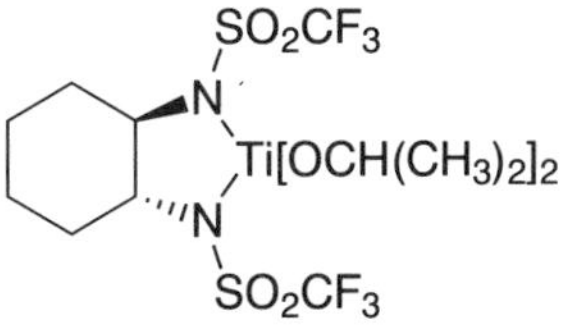

α-(Phenylthio)alkenylzinc reagents,[7] $RCH{=}C(ZnBr)(SC_6H_5)$ Reagents of this type can be obtained by reaction of an alkynyllithium **1** with phenyl benzenethiosulfonate (1 equiv.) at $-10 \rightarrow 20°$ to form a phenylthioalkyne, which on Pd(0)-catalyzed *cis-*

hydrostannylation affords an α-(phenylthio)alkenylstannane (**2**). This tin reagent can undergo Stille coupling, but the corresponding zinc reagent **3** obtained by transmetalation is more reliable.

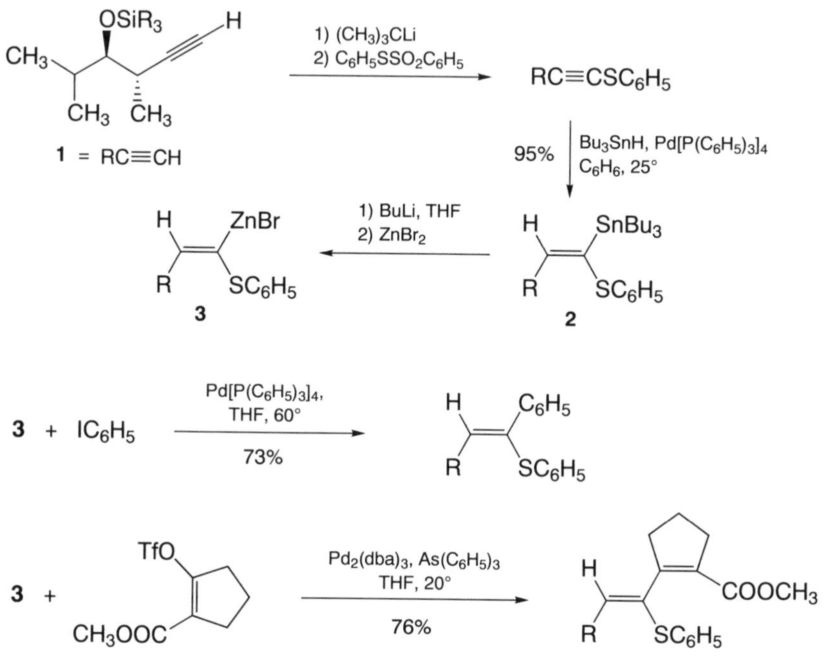

Substituted aryl- and alkenylzinc reagents.[8]   These are difficult to prepare directly by zinc insertion, but the alkenyl- and aryllithiums are readily obtained by halogen–lithium exchange. Reaction of these with ZnI_2 or $CuCN \cdot 2LiCl$ in THF at $-100°$ provide stable solutions of the organozinc or organocopper reagents or Cu/Zn organobimetallics.

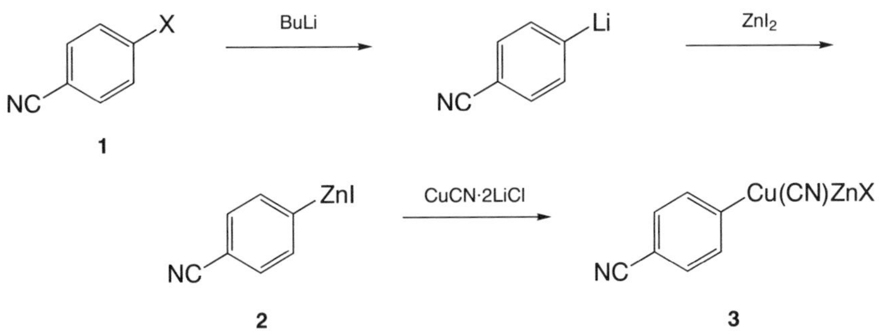

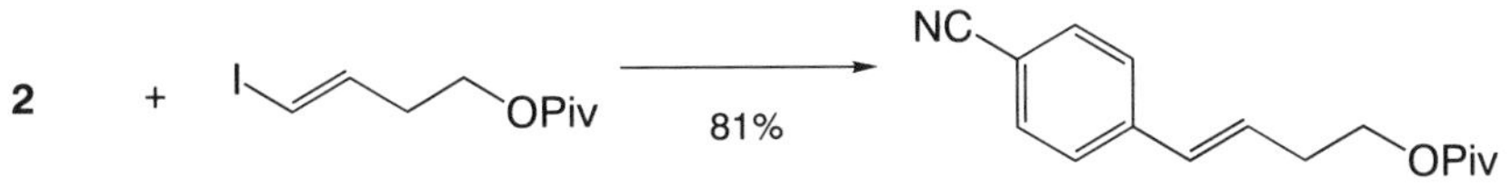

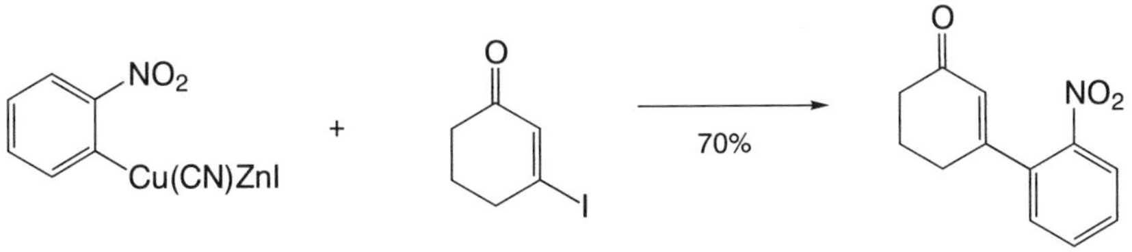

Transmetalation of a vinyl iodide to magnesium is also possible (equation I).

(I)

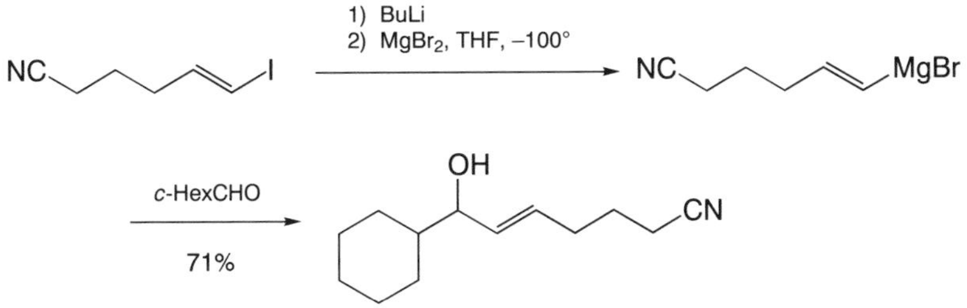

Organozinc carbenoids, $\underset{R^2}{\overset{R^1}{\diagdown}}C{=}Zn.$[9]   These reagents can be generated by reaction of carbonyl compounds with 1,2-bis(chlorodimethylsilyl)ethane and amalgamated zinc in ether. These can be trapped by alkenes with formation of cyclopropanes.

$$C_6H_5CHO + \underset{\substack{CH_3 \quad CH_3 \\ ClSi \\ | \\ ClSi \\ CH_3 \quad CH_3}}{} + Zn \xrightarrow{ether} \left[\underset{H}{\overset{C_6H_5}{\diagdown}} =Zn \right]$$

68%

(cis)   4:1   (trans)

1) [ClSi(CH₃)₂CH₂]₂Zn
2) C₆H₅CH=CH₂

$\xrightarrow{59\%}$

(cis/trans = 11:1)

[1] E. Erdik, *Tetrahedron,* **48,** 9577 (1992).
[2] D. Seebach, L. Behrendt, and D. Felix, *Angew. Chem. Int. Ed.,* **30,** 1008 (1991).
[3] W. Oppolzer and R. N. Radinov, *Helv.,* **75,** 170 (1992).
[4] L. Labaudinière, J. Hanaizi, and J.-F. Normant, *J. Org.,* **57,** 6903 (1992).
[5] E. Negishi, Z. R. Owczarczyk, and D. R. Swanson, *Tetrahedron Letters,* **32,** 4453 (1991).
[6] M. J. Rozema, A. Sidduri, and P. Knochel, *J. Org.,* **57,** 1956 (1992).
[7] A. Pimm, P. Kocieński, and S. D. A. Street, *Synlett,* 886 (1992).
[8] C. E. Tucker, T. N. Majid, and P. Knochel, *Am. Soc.,* **114,** 3983 (1992).
[9] W. B. Motherwell and L. R. Roberts, *J. C. S. Chem. Comm.,* 1582 (1992).

Organozinc/zirconium reagents.

Reagents of this type are readily available by hydrozirconation of alkenyl- or alkynylzinc halides with Schwartz's reagent, H(Cl)ZrCp₂, in CH₂Cl₂ at 25° (equations I and II).

(I)   RCH=CHZnBr $\xrightarrow[\text{CH}_2\text{Cl}_2]{\text{H(Cl)ZrCp}_2}$ $RCH_2 - C \underset{Cp_2}{\overset{Br}{\underset{Zr}{\overset{Zn}{\diagup}}}} Cl$

1

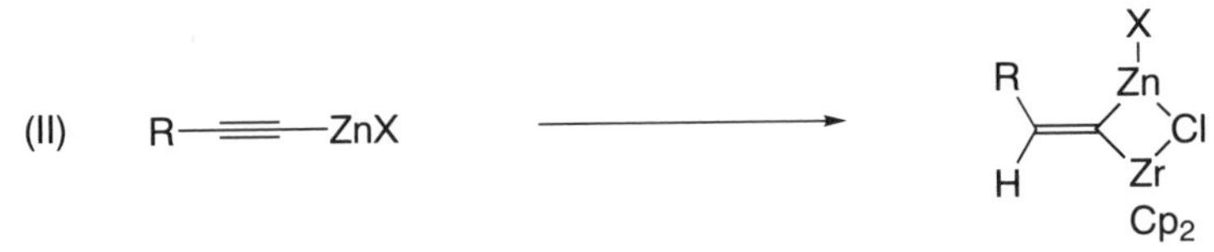

(II)   R————ZnX   ⟶

2

Reagents of the type **1** or **2** are unstable at 25°, but they can be trapped as formed by aldehydes or ketones to give alkenes.[1] The reaction of aldehydes with reagents of type **1** is useful because this olefination provides (E)-alkenes with high selectivity (94–100:6–0) and in fair to good yield (49–89%). A similar reaction with a ketone shows only slight

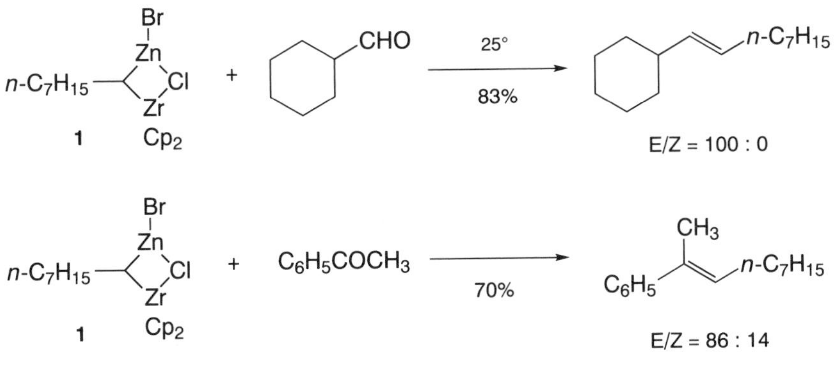

regioselectivity. The reagent does not react with esters. Reagents of type **2** react with aldehydes to form allenes in satisfactory yield (>70%).

[1] C. E. Tucker and P. Knochel, *Am. Soc.*, **113**, 9888 (1991).

Osmium(III) chloride, OsCl₃.

Oxidation of β-lactams.[1]   In the presence of OsCl₃ (catalytic), β-lactams are oxidized by a peracid in acetic acid to give 4-acetoxy-β-lactams. Highest yields obtain with peracetic acid, but C_6H_5IO and $C_6H_5I(OAc)_2$ can be used. This oxidation is

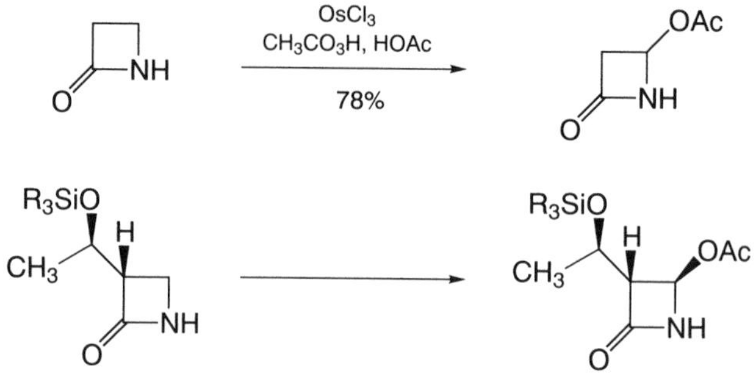

1   **2** (>99% de)

particularly useful for conversion of **1** into the acetate **2**, a key intermediate in the synthesis of carbapenem antibiotics. Presumably an oxoosmium(V) complex is involved.

[1] S. Murahashi, T. Saito, T. Naota, H. Kumobayashi, and S. Akutagawa, *Tetrahedron Letters*, **32**, 2145 (1991).

Osmium tetroxide

Asymmetric catalytic osmylation (**14**, 237–239; **15**, 240–241; **16**, 249).   In the early versions of this reaction the asymmetry was obtained by use of esters of dihydroquinine and dihydroquinidine as ligands. Markedly higher enantioselectivity obtains by use of ligands **1** and **2**, prepared by reaction of 1,4-dichlorophthalazine with dihydroquinidine (ligand 1) and dihydroquinine (ligand 2).[1]

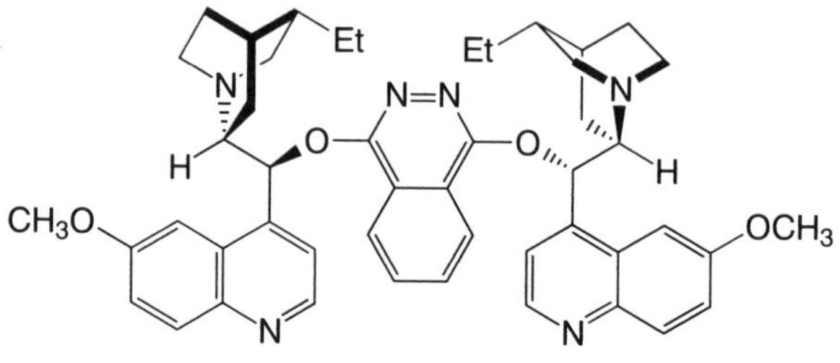

1

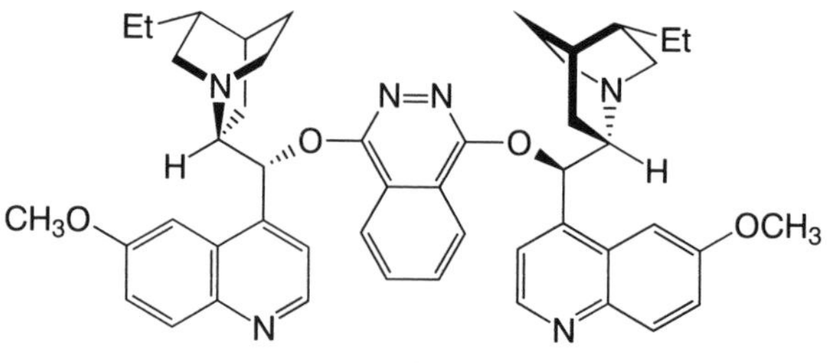

2

Note that the mirror-image reciprocity between **1** and **2** is greater than in the cinchona bases themselves.

A further improvement can be effected by addition of methanesulfonamide (1 equiv. based on olefin), which accelerates hydrolysis of osmate ester intermediates. This catalyst is useful if the alkene is trisubstituted or 1,2-disubstituted, but is not useful in the case of terminal alkenes. Addition of the sulfonamide permits osmylations at 0°, with enhances enantioselectivity.

Use of these phthalazine ligands and addition of a sulfonamide results in enantioselectivities of 95–99.5% in the case of nonterminal alkenes. In general, reactions with ligand 1 are somewhat more enantioselective than those with ligand 2.

However, the use of these dihydroquinidine and dihydroquinine phthalazine ligands does not improve the enantioselectivity of *cis*-disubstituted alkenes.[2] The problem in this case can be solved by use of carbamate ligands such as (9-O-indolinylcarbamoyl)dihydroquinidine, DHQO-IND (**1**).

Asymmetric monodihydroxylation of 1,3-dienes.[3]   This osmylation can be effected with use of [1,4-bis(9-O-dihydroquinidinyl)phthalazine (**1**) as the ligand. In this reaction the products with few exception are ene diols; generally osmylation occurs with the more electron-rich double bond. Some regioselectivity is observed with nonconjugated dienes, with a preference for trisubstituted double bonds over terminal ones.

Asymmetric dihydroxylation of enol ethers.[4]   This reaction can be effected in high yield by reaction with $K_2O_3O_2(OH)_4$, $K_3Fe(CN)_6$, or K_2CO_3 catalyzed by either (DHQD)$_2$-PHAL or (DHQ)$_2$-PHAL. The dihydroxylation can be carried out on the

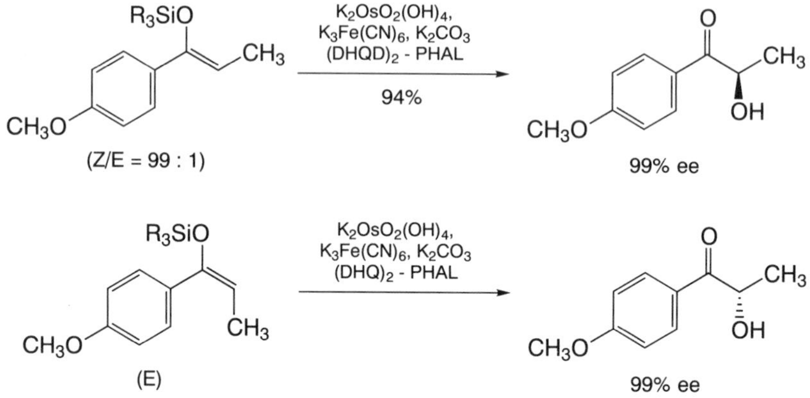

crude E/Z mixture obtained by conversion of ketones into enol methyl ethers or enol silyl ethers.

Asymmetric dihydroxylation of enynes.[5]   Even though alkynes can be oxidized to α-diketones by OsO$_4$, 1,3-enynes can be converted to ynediols exclusively by osmium-catalyzed asymmetric dihydroxylation using a dihydroquinidine catalyst. The enantio-selectivity in this reaction can be markedly enhanced by use of a new ligand **1**, in which two dihydroquinidines are connected by a phthalazine spacer group. Using this new ligand, 1,3-enynes in which the ene group is *trans*-disubstituted are converted into 1,2-diols in 73–97% ee.

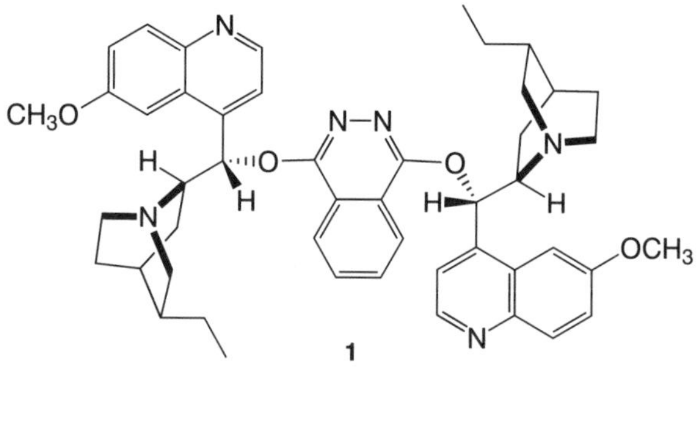

1

Asymmetric synthesis of hydroxy-γ-lactones.[6]   Asymmetric dihydroxylation of β, γ- and γ, δ-unsaturated esters is accompanied by spontaneous lactonization to hydroxy γ-lactones, generally in 96–98% ee, which can be increased to 100% by a single crystallization.

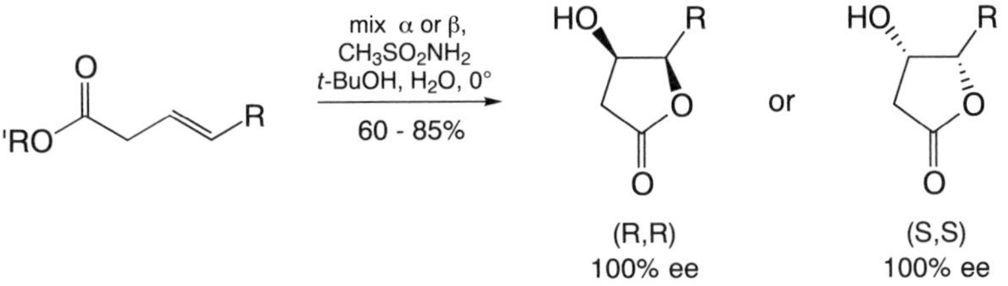

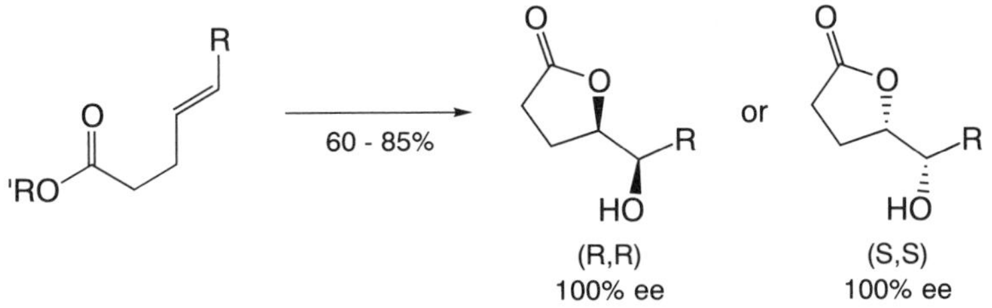

Asymmetric dihydroxylation of alkenes.[7]   DABCO, known to accelerate dihydroxylation, can accelerate OsO_4-dihydroxylations; new C_2-symmetrical 2,3-disubstituted DABCOs have been shown to be useful for asymmetric dihydroxylations. Of several DABCOs of this type available from (S,S)-threitol 1,4-dibenzyl ether, the most effective is **1**.

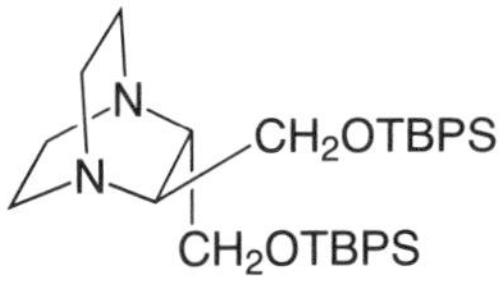

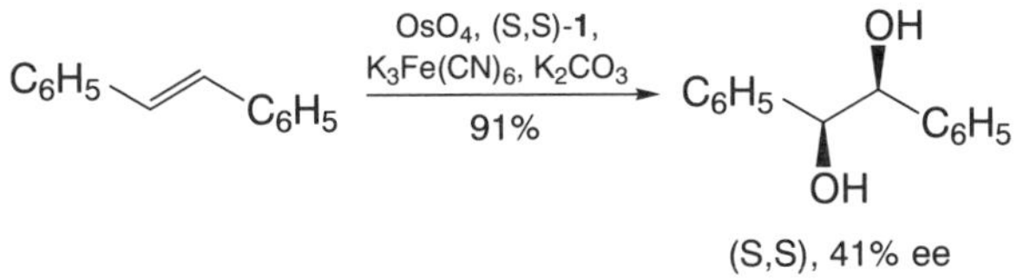

Review of asymmetric dihydroxylation.   Lohray[8] has reviewed this reaction, which was initiated by Criegee's observation in 1936 that pyridine can accelerate the rate of reaction. As a consequence other tertiary amines and diamines were examined as catalysts, and were found to also effect enantioselective oxidation of alkenes by Sharpless in 1980. Further search eventually led to bisdihydroquinidine and bisdihydroquinine 1,4-ethers of phthalazine. Lohray and Bhushan (1992) have found that the bisester (1) of a dicarboxylic acid is an excellent chiral auxiliary for asymmetric dihydroxylation of *trans*-disubstituted alkenes.

1

[1] K. B. Sharpless, W. Amberg, Y. L. Bennani, G. A. Crispino, J. Hartung, K.-S. Jeong, H.-L. Kwong, K. Morikawa, Z.-M. Wang, D. Xu, and X.-L. Zhang, *J. Org.,* **57,** 2768 (1992).

[2] L. Wang and K. B. Sharpless, *Am. Soc.,* **114,** 7568 (1992).

[3] D. Xu, G. A. Crispino, and K. B. Sharpless, *ibid.,* **114,** 7570 (1992).

[4] T. Hashiyama, K. Morikawa, and K. B. Sharpless, *J. Org.,* **57,** 5067 (1992).

[5] K.-S. Jeong, P. Sjö, and K. B. Sharpless, *Tetrahedron Letters,* **33,** 3833 (1992).

[6] Z.-M. Wang, X.-L. Zhang, K. B. Sharpless, S. C. Sinha, A. Sinha-Bagchi, and E.Keinan, *ibid.,* **33,** 6407 (1992).

[7] T. Oishi and M. Hirama, *ibid.,* **33,** 639 (1992).

[8] B. B. Lohray, *Tetrahedron: Asymmetry,* **3,** 1317 (1992).

Osmium tetroxide – Bis(3-methyl-2,4-pentanedionato)nickel(II), $OsO_4/Ni(mac)_2$.

α-Hydroxy ketones.[1]   In the presence of OsO_4 (1 mol %) and $Ni(mac)_2$ (3 mol %), various alkenes are oxidized by molecular O_2 and an aldehyde (usually isobutyraldehyde) to *α*-hydroxy ketones in good yield, which can be improved by addition of 2,6-lutidine (2 mol %) to suppress epoxidation.

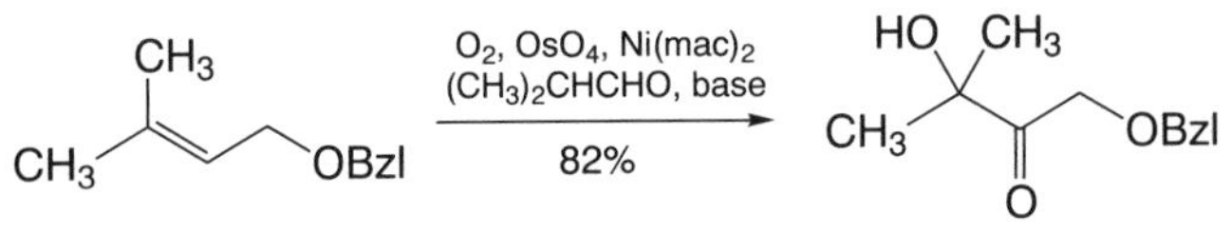

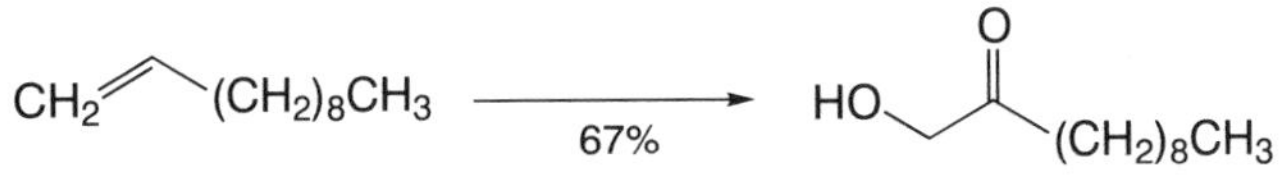

[1] T. Takai, T. Yamada, and T. Mukaiyama, *Chem. Letters,* 1499 (1991).

Oxalyl chloride $\overset{\overset{\displaystyle O}{\|}}{(ClC)_2}$.

N-Alkyl maleimides.[1]   A short route to these compounds involves reaction of maleic anhydride (1) with amines to form maleamic acids (2, >95% yield). On treatment of these products with oxalyl chloride, the hydrochlorides 3 are obtained. Elimination of HCl affords the desired maleimides (4), which are obtained in 55–73% overall yield.

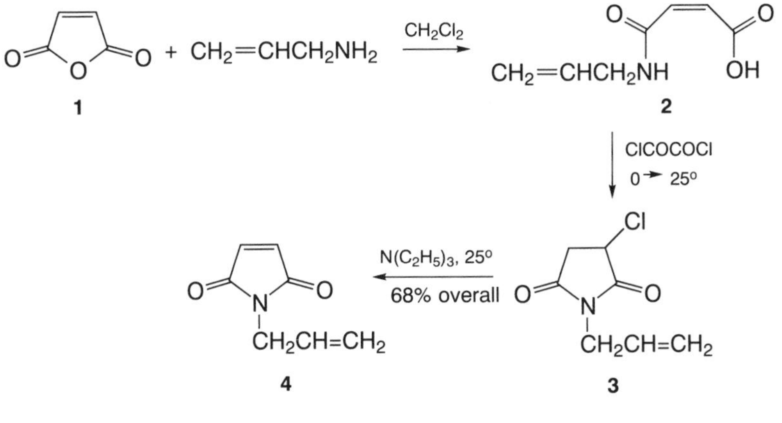

3-Aryl-3,4-dihydroisoquinolines.[2]   Isoquinolines are generally prepared by the Bischler–Napieralski reaction, but this classical route is not useful in the case of 3-aryl-isoquinolines. In a modified procedure, the precursor, (phenylethyl)amide (1), is treated with oxalyl chloride to form **a**, which on treatment with FeCl₃ forms an N-acyl-liminium ion **b**. This ion cyclizes to 2, which is converted into 3,4-dihydroisoquinoline 3 by treatment with sulfuric acid in methanol. Overall yields of 3 are in the range 55–90%.

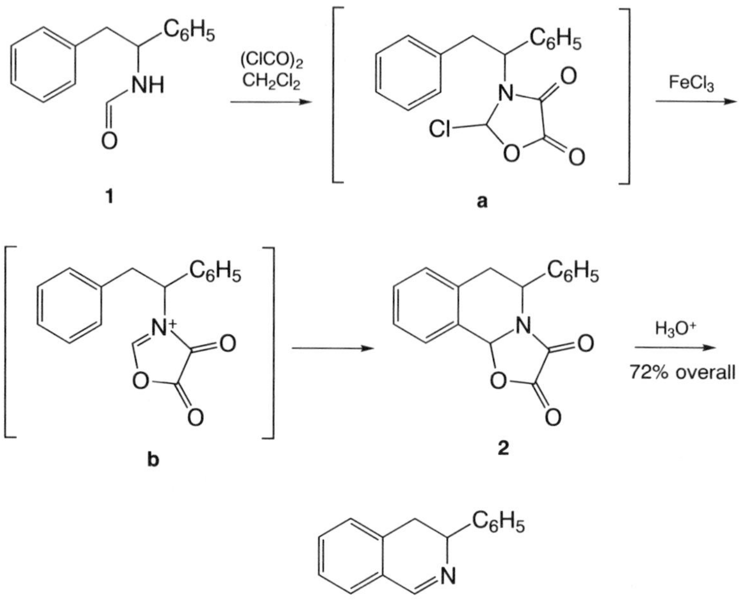

[1] T. F. Braish and D. E. Fox, *Synlett,* 979 (1992).
[2] R. D. Larsen, R. A. Reamer, E. G. Corley, P. Davis, E. J. J. Grabowski, P. J. Reider, and I. Shinkai, *J. Org.,* **56**, 6034 (1991).

Oxazaborolidines, 19, 110–111; **15,** 239–242; **16,** 253–255.

These useful heterocycles were prepared originally by reaction of amino alcohols with $BH_3 \cdot THF$ or boronic acids, but these methods can require lengthy reaction times in the case of hindered pyrrolidines such as **1**, (S)-2-(diphenylhydroxymethyl)pyrrolidine. Reaction of **1** with a bis(trifluoroethyl)alkylboronate **2** at 110° for 30 minutes, generates the desired oxazaborolidine **3** *in situ*, which can be isolated if desired in 95% yield.[1]

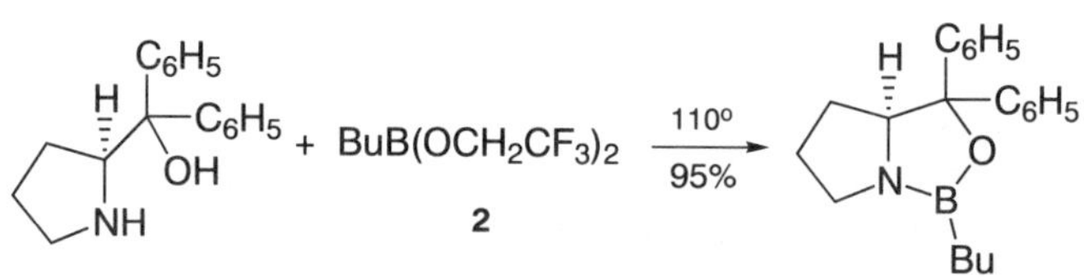

The boronate (**2**) is prepared in two steps in 82% overall yield (equation I).

$$\text{(I)} \quad BH_3 \cdot THF + CF_3CH_2OH \xrightarrow[-H_2]{23°} B(OCH_2CF_3)_3 \xrightarrow[82\%]{\substack{1)\ BBu_3 \\ 2)\ BH_3 \cdot HF}} BuB(OCH_2CF_3)_2$$

2

Enantioselective synthesis of α-amino acids.[2]   A practical and general route to α-amino acids involves enantioselective reduction of alkyl trichloromethyl ketones with catecholborane (CB) in the presence of the oxazaborolidine (S)-**1** (0.1 equiv.) to give (R)-alcohols **2**. The (R)-alcohols (**2**) are converted on treatment with NaOH

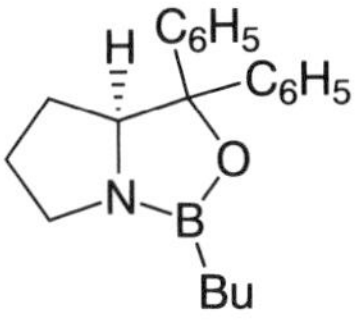

(S) - **1**

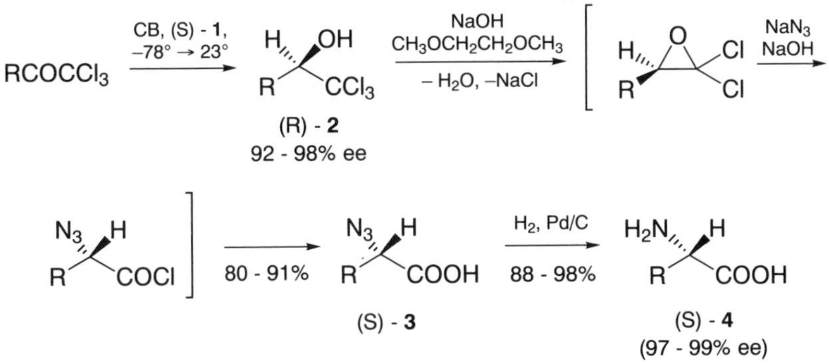

(4 equiv.) and NaN$_3$ (2 equiv.) to (S)-α-azido acids (**3**), which are converted to (S)-α-amino acids (**4**) by catalytic hydrogenation.

Enantioselective aldol reactions.[3]   The tryptophan-derived oxazaborolidine **1** can also effect enantioselective Mukaiyama-aldol reactions of aldehydes with trimethyl-silyl ethers.

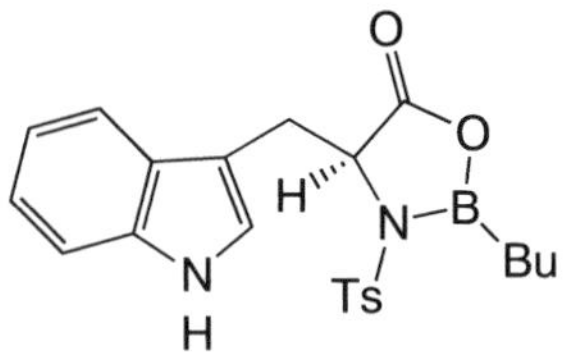

1

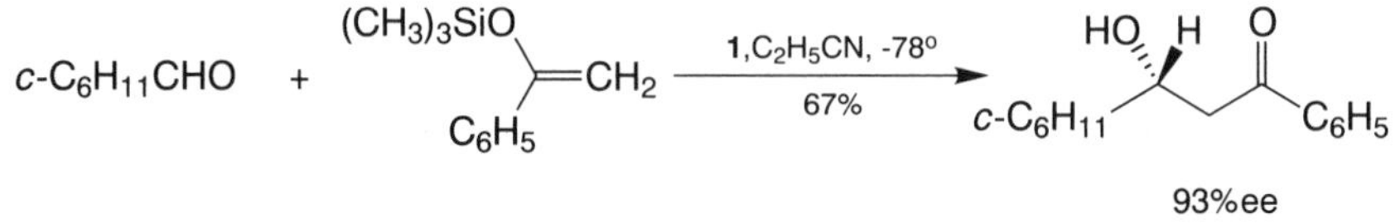

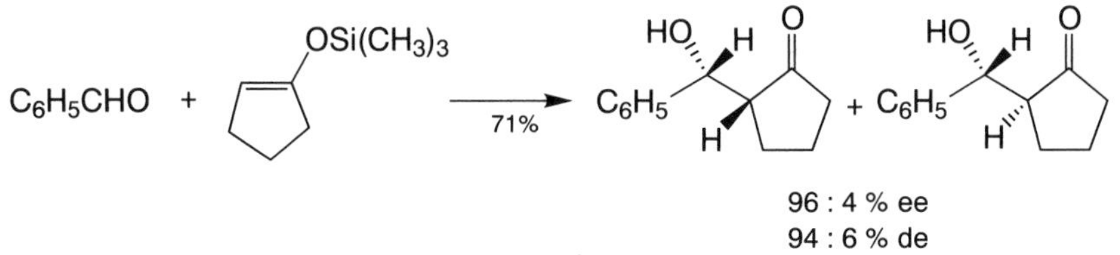

$c\text{-}C_6H_{11}CHO$ + (structure) $\xrightarrow[67\%]{1,C_2H_5CN,\ -78°}$ (product) 93%ee

C_6H_5CHO + (structure) $\xrightarrow{71\%}$ (products)

96 : 4 % ee
94 : 6 % de

Reaction of aldehydes with 1-methoxy-3-trimethylsilyloxy-1,3-butadiene catalyzed by **1** provides an aldol as the major product.

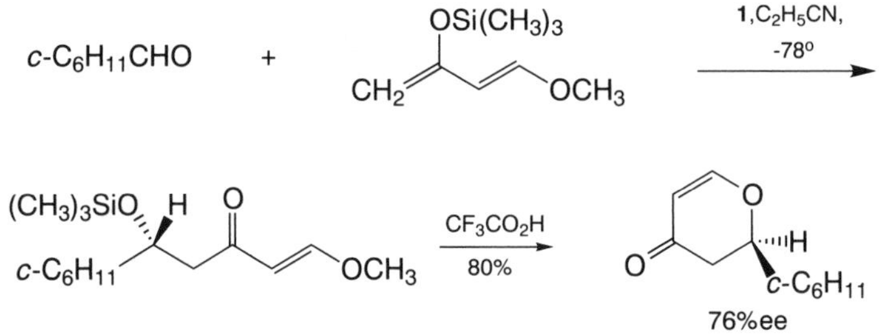

$c\text{-}C_6H_{11}CHO$ + (structure with $OSi(CH_3)_3$, CH_2, OCH_3) $\xrightarrow{1,C_2H_5CN,\ -78°}$

(product with $(CH_3)_3SiO$, $c\text{-}C_6H_{11}$, OCH_3) $\xrightarrow[80\%]{CF_3CO_2H}$ (product $c\text{-}C_6H_{11}$) 76%ee

(3S)-2,3-Oxidosqualene, (S)-1.[4]   This biosynthetic precursor to steroids and triter-penes can be prepared from (±)-oxidosqualene (**1**). The first steps involves conversion of (±)-**1** to a fluorohydrin (**2**), which is converted to an α-fluoro ketone **3** by Swern oxidation. This α-fluoro ketone is reduced by catecholborane (CB) catalyzed by the (R)-oxazaborolidine **4** to the (S)-fluorohydrin **5** in 92% ee. The final step involves conversion to the (3S)-oxide **1** by base.

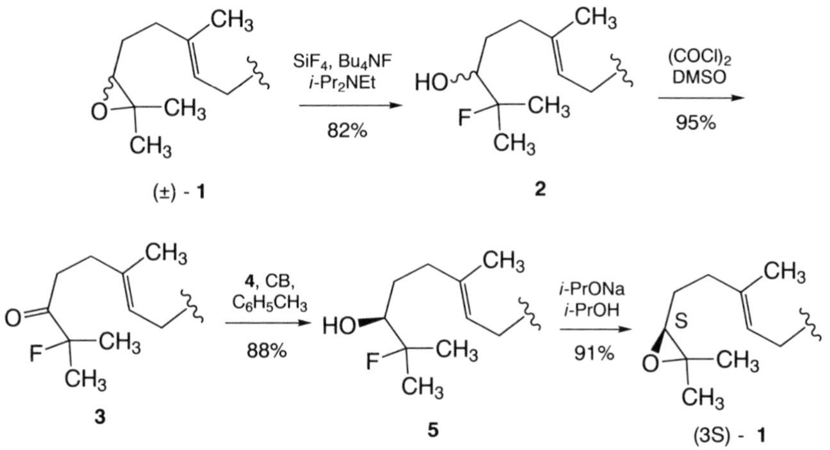

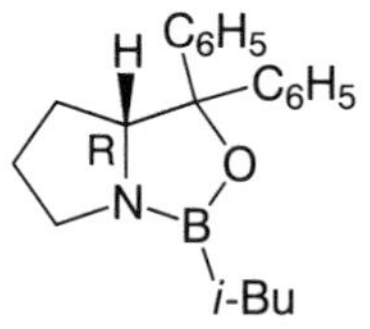

Enantioselective reduction of trichloromethyl ketones.[5] These ketones are reduced by catecholborane (CB) in the presence of 0.1 equiv. of the chiral oxazaborolidine **1** to trichloromethylcarbinols **2** in 92–98% ee. These products are useful in themselves,

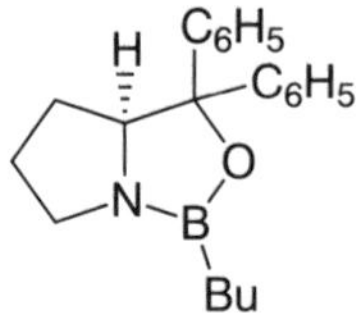

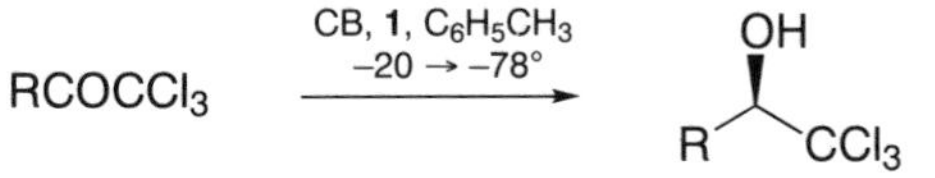

but particularly because they can be converted into optically active α-amino acids, α-aryloxy acids (**3**), and α-hydroxy esters (**4**), with inversion of configuration.

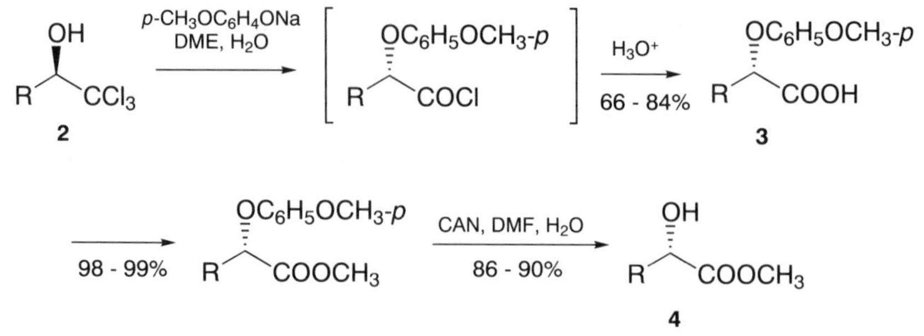

Asymmetric Diels–Alder reactions.[6]   The oxazaborolidine **1a**, prepared by reaction of N-tosyl-(S)-tryptophan with BuB(OH)$_2$ in toluene-THF (2:1) with removal of H$_2$O, is a highly effective catalyst for enantioselective Diels–Alder reactions. The same reaction but with BH$_3$ · THF at 23° provides **1b**.

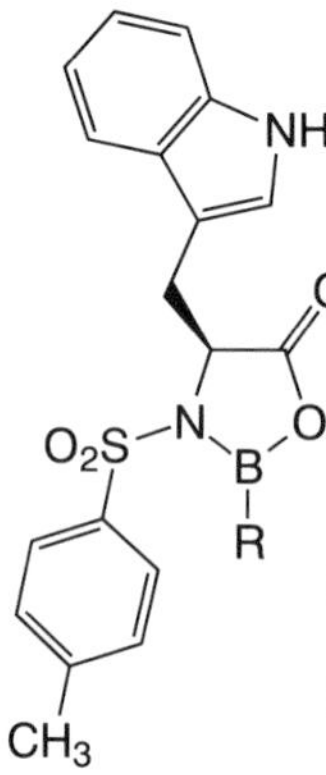

1a, R = Bu

1b, R = H

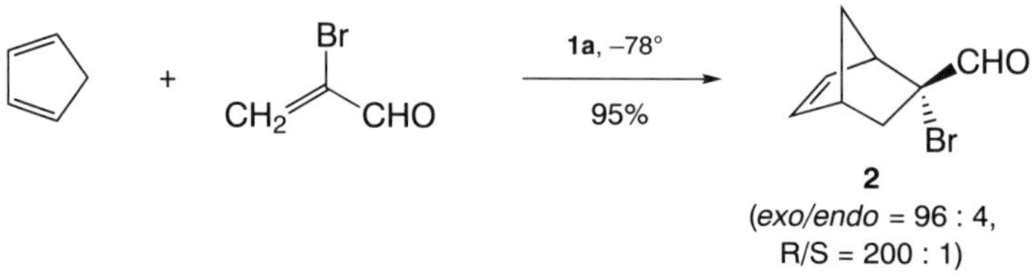

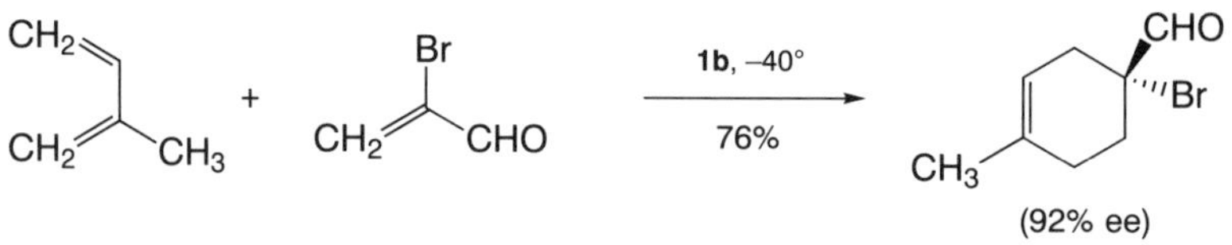

2
(*exo/endo* = 96 : 4,
R/S = 200 : 1)

(92% ee)

Significantly, a catalyst prepared from N-tosyl-(S)-valine provides the enantiomer of the adducts formed by use of **1a** or **1b**.

[1] E. J. Corey and J. O. Link, *Tetrahedron Letters,* **33**, 4141 (1992).
[2] E. J. Corey and J. O. Link, *Am. Soc.,* **114**, 1906 (1992).
[3] E. J. Corey, C. L. Cywin, and T. D. Roper, *Tetrahedron Letters,* **33**, 6907 (1992).
[4] E. J. Corey, K. Y. Yi, and S. P. T. Matsuda, *ibid.,* **33**, 2319 (1992).
[5] E. J. Corey and J. O. Link, *Tetrahedron Letters,* **33**, 3431 (1992).
[6] E. J. Corey and T.-P. Loh, *Am. Soc.,* **113**, 8966 (1991).

Oxazolidinones, chiral.

Stereoselective anti-aldol reactions.   As part of a synthesis of polypropionate natural products, Evans *et al.*[1] have studied the stereoselectivity of the reaction of isobutyraldehyde with the chiral β-ketoimide **1a**, which has been shown to undergo *syn*-selective aldol reactions.[4] Surprisingly, the (E)-boron enolate, generated in ether from dicyclohexylchloroborane and ethyldimethylamine, reacts with isobutyraldehyde to give the *anti, anti*-aldol **2** and the *syn, anti*-aldol **2** in the ratio 84:16. Similar diastereoselectivity obtains with the reaction of the isomeric β-ketoimide **1b**.

Under the same conditions, similar *anti*-selective aldol reactions obtain with several other chiral ethyl ketones such as **3**.[4]

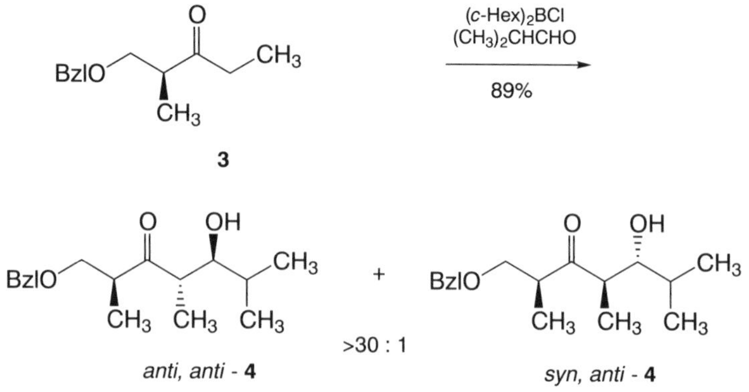

Asymmetric aldol reactions.[3]   The chiral N-propionyloxazolidinone (**1**), prepared in several steps from (1R)-(−)-camphorquinone, undergoes highly diastereoselective aldol reactions with the additional advantage of high crystallinity for improving the optical purities of crude aldols. Either the lithium enolate or the titanium enolate, prepared by transmetalation with ClTi(O-*i*-Pr)$_3$, reacts with aldehydes to form *syn*-adducts with diastereomeric purities of 98–99% after one crystallization. The observed facial selectivity is consistent with metal chelation of intermediate (Z)-enolates (supported by an X-ray crystal structure of the trapped silyl enol ether). The lithium enolate also exhibits

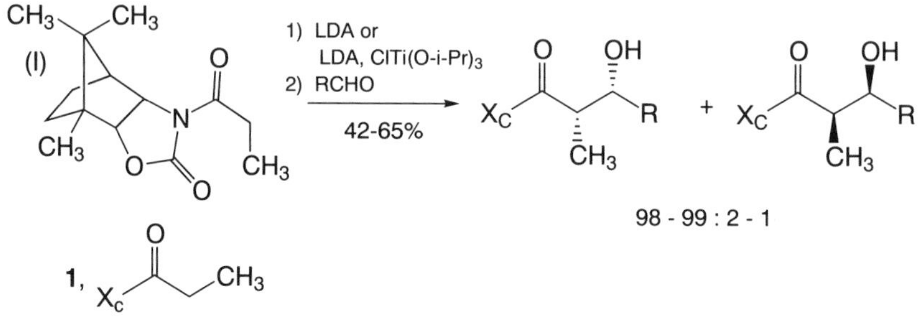

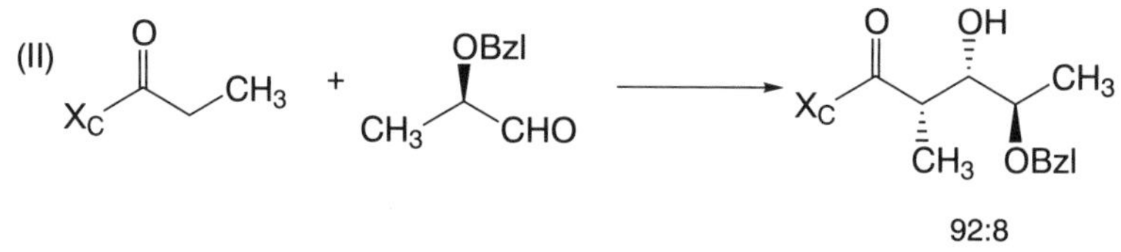

highly double asymmetric induction on reaction with (R)-2-benzyloxypropanol (matching pair, equation II). Hydrolytic or reductive removal of the chiral auxiliary provides chiral carboxylic acids or alcohols.

Asymmetric reduction of ketimines to sec-amines.[4]   Of the various hydride reagents found to achieve high enantioselective reduction of ketones, the oxazaborolidine **1** of Itsuno, prepared from BH_3 and (S)-(−)-2-amino-3-methyl-1,1-diphenylbutane-1-ol, derived from (S)-valine, (**12,31**), is the most effective in terms of asymmetric induction. Like Corey's oxazaborolidines derived from (S)-proline, **1** can also be used in catalytic amounts. The highest enantioselectivities obtain in reduction of N-phenylimines of aromatic ketones (as high as 88% ee). The enantioselectivities are lower in the case of N-*t*-butylimines of aryl ketones (80% ee). Reduction of N-phenylimines of prochiral dialkyl ketones with **1** results in 10–25% ees.

[1] D. A. Evans, H. P. Ng, J. S. Clark, and D. L. Rieger, *Tetrahedron,* **48**, 2127 (1992).
[2] D. A. Evans, J. S. Clark, R. Metternich, V. J. Novack, and G. S. Sheppard, *Am. Soc.,* **112**, 866 (1990).
[3] M. P. Bonner and E. R. Thornton, *Am. Soc.,* **113**, 1299 (1991).
[4] B. T. Cho and Y. S. Chun, *Tetrahedron: Asymmetry,* **3**, 1583 (1992).

Oxodiperoxymolybdenum(pyridine)(hexamethylphosphoric triamide).
(Vedejs reagent, MoOPH).

(3R)- or (3S)-3-Hydroxyaspartates.[1]   These amino acids can be prepared in high diastereoselectivity by hydroxylation of the enolate of N-(9-phenylfluorenyl)aspartates. Thus this derivative **1** of L-aspartic acid on treatment with a base followed by reaction with MoOPH provides the 3-hydroxy derivative, *syn-* and *anti-***2**. The stereoselectivity of this hydroxylation is highly dependent on the base and the solvent. The highest *syn*-selectivity is obtained with LHMDS, LiN[Si$(CH_3)_3$]$_2$, as base in THF-DMPU or THF-HMPA. The highest *anti*-selectivity is obtained by treatment first with BuLi and then with LHMDS in THF.

Pf =

(structure: 9-phenyl-9-methylfluorene with C₆H₅)

CH_3O_2C ⎯⎯ CO_2CH_3
         |
       NHPf

L - 1

1) base
2) MoOPH
⟶

OH
|
CH_3O_2C ⎯⎯ CO_2CH_3
            |
          NHPf

(2S, 3S) - 2

+

OH
|
CH_3O_2C ⎯⎯ CO_2CH_3
            |
          NHPf

(2S, 3R) - 2

LHMDS, HMPA	95%	11 : 1
BuLi, LHMDS, THF	70%	1 : 20

[1] F. J. Sardina, M. M. Paz, E. Fernańdez-Megia, R. F. de Boer, and M. P. Alvarez, *Tetrahedron Letters*, **33**, 4637 (1992).

Oxygen

Epoxidation.[1]   The epoxidation of alkenes can be effected with molecular oxygen and an aldehyde. Isobutyraldehyde and pivaldehyde are the most effective. Note that a metal catalyst is not involved. Yields are highest when the ratio of alkene to aldehyde is 1 : 3.

C_6H_{13}⎯⎯=CH₂ + O₂ + $(CH_3)_3CCHO$ ⎯⎯$\xrightarrow{ClCH_2CH_2Cl, 40°C}$⎯⎯ 100%

C_6H_{13}⎯⎯(epoxide) + $(CH_3)_3CCO_2H$

(cyclohexene with CH₂ substituent) ⎯⎯$\xrightarrow{O_2, (CH_3)_3CCHO}$⎯⎯ 78% → (epoxycyclohexane with CH₂ substituent)

Epoxidation of cholesterol.[2]   This reaction can be effected with O₂ and isobutyraldehyde as reductant in quantitative yield in CH_2Cl_2 at room temperature. This reaction provides the α- and β-oxides as a 1 : 1 mixture. Addition of a (tetraphenylporphrinato)nickel(II) catalyst does not affect the rate, but enhances β-stereoselectivity (74 : 26).

[1] K. Kaneda, S. Haruna, T. Imanaka, M. Hamamoto, Y. Nishiyama, and Y. Ishii, *Tetrahedron Letters,* **33,** 6827 (1992).

[2] R. Ramasseul, M. Tavares, and J.-C. Marchon, *J. Chem. Res. (S),* 104 (1992).

Oxygen, singlet.

Stereoselective photooxygenation of allylic alcohols.[1]   The ene reaction of singlet oxygen with allylic alcohols can show high regioselectivity and also high diastereoselectivity. Thus the allylic alcohol (E)-**1** reacts to form **2** and **3** in the ratio 96:4, but **2** is

formed with 93:7 diastereoselectivity. In contrast, the acetate of (E)-**1** shows high regioselectivity (82:18) but only 32:68 diastereoselectivity.

Photooxygenation of Ge—Ge bonds.[2]   Photooxygenation of the digermirane **1** or the azadigermiridine **2** with tetraphenylporphine (TPP) as sensitizer affords cyclic peroxides (**3** and **4**, respectively).

Both **3** and **4** are reduced by triphenylphosphine to **5** and **6**, respectively.

The 1,2-digermetene **7** undergoes photooxygenation in the presence of 9,10-dicyanoanthracene (DCA) to form the 1,2,3,6-dioxadigermine **8** in 80% yield.

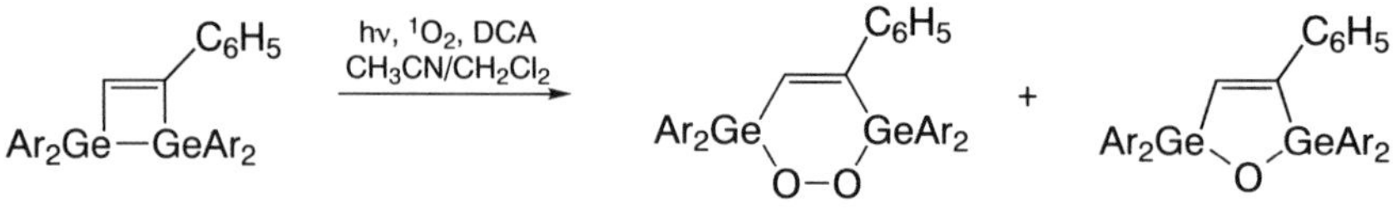

Oxidative decarboxylation of pyrrole-2-carboxylic acids.[3]   Reaction of pyrrole-2-carboxylic acids such as **1** and **2** with singlet oxygen in *i*-PrOH or CH_3CN and water (3 : 1) results in 5-hydroxy-3-pyrrolin-2-ones in high yield.

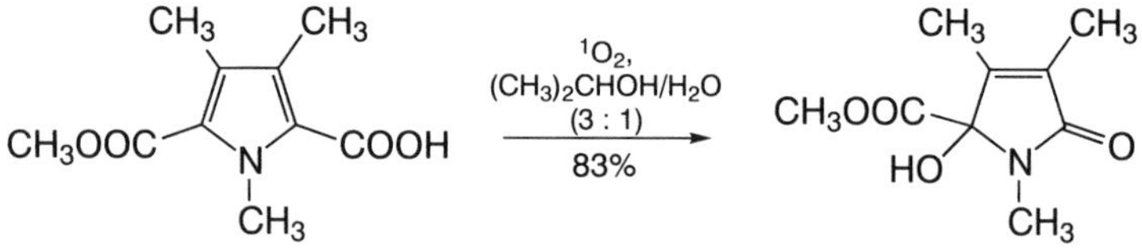

[1] W. Adam and B. Nestler, *Am. Soc.,* **114**, 6549 (1992).
[2] M. Kako, T. Akasaka, and W. Ando, *J. C. S. Chem. Comm.,* 457, 458 (1992).
[3] D. L. Boger and C. M. Baldino, *J. Org.,* **56**, 6942 (1991).

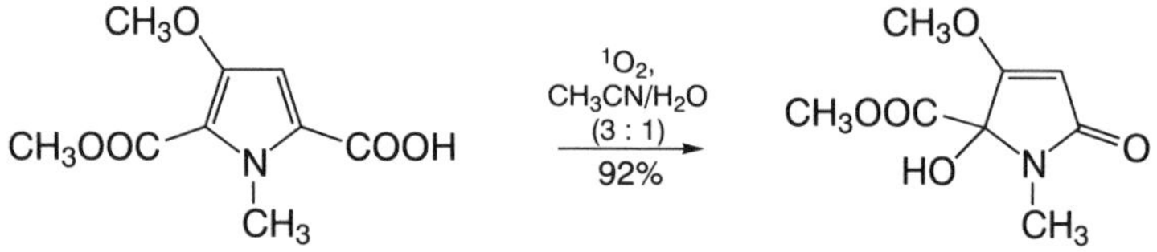

Oxygen, singlet – Titanium(IV) isopropoxide, 15, 322.

Vinylsilanes → silyl epoxy alcohols.   Photooxygenation of vinylsilanes in the presence of Ti(O-*i*-Pr)$_4$ affords silyl epoxy alcohols with high regio- and diastereoselectivity. The conversion involves as the first step an ene reaction with O_2 to provide a β-silyl allylic alcohol, which then undergoes epoxidation.

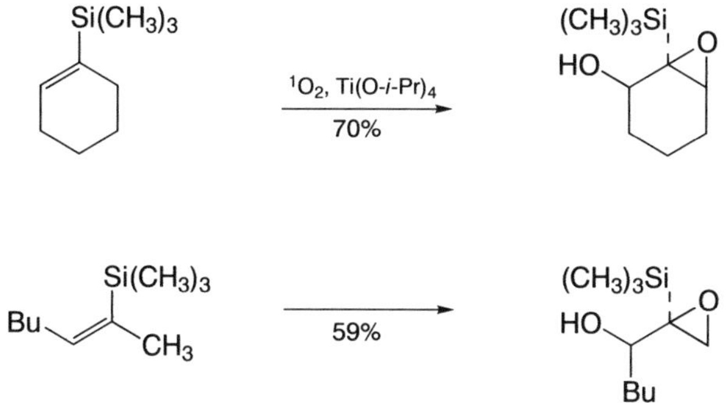

[1] W. Adam and M. Richter, *Tetrahedron Letters,* **33**, 3461 (1992).

Ozone.

Oxidation of R¹CH(OH)SiR₃ to R¹COOH.[1] α-Hydroxytrialkylsilanes and acylsilanes are oxidized by ozone to the corresponding carboxylic acids, even when one group attached to silicon is an allyl group.

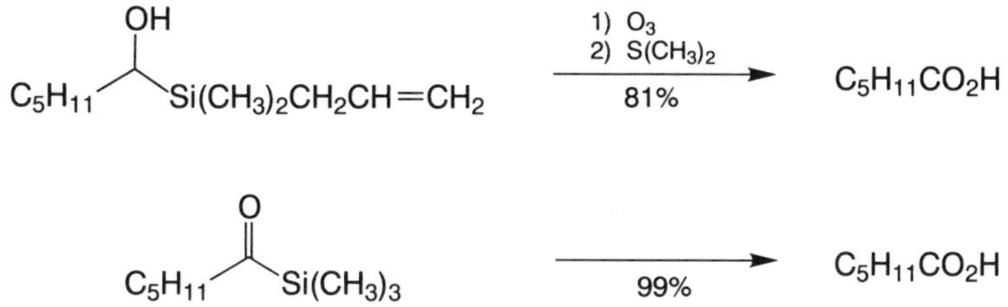

This reaction can be used to convert aldehydes to acids.

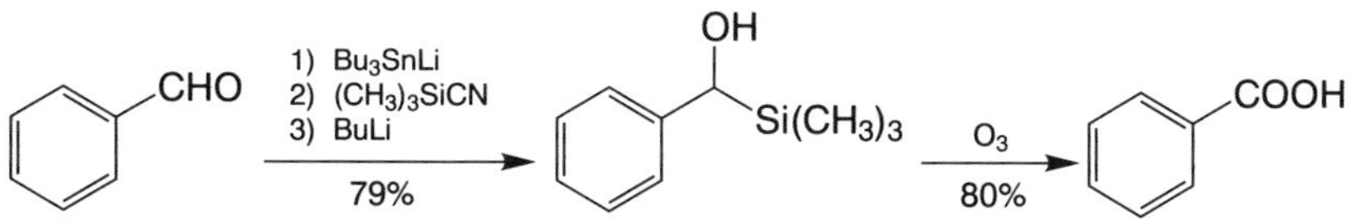

3-Alkoxy-1-alkenes → α-alkoxy esters.[2]   This conversion can be effected by ozone in the presence of $CH_3OH/NaOH$. Presumably an intermediate α-alkoxy aldehyde is converted into a dimethyl acetal, which is converted to the α-alkoxy ester by further reaction with ozone.

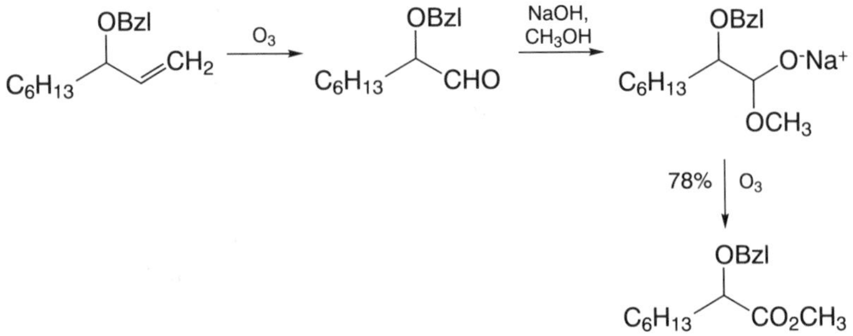

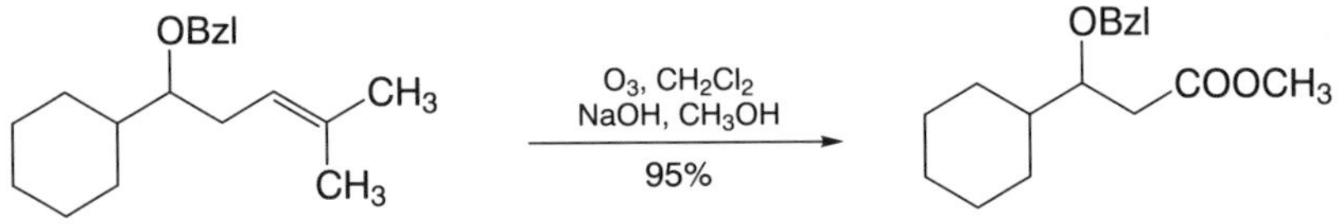

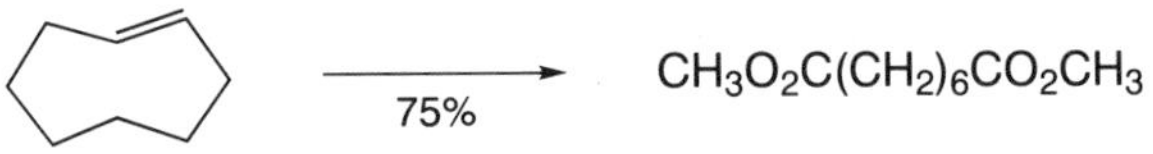

[1] R. J. Linderman and K. Chen, *Tetrahedron Letters,* **33**, 6767 (1992).
[2] J. A. Marshall, A. W. Garofalo, and R. C. Sedrani, *Synlett,* 643 (1992).

P

Palladium(II) acetate.

Cyclization of 4,6- and 5,7-dieneamides.[1]   (cf., **16**,261). These dieneamides (**1** and **3**) cyclize in the presence of Pd(OAc)$_2$ and an oxidant (CuCl$_2$/O$_2$) to pyrrolizidinone **2** and indolizidinone **4**, respectively, in 85–90% yield.

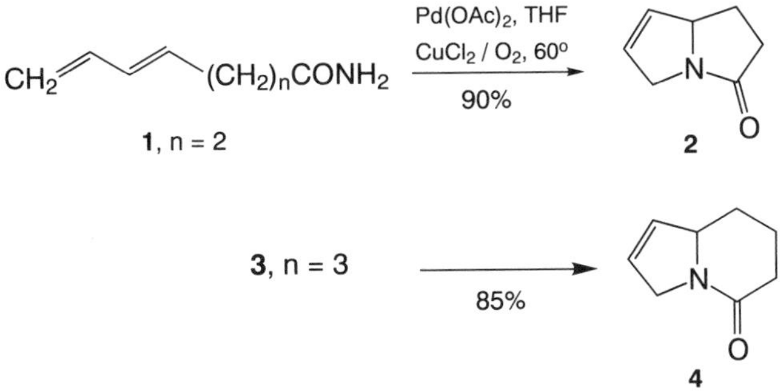

CH$_2$⟍⟍⟍(CH$_2$)$_n$CONH$_2$ → [Pd(OAc)$_2$, THF / CuCl$_2$ / O$_2$, 60°] → 90%

1, n = 2   **2**

3, n = 3 → 85% **4**

Oxaspirocyclic addition to 1,3-dienes.[2]   Pd(II) catalyzed oxidation of the diene **1** with *p*-benzoquinone in acetone–acetic acid in the presence of a base, Li$_2$CO$_3$, results in a spirocyclic ether **2**, formed by an overall *trans*-addition to the diene. Overall *cis*-addition of the oxygen function can be effected by replacement of Li$_2$CO$_3$ by LiCl, which results in the spirocyclic ether **3**.

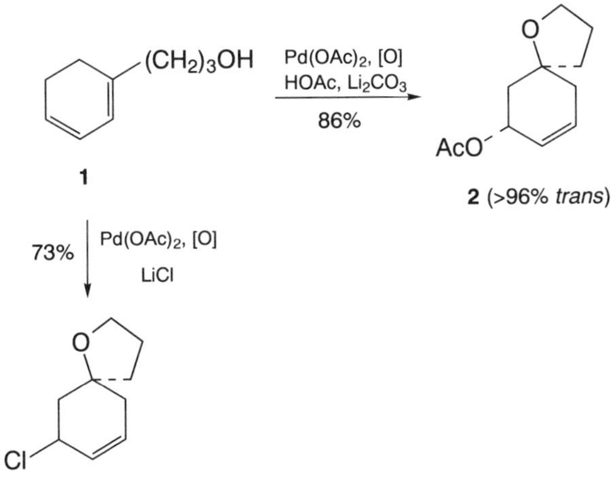

(CH$_2$)$_3$OH → [Pd(OAc)$_2$, [O] / HOAc, Li$_2$CO$_3$] → 86%

1   **2** (>96% *trans*)

73% | [Pd(OAc)$_2$, [O] / LiCl]

3 (>99% *cis*)

$RCH_2CHO \rightarrow RCH=CHOCH_3 \rightarrow RCH=CHCHO.$[3]   Methyl enol ethers readily prepared by reaction of aldehydes with (methyloxymethylene)triphenylphosphorane, $[(C_6H_5)_3P=CHOCH_3]$, when treated with $Pd(OAc)_2$ in CH_3CN and then with $NaHCO_3$ and $Cu(OAc)_2 \cdot H_2O$ (1 equiv.) are converted into α, β-unsaturated aldehydes.

$$C_6H_5(CH_3)_4CHO \xrightarrow[94\%]{} C_6H_5(CH_3)_4CH=CHOCH_3 \xrightarrow[92\%]{\substack{Pd(OAc)_2, NaHCO_3 \\ Cu(OAc)_2 \cdot H_2O}}$$

$$C_6H_5(CH_3)_4CH=CHCHO$$

Heck cyclization.[4]   A short, efficient, and asymmetric synthesis of the anticancer alkaloid (S)-camptothecin (**4**) involves N-alkylation of the optically active hydroxylactone **1** with the bromoquinoline **2** to provide **3**. This product undergoes Heck cyclization to the alkaloid **4** in 59% yield.

(S)-4, α_D +42°

Heck-type coupling of allylic alcohols and enol triflates.[5]   This vinylation of allylic alcohols can be effected by catalysis with $Pd(OAc)_2$ and tri-o-tolylphosphine (1:2) and triethylamine (excess) as base. The major products are conjugated dienols.

$$CH_2 = CHCH_2OH$$
$$+$$

$$\xrightarrow[\text{55\%}]{\begin{array}{c}\text{Pd(OAc)}_2,\ \text{Ag}_2\text{CO}_3\\ \text{Bu}_4\text{NHSO}_4,\ \text{CH}_3\text{CN}\end{array}}$$

Coupling of vinyl halides with allyl alcohols.[6]   Palladium-catalyzed coupling of vinyl halides with allylic alcohols under usual Heck conditions results in a complex mixture of products. Addition of silver carbonate and a phase-transfer catalyst can result in a palladium-catalyzed coupling of vinyl halides with a primary allylic alcohol to give a γ, δ-enal, with retention of configuration of the vinyl halide.

Palladium-catalyzed coupling of a vinyl halide with a secondary allylic alcohol in the presence of silver acetate or carbonate results in a conjugated dienol with retention of configuration of the vinyl halide.

Arylation of allylic alcohols.[8]   Aryl iodides and allylic alcohols undergo a Heck-type reaction to form β-aryl-α,β-unsaturated alcohols in the presence of Pd(OAc)$_2$ and P(C$_6$H$_5$)$_3$ as catalysts and 1 equiv. of AgOAc to prevent isomerization to carbonyl compounds.

Reaction of aryl iodides and allylic alcohols catalyzed by Pd(OAc)$_2$ under phase-transfer conditions (NaHCO$_3$/Bu$_4$NCl) results in β-aryl aldehydes or ketones.

ArI + CH$_2$ <OH / R>   → [Pd(OAc)$_2$ - P(C$_6$H$_5$)$_3$ / AgOAc, DMF / 60 - 90%] → Ar <OH / R>

(E)

ArI + CH$_2$ <OH / R>   → [Pd(OAc)$_2$, NaHCO$_3$ / Bu$_4$NCl, DMF / 75 - 90%] → ArCH$_2$CH$_2$ <O / R>

Hydroacetoxylation of 2-alkynoates.[9] Under catalysts with Pd(OAc)$_2$, lithium acetate adds to 2-alkynoates to form (Z)-3-acetoxyalkenoates with high regio- and stereospecificity.

CH$_3$≡CO$_2$C$_6$H$_5$ + LiOAc → [Pd(OAc)$_2$ / CH$_3$CO$_2$H / 83%] → <CH$_3$ H / AcO CO$_2$C$_6$H$_5$>

Three-component coupling of alkenes. Larock and Lee[10] have reported an efficient one-step intermolecular coupling of three alkenes to form an intermediate (3) to prostaglandins. The starting material is the chiral cyclopentenol 1, which couples with ethyl vinyl ether and 1-octen-3-one (2) in the presence of Pd(OAc)$_2$, NaOAc, and NaI at room temperature to afford the bicyclic enone 3 in 72% yield as a 2–3:1 mixture of *exo-* and *endo*-isomers, which can be readily separated. A number of analogs of this bicyclic system can be prepared by substitution of 2 by other 1-alkenes. Pure *exo-3* was

1

+ C$_2$H$_5$OCH=CH$_2$ + CH$_2$=CHCOC$_5$H$_{11}$ → [Pd(OAc)$_2$, / NaOAc, NaI / 72%]

2

3, α$_D$ -51°, 2-3:1

used for synthesis of 12-epi-PGF$_{2\alpha}$ (**5**) in 54% yield. Thus reduction of **3** with (S)-BINAL-H (**12**,190) provides **4** which is converted into **5** by a Wittig reaction.

[1] P. G. Andersson and J.-E. Bäckvall, *Am. Soc.*, **114**, 8696 (1992).

[2] J.-E. Bäckvall and P. G. Andersson, *J. Org.*, **56**, 2274 (1991).

[3] H. Takayama, T. Koike, N. Aimi, and S. Sakai, *J. Org.*, **57**, 2173 (1992).

[4] D. L. Comins, M. F. Baevsky, and H. Hong, *Am. Soc.*, **114**, 10971 (1992).

[5] E. Bernocchi, S. Cacchi, P. G. Ciattini, E. Morera, and G. Ortar, *Tetrahedron Letters*, **33**, 3073 (1992).

[6] T. Jeffery, *Tetrahedron Letters*, **31**, 6641 (1990).

[7] *Idem, J. C. S. Chem. Comm.*, 324 (1991).

[8] *Idem, Tetrahedron Letters*, **32**, 2121 (1991).

[9] X. Lu, G. Zhu, and S. Ma, *ibid.*, **33**, 7205 (1992).

[10] R. C. Larock and N. H. Lee, *Am. Soc.*, **113**, 7815 (1991).

Palladium(II) acetate – 1,3-Bis(diphenylphosphino)propane, Pd(OAc)$_2$ – dppp (1:1), **1**.

α-Regioselective arylation of enol ethers.[1] This Heck reaction usually shows only slight regioselectivity. However use of a bidentate phosphine (a chelating ligand) greatly favors α-regioselectivity in coupling of aryl triflates with butyl vinyl ether (**2**) in DMF.

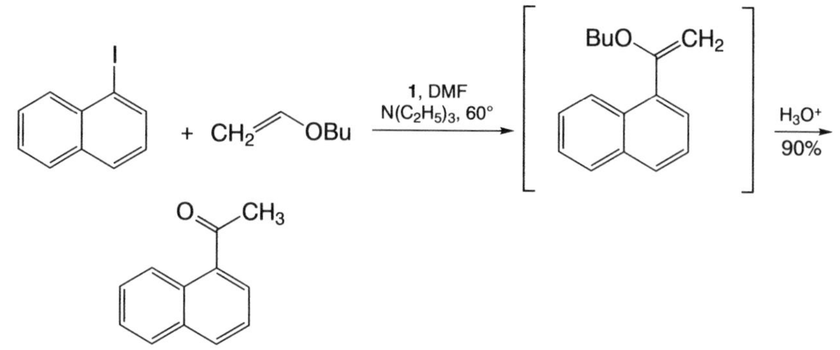

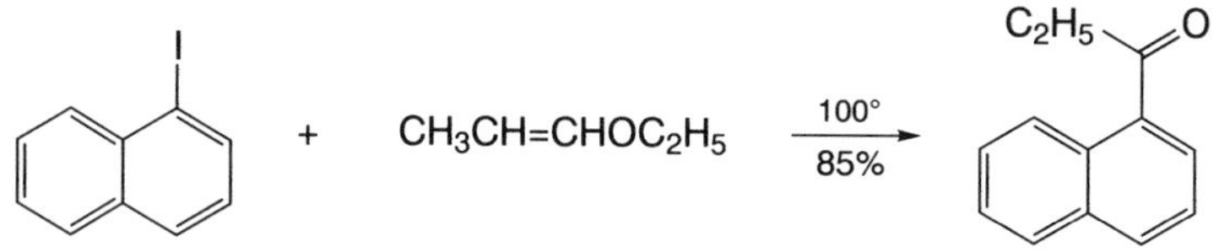

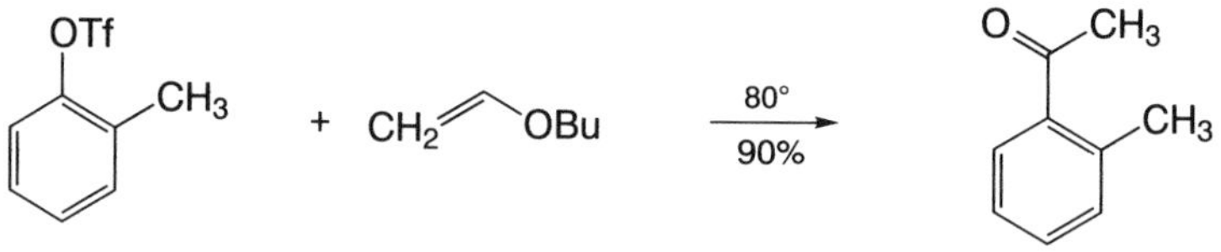

[1] W. Cabri, I. Candiani, A. Bideschi, and S. Penco, *J. Org.*, **57**, 1481 (1992).

Palladium(II) acetate–*t*-Octyl isocyanide (1,1,3,3-tetramethylbutyl isocyanide).

Bis-silylation of alkynes and alkenes.[1]   Palladium(II) acetate in combination with a large excess of a *t*-alkyl isocyanide can effect bis-silylation of terminal alkynes by otherwise unreactive disilanes (equation I). Intramolecular bis-silylation of an alkyne is also possible with this catalyst (equation II).

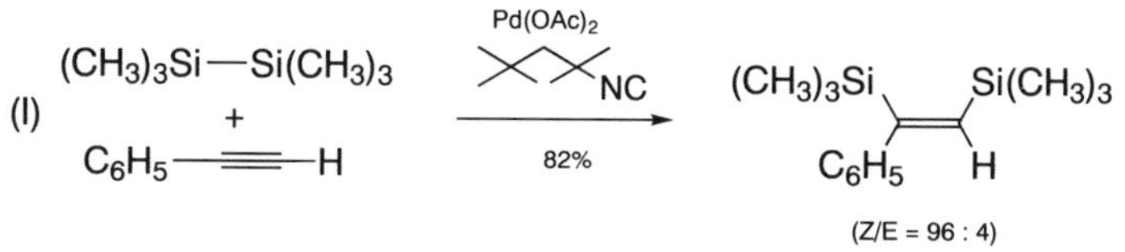

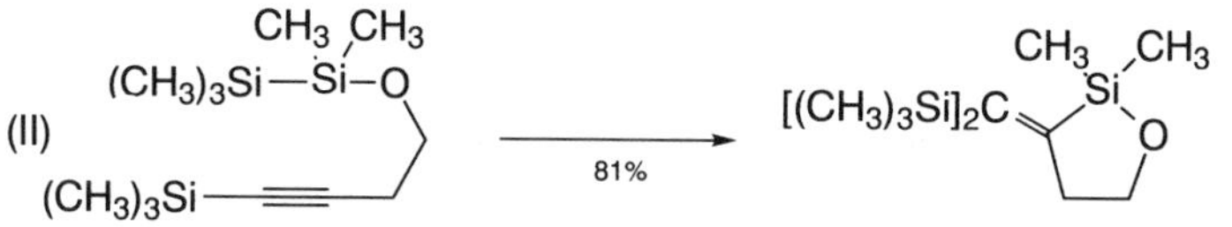

This intramolecular bis-silylation has been extended to terminal and 2,2-disubstituted alkenes. 1,2-Disubstituted alkenes do not undergo this reaction. The substrates, unsaturated disilanyl ethers, are prepared by silylation of allylic and homoallylic alcohols.

Intramolecular bis-silylation does not occur if the disilane is tethered to the double bond by chains of more than four atoms. Therefore, it is not surprising that alkenes do not undergo intermolecular addition of disilanes. The bis-silylation is stereoselective when the alkene bears an α- or β-group (equations III and IV).[2]

(III)

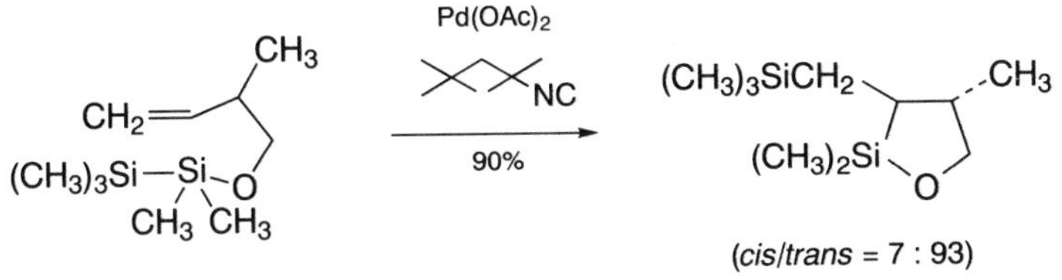

(cis/trans = 7 : 93)

(IV)

(cis/trans = 96 : 4)

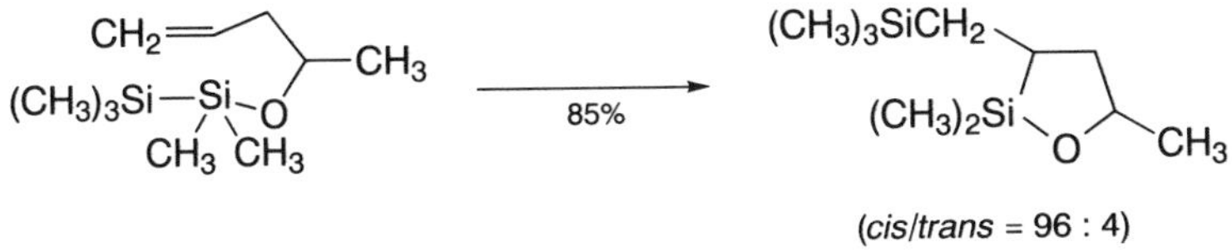

This reaction can provide a stereoselective synthesis of 1,2,4-triols.[4] Thus use of isopropoxydisilyl ethers of allylic or homoallylic alcohols results in adducts that can be oxidized to triols.

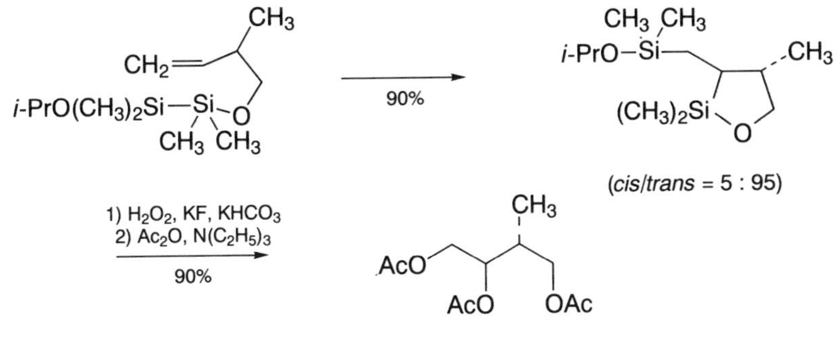

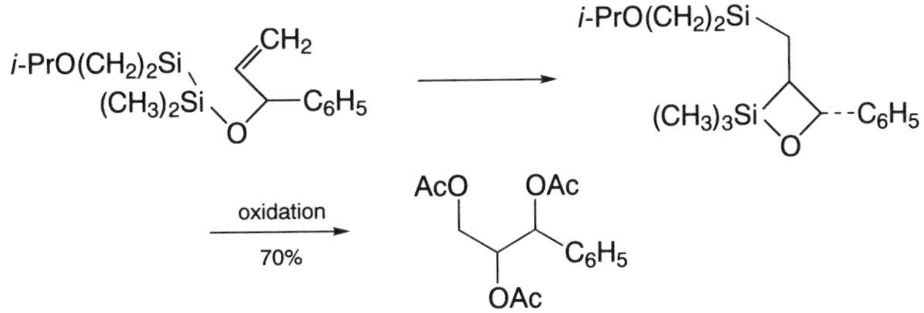

(cis/trans = 5 : 95)

[1] Y. Ito, M. Suginome, and M. Murakami, *J. Org.*, **56**, 1948 (1991).
[2] M. Murakami, P. G. Andersson, M. Suginome, and Y. Ito, *Am. Soc.*, **113**, 3978 (1991).

Palladium(II) acetate-3,3′,3″-Phosphinetriyltribenzenesulfonate (TPPTS).
Cross-coupling in homogeneous aqueous medium.[1] A catalyst **1** prepared from Pd(OAc)$_2$ and TPPTS (5:1) is soluble in H$_2$O–CH$_3$CN or H$_2$O–ethanol. It is highly effective in various reactions that have been effected with Pd(OAc)$_2$. It is effective for inter- and intramolecular Heck reactions, for coupling of vinyl or aryl iodides with terminal alkynes, and for allylic substitution.

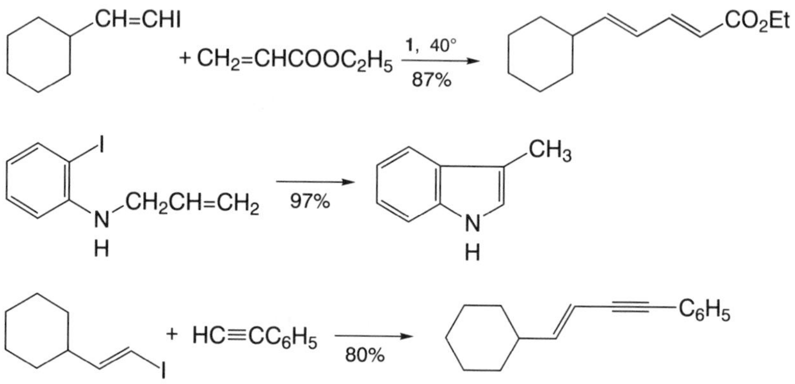

[1] J. P. Genet, E. Blart, and M. Savignac, *Synlett*, 715 (1992).

Palladium(II) acetate–Triphenylarsine.
ArBr → Ar-Ar.[1] Bromoarenes couple to biaryls in DMF at 140° in the presence of Pd(OAc)$_2$ and triphenylarsine as catalysts and 1 equiv. of a trialkylamine. Iodoarenes couple under the same conditions but do not require triphenylarsine.

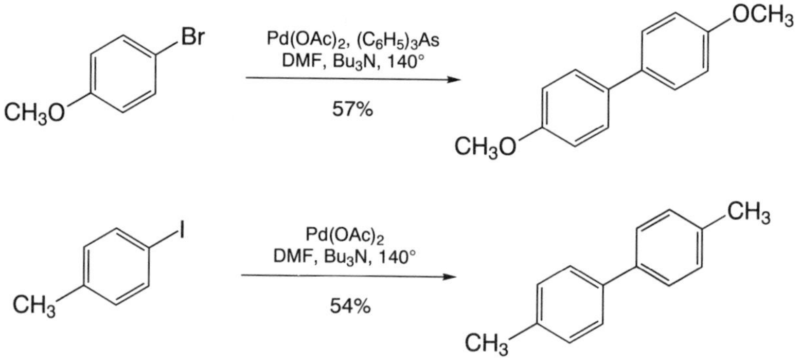

[1] M. Brenda, A. Knebelkamp, A. Greiner, and W. Heitz, *Synlett*, 809 (1991).

Palladium(II) acetate–Triphenylphosphine.

Indole synthesis.[1] A new approach to indoles involves annelation of *o*-iodoanilines with internal alkynes catalyzed by Pd(OAc)$_2$–(C$_6$H$_5$)$_3$P (1:1) with 1 equiv. of LiCl (or Bu$_4$NCl) and 5 equiv. of K$_2$CO$_3$ or Na$_2$CO$_3$ (equation I). R^1 can be H, CH$_3$, Ac,

(I)

or tosyl. The alkyne can bear a variety of groups: alkyl, aryl, alkenyl, or trimethylsilyl. In the case of unsymmetrical alkynes, the more bulky group is located nearer the nitrogen atom. The reaction shows 2:1 regioselectivity even with 2-pentyne. The facile reaction with silylalkynes followed by desilylation, provides a useful route to 3-substituted indoles.

Bicyclization-anion capture.[2] Strained bicyclic systems can be prepared by a tandem cyclization-anion capture process using a Pd catalyst in combination with sodium formate, triethylamine, or K$_2$CO$_3$

α-Alkylidene cyclic ethers.[3] A route to substituted tetrahydrofurans or -pyrans involves conversion of acetylenic alcohols such as **1** to the lithium alkoxide (BuLi) followed by Pd(II)-catalyzed cross-coupling with an alkyl or aryl halide.

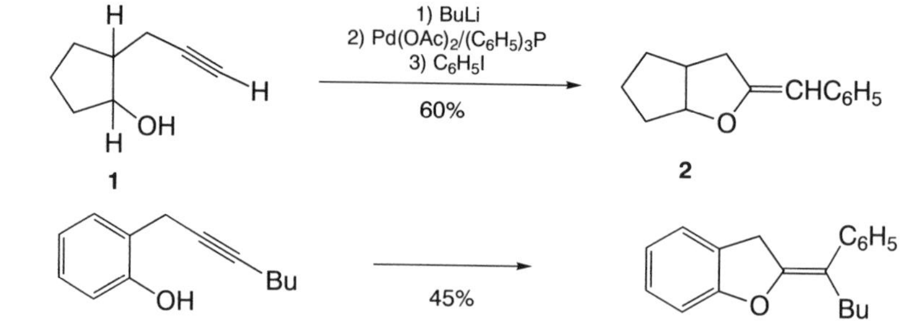

Decarboxylation of allylic aryl carbonates.[4]   These carbonates (**1**) are available by reaction of a phenol with triphosgene in pyridine followed by addition of an allylic alcohol (71–90% yield). They undergo decarboxylation with Pd(OAc)$_2$ (5 mol %) and (C$_6$H$_5$)$_3$P (20 mol %) in THF or CH$_2$Cl$_2$ at 25° to form allylic aryl ethers (**2**).

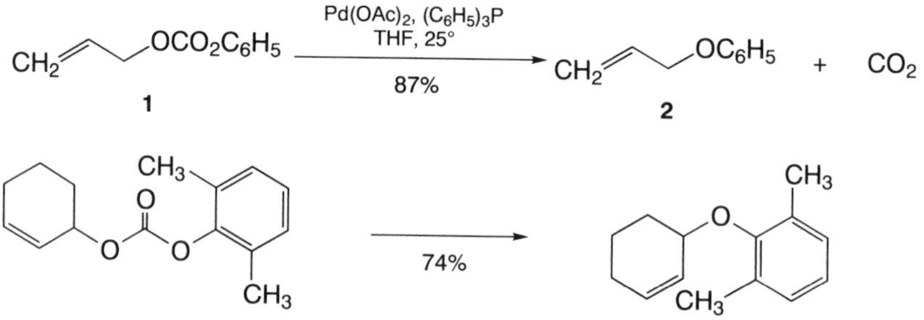

Arylation of 1,3-dienes.[5]   Pd-catalyzed coupling of aryl iodides with 1,3-dienes results in (E,E)-4-aryl-1,3-dienes regardless of the geometry of the original 1,3-diene.

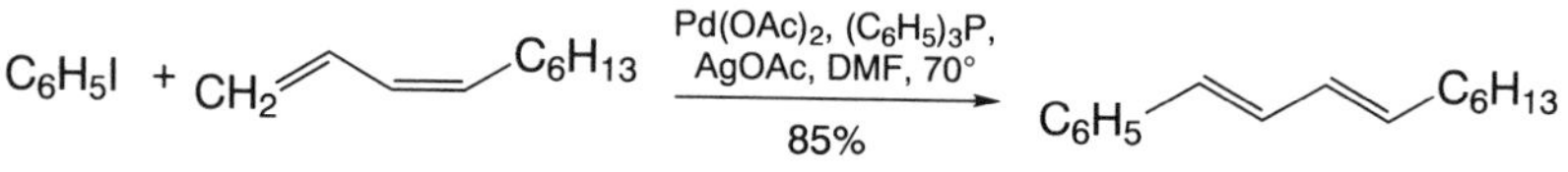

Alkylbenzoquinones via a Heck-type coupling.[6]   A palladium-catalyzed coupling of bromobenzene with a terminal alkene provides a convenient route to benzoquinones bearing a long-chain alkyl group, such as masanin (**5**). Thus Heck-type coupling of the bromobenzene **1** with 9-decene-1-ol furnishes **2** in 49% yield. Hydrogenation of **2** followed by Swern oxidation furnishes an aldehyde (**3**), which undergoes Wittig olefination to (Z)-**4**. This product was converted by deprotection and oxidation (Ag$_2$CO$_3$) into the quinone **5**.

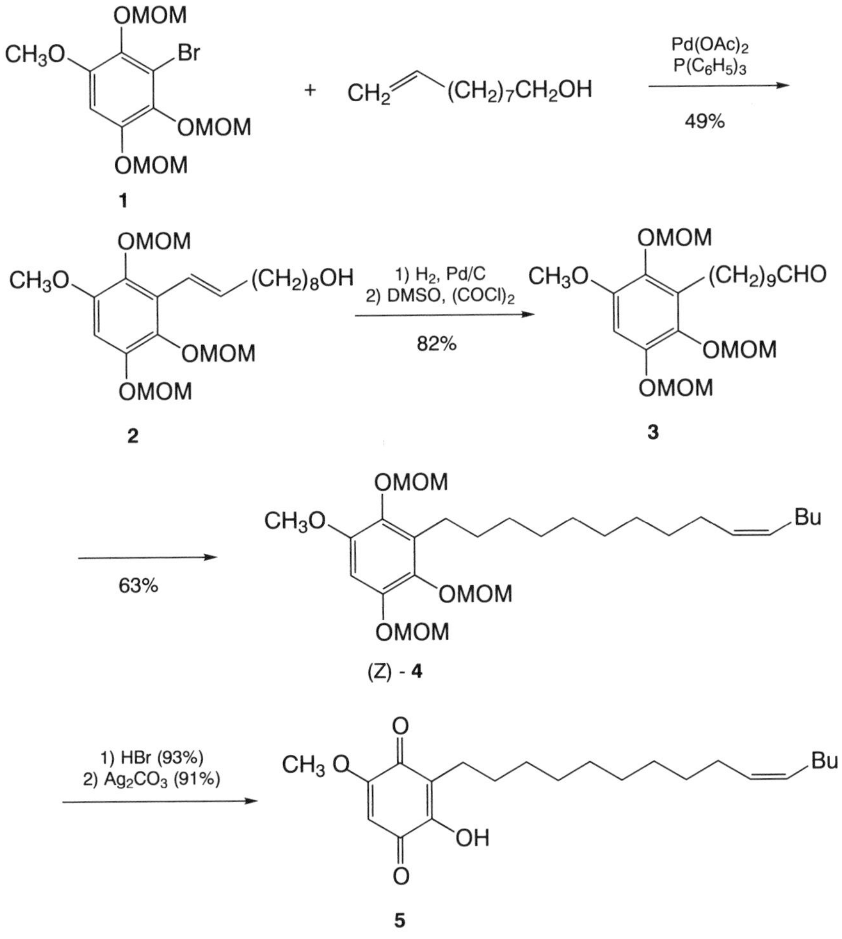

1   + CH_2=CH-$(CH_2)_7CH_2OH$ $\xrightarrow[49\%]{\text{Pd(OAc)}_2 \quad \text{P(C}_6\text{H}_5)_3}$

2 $\xrightarrow[82\%]{\text{1) H}_2, \text{Pd/C} \quad \text{2) DMSO, (COCl)}_2}$ **3**

$\xrightarrow{63\%}$ **(Z) - 4**

$\xrightarrow{\text{1) HBr (93\%)} \quad \text{2) Ag}_2\text{CO}_3 \text{(91\%)}}$ **5**

Heck tetracyclization of dienynes.[7]   Treatment of the dienyne **1** with Pd(OAc)$_2$, P(C$_6$H$_5$)$_3$, K$_2$CO$_3$ (2 equiv.), and AgNO$_3$ (1 equiv.) in CH$_3$CN results in the tricyclic product **2** (52% yield) together with the tetracyclic product **3** (10% yield.)

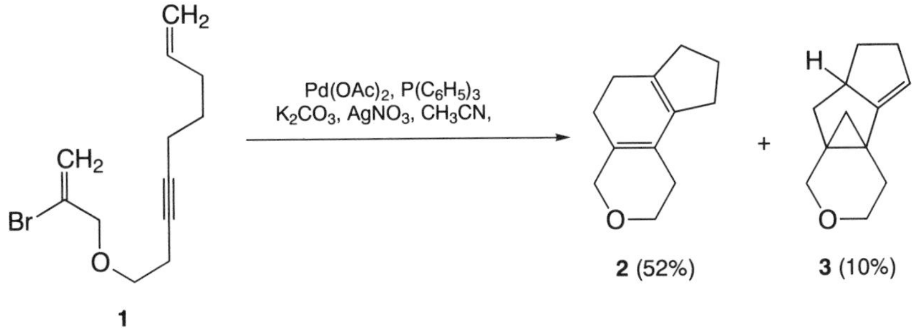

Treatment of the related dienyne **4** under the same conditions provides tetracycle **5** in 71% yield (isolated).

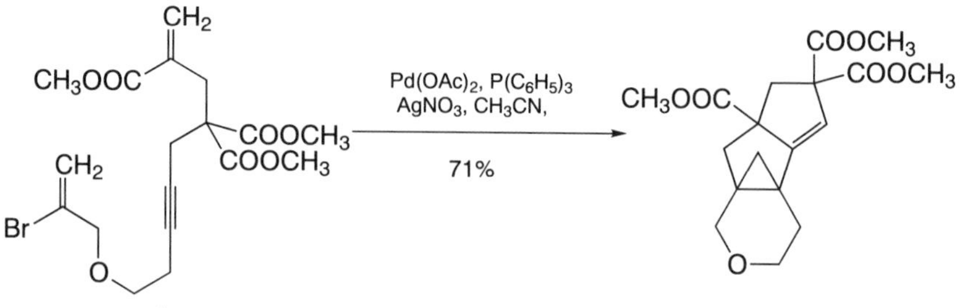

Heck polycyclization of dienynes.[8]   Treatment of dienyne **1** bearing a vinyl bromide group with catalytic amounts of Pd(OAc)$_2$ and P(C$_6$H$_5$)$_3$ and 2 equiv. of silver carbonate in CH$_3$CN at 80° results in the tricyclic cyclohexadiene **2** in 60% yield. A similar reaction occurs if the dienyne group is attached to a cyclohexane ring (**3 → 4**, equation I).

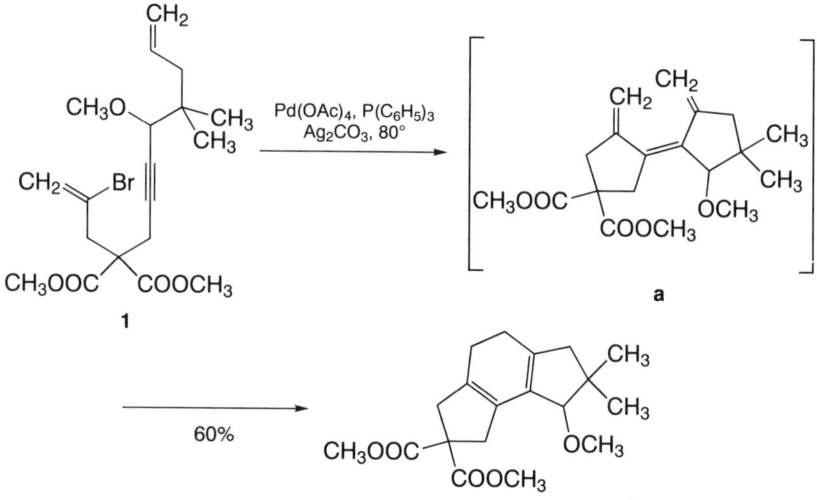

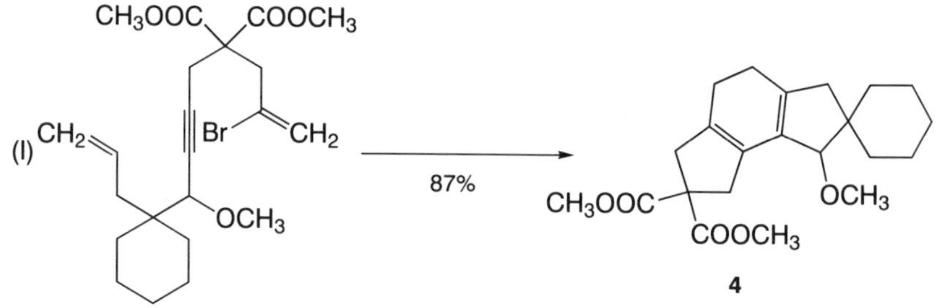

Polycyclization of dienynes.[9] Treatment of the enyne **1** with Pd(OAc)$_2$/ triphenylphosphine and a base (Heck conditions) results in cyclization to a mixture of *cis-* and *trans-*trienes **2**. At 130° and with K$_2$CO$_3$, only *trans-*2 undergoes an intramolecular Diels–Alder reaction to **3**.

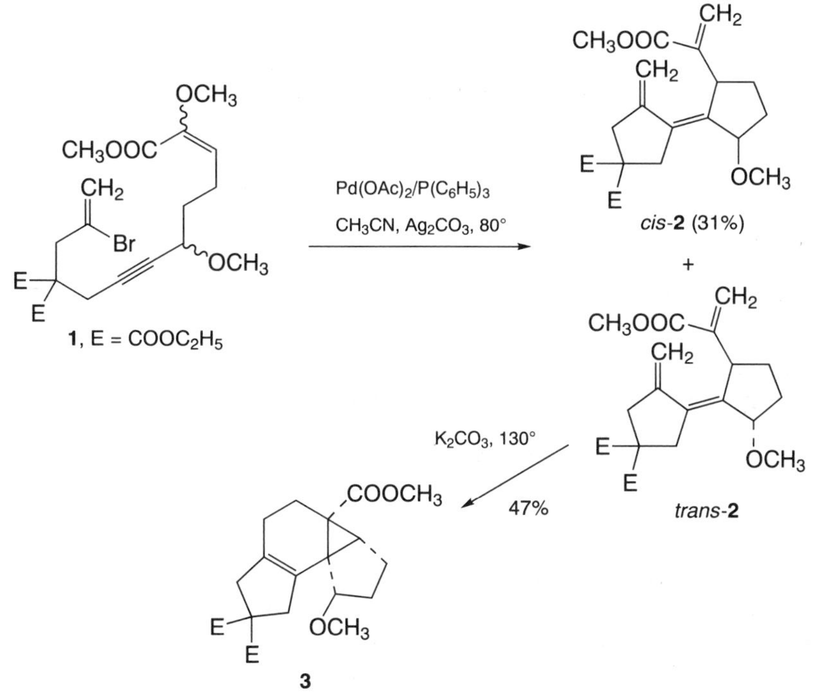

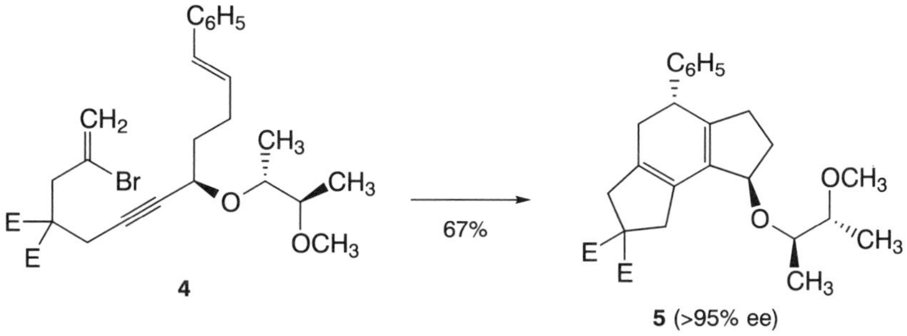

Under the same conditions, the enantiomerically pure dienyne **4** cyclizes to the tricyclic **5** in >95% ee and in 67% isolated yield.

Stille coupling of 2-carboxyethyl enol triflates with organostannanes.[10] In the presence of Pd(OAc)$_2$/P(C$_6$H$_5$)$_3$ (1:2) or Pd[P(C$_6$H$_5$)$_3$]$_4$, the enol triflates of β-keto esters couple with vinyl-, allyl-, and alkynyl(tributyl)stannanes, but not with phenyltributyltin.

[1] R. C. Larock and E. K. Yum, *Am. Soc.*, **113**, 6689 (1991).
[2] R. Grigg, V. Sridharan, and S. Sukirthalingam, *Tetrahedron Letters*, **32**, 3855 (1991).
[3] F.-T. Luo, I. Scheyder, and R.-T. Wang, *J. Org.*, **57**, 2213 (1992).
[4] R. C. Larock and N. H. Lee, *Tetrahedron Letters*, **32**, 6315 (1991).
[5] T. Jeffery, *ibid.*, **33**, 1989 (1992).
[6] J. S. Yadav, V. Upender, and A. V. Rama Rao, *J. Org.*, **57**, 3242 (1992).
[7] F. E. Meyer, P. J. Parsons, and A. de Meijere, *ibid.*, **56**, 6487 (1991).
[8] F. E. Meyer, J. Brandenburg, P. J. Parsons, and A. de Meijere, *J. C. S. Chem. Comm.*, 390 (1992).
[9] F. E. Meyer, H. Henniges, and A. de Mierjere, *Tetrahedron Letters*, **33**, 8039 (1992).
[10] I. N. Houpis, *ibid.*, **32**, 6675 (1991).

Palladium(II) acetate/Triphenylphosphine – Hexaalkylditin.

Coupling of aryl and benzylic halides. Grigg[1] has extended the inter- and intramolecular coupling of vinyl bromides to form 1,3-dienes catalyzed by Pd(OAc)$_2$–triphenylphosphine (**13,234**) to coupling of aryl and benzylic halides by the combination of a Pd(0) catalyst and [(CH$_3$)$_3$Sn]$_2$ or (Bu$_3$Sn)$_2$ (1 equiv.).

This coupling has been used for synthesis of the alkaloid hippadine (**1**).

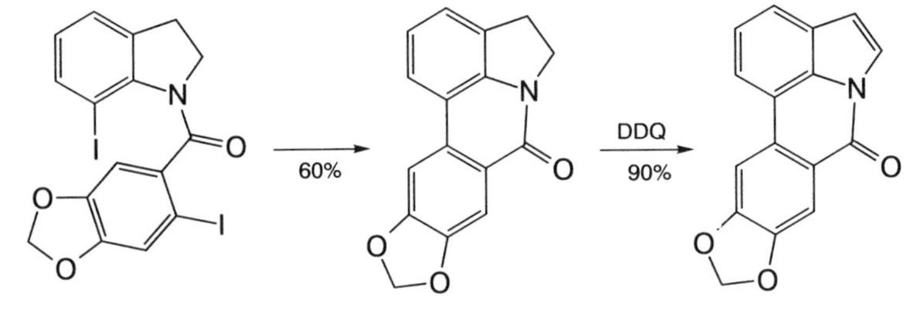

1

[1] R. Grigg, A. Teasdale, and V. Sridharan, *Tetrahedron Letters,* **32**, 3859 (1991).

Palladium(II) acetylacetonate–Tributylphosphine, Pd(acac)$_2$–Bu$_3$P (1:1).

Hydrogenolysis of allylic formates.[1]   The Pd(0) catalyst generated from Pd(OAc)$_2$ and Bu$_3$P effects hydrogenolysis of allylic acetates or carbonates with ammonium formate to afford 1-alkenes. Allylic formates are also converted to 1-alkenes by this Pd(0) catalyst without use of ammonium formate.

$$RCH=CHCH_2OR^1 \ + \ HCOONH_4 \ \xrightarrow{\text{Pd(0)}} \ RCH_2CH=CH_2 \ + \ CO_2$$

This hydrogenolysis can be used to control the ring junction in hydrindane and decalin systems.

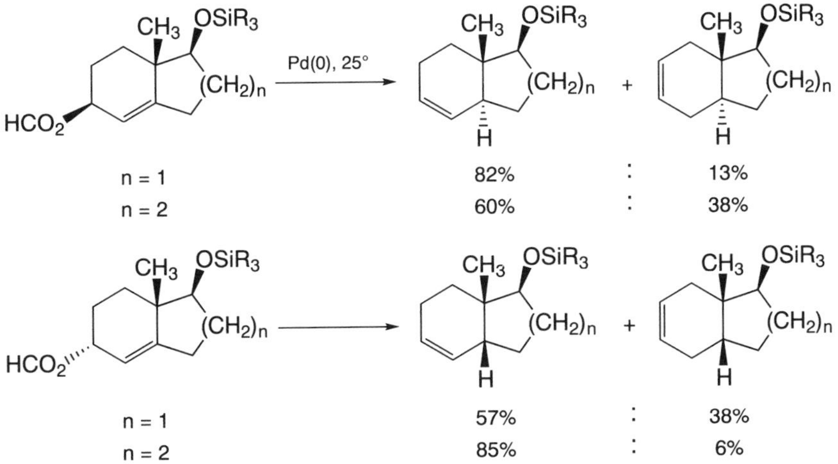

n = 1	82%	13%
n = 2	60%	38%

n = 1	57%	38%
n = 2	85%	6%

[1] T. Mandai, T. Matsumoto, M. Kawada, and J. Tsuji, *J. Org.,* **57**, 1326 (1992).

(R)-(−)-Pantolactone, 16,269−270.

Asymmetric cyclopropanation.[1]   Asymmetric cyclopropanation of styrene can be effected with chiral vinyl α-diazo esters with $Rh_2(OAc)_4$ or rhodium(II) octanoate as catalyst. The products can be converted into optically active cyclopropylamino acids. Of several chiral auxiliaries, (R)-(−)-pantolactone is the reagent of choice.

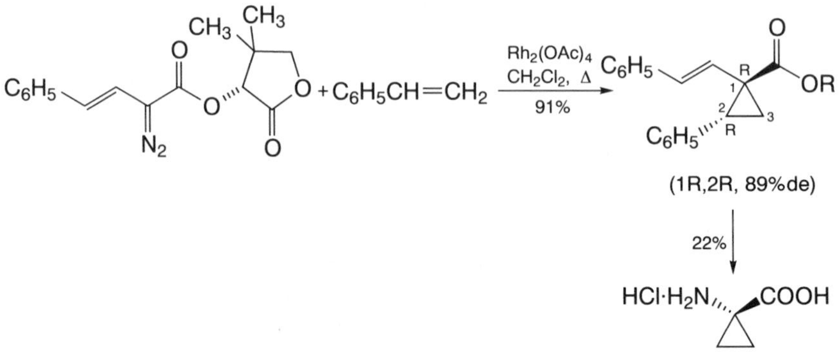

(1R,2R, 89%de)

(S)-α-Halo esters.[2]   These α-halo esters (**3**) can be prepared in 75−95% de by reaction of α-haloketenes (**1**) with (R)-pantolactone (**2**).

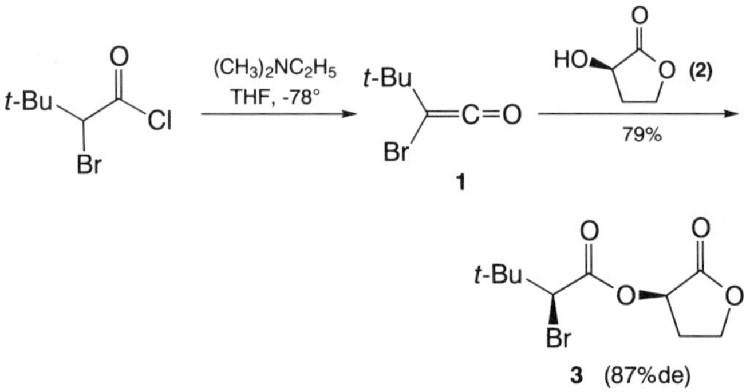

3 (87%de)

The products (**3**) can be converted into α-amino acids by azidation (inversion, 70% yield) and saponification (72% ee).

[1] H. M. L. Davies and W. R. Cantrell, Jr., *Tetrahedron Letters*, **32**, 6509 (1991).
[2] T. Durst and K. Koh, *Tetrahedron Letters*, **33**, 6799 (1992).

Pentacarbonyliron.

[4+1]Cycloaddition; 2,5-dialkylcyclo-3-pentenones.[1]   In the presence of $Fe(CO)_5$ or $Fe_2(CO)_9$, diallenes (**1**) react with carbon monoxide to give 2,5-dialkylidene-3-cyclopentenones **2** in surprisingly good yield.

$$\mathbf{1} + CO \xrightarrow[79\%]{Fe(CO)_5} \mathbf{2}$$

[1] B. E. Eaton, B. Rollman, and J. A. Kaduk, *Am. Soc.*, **114**, 6245 (1992).

Pentafluorophenyl diphenylphosphinate,   **(1).**

Prepared by reaction of diphenylphosphinic chloride and pentafluorophenol with imidazole.

Coupling of amino acids with α-amino esters.[1]   This coupling to peptides can be effected with the phosphinate (1 equiv.) and diisopropylethylamine in DMF at 25°. Yields are generally >90%, and the reaction is racemization free. The reagent can also be used for solid-phase peptide synthesis.

[1] S. Chen and J. Xu, *Tetrahedron Letters,* **32**, 6711 (1991).

Periodinane (1) of Dess-Martin, 12,378–379.

Full details for preparation of this tetravalent iodine compound are now available.[1] Since explosions have been reported in the preparation, the directions should be followed with care. Moreover, the reagent is sensitive to moisture, which converts it into iodosylbenzoic acid, also reported as explosive. The periodinane oxidizes primary and secondary alcohols rapidly and efficiently to carbonyl compounds without further oxidation to acids. Benzylic and allylic alcohols can be selectively oxidized. It does not react with sulfides or vinyl ethers. The iodine-containing by-product can be removed by hydrolysis to 2-iodosylbenzoic acid.

[1] D. B. Dess and J. C. Martin, *Am. Soc.*, **113**, 7277 (1991).

(−)-8-(4-Phenoxyphenyl)menthol.

Asymmetric synthesis of dihydropyridones.[1]   8-Phenylmenthol has been used frequently as a chiral auxiliary, but this derivative (1) is more effective as the chiral auxiliary for enantioselective reactions of N-acylpyridinium salts with organometallic reagents. Thus 2, prepared by reaction of 4-methoxy-3-(triisopropylsilyl)pyridine with the chloroformate of 1 reacts with the Grignard reagent 3 to give an adduct that on acidic deprotection provides the dihydropyridone 4 in 94% de. Conversion of the alcohol to a chloride [P(C₆H₅)₃ and NCS] followed by treatment with sodium methoxide in methanol cleaves the chiral auxiliary as the methyl carbonate (94% yield) and at the same time effects cyclization to form the chiral bicyclic dihydropyridone 5 in 84% yield.

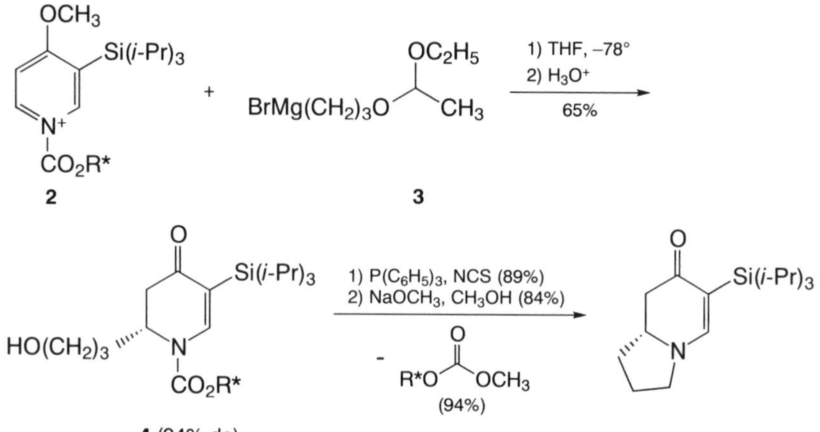

2   +   **3**   →   **4 (94% de)**   →   **5**

This chiral product **5** was used for asymmetric synthesis of the alkaloids (+)-elaeokanine A (**9**) and (+)-elaeokanine C (**8**). Thus the anion of **5** reacts with dimethyl-carbamyl chloride to form mainly the *trans*-amide **6** (97:3). Desilylation (oxalic acid, CH$_3$OH, 96%) and catalytic hydrogenation (96%) provides **7**. The dimethylamide group is converted to a propyl ketone by reaction with PrMgCl catalyzed by CeCl$_3$ (2.5 equiv.) to give (+)-**8**. Finally, treatment of (+)-**8** with NaOH in CH$_3$OH provides (+)-**9** in 30% yield.

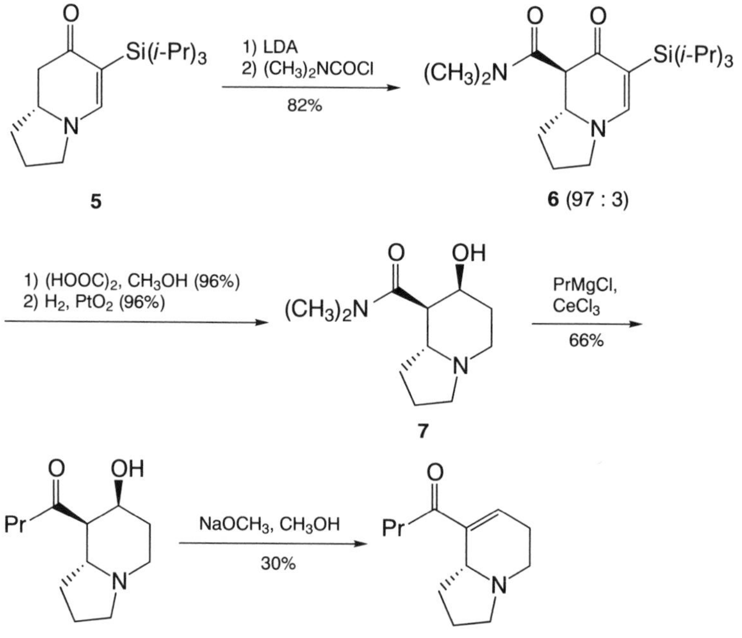

5   →   **6 (97 : 3)**   →   **7**   →   **(+) - 8**   →   **(+) - 9**

[1] D. L. Comins and H. Hong, *Am. Soc.*, **113**, 6672 (1991).

(S)-Phenylalanine dimethylamide,

$$\underset{H_2N}{\overset{C_6H_5}{\diagup}}\!\!\!\diagdown\!\!\!\underset{O}{\overset{}{\diagdown}}N(CH_3)_2 \quad (1)$$

(R)- and (S)-α,α-Disubstituted α-amino acids.[1]   Although only a few α-amino acids of this type are known, they are present in some microbial peptide antibiotics. A number of routes are known to racemic or to one enantiomer of amino acids of this type, but the most practical route to both enantiomers appears to be resolution of racemic N-acylated derivatives of these acids by a chiral amide containing one or two (S)-phenylalanine residues. The diastereomers obtained in this way can be separated by crystallization or by flash chromatography on silica gel. Thus the racemic amino acid (**2**) is converted to the 1,3-oxazol-5(4*H*)-one (**3**) by an activating agent such as 1,1′-carbonyldiimidizole (CDI), which without isolation is coupled with the chiral amine **1** (DMF, 50°) to form two diastereomeric dipeptides (**4**) which are separated by crystallization or chromatography on silica gel. Several methods are available for cleaving the chiral amide, but the most useful is reaction with CF_3SO_3H in CH_3OH at 80° (Scheme I).

Scheme I

In some cases, better results were obtained by use of L-phenylalanyl-L-phenylalanine dimethylamide in place of **1**.

[1] D. Obrecht, C. Spiegler, P. Schönholger, K. Müller, H. Heimgartner, and F. Stierli, *Helv.,* **75,** 1666 (1992).

(1S,2R)-(+)-Phenylcyclohexanol (13,244; 14,128–129, 16,113–114)

Use as chiral auxiliary. Whitesell[1] has reviewed use of this chiral alcohol (**1**), particularly as compared with that of (−)-menthol and (1R)-(+)-8-phenylmenthol. One advantage is that both enantiomers of **1** are available by resolution of *trans*-2-phenyl-cyclohexanol by means of enzymatic hydrolysis of the esters and that 2-substituted cyclohexanols are readily available. Although this chiral auxiliary was used originally for control of ene reactions of glyoxylates, it is also useful for asymmetric alkylation of enolates, and for control of various cycloaddition reactions.

Asymmetric [4+2]//[3+2]cycloadditions.[2]   The reaction of a chiral vinyl ether (**2**) and a nitroalkene (**1**) catalyzed by a Lewis acid results in an intermediate adduct (**a**) which undergoes an intramolecular [3+2]cycloaddition. Both (+)-camphor and (−)-*trans*-2-phenylhexanol have been used as the chiral auxiliary. Thus the vinyl ether (**2**) formed from the latter auxiliary reacts with **1** to give **3** in 73% yield. Hydrogenation cleaves the chiral auxiliary and provides a tricyclic α-hydroxy lactam **4**, in which

C_1 has the (S)-configuration. If the reaction of **1** and **2** is catalyzed by methylaluminum bis(2,6-diphenylphenoxide), MAPh, (1R)-**4** is obtained in 79% ee. Surprisingly, use of methylaluminum bis(2,6-di-*t*-butyl-4-methylphenoxide), MAD, provides (1S)-**4** in 75% ee. When (+)-camphor is used as the chiral auxiliary, the enantioselectivity can also be controlled by the choice of Lewis acid, but in this case MAPh provides (1R)-**4** in 99% ee, and $Cl_2Ti(O\text{-}i\text{-}Pr)_2$ provides (1S)-**4** in 98% ee. This result indicates that a change in the Lewis acid catalyst can be as effective as use of enantiomeric chiral auxiliaries.

Chiral α-methylene-γ-butyrolactones.[3]   An asymmetric synthesis of α-methylene-γ-butyrolactones such as **7** utilizes the benzoate (**2**) of (1R,2S)-**1**, which on reaction with 1,1-dibromohexane and $Zn-TiCl_4 \cdot TMEDA$ forms the (Z)-enol ether **3**. Thus this product reacts with dichloroketene to form the cyclobutanone (−)-**4**. Baeyer–Villiger oxidation of **4** followed by oxidation with chromium(II) perchlorate provides the α-chlorobutenolide (+)-**5**. Hydrogenation and oxidation provides the β-carboxy-γ-butyrolactone (−) **6**. The final step involves introduction of a methylene group to provide (−)-**7**, methylenolactocin, a natural antitumor antibiotic.

3-Hydroxy-4-phenyl-β-lactams.[4]   The lithium enolate of (silyloxy)acetates (**2**) couple with the N-(trimethylsilyl)imine **3** to give 3-hydroxy-4-aryl-β-lactams (**4**). The stereoselectivity depends in part on the size of the silyloxy group but mainly on the ester group. Use of either (+)- or (−)-trans-2-phenyl-1-cyclohexyl (**1**) as the chiral auxiliary results entirely in a cis-β-lactam (**4**) in 80% yield and 96–98% ee. Use of (−)-menthyl or of Oppolzer's D-isobornyl auxiliary (**12**,103–104) results in lower yields and enantioselectivity. The cyclocondensation with the ester **2** from (−)-**1** results in (3R,4S)-**4** in 96% ee. On desilylation and acid hydrolysis, this lactam provides (2R,3S)-

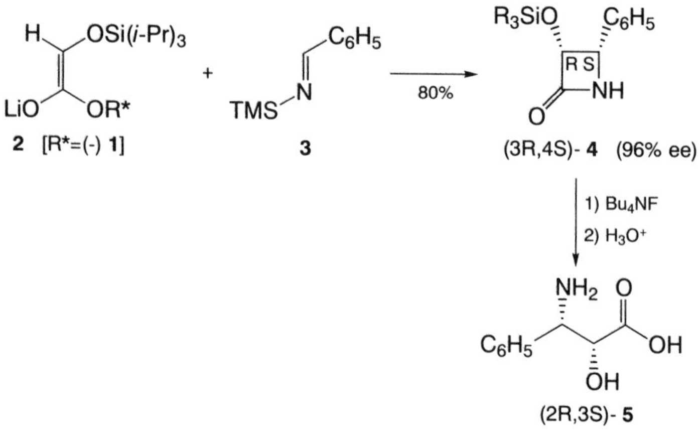

3-phenylisoserine. The N-benzoate of this unnatural amino acid provides the side chain of the diterpine taxol and is essential for the potent antitumor activity of the taxol.

[1] J. K. Whitesell, *Chem. Rev.*, **92**, 953 (1992).
[2] S. E. Denmark and M. E. Schnute, *J. Org.*, **56**, 6738 (1991).
[3] M. B. M. de Azevedo, M. M. Murta, and A. E. Greene, *J. Org.*, **57**, 4567 (1992).
[4] I. Ojima, I. Habus, M. Zhao, G. I. Georg, and L. R. Jayasinghe, *J. Org.*, **56**, 1681 (1991).

(R)- and (S)-1-Phenylethylamine, $CH_3\overset{*}{C}H(C_6H_5)NH_2$ **(1), 14,257.**

Asymmetric Michael reactions via enamines.[1]   Imines (*via* the tautomeric enamines) undergo conjugate addition to electrophilic alkenes, but the first asymmetric reaction of this type was accomplished in 1985 using chiral imines derived from (+)- or (−)-**1** (equation I). Of a large number of chiral amines examined as the chiral auxiliary, the

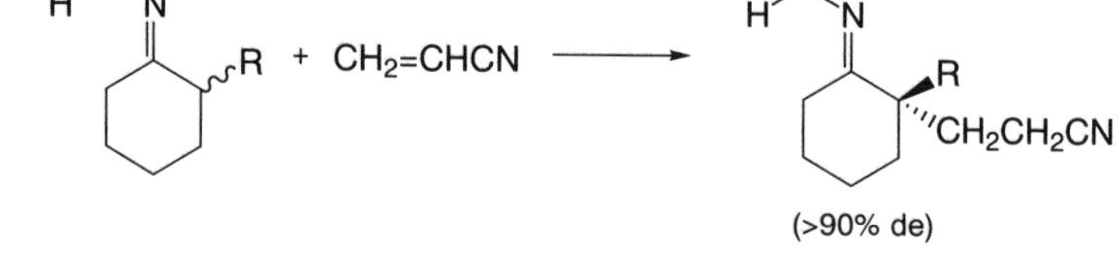

most useful all possess an aryl group α to the amine group. Modern extensions of this reaction including unpublished results are discussed in detail (77 references, 1985–1991).

Chiral tetronic acids.[2]   The key step in a synthesis of chiral 5,5-disubstituted tetronic acids (4-hydroxybutenolides) involves Michael addition to acrylates of a chiral imine (**2**) derived from a dihydrofuran-3-one and (R)-1-phenylethylamine (**1**). After

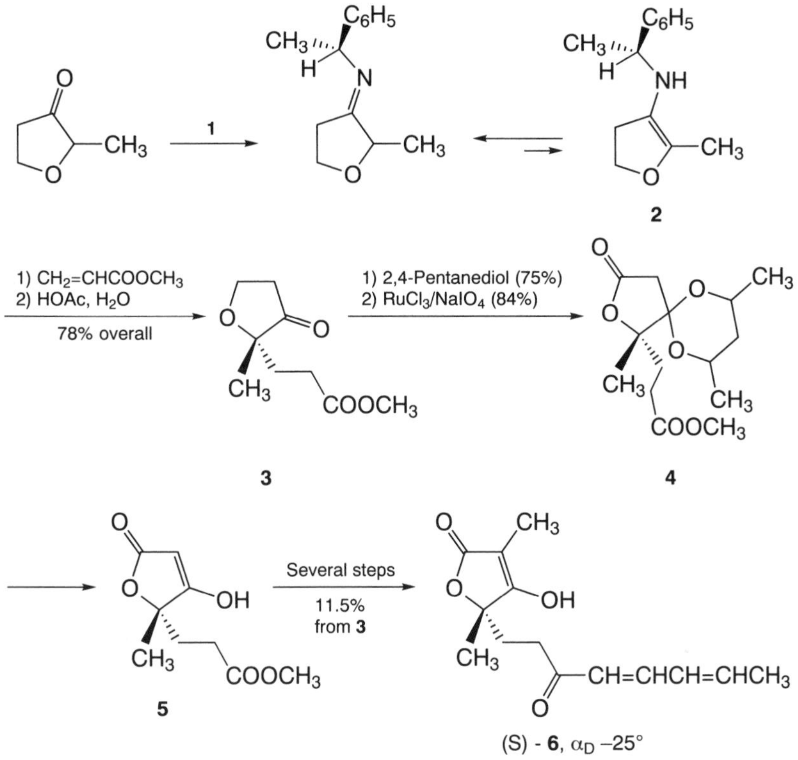

protection of the carbonyl group at C_4, the ketone group at C_2 was introduced by oxidation with ruthenium tetroxide (84% yield) to give **4**. This compound can be converted into vertinolide (**6**) in 12 steps in 11.5% yield.

Asymmetric Diels–Alder reaction of imines.   Diels–Alder reactions of imines generally require an acyl or tosyl activating group on nitrogen, but can be effected in the presence of trifluoroacetic acid (1 equiv.) and a catalyst such as BF_3 etherate[3] or water (0.03 equiv.).[4] Under these conditions, the imine BzlN=CHCOOR, formed from benzylamine and an alkyl glyoxylate, reacts with simple dienes to form derivatives of pipecolic acid in high yield (equation I). This aza-Diels–Alder reaction has been used

(I)

for an enantio- and diastereoselective synthesis of 4-methylpipecolic acid (equation II). Thus reaction of the imine (R)-**1**, prepared from (R)-1-phenylethylamine and ethyl glyoxylate, with 2-methylbutadiene provides the adduct **2** in 44% yield and 70% de. This adduct is converted by hydrogenation and removal of the chiral auxiliary to ethyl (2S,4R)- and (2S,4S)-4-methylpicolate (**3**). Note that reaction of (R)-**1** with cyclopentadiene gives the *exo*-adduct in 82% yield with 89% de.

(II)

[1] J. d'Angelo, D. Desmaële, F. Dumas, and A. Guingant, *Tetrahedron: Asymmetry,* **3**, 459 (1992).
[2] D. Desmaële, *Tetrahedron,* **48**, 2925 (1992).
[3] L. Stella, H. Abraham, J. Feneau-Dupont, B. Tinant, and J. P. Declercq, *Tetrahedron Letters,,* **31**, 2603 (1990).
[4] P. D. Bailey, R. D. Wilson, and G. R. Brown, *J. C. S. Perkin I,* 1337 (1991).

(S)-Phenylglycine.

Asymmetric synthesis of α-methylene-γ-lactams.[1]   The organozinc reagents de-rived from 2-(bromomethyl)acrylates react with imines to form α-methylene-γ-lactams. Use of (S)-1-phenylglycine as the chiral auxiliary for the imine provides these products in >95% ee.

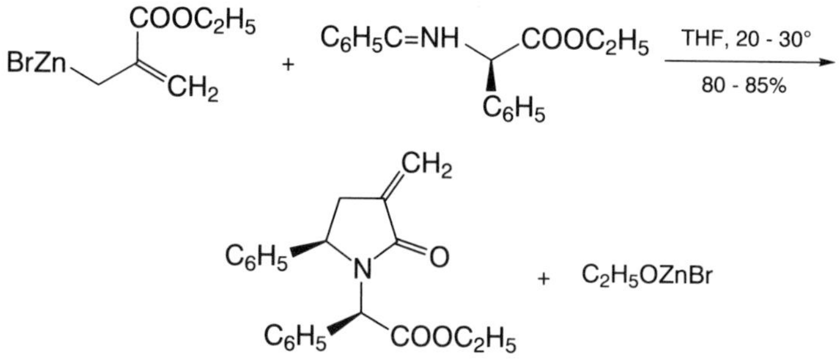

(>95% ee)

[1] Y. A. Dembélé, C. Beland, P. Hitchcock, and J. Villieras, *Tetrahedron: Asymmetry*, **3**, 351 (1992).

(R)-(−)-2-Phenylglycinol (1), 15,256

Asymmetric synthesis of pyrrolidines and pyrrolinones.[1]   This chiral β-hydroxy-amine (1) reacts with a γ-keto acid in refluxing toluene to form a chiral bicyclic lactam 2 in 75–98% yield. Treatment of 2 with excess alane, generated from AlCl$_3$ and LiAlH$_4$,

cleaves the C–O bond and also reduces the carbonyl group to provide **3**, which on transfer hydrogenolysis provides the 1-substituted pyrrolidine **4**.

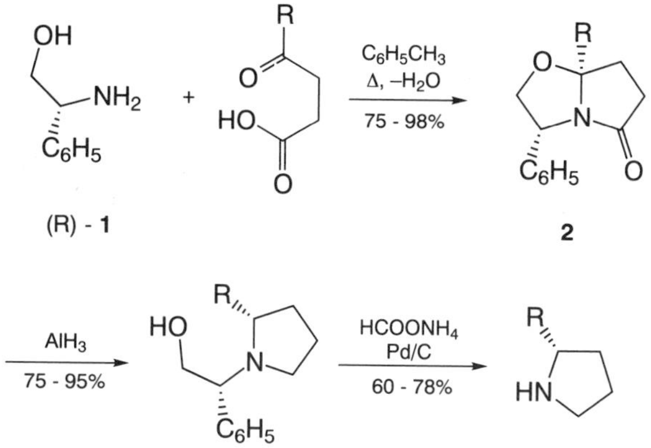

Reduction of **2** with triethylsilane catalyzed by TiCl$_4$ effects only cleavage of the C–O bond to give **5** in >94% de. This product is reduced to the lactam **6** by Li/NH$_3$. Unfortunately the chiral auxiliary is not recovered.

5   **6**

[1] L. E. Burgess and A. I. Meyers, *J. Org.*, **57**, 1656 (1992).

Phenyliodine(III) diacetate, C₆H₅I(OAc)₂.

Oxidation of N-protected tyrosine.[1]   Oxidation of an N-protected tyrosine **1** with C₆H₅I(OAc)₂ in methanol provides a spirolactone (**2**), which is converted by treatment with NaHCO₃ in CH₃OH into the *exo*-hydroindole **3**, a bicyclic system common to various *Stemona* alkaloids. Actually **1** can be converted directly into **3** by oxidation of **1** in the presence of sodium bicarbonate. The conversion of **1** to **3** evidently involves *para*-hydroxylation of a phenol followed by a highly diastereoselective intramolecular

1, R = Cbz or Boc

conjugate addition.

[1] P. Wipf and Y. Kim, *Tetrahedron Letters*, **33**, 5477 (1992).

Phenyliodine(III) diacetate – Iodine.

Hypoiodite-type reaction.[1]   Treatment of the allylic alcohol **1** with C₆H₅I (OAc)₂ and iodine generates an alkyloxy radical, which undergoes β-scission to **a** followed by cyclization to a *cis*-hydroazulenone ring system (**2**). The CH₂I group of **2** can be reduced to a methyl group by a catalytic amount of Bu₃SnH (60% yield). Treatment of **2** with Bu$_3$SnH–AIBN under high dilution results in ring expansion to **3**.

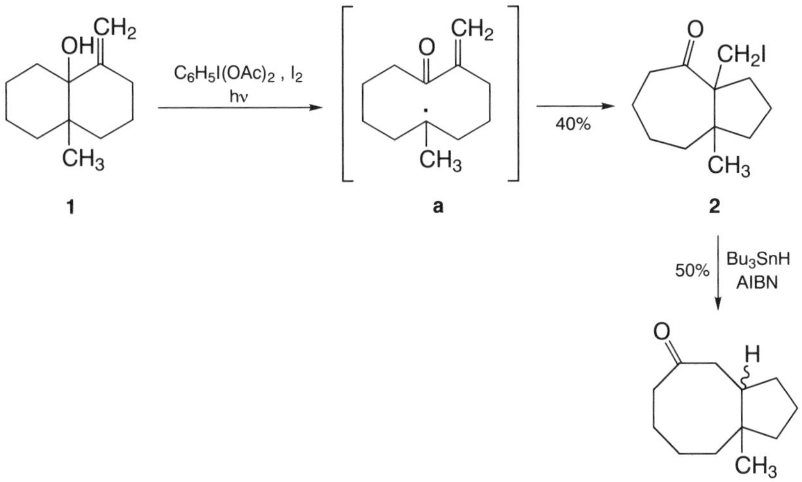

Iodoepoxides from allylic alcohols; cyclopentanols.[2]   Reaction of a cyclic allylic alcohol (**1**) with 1 equiv. each of C$_6$H$_5$I(OAc)$_2$ and I$_2$ under irradiation with a sunlamp results in the iodoepoxide **2**. The product undergoes an atom transfer cyclization on irradiation with (Bu$_3$Sn)$_2$ in refluxing benzene.

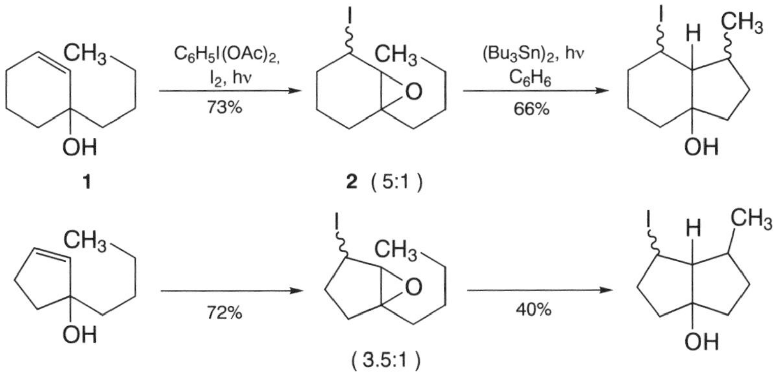

[1] C. W. Ellwood and G. Pattenden, *Tetrahedron Letters,* **32**, 1591 (1991).
[2] V. H. Rawal and S. Iwasa, *ibid.,* **33**, 4687 (1992).

8-Phenylmenthol.

Chiral 1-acyl-2-alkyl-1,2-dihydropyridines.[1]   The useful 1,2-dihydropyridines can be prepared by reaction of Grignard reagents with the chloroformate **1** derived from reaction of (−)-8-phenylmenthol with 4-methoxy-3-(triisopropylsilyl)pyridine. The bulky group at C_3 is essential for high diastereoselectivity in this reaction, and the C_4-methoxy group facilitates removal of the chiral auxiliary. Reaction of Grignard reagents with **1**

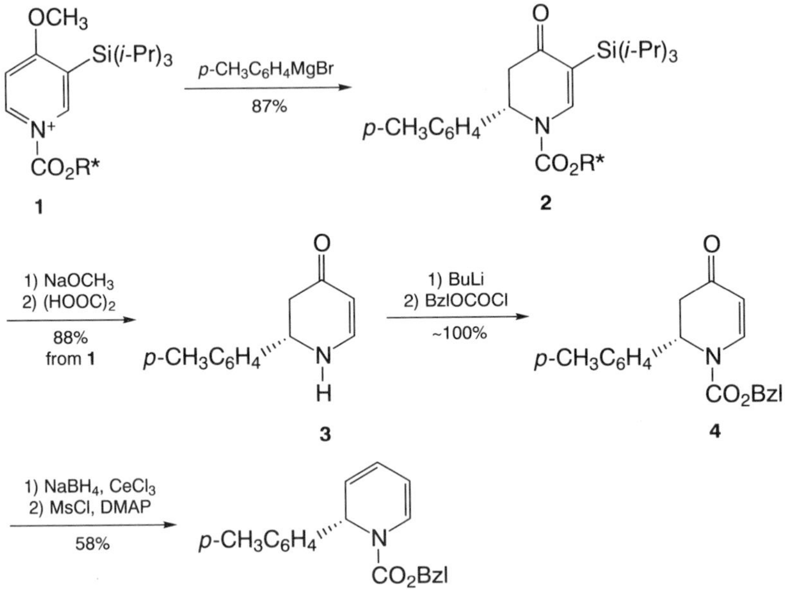

provides **2** in high diastereoselectivity, which can be converted to enantiomerically pure **5** as shown.

[1] D. L. Comins, H. Hong, and J. M. Salvador, *J. Org.*, **56**, 7197 (1991).

(2R,4S)-2-Phenyl-1,3,2-oxazaphospholidine (1).

Prepared from (S)-(+)-prolinol and bis(dimethylamino)phenylphosphine.[1]

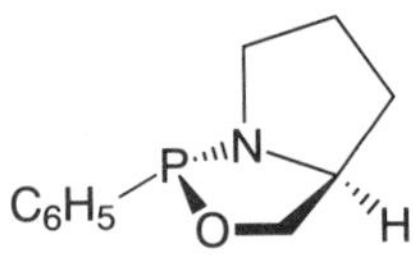

Enantioselective reduction of ketones with a borane complex.[2]   The complex (**2**), obtained by reaction of $BH_3 \cdot THF$ or $BH_3 \cdot S(CH_3)_2$ with **1** at 25°, is an effective catalyst for reduction of ketones by diborane in toluene, but only at temperatures above $-20°$. Highest enantioselectivity obtains from reactions at $110°$. At that temperature, isopropyl methyl ketone is reduced with 92% ee by $BH_3 \cdot THF$ in the presence of 2 mol % of **2**. Use of 1 equiv. of **2** permits reduction of dialkyl and alkyl aryl ketones at $110°$ in >99% ee. This oxazaphospholidine resembles Corey's chiral oxazaborolidine (**16**,254).

[1] W. J. Richter, *Ber.,* **117**, 2328 (1984).
[2] J. M. Bruneli, O. Pardigon, B. Faure, and G. Buono, *J. C. S. Chem. Comm.,* 287 (1992).

N-(Phenylseleneno)phthalimide.

Cyclization of homoallylic alcohols to tetrasubstituted tetrahydrofurans.[1]   This cyclization can be effected with almost complete stereocontrol by phenylselenenylation with Nicolaou's reagent **1**. The products are reduced with $Bu_3SnH/AIBN$ to trialkylated tetrahydrofurans.

In the last two examples higher regioselectivity can be obtained by use of C_6H_5SeCl in CH_3CN. But in any case the favored product has an *anti*-relationship between the SeC_6H_5 group and the adjacent methyl.

[1] E. D. Mihelich and G. A. Hite, *Am. Soc.,* **114**, 7318 (1992).

Phenylsilane.

Deoxygenation of alcohols.[1] This reaction can be effected by conversion of an alcohol to a derivative that undergoes radical reduction by a hydrogen atom donor (Barton–McCombie reaction). Xanthates are usually employed as the intermediate but various thionocarbonates can be used. They are easily prepared by acylation of the alcohol by a phenoxythiocarbonyl chloride, $ArOCCl$, prepared by reaction of a phenol with thiophosgene. Originally $R_3SnH/AIBN$ was used for the reduction, but the most convenient reagent is phenylsilane in combination with an initiator, AIBN or dibenzoyl peroxide.

[1] D. H. R. Barton, D. O. Jang, and J. C. Jaszberenyi, *Synlett,* 435 (1991).

(+)- or (−)-(N-Phenylsulfonyl-3,3-dichlorocamphoryl)oxaziridine (1), 16,119–120.

Asymmetric oxidation of sulfenimines to sulfinimines; β-amino acids.[1] This oxidation can be effected by use of (+)- or (−)-**1** (1 equiv.) in CCl_4 in 80–90% yield and in 85–90% ee.

The products are useful for preparation of optically active β-amino acids. Thus addition of the enolate (LDA) of methyl acetate to (−)-(R$_s$)-**2** affords sulfinamides **3**, separated by flash chromatography. Pure (3S)-**3** is converted by hydrolysis into the (S)-

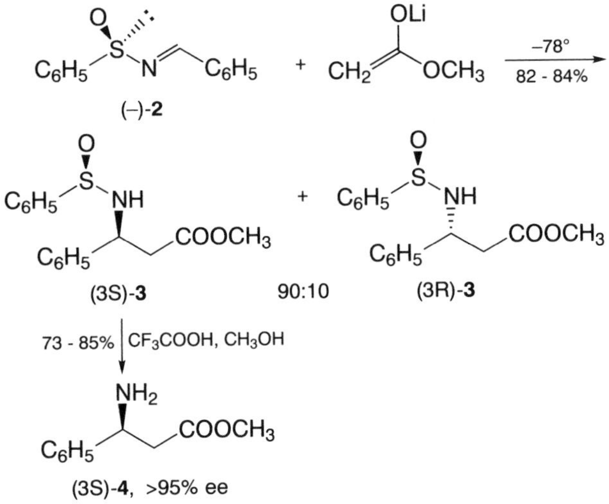

β-amino acid **4**.

Note that the enolate of a β-amino acid such as (3R)-**6** is oxidized by the (10-

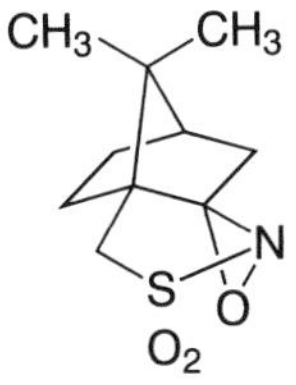

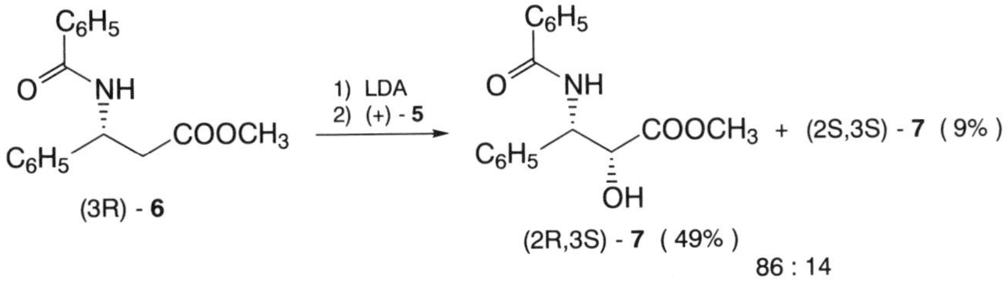

camphorylsulfonyl)oxaziridine (+)-**5** to the *syn-α*-hydroxy-*β*-benzoylamino acid **7** in 49% yield with >93% ee.

[1] F. A. Davis, R. T. Reddy, and R. E. Reddy, *J. Org.*, **57**, 6387 (1992).

(Phenylthio)methyl azide, (azidomethyl phenyl sulfide), $C_6H_5SCH_2N_3$ **(1, 10,14).**

N-Methylaziridines.[1]   A key step in a total synthesis of the alkaloid ($\pm$)-mitomycin K (**7**) is the introduction of the N-methylaziridine ring, which was effected with this reagent. Thus **1** adds to the imide **2** at 80° to give the triazoline **3** in 90% yield. After elimination of the carbonyl group to form **4**, irradiation of the triazoline group

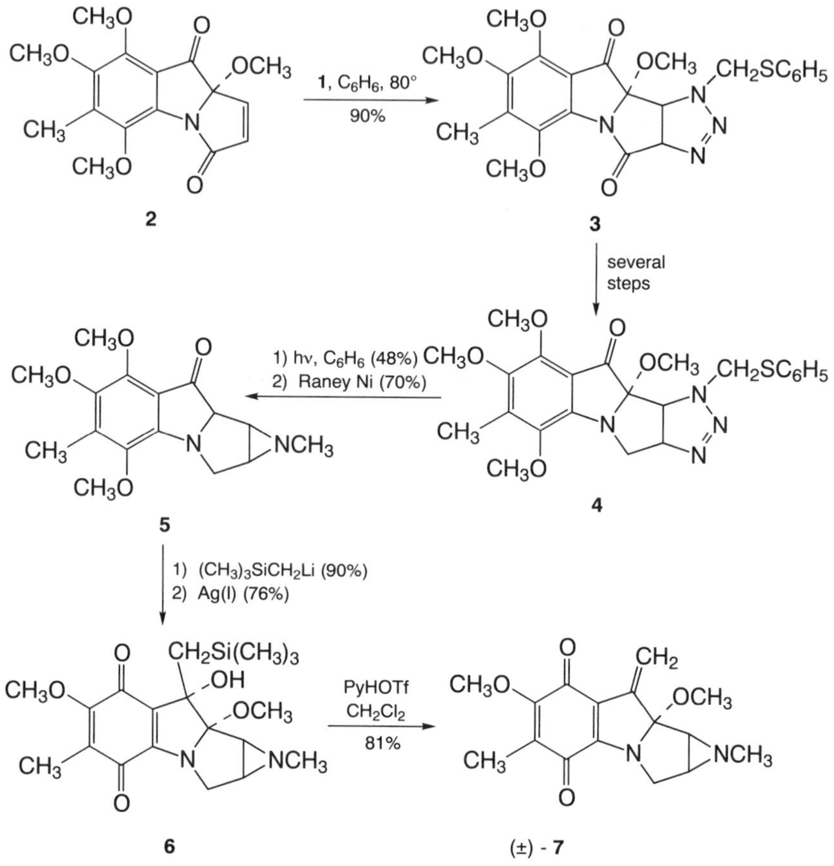

results in a (phenylthio)methylaziridine (48% yield), which is reduced by Raney nickel to an N-methylaziridine (**5**). Remaining steps to **7** involved conversion of the keto group at C_9 to methylene and oxidation to a quinone **6**, the immediate precursor to ($\pm$)-mitomycin K (**7**).

[1] J. W. Benbow, G. K. Schulte, and S. J. Danishefsky, *Angew. Chem. Int. Ed.,* **31,** 915 (1992).

3-Phenyl-3-(trifluoromethyl)diazirine, $\begin{array}{c}CF_3\\C_6H_5\end{array}\!\!>\!\!<\!\!\begin{array}{c}N\\||\\N\end{array}$ (1)

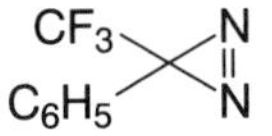

$RCOOH \rightarrow RHN_2$.[1] Acyl derivatives (2) of N-hydroxy-2-thiopyridone (12,417; 14,268) when photolyzed in the presence of this diazirine (1) transfer the alkyl group to 1 to form the imine 3. These imines are hydrolyzed by boric acid in refluxing ethanol to primary amines (4).

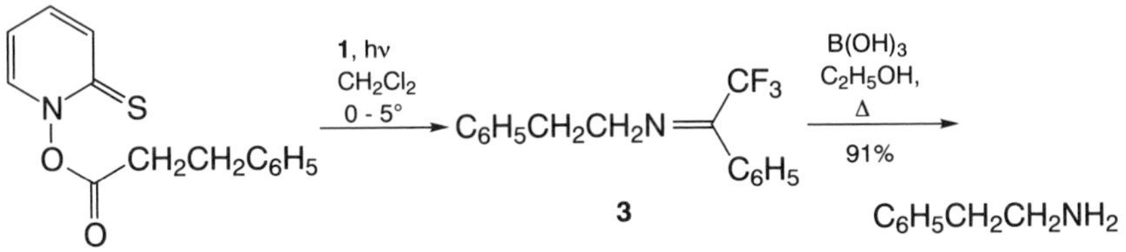

[1] D. H. R. Barton, J. C. Jaszberenyi, and E. A. Theodorakis, *Am. Soc.*, **114**, 5904 (1992).

Phosphazenium fluoride (1).

Preparation:

The value of this salt is that it is anhydrous as obtained and is soluble in THF. It should be as useful as $(CH_3)_4N^+F^-$ as a form of fluoride ion. Thus it converts 2 to 3 in THF at 25° in 93% yield.[1]

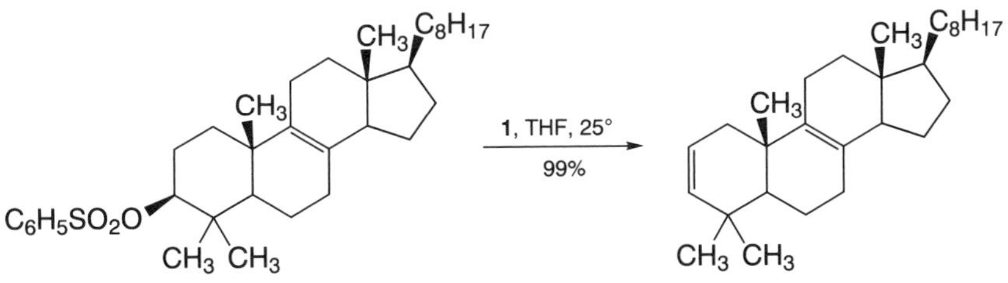

[1] R. Schwesinger, R. Link, G. Thiele, H. Rotter, D. Honert, H.-H. Limbach, F. Männle, *Angew. Chem. Int. Ed.*, **30**, 1372 (1991).

Phosphoryl chloride.

Preparation of cyanides.[1] An attractive route to cyanides involves as the first step dehydration of alkylformamides with phosgene–triethylamine (**1**,857) or phosphoryl chloride–diisopropylamine (**13**,249) to form isocyanides followed by isocyanide–cyanide rearrangement. This rearrangement traditionally was conducted in the gas phase, but proceeds in almost quantitative yield when carried out by flash pyrolysis at 600°. This route to cyanides is attractive because allyl isocyanides rearrange without allylic rearrangement. Moreover, optically active carboxylic acids can be obtained from optically active amines without racemization.

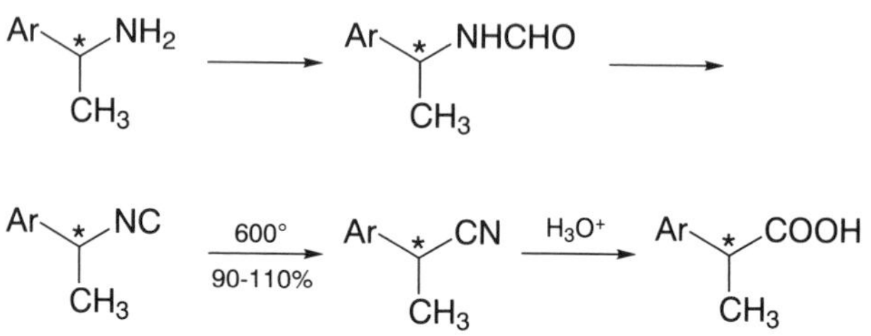

This sequence can convert optically active amino acids to optically active β-acyloxy cyanides (equation I).

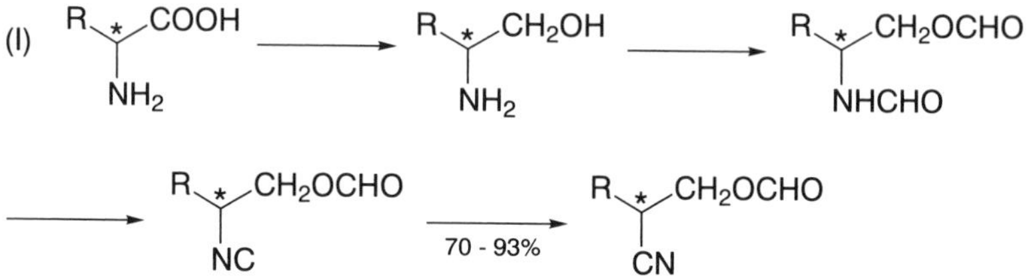

[1] C. Rüchardt, M. Meier, K. Haaf, J. Pakusch, E. K. A. Wolber, and B. Müller, *Angew. Chem. Int. Ed.*, **30**, 893 (1991).

Pinacolborane (1).

Preparation:

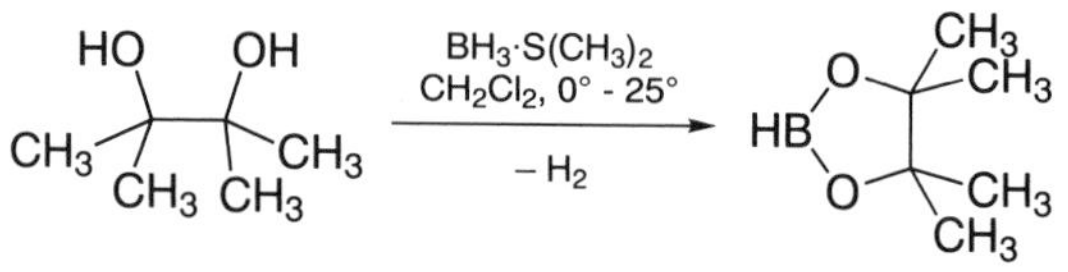

1

Hydroboration of alkynes and alkenes.[1]   This borane is more efficient than catecholborane for hydroboration of alkynes and alkenes in terms of regio- and stereochemistry and in tolerance for various functional groups.

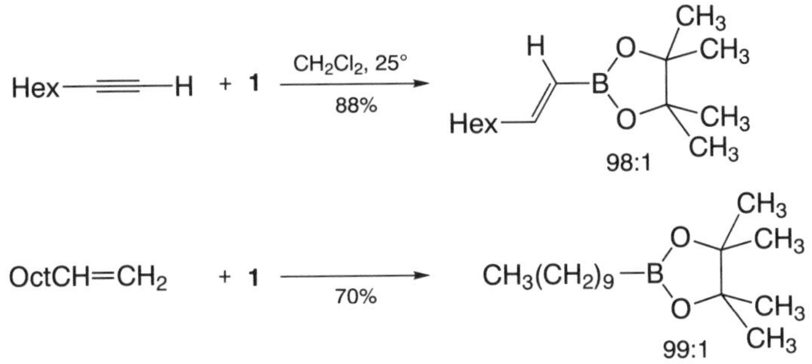

[1] C. E. Tucker, J. Davidson, and P. Knochel, *J. Org.*, **57**, 3482 (1992).

Potassium *t*-butoxide.

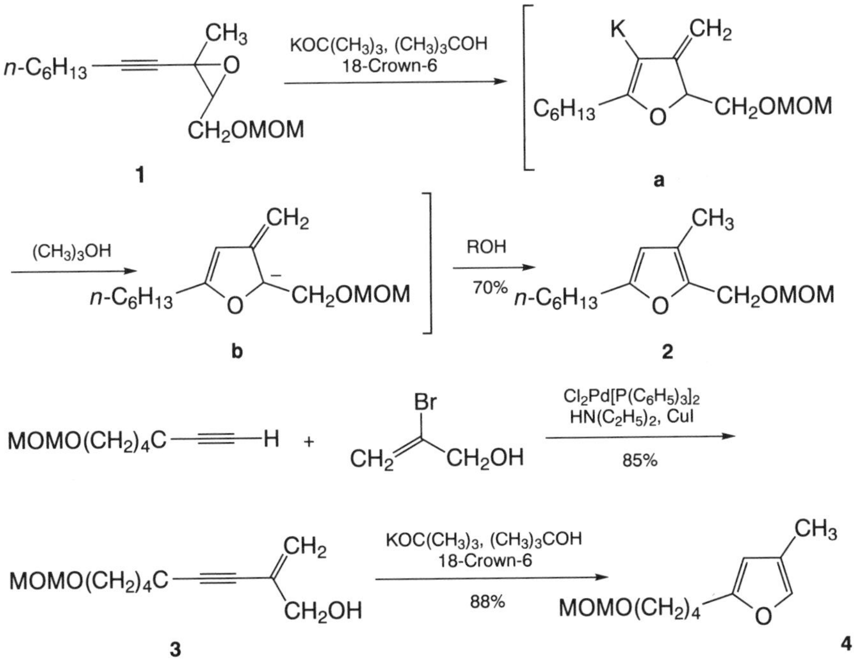

Furan synthesis.[1]   Reaction of an alkynyloxirane such as **1** with $KOC(CH_3)_3$ in $(CH_3)_3COH$/18-crown-6 results in isomerization to a furan **2**, via an α-methylene homopropargylic alkoxide. Under the same conditions the α-methylene homoproparglyic alcohol **3** cyclizes to the furan **4**.

[1] J. A. Marshall and W. J. DuBay, *Am. Soc.*, **114**, 1450 (1992).

Potassium hydride–18-Crown-6.

Asymmetric oxy-Cope rearrangement.[1]   The anionic rearrangement of 1,5-dien-3-ols (**3**) can proceed with high diastereoselectivity. The *trans*- and *cis*-isomers of **3** are prepared by [2,3]-Wittig rearrangement of **1**, followed by (E)- and (Z)-reduction of the

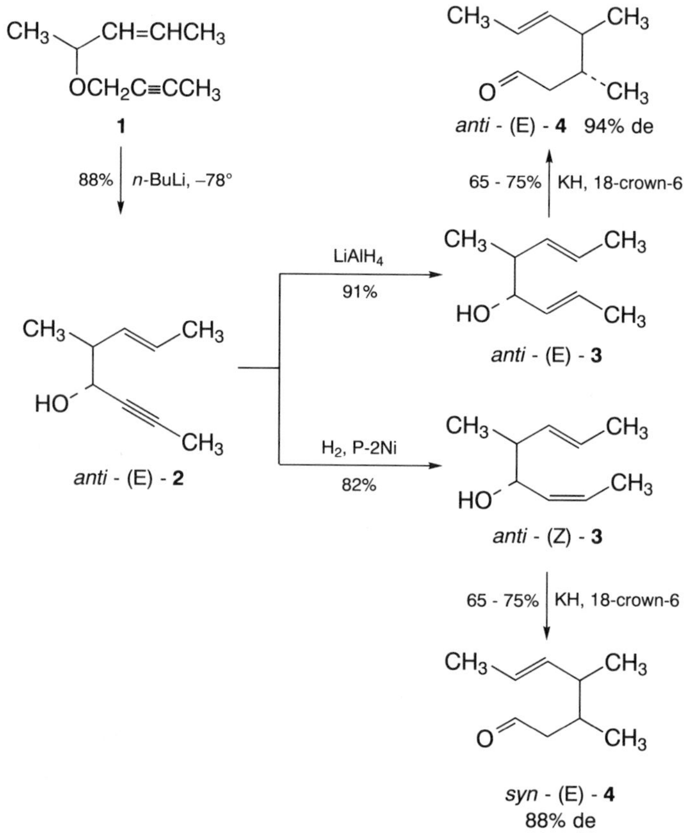

CH3_/CH=CHCH3
OCH2C≡CCH3
1

88% | *n*-BuLi, −78°

CH3_/=_CH3
HO⁻
≡
CH3
anti - (E) - **2**

LiAlH4
91%

H2, P-2Ni
82%

CH3_/_CH3
O⁼
⁼CH3
anti - (E) - **4**   94% de

65 - 75% | KH, 18-crown-6

CH3_/=_CH3
HO⁻   /=_CH3
anti - (E) - **3**

CH3_/=_CH3
HO⁻   _/=_CH3
anti - (Z) - **3**

65 - 75% | KH, 18-crown-6

CH3_/=_CH3
O⁼
⁼CH3
syn - (E) - **4**
88% de

triple bond to provide *anti*-(E)- and -(Z)-**3**. Rearrangement of (E)-**3** provides *anti*-(E)-**4** in 94% de, whereas *anti*-(Z)-3 rearranges to *syn*-(E)-**4** in 88% de.

[1] S.-Y. Wei, K. Tomooka, and T. Nakai, *J. Org.*, **56**, 5973 (1991).

Potassium nitrosodisulfonate [Fremy's salt, $(KSO_3)_2NO$, **1**).

4,7-Indoloquinones. A route to these quinones (**5**) involves ozonolysis of benzox-azoles such as **1** to provide **2**, which undergoes ready dehydration to 1-acetyl-4-formyl-7-hydroxyindoles (**3**). These products undergo deacylation on chromatography on silica gel to provide **4**. This 4-formyl-7-hydroxyindole (**4**) is oxidized by buffered Fremy's

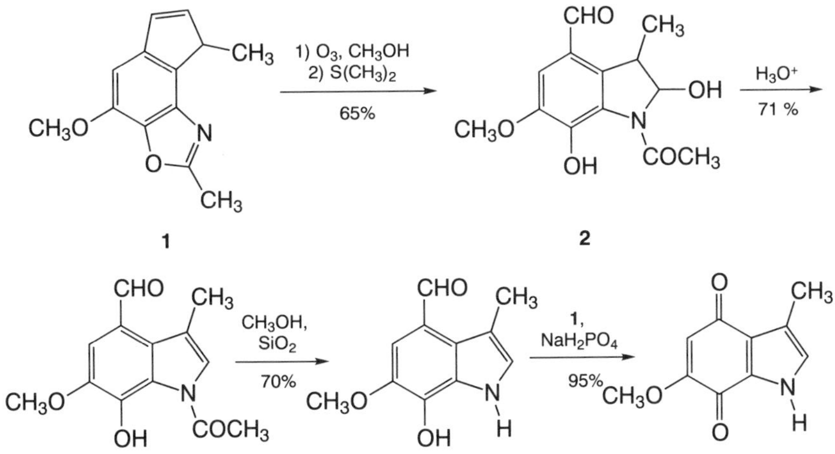

salt to the indoloquinone **5** in 95% yield. This ready oxidation is unexpected, particularly since **3** is inert to this oxidant.

[1] J. M. Saá, C. Martí, and A. García-Raso, *J. Org.,* **57**, 589 (1992).

Potassium triethylmethoxide, $(C_2H_5)_3COK$ (2,349).

Ramberg-Bäcklund rearrangement.[1] The key step in a synthesis of (+)-eremantholide A (**4**), involves ring contraction of the chloro sulfone **2** by the Ramberg–Bäcklund rearrangement. Thus treatment of **2** with this highly hindered base (**1**), 2.2 equiv., and HMPA (10 equiv.) in DME at 70° provides **3** in 82% yield. Use of less hindered bases results mainly in reduction of the chloro group.

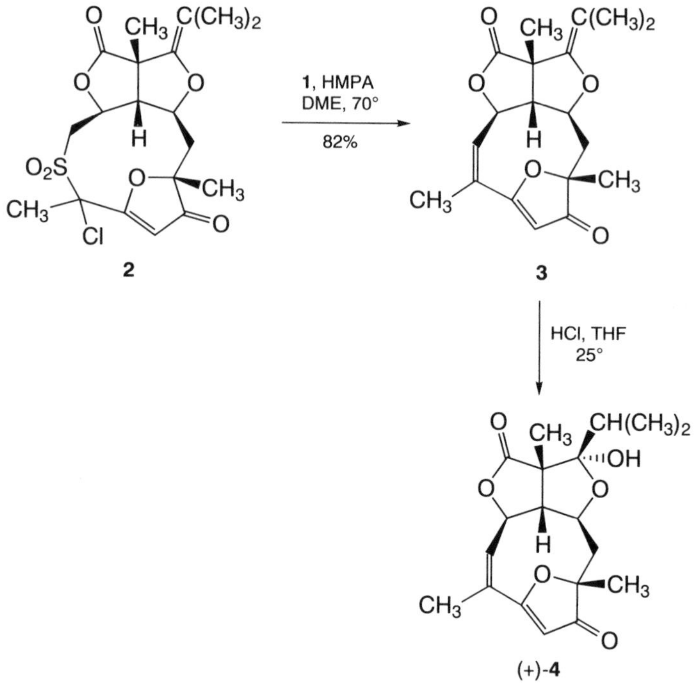

[1] R. K. Boeckman, Jr., S. K. Yoon, and D. K. Heckendorn, *Am. Soc.*, **113**, 9682 (1991).

2-Pyridinethiol 1-oxide, 12,417.

Julia alkene synthesis.[1]   The Julia alkene synthesis involves deprotonation of an alkyl sulfone followed by reaction with a carbonyl compound to form a β-hydroxy sulfone, which is converted to an alkene on reduction with sodium amalgam. Although this synthesis has proved useful (**11,**474), yields in the second step can be low. An alternative method is a radical reduction of the xanthate of the secondary alcohol, known to result in an alkene (**8,**497–498). Tributyltin hydride/AIBN was originally used as the source of radicals, but $(C_6H_5)_2SiH_2$ in combination with AIBN or $B(C_2H_5)_3$ is also effective. Generally the highest yields obtain with acyl derivaties of N-hydroxy-2-thiopyridine, prepared by acylation of 2-pyridinethiol 1-oxide. Yields are generally around 85% for this radical process, and only the (E)-isomer is formed.

CH₃(CH₂)₆CH₂SO₂C₆H₅ $\xrightarrow{\text{1) BuLi}}_{\text{2) RCHO}}$

Note that samarium diiodide can also effect this reductive elimination, also by a radical process (**16**,297).

[1] D. H. R. Barton, J. C. Jaszberenyi, and C. Tachdjian, *Tetrahedron Letters,* **32**, 2703 (1991).

N-(2-Pyridyl)triflamide (1); N-(5-chloro-2-pyridyl)triflimide (2).

Prepared by reaction of the aminopyridine with triflic anhydride and pyridine in CH₂Cl₂ at 20°.

Vinyl triflates.[1]   These triflates are generally prepared by reaction of lithium enolates with N-phenyltriflimide, $(CF_3SO_2)_2NC_6H_5$, **12**,395. These new pyridine-based triflimides are considerably more reactive than N-phenyltriflimide.

[1] D. L. Comins and A. Dehghani, *Tetrahedron Letters,* **33**, 6299 (1992).

Pyridylzinc halides, PyZnX.

These reagents can be prepared by reaction of halopyridines with activated Zn in THF.

Coupling with ArX or (ArCO)₂O.[1]   In the presence of Pd[P(C₆H₅)₃]₄ these zinc reagents couple with aryl halides or aroyl anhydrides. It is particularly useful for preparation of unsymmetrical bipyridines.

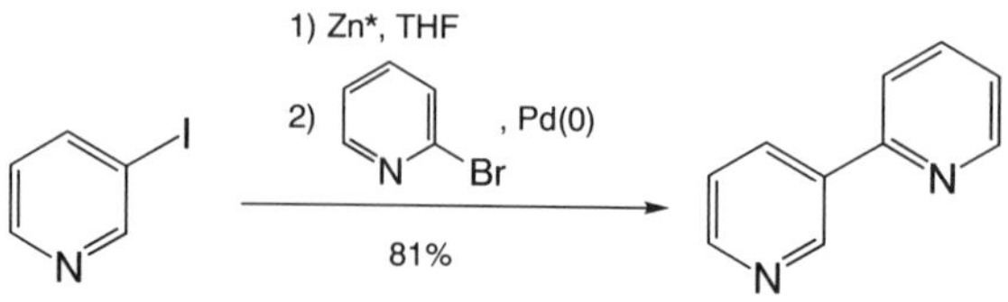

$$1) \ Zn^*, \ THF$$
81%

[1] T. Sakamoto, Y. Kondo, N. Murata, and H. Yamanaka, *Tetrahedron Letters*, **33**, 5373 (1992).

Pyran-2-ones.

Diels–Alder reaction of pyran-2-ones. Diels–Alder reaction of 2-pyrones, if successful, can provide unusual cyclohexenecarboxylic acids, but thermally promoted cycloadditions with these electron-deficient dienes usually result in decarboxylation and aromatization of the adducts as a result of the required high temperatures (6,291–292). Successful Diels–Alder reactions of 3-bromo-2-pyrone (**1**) with the electron-rich dioxole **2** can be effected with a catalytic amount of ethyldiisopropylamine at 90° (4 days) to give the major adduct (*endo*-**3**) in 63% yield. The adduct is hydrolyzed by *p*-toluenesulfonic acid in methanol to **4** as the only diastereomer. The trisilyl ether of **4** was transformed to the α,β-unsaturated ester **5** by radical debromination and DBU isomerization.[1]

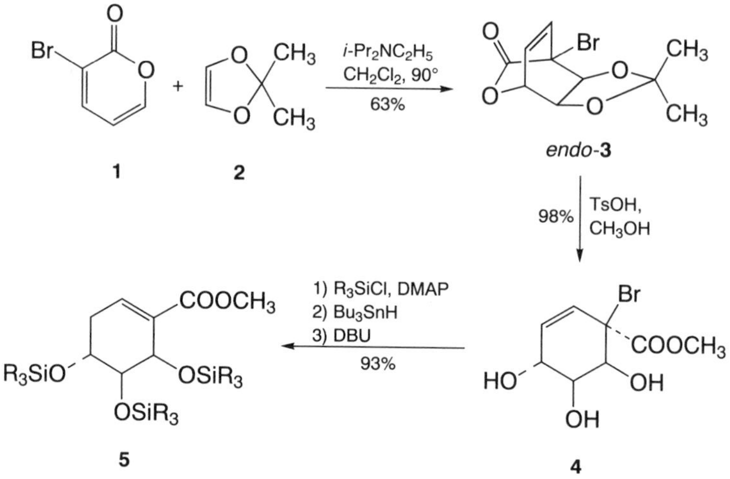

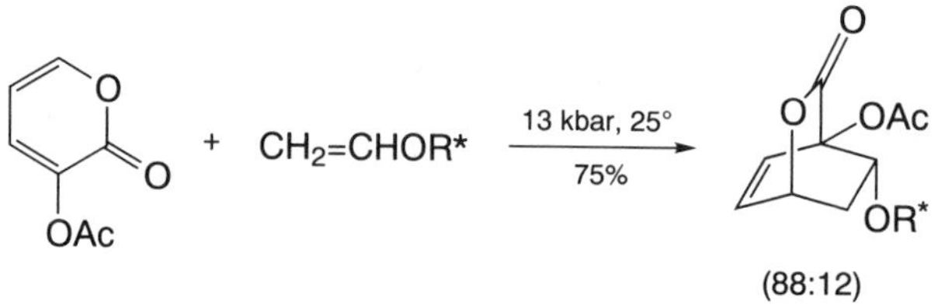

Diels–Alder reactions of 2-pyrones with reactive dienophiles can also be effected without decomposition by use of high pressure (**8**,254). This version has been used to effect cycloaddition of 3-acyloxy-2-pyrones with vinyl ethers at 25° to give bicyclic adducts in good yield.[2] Use of chiral vinyl ethers can result in diastereoselective reactions. The highest levels of induction are observed with 8-(β-naphthyl)menthyl vinyl ether and 8-(3,5-dimethylphenyl)menthyl vinyl ether (88:12).

[2+4]Cycloadditions.   Posner et al.[3] have reviewed Diels–Alder reactions with 2-pyrones and 2-pyridones in detail, with special attention to the selectivity (213 references).

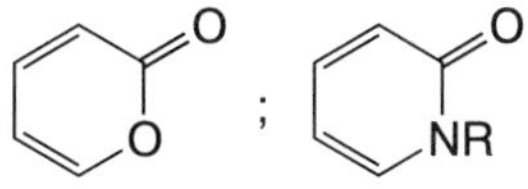

[1] G. H. Posner and T. D. Nelson, *J. Org.,* **56**, 4339 (1991).

[2] V. Propansiri and E. R. Thorton, *Tetrahedron Letters,* **32**, 3147 (1991).

[3] K. Afarinkia, V. Vinader, T. Thornton, D. Nelson, and G. H. Posner, *Tetrahedron,* **48**, 9111 (1992).

R

Raney nickel.

Reductive N–N cleavage of hydrazines. A diastereoselective synthesis of chiral α-amino aldehydes (**4**) from glyoxal involves an intermediate dimethylhydrazone (**1**), which reacts with an alkyllithium to provide an α-substituted N',N'-dimethylhydrazine (**2**). The mildest method for cleavage of hydrazines to primary amines (**3**) involves hydrogenolysis catalyzed by Raney nickel at 30–50°. This Raney nickel reductive cleavage is markedly improved by sonication and proceeds at atmospheric pressure of H_2 at 20° in CH_3OH in yields of 66–84% with no racemization or debenzylation.[1]

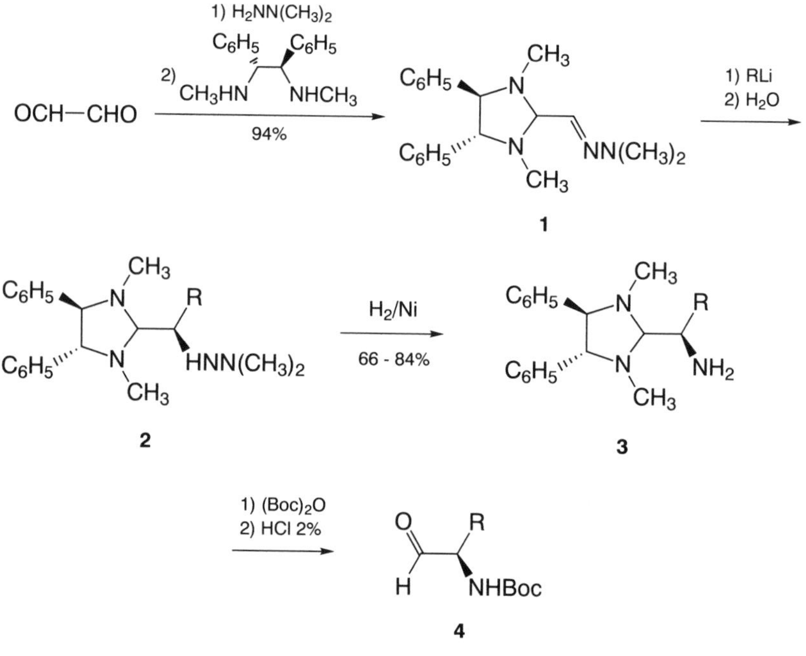

[1] A. Alexakis, N. Lensen, and P. Mangeney, *Tetrahedron Letters,* **32**, 1171 (1991); *idem, Synlett,* 625 (1991).

Rhenium(VII) oxide, Re_2O_7.

Oxidative cyclization of 5-hydroxy alkenes.[1] Reaction of these substrates with Re_2O_7 and a base (preferably 2,6-lutidine) in CH_2Cl_2 at 25° provides 2-hydroxytetrahydrofurans. These products could be formed by dihydroxylation followed by cyclization. And indeed the product in the first example is also obtained by epoxidation of the (Z)-isomer of this alkene followed by cyclization.

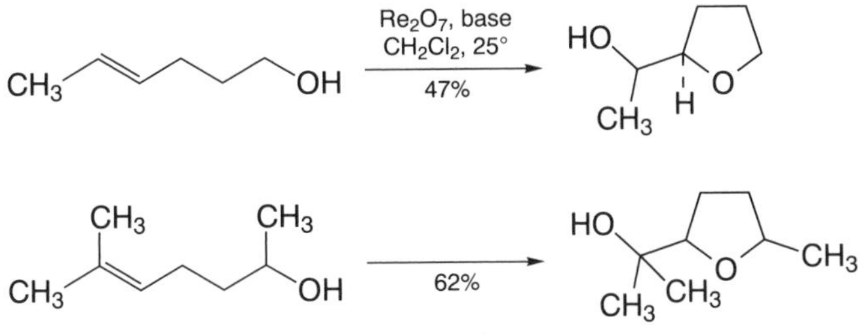

(trans/cis = 33 : 1)

Oxidation of chiral ketals.[2]   The chiral ketal **1** is oxidized by Re_2O_7 and 2,6-lutidine to 2-hydroxy ketal **2** with high (>99:1) enantioselectivity. This reaction is believed to proceed via a perrhenate ester of a 2-hydroxy enol ether, since the 2-hydroxyethyl enol ether **3** is oxidized to the 2-hydroxy ketal **2** by Re_2O_7 in similar enantioselectivity and comparable yield.

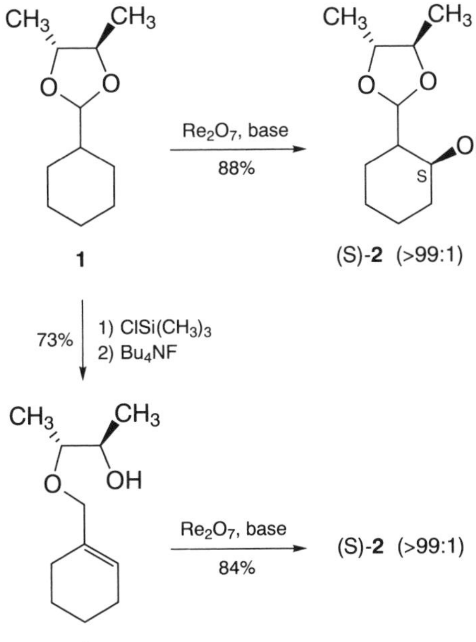

[1] R. M. Kennedy and S. Tang, *Tetrahedron Letters,* **33**, 3729 (1992).
[2] S. Tang and R. M. Kennedy, *Tetrahedron Letters,* **33**, 7823 (1992).

Rhodium(II) carboxylates.

Tropones.[1]   Reaction of vinyldiazomethanes (**1**) with 1-methoxy-1-(trimethyl-silyloxy)butadiene (**2**) catalyzed by rhodium(II) acetate or rhodium(II) pivalate results in [3+4]cycloaddition via cyclopropanation/Cope rearrangement to form a cycloheptadiene (**3**). Short exposure of **3** to citric acid followed by oxidation (DDQ) provides the

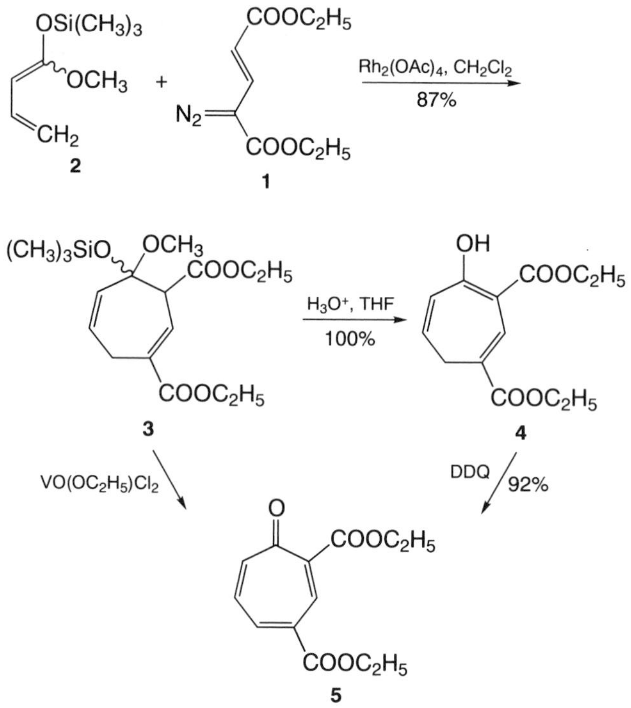

tropone **5** in high yield. The conversion of **3** to **5** can be effected in one step with dichloroethyloxyoxovanadium, VO(OC$_2$H$_5$)Cl$_2$.[2]

This tropone synthesis was used for a short synthesis of nezukone (**7**) (equation I).

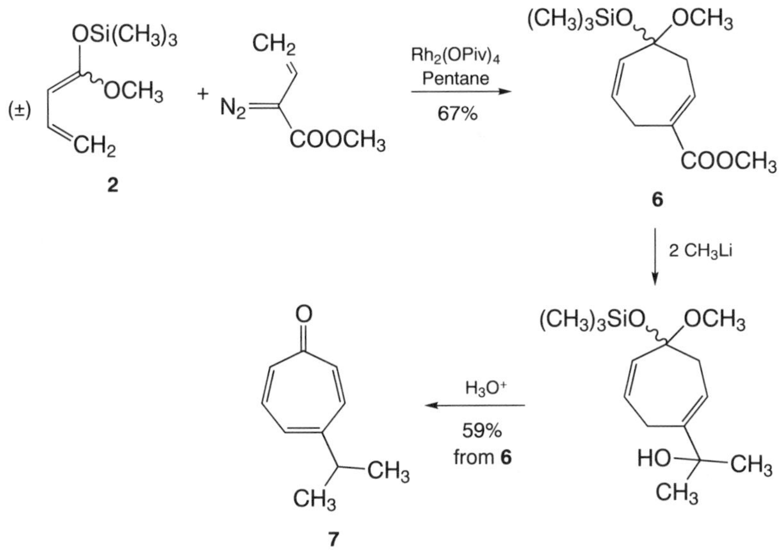

Reaction of α-diazo ketones with alkynes.[3]   Reaction of the acetylenic α-diazo ketone **1** with an alkyne in the presence of Rh₂(OAc)₄ at 25° results in an intermediate **a**, formed by an intramolecular alkyne insertion, that reacts with the external alkyne to form a cyclopropene (**2**). Rearrangement of the cyclopropene group by Rh₂(OAc)₄[4]

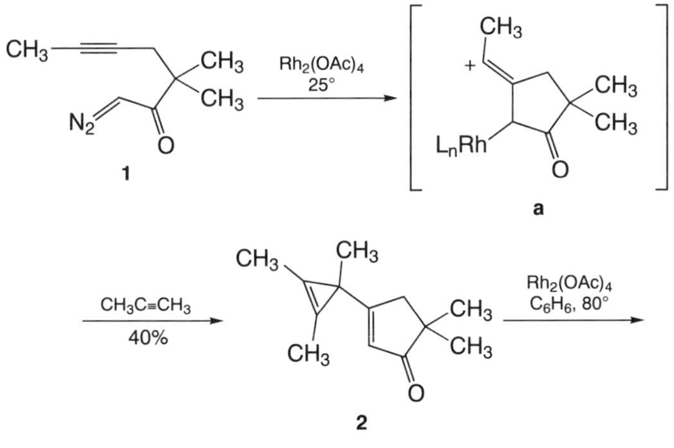

provides an intermediate **b** that cyclizes to form the dihydropentalenone **3**.

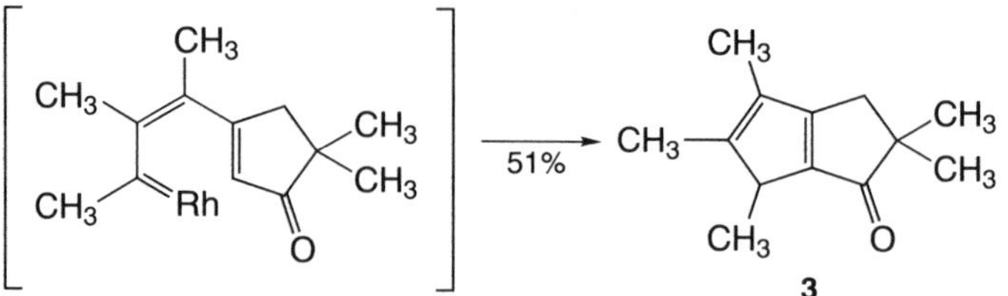

Synthesis of (+)-griseofulvin (3).[5]   The key step in a recent asymmetric synthesis of this antifungal reagent is the decomposition of the diazo ketone **1** with rhodium(II) pivalate to provide **2** in 62% yield as the only isolable product.

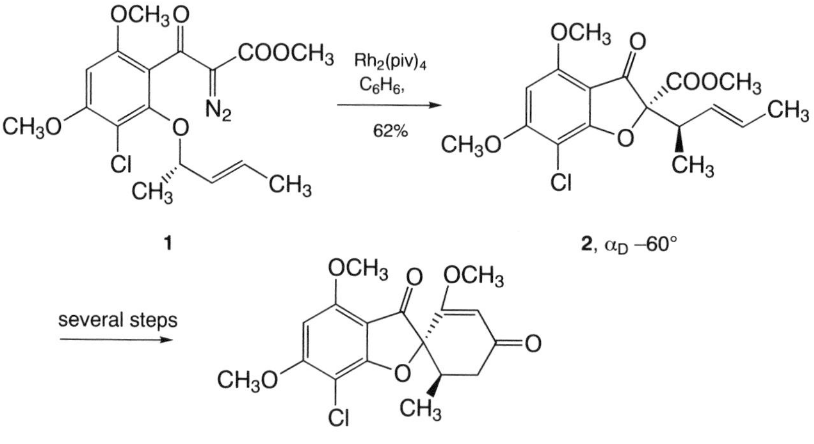

Reaction of cyclic diazo-1,3-dicarbonyls with aryl heterocycles.[6]   In the presence of 2 mol % of $Rh_2(OAc)_4$, 2-diazo-1,3-cyclohexanedione (**1**) reacts with dihydrofuran to form the dipolar cycloadduct **2** in 70% yield. The reaction of **1** with 2,5-dimethylfuran

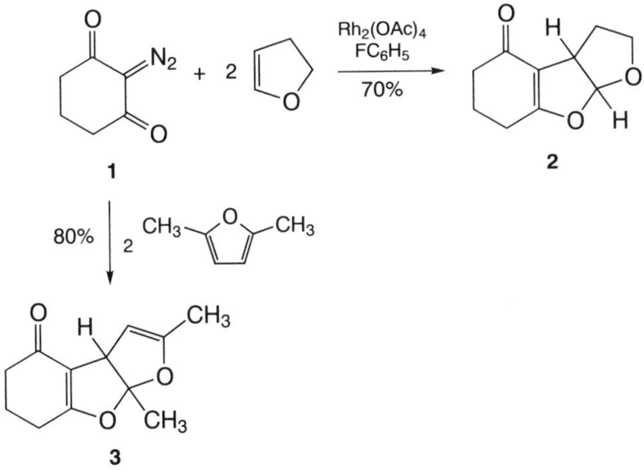

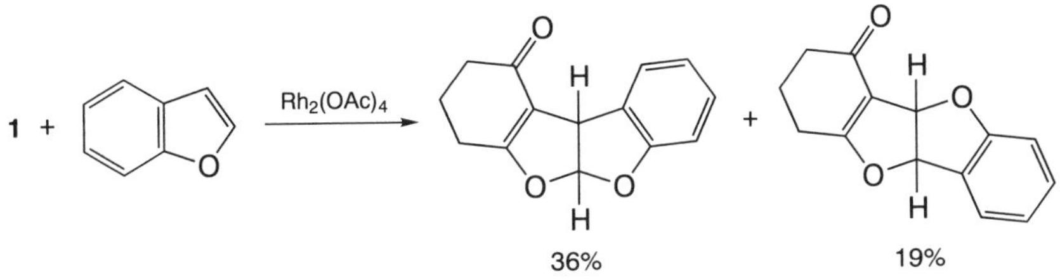

proceeds in even higher yield to give **3**.

Tricyclic dihydropyrrolizines.[7]   The reaction of N-acyl-2-(1-diazoacetyl)-pyrrolidines (**1**) with a catalytic amount of a rhodium(II) carboxylate, particularly rhodium(II) octanoate and a dipolarophile (DMAD), results in a tricyclic dihydropyrrolizine **3** as the major product. The expected product (**2**) is formed only in traces (10%). Apparently the intermediate carbonyl ylide **a** rearranges to the more stable azomethine ylide **b**.

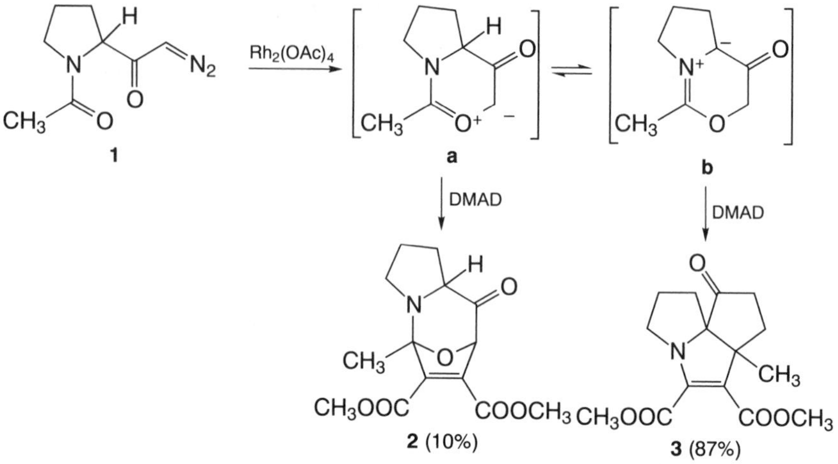

Reaction with cyclopropenes.[8]   Reaction of the cyclopropene **1** with $Rh_2(OAc)_4$ in C_6H_6 (80°) results in two furans, **2** and **3**, in 78% and 3% yield, respectively. Thus the less substituted bond is cleaved selectively. In contrast, treatment of **1** with a Rh(I) catalyst in CH_2Cl_2 at 25° gives only furan **3** in 86% yield.

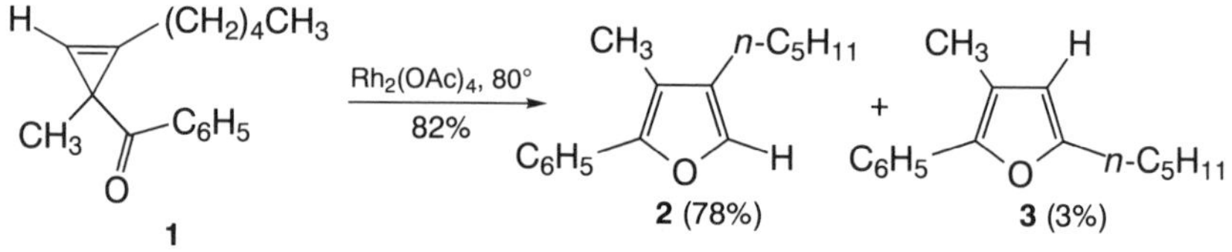

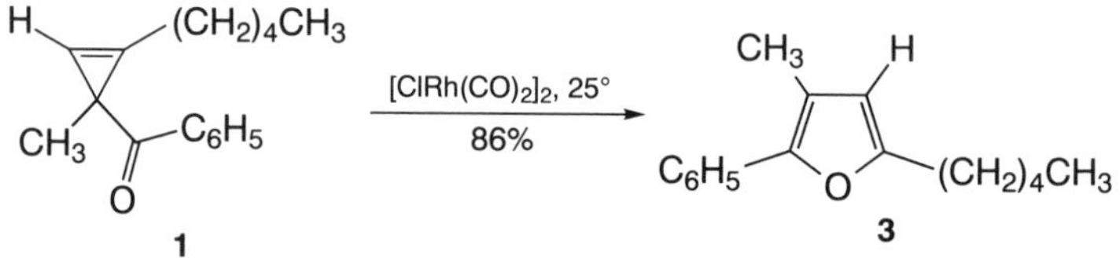

[1] H. M. L. Davies, T. J. Clark, and G. F. Kimmer, *J. Org.*, **56**, 6440 (1991).
[2] T. Hirao, M. Mori, and Y. Ohshiro, *ibid.*, **55**, 358 (1990).
[3] T. R. Hoye and C. J. Dinsmore, *Tetrahedron Letters*, **32**, 3755 (1991).
[4] P. Müller, N. Pautex, M. P. Doyle, and V. Bagheri, *Helv.*, **73**, 1233 (1990).
[5] M. C. Pirrung, W. L. Brown, S. Rege, and P. Laughton, *Am. Soc.*, **113**, 8561 (1991).
[6] M. C. Pirrung, J. Zhang, and A. T. McPhail, *J. Org.*, **56**, 6269 (1991).
[7] A. Padwa, D. C. Dean, and L. Zhi, *Am. Soc.*, **114**, 593 (1992).
[8] A. Padwa, J. M. Kassir, and S. L. Xu, *J. Org.*, **56**, 6971 (1991).

Rhodium(II) pyroglutamates.

Asymmetric cyclopropanation.[1] The ability to effect ligand exchange between rhodium(II) acetate and various amides has lead to a search for novel, chiral rhodium(II) catalysts for enantioselective cyclopropanation with diazo carbonyl compounds. The most promising to date are prepared from methyl (S)- or (R)-pyroglutamate (1), [dirhodium(II) tetrakis(methyl 2-pyrrolidone-5-carboxylate)]. Thus these complexes, $Rh_2[(S)$- or (R)-1$]_4$, effect intramolecular cyclopropanation of allylic diazoacetates (2) to give the cyclo-propanated γ-lactones 3 in 65 $\geqslant$ 94% ee (equation I). In general, the enantioselectivity is higher in cyclopropanation of (Z)-alkenes.

(I)

$R^1,R^2=H$	74%	88% ee
$R^1=H, R^2=C_2H_5$	88%	> 94% ee
$R^1=Pr, R^2=H$	74%	75% ee

Enantioselective cyclopropenation.[2] This chiral rhodium(II) catalyst can effect highly enantioselective cyclopropenation of 1-alkynes with alkyl diazoacetates. The enantioselectivity increases with the steric size of the ester group, and the size and polarity of the alkyne substituent also affects the enantioselectivity. The highest selectivity is observed with d-menthyl diazoacetates as a result of double diastereoselection.

$$HC\equiv CR \ + \ N_2CHCOOR' \ \xrightarrow{\ 1\ }$$

$R = CH_2OCH_3$	$R' = C_2H_5$	73%	69% ee
$R = CH_2OCH_3$	$R' = d$ - Men*	43%	98% ee
$R = Bu$	$R' = t$-Bu	69%	53% ee
$R = Bu$	$R' = d$ - Men*	46%	86% ee

Yields and enantioselectivities are low in reactions with disubstituted alkynes.

Asymmetric synthesis of lactams.[3] A rhodium catalyst with a chiral oxazolidinone ligand, dirhodium(II) tetrakis[methyl 2-oxazolidinone-4 (S)-carboxylate] (1), can effect cyclization of N-2-alkoxyethyl-N-(t-butyl)diazoacetamide (**2**) to N-(t-butyl)-4-ethoxy-2-pyrrolidone **3** in 78% ee and 97% yield.

$$\text{2}$$ **3**, 78% ee

Enantioselective cyclopropanation of homoallylic diazoacetates.[4]   In the presence of this chiral catalyst, homoallylic diazoacetates undergo asymmetric intramolecular cyclopropanation to form oxabicyclo[4.1.0]heptanes in 71–90% ee.

79% ee

[1] M. P. Doyle, R. J. Pieters, S. F. Martin, R. E. Austin, C. J. Oalmann, and P. Müller, *Am. Soc.*, **113**, 1423 (1991).

[2] M. N. Protopopova, M. P. Doyle, P. Müller, and D. Ene, *ibid.*, **114**, 2755 (1992).

[3] M. P. Doyle, M. N. Protopopova, W. R. Winchester, and K. L. Daniel, *Tetrahedron Letters*, **33**, 7819 (1992).

[4] S. F. Martin, C. J. Oalmann, and S. Liras, *ibid.*, **33**, 6727 (1992).

Ruthenium(IV) dioxide, RuO$_2$.
Oxidative phenolic coupling.[1]   RuO$_2$ · 2H$_2$O (2 equiv.) in a TFA/TFAA medium is an excellent reagent for *para–para* coupling of **1** to afford the lignan prostegone A (**2**) in 80% yield. Substitution of RuO$_2$ by Tl$_2$O$_3$ also converts **1** into **2**, but the yield is 45%.

1   **2**

[1] J.-P. Robin and Y. Landais, *Tetrahedron*, **48**, 819 (1992).

S

Salicyl alcohol (2-hydroxybenzyl alcohol).

Unsymmetrical biphenyls.[1]   This alcohol can be used as a template for synthesis of unsymmetrical biphenyls with substituents at *ortho*-positions by an intramolecular Ullmann coupling. The two ester bonds of the coupling product can be cleaved selectively by $NaOCH_3$ or a primary amide.

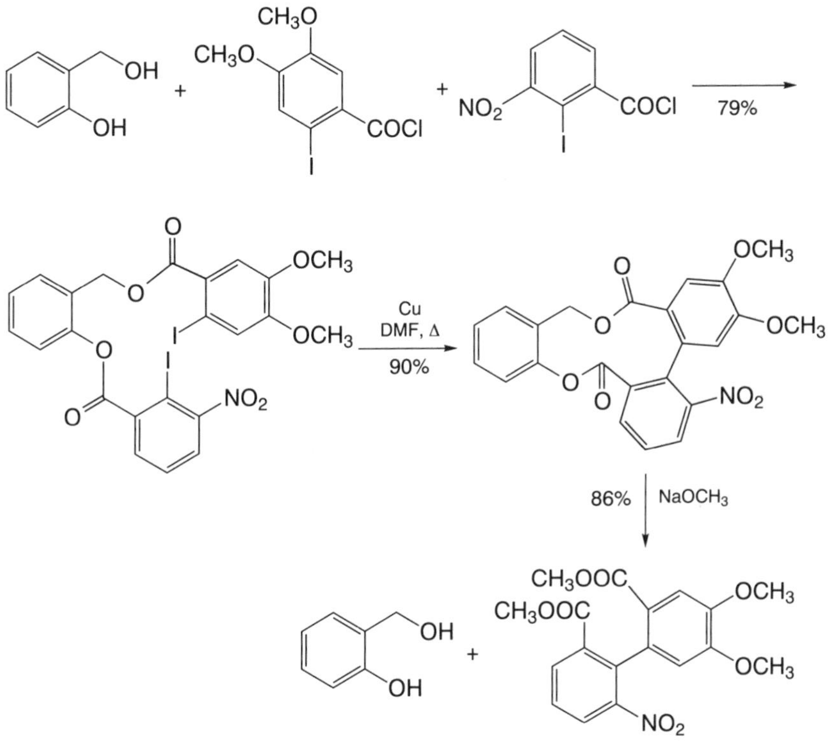

[1] M. Takahashi, Y. Moritani, T. Ogiku, H. Ohmizu, K. Kondo, and T. Iwasaki, *Tetrahedron Letters*, **33**, 5103 (1992).

Samarium/Methylene iodide (14,275).

Cyclopropanation of vinylorganometallic compounds.[1]   Molander cyclopropanation of allylic alcohols substituted by silyl or stannyl groups can show high diastereoselectivity particularly when carried out with a large excess of the samarium reagent.

The diastereoselectivity of an (E)-vinylsilane (**1**) depends on the steric bulk of the R group (equation I). Higher diastereoselectivity obtains in the cyclopropanation of a (Z)-vinylsilane (**2**), and is not affected by size of the R group (equation II).

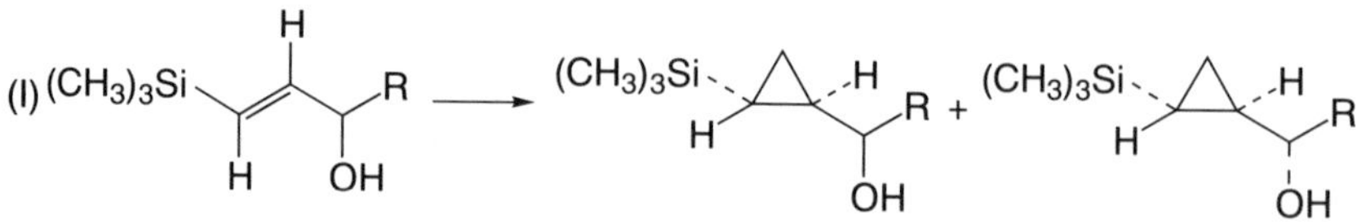

1a, R = c-C$_6$H$_{11}$
1b, R = CH$_3$

81% 46:1
76% 1:10

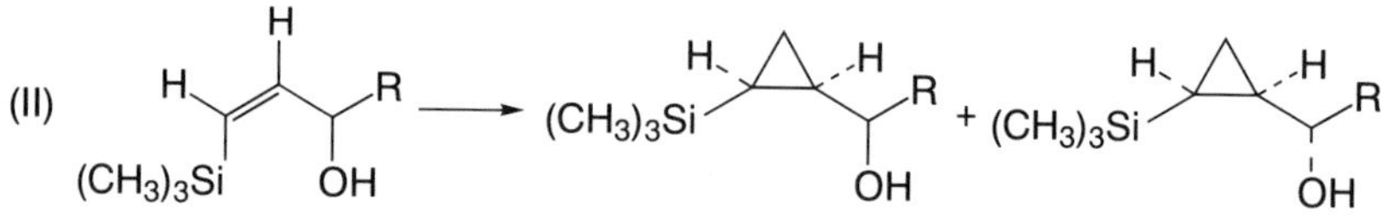

2a, R = Pr
2b, R = CH$_3$

67% >100:<1
67% >100:<1

Cyclopropanation of allylic alcohols substituted by a tin group gives a single diastereomer in good yield. These tin-substituted cyclopropanes can undergo *trans*-metalation (CH$_3$Li, THF, 0°).

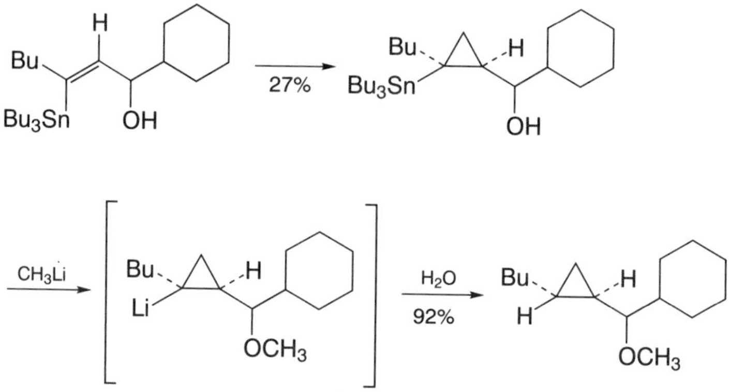

[1] M. Lautens and P. H. M. Delanghe, *J. Org.*, **57**, 798 (1992).

Samarium(0)/Samarium(II) iodide.

Deoxygenative coupling of amides; vic-diamines.[1] This combination (1:2) can effect deoxygenative coupling of amides to form *vic*-diaminoalkenes.

$$C_6H_5CON(C_2H_5)_2 \xrightarrow[72\%]{\substack{Sm/SmI_2 \\ THF}}$$

E/Z = 38:62

[1] A. Ogawa, N. Takami, M. Sekiguchi, I. Ryu, N. Kambe, and N. Sonoda, *Am. Soc.*, **114**, 8729 (1992).

Samarium(II) iodide.

Coupling of lactones and ketones.[1] SmI$_2$ in THF/HMPA effects coupling of a carbohydrate-derived lactone such as **1** and a ketone that involves oxygenation of **1** followed by a carbonyl addition reaction with high diastereoselectivity.

(35 : 1)

Reduction of ArCOOH and derivatives.[2] Aryl carboxylic acids can be reduced to ArCH$_2$OH by SmI$_2$ (excess) in the presence of an acid, particularly phosphoric acid (85%). In contrast pyridinecarboxylic acids are reduced to the corresponding methylpyridine in 43–48% yield. Benzoic anhydride and benzoyl chloride are also reduced to benzyl alcohol, but in lower yield. Benzonitrile is reduced under the same conditions to benzylamine in 99% yield. This behavior is characteristic for aryl nitriles.

Unexpectedly, SmI$_2$ reduction of amides provides aldehydes in almost quantitative yield.

Aryl radical cyclization.[3] SmI$_2$ can promote an intramolecular cyclization of aryl bromides with double or triple bonds. This electron-transfer reaction is a useful route to heterocycles.

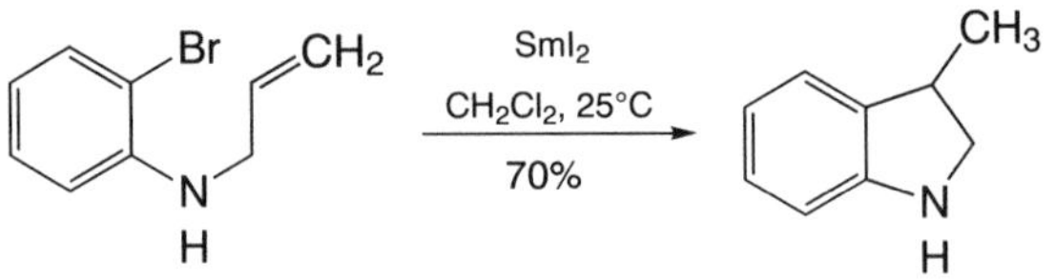

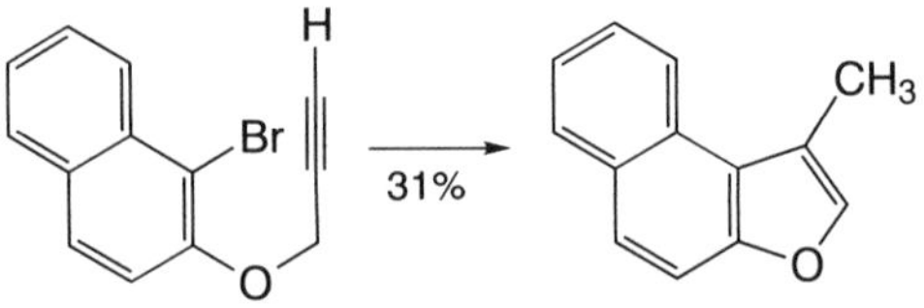

Barbier/aldol reactions of enones.[4]   The iodo enone **1** on treatment with SmI$_2$ (2 equiv.) in THF containing DMPU (10 equiv.) undergoes reductive cyclization to the bicyclo[3.3.0]octane-3-one (**2**) in 70% yield. Such reductive cyclizations have been effected with a tin hydride. The advantage of the SmI$_2$ reaction is that it proceeds

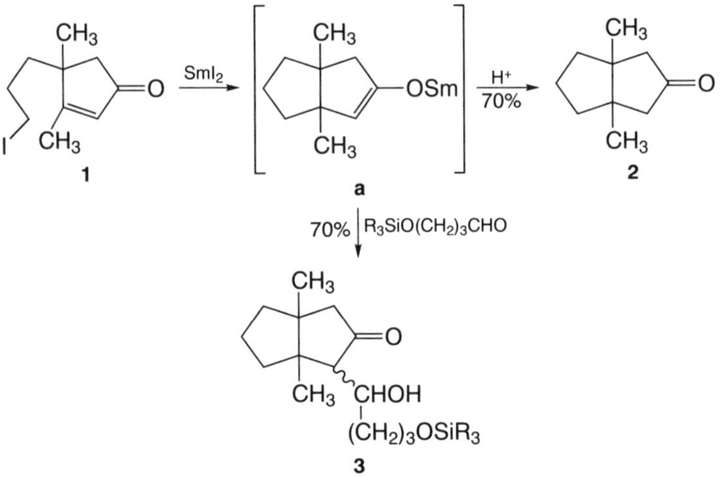

through a samarium enolate (**a**), which can undergo an aldol reaction with an aldehyde.

α-Ketols.[5]   In the presence of 2 equiv. of SmI$_2$ or dicyclopentadienylsamarium (SmCp$_2$), acid chlorides couple to α-ketols via an acylsamarium intermediate. In the case of benzoyl chloride, benzil is obtained in high yield because of ready air oxidation of the intermediate enediol. α-Ketols are also formed by addition of a mixture of RCOCl and RCHO or RCOR (1:1) to a solution of SmI$_2$ (2 equiv.).

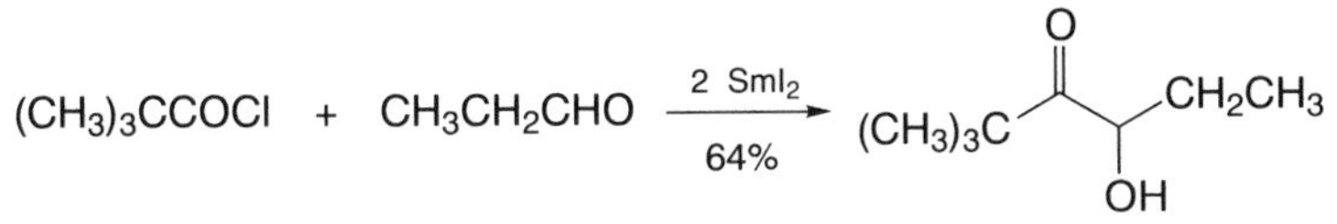

$(CH_3)_3CCOCl$ + CH_3CH_2CHO $\xrightarrow[64\%]{2\ SmI_2}$ (reaction)

Addition of $ClCH_2OCH_2C_6H_5$ to $\rangle C=O$.[6] In the presence of SmI_2, benzyl chloromethyl ether adds to carbonyl compounds to effect benzyloxymethylation. These adducts on hydrogenolysis provide 1,2-diols.

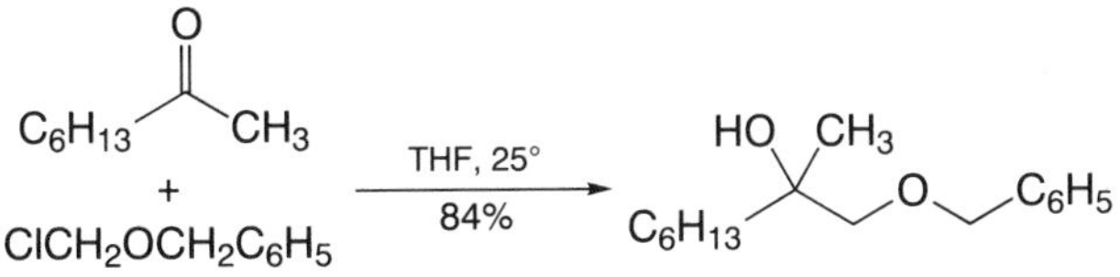

This Barbier-type reaction with CH_2I_2 can be used to obtain iodohydrins and cyclopropanols. Highest yields of iodohydrins obtain when the ratio of ketone/CH_2I_2/SmI_2 = 1:3:2.

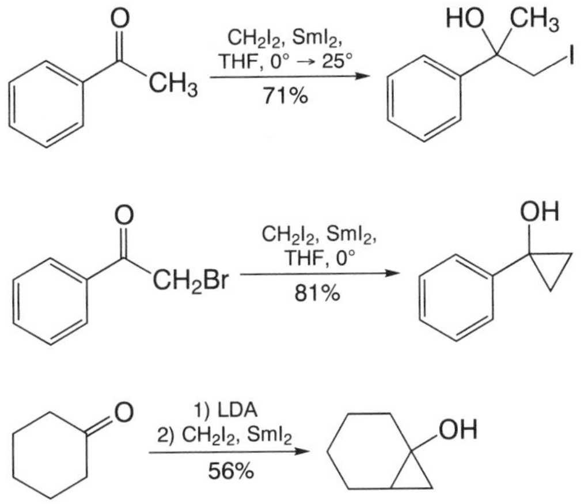

Cyclization of alkynyl halides; methylenecyclopentanes.[7] SmI_2 is as effective as Bu_3SnH for cyclization of 6-halo-1-ynes, particularly when used in excess (3 equiv.) and with DMPU as a cosolvent with THF.

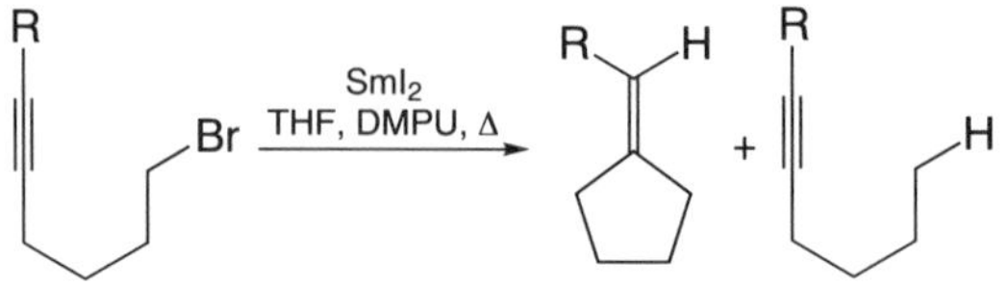

R = C$_6$H$_5$ 83%

R = Bu   81%   2%

R = Si(CH$_3$)$_3$ 67%   8%

Intramolecular Reformatsky reaction.[8]   Medium- to large-sized rings can be obtained by reaction of an α-bromo-ω-oxo ester (**1**) with SmI$_2$ in THF followed by acetylation. The precursors of **1** can be 1,ω-diols or cycloalkanones.

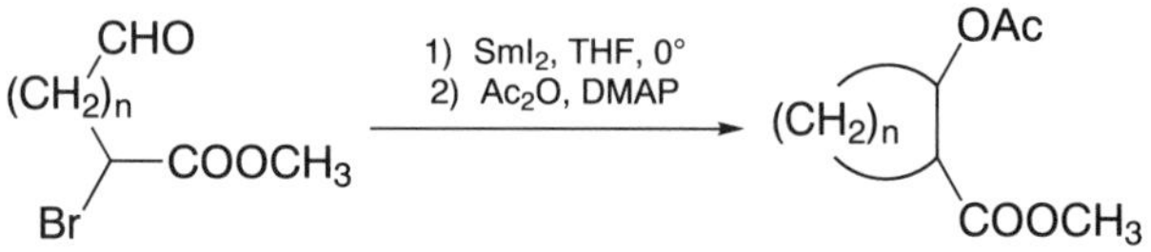

n = 6   69%

= 9   75%

= 13   82%

Reductive cyclization of unsaturated ketones.[9]   Treatment of unsaturated ketones with SmI$_2$ (2 equiv.)/HMPA (8 equiv.) can lead to 5- and 6-membered carbocycles. The intermediate organosamarium species can be trapped with various electrophiles to provide elaborate carbocycles. Both processes are highly diastereoselective.

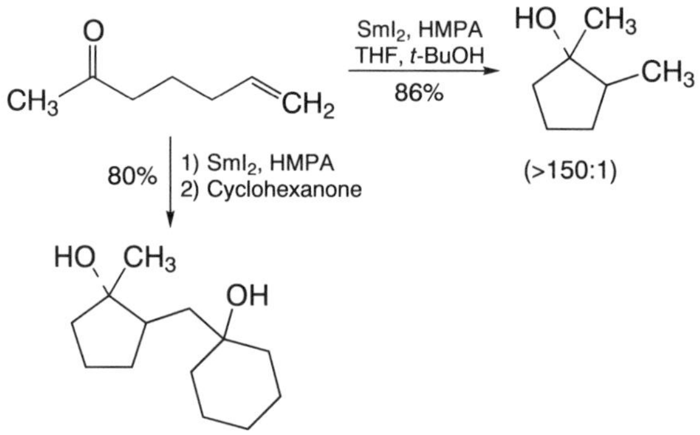

Vinylogous Barbier reaction.[4]   SmI$_2$ (2 equiv.) can effect cyclization of the unsaturated iodo ketone **1** in THF containing DMPU (or HMPA) to give a bicyclic samarium enolate (**a**) that can be trapped by an aldol reaction.

Reduction of RNO₂.[10]   Primary, secondary, or tertiary nitroalkanes can be reduced by SmI_2 with CH_3OH as the proton source to either alkyl hydroxylamines or amines depending on the amount of SmI_2. Reactions with 4 equiv. of SmI_2 at 25° in THF/CH_3OH (2:1) for less than five minutes provides hydroxylamines in yields of 60–93%. Reactions with 6 equiv. of SmI_2 for eight hours provides amines, identified as the 4-phenylbenzamides, in 50–80% yield.

Review of SmI₂ reduction of halides and radicals.   This report (69 references) reviews various proposed mechanisms for these reactions, and suggests that both reactions may involve organosamarium(III) intermediates.[11]

[1] E. J. Enholm and S. Jiang, *Tetrahedron Letters,* **33,** 6069 (1992).
[2] Y. Kamochi and T. Kudo, *Tetrahedron,* **48,** 4301 (1992).
[3] J. Inanaga, O. Ujikawa, and M. Yamaguchi, *Tetrahedron Letters,* **32,** 1737 (1991).
[4] D. P. Curran and R. L. Wolin, *Synlett,* 317 (1991).
[5] J. Collin, J.-L. Namy, F. Dallmer, and H. B. Kagan, *J. Org.,* **56,** 3118 (1991).
[6] T. Imamoto, T. Hatajima, N. Takiyama, T. Takeyama, Y. Kamiya, and T. Yoshizawa, *J. C. S. Perkin I,* 3127 (1991).
[7] S. M. Bennett and D. Larouche, *Synlett,* 805 (1991).
[8] J. Inanaga, Y. Yokoyama, Y. Handa, and M. Yamaguchi, *Tetrahedron Letters,* **32,** 6371 (1991).
[9] G. A. Molander and J. A. McKie, *J. Org.,* **57,** 3132 (1992).
[10] A. S. Kende and J. S. Mendoza, *Tetrahedron Letters,* **32,** 1699 (1991).
[11] D. P. Curran, T. L. Fevig, C. P. Jasperse, and M. J. Totleben, *Synlett,* 943 (1992).

Samarium(II) iodide–Lithium amide, SmI_2–$LiNH_2$.

Reduction of esters, anhydrides, or carbonyl groups.   These groups can be reduced by SmI_2 in combination with a base ($LiNH_2$, $LiOCH_3$, or KOH) and methanol (hydrogen source) in THF at 25° to alcohols.[1]

$$C_6H_5COOCH_3 \xrightarrow[\text{64\%}]{\begin{array}{c}\text{SmI}_2\text{, LiNH}_2,\\ \text{CH}_3\text{OH, THF}\end{array}} C_6H_5CH_2OH$$

$$C_6H_5CONH_2 \xrightarrow{\hspace{3cm}} C_6H_5CH_2OH \;+\; C_6H_5CH_2NH_2$$

$$81\% \qquad\qquad 8\%$$

$$C_6H_5CH=NOH \xrightarrow{\hspace{3cm}} C_6H_5CH_2NH_2 \;+\; C_6H_5CH_2OH$$

$$45\% \qquad\qquad 3\%$$

[1] Y. Kamochi and T. Kudo, *Tetrahedron Letters*, **32**, 3511 (1991).

Selenium dioxide, SeO₂.

Cyclic diacetylenes. A useful route to cyclic diacetylenes involves reaction of a cyclic diketone such as **1** with semicarbazide acetate in ethanol to afford **2**. Reaction of **2** with SeO₂ forms selenadiazoles (**3** and **4**), which on thermolysis affords cyclic diacetylenes **5** and **6**.

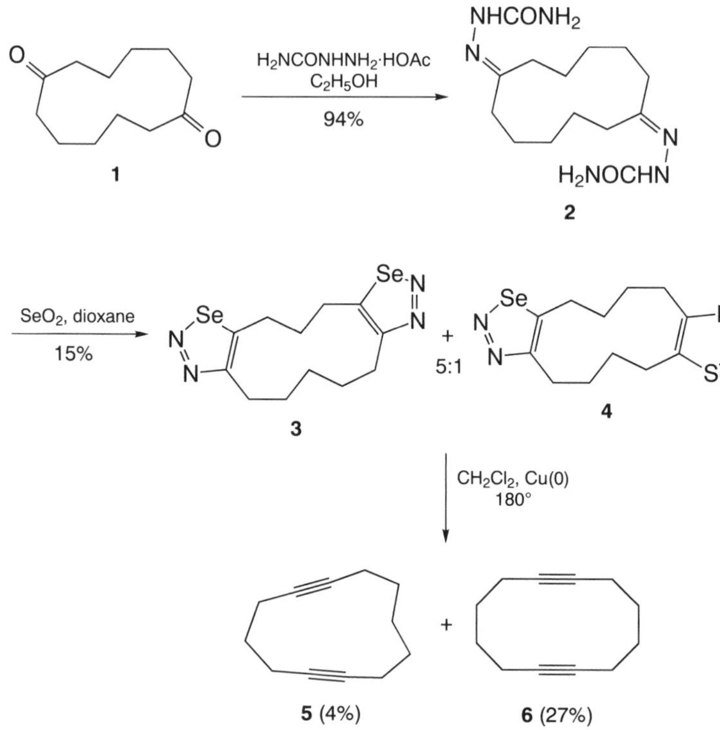

Cyclic diacetylenes can also be obtained by a similar strategy from a cyclic ynone such as 5-cyclodecynone **7**. In this case only one selenadiazole intermediate (**9**) was formed.[1]

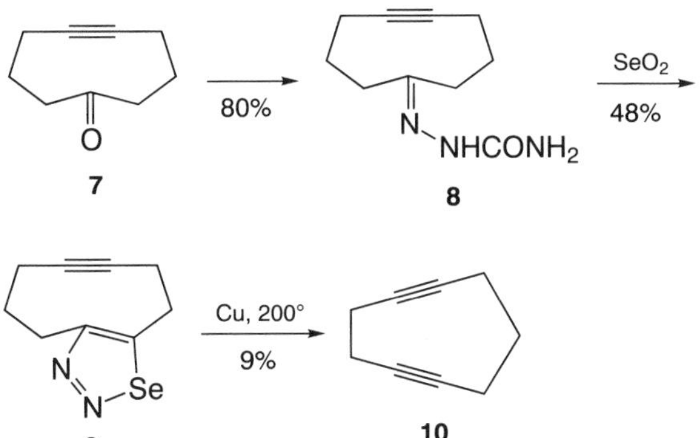

[1] R. Gleiter, D. Kratz, W. Schäfer, and V. Schehlmann, *Am. Soc.*, **113**, 9258 (1991).

Selenium dioxide–Trimethylsilyl polyphosphate (PPSE).

Aromatization of cyclohexenes and cyclohexadienes.[1] This reaction can be effected in refluxing CCl_4 by treatment with SeO_2 (1 eq.) and PPSE (10 equiv.; **10**,437; **11**,427; **12**,543). Note that a cyclohexane fused to an aromatic ring is not affected.

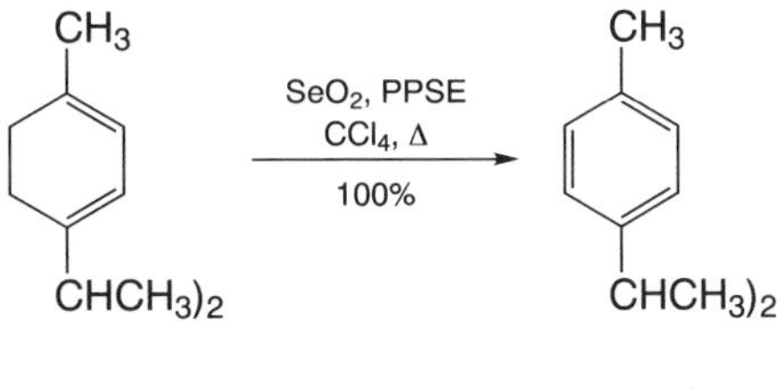

[1] J. G. Lee and K. C. Kim, *Tetrahedron Letters,* **33**, 6363 (1992).

Silver trifluoroacetate, CF_3COOAg. **14**,331; **16**,301–302.

α-Alkylation of ketones.[1] Silyl enol ethers can be alkylated by *n*-alkyl iodides (but not bromides) activated by silver trifluoroacetate (1 equiv.). Although the yields are generally moderate, this reaction is a useful route to α-alkyl ketones.

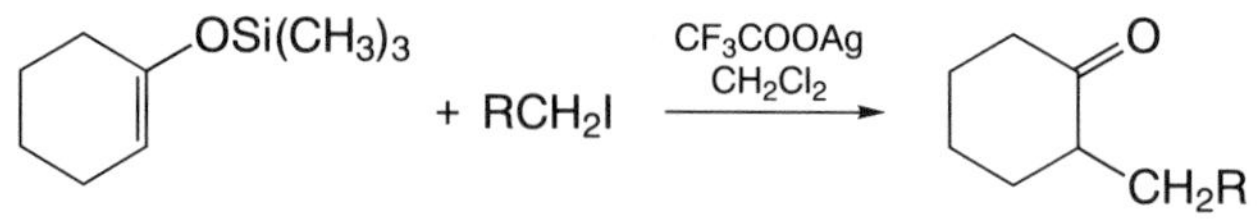

R = H	83%
R = CH₃CH₂	56%
R = C₂H₅O₂C	50%

[1] C. W. Jefford, A. W. Sledeski, P. Lelandais, and J. Boukouvalas, *Tetrahedron Letters,* **33,** 1855 (1992).

Silver trifluoromethanesulfonate, AgOTf.

Iodocyclization of carbamates.[1] Silver triflate (3 equiv.) markedly promotes iodocyclization with *trans*-stereoselectivity of carbamates (**1**) derived from amino acids.

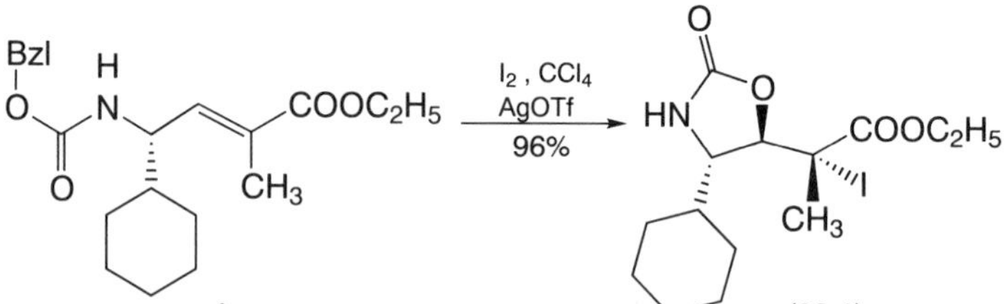

[1] Y. Guindon, A. Slassi, E. Ghiro, G. Bantle, and G. Jung, *Tetrahedron Letters,* **33,** 4257 (1992).

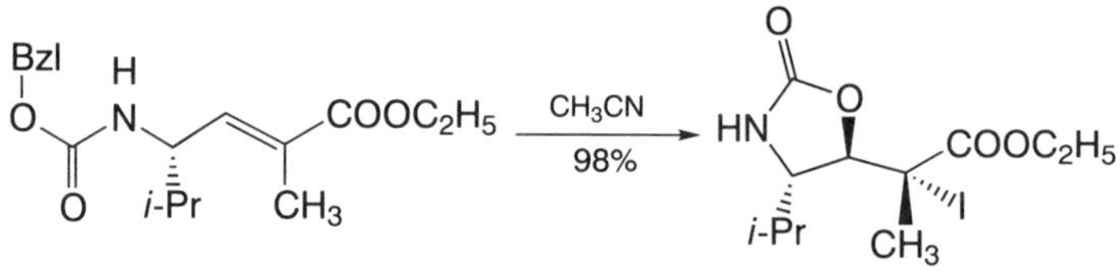

Sodium borohydride–Cerium(III) chloride (Luche reagent).

Stereoselective reduction of ketones.[1] The Luche reagent is useful for selective reduction of ketones in the presence of aldehydes, and for selective reduction of enones and enals to alcohols. The reagent can also invert the selectivity of hydride reduction of cyclic ketones such as **1a**. Reduction of **1a** with a metal boron and aluminum hydrides,

1a , R¹,R² = CH₃   exo-2   endo-2

NaBH₄ , CH₃OH, -78°   7:93
NaBH₄ , CeCl₂, CH₃OH , -78°   96:4

including NaBH₄, results in highly selective attack from the least hindered face to give the *endo*-alcohol (**2**). Reaction with NaBH₄–CeCl₃ (1:1) results in reduction to provide *exo*- and *endo*-2 in the ratio 96:4. Reduction to the *exo*-alcohol predominates (85/15) even when only 0.1 equiv. of CeCl₃ is present, but the reaction is slow. The order of admixture of the reagents has no effect at −78°, but no reduction at all is observed if the ketone is added to the reductant at 20°. The actual reagent in this case may be the unreactive sodium tetramethylboronate. Apparently at −78° the reaction of NaBH₄ with the solvent is suppressed. Lithium borohydride in combination with CeCl₃ is comparable to NaBH₄/CeCl₃, but chlorocerium borohydride and cerium borohydride show only moderate stereoselectivity with **1a** (*exo/endo* = 65:35). The same stereoselective reduction obtains in the absence of the additional keto group of **1a**. However, the *endo*-methyl group at C_6 of **1a** is essential. When it is absent the ketone is reduced to a 1:1-mixture of alcohols.

anti-**Selective reduction of epoxy ketones.**[2]   α, β-Epoxy ketones are reduced to epoxy alcohols with high *anti*-selectivity under Luche conditions.

[1] A. Krief and D. Surleraux, *Synlett,* 273 (1991).
[2] K. Li, L. G. Hamann, and M. Koreeda, *Tetrahedron Letters,* **33**, 6569 (1992).

Sodium borohydride–Iodine, $NaBH_4$-I_2 (12:5).

Reduction of amides, nitriles, esters, and acids; hydroboration.[1]   This system (12:5) in THF reduces amides or nitriles to the corresponding amines in 70–75% yield. Carboxylic acids or esters are reduced to alcohols in 60–90% yield. The actual reagent is diborane, and indeed $NaBH_4$–I_2 can effect hydroboration of alkenes.

$$C_6H_5CH{=}CH_2 \xrightarrow[\substack{90\%}]{\substack{1)\ I_2,\ NaBH_4,\ THF \\ 2)\ H_2O_2,\ ^-OH}} C_6H_5CH_2CH_2OH$$

[1] A. S. B. Prasad, J. V. B. Kanth, and M. Periasamy, *Tetrahedron,* **48**, 4623 (1992).

Sodium hypochlorite.

Review.[1]   Use of NaOCl and, to a lesser extent, $Ca(OCl)_2$ as oxidants during the period 1980–1992 has been reviewed (104 references). NaOCl is readily available; yields can be improved somewhat by addition of catalytic amounts of $RuCl_3 \cdot 3H_2O$ (1 mol %) or $RuO_2 \cdot 2H_2O$. Oxidation can be conducted in a two-phase system by use of a phase-transfer catalyst, usually $Bu_4NBr(Cl)$, 5 mol %. One significant development is that primary hydroxy groups can be oxidized in the presence of primary–secondary diols to give hydroxy aldehydes in 70–98% yield by using NaOCl (1.3 equiv.) and catalytic amounts of 2,2,6,6-tetramethylpiperidine-1-oxyl (TEMPO) and of Bu_4NCl. This system, but with 2.5 equiv. of NaOCl, can also oxidize primary alcohols to carboxylic acids, $RCH_2OH \rightarrow RCOOH$, in 80–98% yield; $Ca(OCl)_2$ is preferred over NaOCl for oxidative cleavage of *vic*-diols. The reaction is effected in aqueous acetonitrile and acetic acid at 20°.

[1] J. Skarzewski and R. Siedlecka, *Org. Prep. Proc. Int.,* **24**, 623 (1992).

Sodium phenyl selenide, C_6H_5SeNa.

Enamides.   One of the last steps in a synthesis of the 14-membered cyclopeptide alkaloid nummularine F (**4**) involves dehydration of the alcohols **1** to form the enamide group of **3**. The standard route through the selenoxide fails, but the mesylates of **1** are converted by reaction with C_6H_5SeNa into the corresponding phenyl selenides (**2**), which are oxidized (H_2O_2) to the selenoxides. One of these decomposes at 25° to **3**; the other, when heated in benzene at 60°. The last steps in the synthesis of **4** involve removal of the Boc group and coupling with N,N-dimethylglycine (DCC, 59% yield).

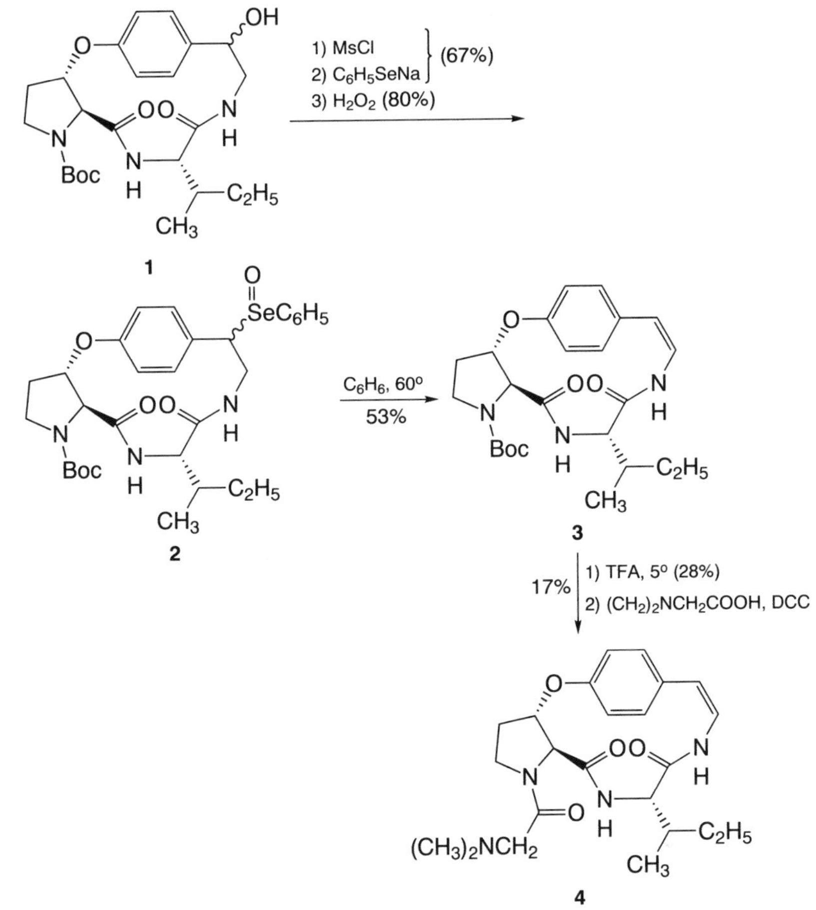

[1] R. J. Heffner, J. Jiang, and M. M. Joullié, *Am. Soc.*, **114**, 10181 (1992).

N-Sulfinyloxazolidinones, chiral.

Two of these reagents, **1** and **2**, have been studied. One is derived from (4R,5S)-norephedrine (HX$_N$), the other (**2**) is derived from (4S)-phenylalanine (HX$_p$).

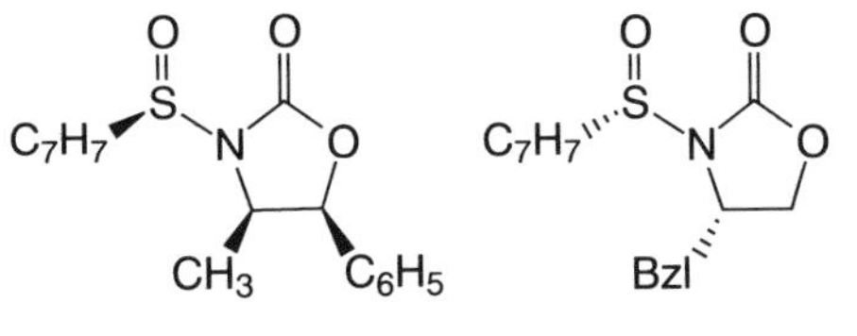

(R)-**1**   (S)-**2**

Asymmetric synthesis of sulfoxides, sulfinates, and sulfonamides.   These sulfiny-
lating reagents react with a wide range of nucleophiles with inversion at sulfur.[1]

(S)-**2** + CH₃MgBr $\xrightarrow[90\%]{\text{THF, -78°}}$

$$\underset{\substack{\\ \text{99\%ee}}}{C_7H_7 \overset{\displaystyle O}{\underset{\displaystyle \|}{\underset{S}{}}} CH_3}$$

(S)-**2** + BrCH₂COOC(CH₃)₃ $\xrightarrow[81\%]{\text{Zn}}$

$$\underset{\substack{\\ > 98\%ee}}{C_7H_7 \overset{\displaystyle O}{\underset{\displaystyle \|}{\underset{S}{}}} CH_2COOC(CH_3)_3}$$

(S)-**2** + (C₂H₅)₂NMgBr $\xrightarrow[91\%]{\text{THF, -78°}}$

$$\underset{\substack{\\ > 98\%ee}}{C_7H_7 \overset{\displaystyle O}{\underset{\displaystyle \|}{\underset{S}{}}} N(C_2H_5)_2}$$

[1] D. A. Evans, M. M. Faul, L. Colombo, J. J. Bisaha, J. Clardy, and D. Cherry, *Am. Soc.,* **114,** 5977 (1992).

(−)-**Sparteine,**   (-)-**1**

Asymmetric deprotonation.[1]   Treatment of Boc-pyrrolidine (**2**) with a slight excess
of a mixture of (−)-**1** and *sec*-BuLi (1:1) in ether at −78° for several hours provides a
chiral 2-lithio-Boc-pyrrolidine (**a**), which reacts with various electrophiles to provide
2-substituted Boc-pyrrolidines in 88–96% ee. A typical synthesis is illustrated in
equation (I).

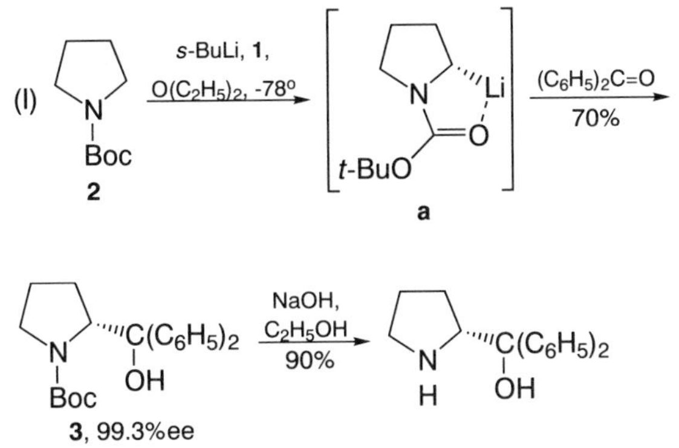

Enantioselective deprotonation of carbamates.[2] 1,3- or 1,4-Diols protected as carbamates of a sterically demanding oxazolidine group undergo highly enantioselective deprotonation with *sec*-BuLi/(−)-sparteine. The protecting group can be removed by sequential reaction with $CH_3SO_3H/Ba(OH)_2$.

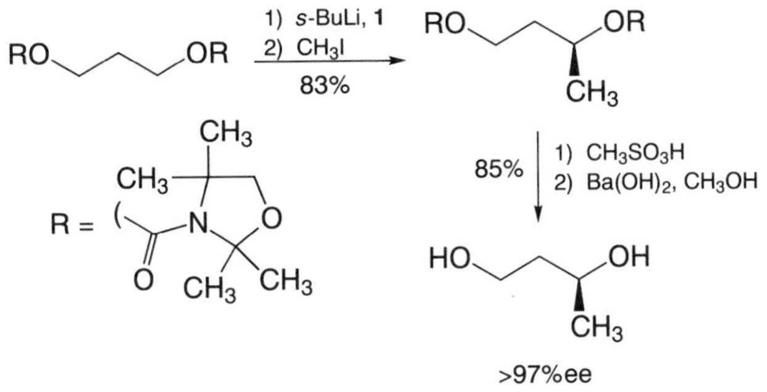

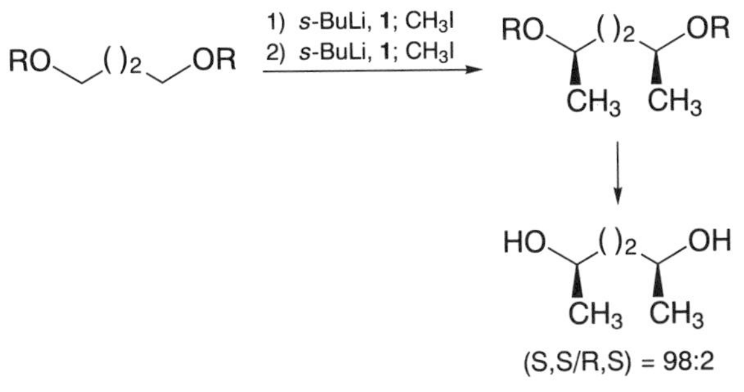

[1] S. T. Kerrick and P. Beak, *Am. Soc.*, **113**, 9708 (1991).
[2] M. Paetow, H. Ahrens, and D. Hoppe, *Tetrahedron Letters*, **33**, 5323, 5327 (1992).

N-Sulfonyloxaziridines.

Asymmetric hydroxylation of enolates. Davis and Chen[1] have reviewed this reaction using in particular (R,R)- and (S,S)-2-phenylsulfonyl)-3-phenyloxaziridene (**1**) and (camphorylsulfonyl)oxaziridine (**2**). Of these reagents, **1** and (+)- and (−)-2, derived from (1R)-10-camphorsulfonic acid, provide highest enantioselectivity and in addition are easy to prepare. They are effective for hydroxylatation of ketones, esters, β-keto esters, amides, lactones, and lactams.

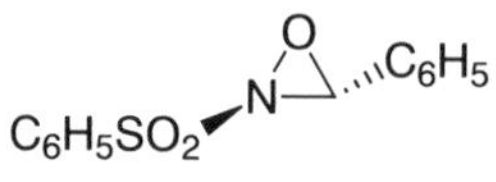

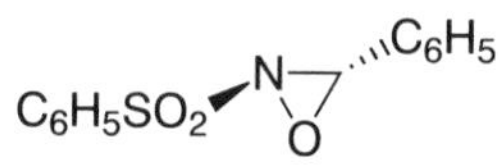

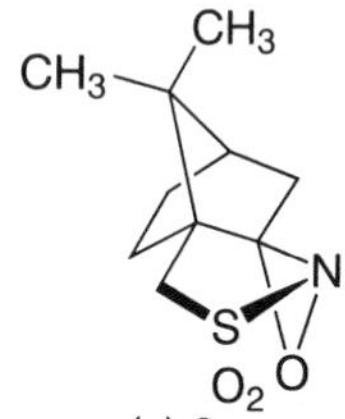

[1] F. A. Davis and B.-C. Chen, *Chem. Rev.*, **92**, 919 (1992).

T

Tantalum(V) chloride/Zinc, 16, 312.

Reaction of Ta–alkyne complexes with $R^1R^2C=O.$[1]   The Ta complexes formed from $TaCl_5/Zn$ with unsymmetrical alkynes react with carbonyl compounds to form two regioisomeric allylic alcohols with a ratio depending on the substituents on the alkyne (both steric and electronic effects) as well as the size of the substituents in the carbonyl group. The complexes from acetylenic esters react with carbonyl compounds mainly at the position α to the ester group, whereas complexes from acetylenic amides react mainly at the position β to the amide.

$$C_{10}H_{21}C \equiv CCO_2C_2H_5 \xrightarrow[76\%]{\substack{1)\ TaCl_5/Zn \\ 2)\ PrCHO}}$$

$$\begin{array}{c} C_{10}H_{21} \quad CO_2C_2H_5 \\ \diagup\diagdown \\ H \qquad -Pr \\ HO \\ (95:5) \end{array}$$

$$c\text{-}C_6H_{11}C \equiv CCON(CH_3)_2 \xrightarrow[79\%]{\substack{1)\ TaCl_5/Zn \\ 2)\ PrCHO}}$$

$$\begin{array}{c} c\text{-}C_6H_{11} \quad CON(CH_3)_2 \\ \diagup\diagdown \\ Pr- \qquad H \\ OH \\ (90:10) \end{array}$$

[1] Y. Kataoka, J. Miyai, M. Tezuka, K. Takai, and K. Utimoto, *J. Org.*, **57**, 6796 (1992).

R,R-Tartaric acid, 16, 312–315.

Asymmetric allyltitanation of RCHO.[1]   A complex, (R,R)-**3**, prepared in two steps from the chiral diol R,R-**1** derived from L-tartaric acid (**14**,232; **16**,314), effects allylation of aldehydes in >90% ee.

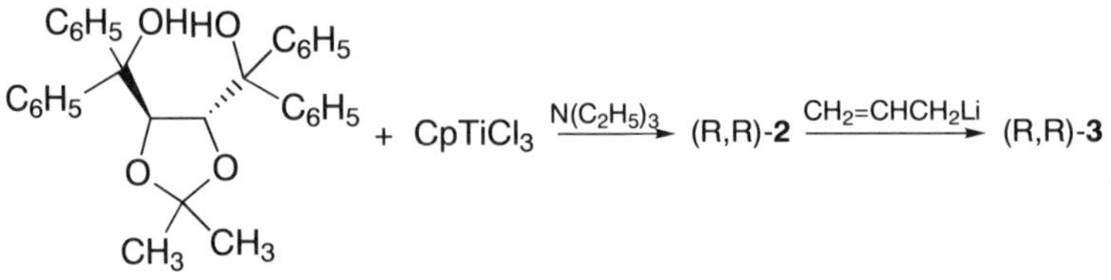

Asymmetric synthesis of β-lactams.[2]   The key reagent for a new asymmetric synthesis of β-lactams is the chiral imine **1**, prepared from (2S,3S)-tartaric acid. Reaction of **1** with an ester enolate (**2**) provides a β-lactam **3**, the configuration of which

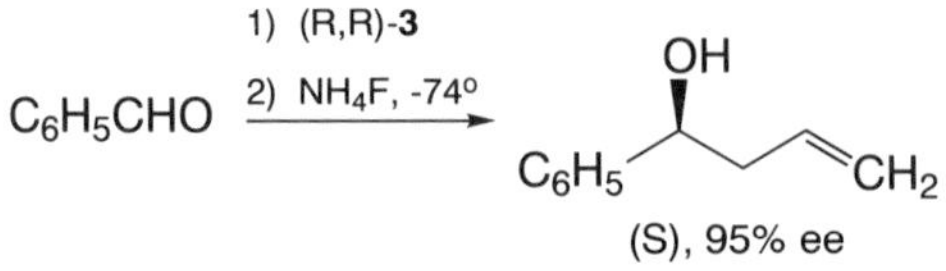

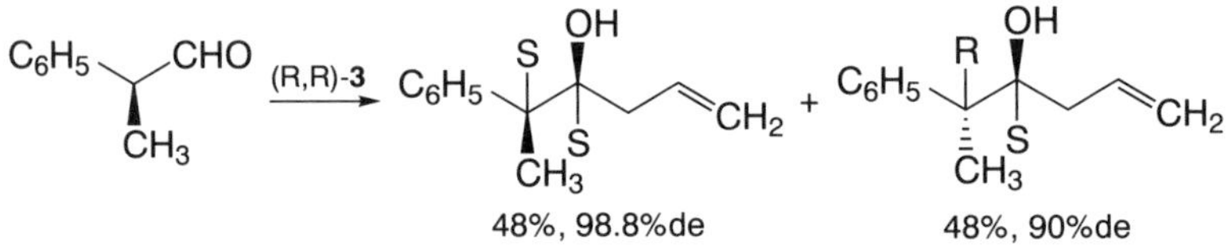

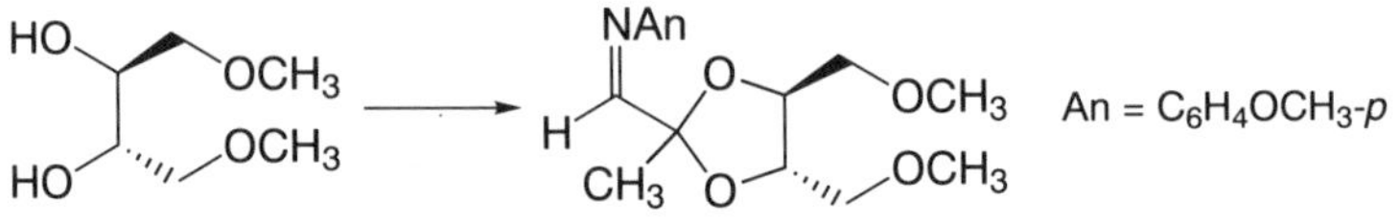

M = Li, DME	96%	S:R = 99:1
M = ZnCl₂, THF	84%	S:R = 85:15
M = Ti(O-*i*-Pr)₃	96%	S:R = 11:89

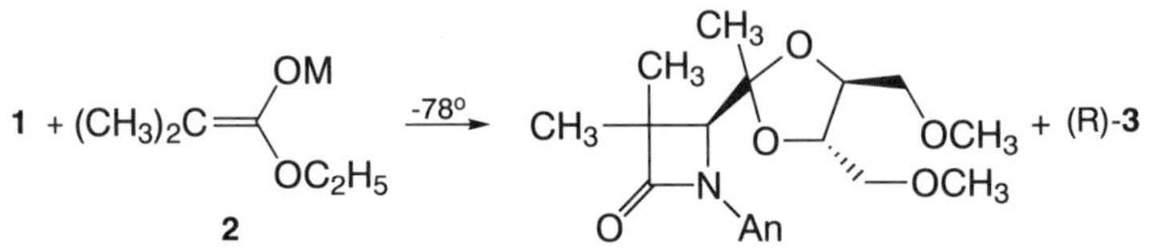

depends on the metal of the enolate. The lithium enolate reacts to form mainly (S)-**3**, whereas (R)-**3** is the major product from the titanium enolate.

Acyloxyborane catalyzed allylation.[3] (cf. **16**,314).   The acyloxyborane **1**, obtained by reaction of borane · THF with (2R,3R)-2,6-diisopropoxybenzoyltartaric acid, catalyzes the reaction of allylsilanes with aldehydes to afford *syn*-homoallylic alcohols with >80% ee. The geometry of the silane has no effect on the *syn*-selectivity.

[1] A. Hafner, R. O. Duthaler, R. Marti, G. Rihs, P. Rothe-Streit, and F. Schwarzenbach, *Am. Soc.,* **114**, 2321 (1992).
[2] T. Fujisawa, Y. Ukaji, T. Noro, K. Date, and M. Shimizu, *Tetrahedron,* **48**, 5629 (1992).
[3] K. Furuta, M. Mouri, and H. Yamamoto, *Synlett,* 561 (1991).

$\alpha, \alpha, \alpha', \alpha'$-Tetraaryl-1,3-dioxolane-4,5-dimethanols (TADDOLS).

Reviews.   Seebach *et al.*[1] provide a general review of these reagents, for which he has coined the abbreviation TADDOL. At the present time there are 26 known reagents of type **1** (C_2-symmetrical) and 12 of type **2** (C_1-symmetrical). In addition there are analogous compounds with heteroaryl, alkenyl, and alkyl substituents in place of the aryl group of **1** and **2**. All are prepared by reaction of (R,R)-tartrate ester acetals with aryl

Grignard reagents. The paper discusses ways to convert **1** and **2** into cyclic titanates; at least 25 titanates (**3**) derived from **1** and **2** have been isolated and characterized, including some in which $R^1 = R^2$ and X = Y. When X≠Y and $R^1 \neq R^2$, two diastereomers exist.

When used to catalyze the addition of diethylzinc to aldehydes, all complexes bearing phenyl groups but differing groups at C_2 of the dioxolane ring show similar enantioselectivity. A complex with a β-naphthyl group shows higher selectivity than that with an α-naphthyl group. When the *gem*-diaryl group of **3** is replaced by *gem*-dimethyl or -benzyl groups only slight enantioselectivity obtains.

A review suggests a mechanism for TADDOL-catalyzed addition of R_2Zn as well as other asymmetric reactions such as [2+2] and [4+2]cycloadditions, aldol reactions, allylic transfer, and others.[2]

Asymmetric addition of RCH$_2$MgBr to ketones.[3]   In the presence of **1** (1 equiv.), primary Grignard reagents (3 equiv.) react with ketones in THF to form tertiary alcohols in high enantioselectivity, usually 95–99:5–1.

1

(R) 99 : 1

---

[1] D. Seebach, D. A. Plattner, A. K. Beck, Y. M. Wang, D. Hunziker, and W. Petter, *Helv.,* **75**, 2171 (1992).

[2] K. Mikami and M. Shimizu, *Chem. Rev.,* **92**, 1021 (1992).

[3] B. Weber and D. Seebach, *Angew. Chem. Int. Ed.,* **31**, 84 (1992).

Tetrabutylammonium fluoride.

Desilylation of siloxanes.[1]   The siloxane **a**, obtained by hydrosilylation of a substrate such as **1**, can undergo protiodesilylation on treatment with Bu_4NF (3 equiv.) in DMF at 25° to give an alcohol (**2**). No reaction occurs if the hydroxyl group is not protected as an ether.

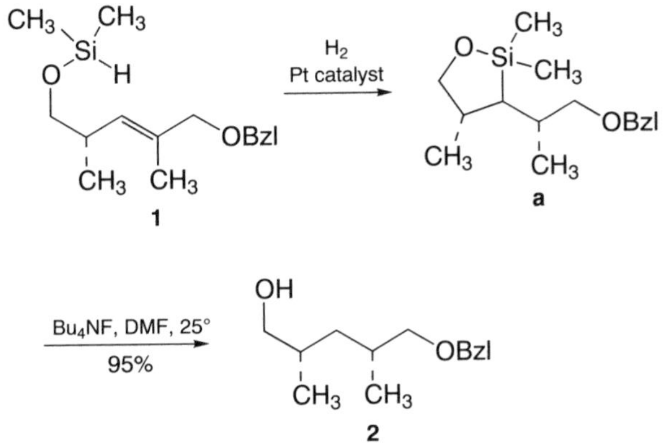

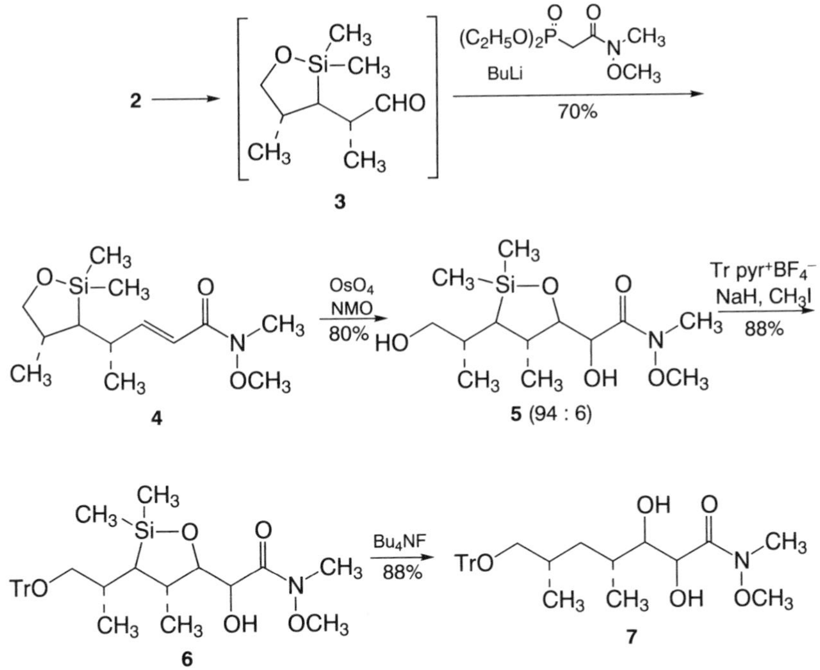

This desilylation provides a key step in an asymmetric osmylation of a double bond joined by one carbon to a five-memberd siloxane. Thus the aldehyde **3**, obtained by deprotection of **a** (H₂,Pd/C) and Swern oxidation, undergoes a Wittig–Horner reaction to provide the unsaturated amide **4** in 70% overall yield. This product undergoes osmylation to give essentially one product (**5**). After protection of the two hydroxy groups, the siloxane ring is removed to give **8**, with a *syn*-1,3-dimethyl group. Thus the siloxane ring can be used to control the diastereoselectivity of osmylation of an adjacent double bond.

[1] M. R. Hale and A. H. Hoveyda, *J. Org.*, **57**, 1643 (1992).

Tetracarbonyl-μ-dichlorodirhodium, $[Rh(CO)_2Cl]_2$ (1).

Phenol synthesis.[1]   Reaction of **1** with a cyclopropene (**2**) forms a rhodium vinyl carbenoid (**a**) that reacts with a terminal alkyne (**3**) to form an oxepine **4** and a phenol **5**. On treatment with HCl **4** rearranges to the phenol **5** in practically quantitative yield.

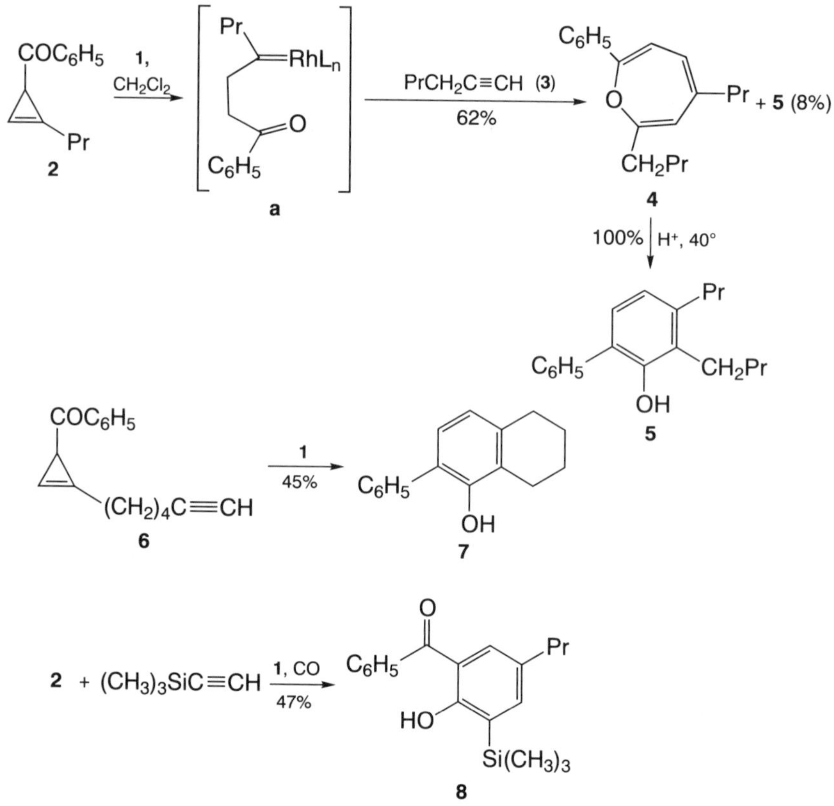

This reaction can also be used to obtain an annelated phenol **7**. The reaction of **2** with **1** under a CO atmosphere provides the phenol **8** via a rhodacycle.

[1] A. Padwa and S. L. Xu, *Am. Soc.*, **114**, 5881 (1992).

Tetrafluorosilane.

Selective cleavage of silyl ethers.[1]   This gaseous reagent can cleave silyl ethers in CH_2Cl_2 or CH_3CN at 25° in rates that are strongly dependent on the substituents at Si. The relative rates are in the order: $Et_3Si > t\text{-}BuMe_2Si \gg t\text{-}Bu(C_6H_5)_2Si$. Silyl ethers of phenols are stable to SiF_4 in CH_2Cl_2 at 23° for 24 hours. SiF_4 is an excellent reagent for selective cleavage of primary or secondary silyl ethers in the presence of tertiary ones.

[1] E. J. Corey and K. Y. Yi, *Tetrahedron Letters,* **33,** 2289 (1992).

Tetrakis(triphenylphosphine)palladium.

Coupling of 2-bromonaphthoquinones with stannanes.[1]   The 2-bromonaphthoquinone **1** undergoes Stille coupling with tetraalkyl-, alkenyl-, and aryltrialkylstannanes in good yield under catalysis with $Pd[P(C_6H_5)_3]_4$ or $Cl_2Pd(1,1'\text{-bisdiphenylphosphino})$-ferrocene [$PdCl_2(dppf)$]. Addition of CuBr usually improves the rate and/or yield. Thus coupling of **1** with the stannane **2** provides **3** in 82% yield. This product was converted into the antibiotic WS 5995C (**4**) in 2 steps.

Addition of $(ArS)_2$ and $(ArSe)_2$ to 1-alkynes.[2] $Pd[P(C_6H_5)_3]_4$ is the most effective catalyst for addition of diaryl disulfides and diselenides to terminal alkynes to provide (Z)-1,2-adducts in high yield (equation I). Note that photoinitiated addition of $(C_6H_5Se)_2$

(I) $CH_3(CH_2)_5-C{\equiv}CH$ + $(C_6H_5Y)_2$ $\xrightarrow{Pd(0),\ 80°}$

Y = S	91%	Z, 100%
Y = Se	82%	Z, 100%

(II) $R-C{\equiv}CH$ + $(C_6H_5Se)_2$ $\xrightarrow{h\nu}$

(E), 95 : 5

to 1-alkynes results mainly in the (E)-1,2-adducts (E/Z = 95:5), equation (II).

Carbonylative addition of diaryl disulfides and diselenides can also be effected with the Pd(0) catalyst to form (Z)-1,3-bis(arylthio)-2-alkene-1-ones and (Z)-1,3-bis(arylseleno)-2-alkene-1-ones, respectively (equation III).

(III) $CH_3(CH_2)_5-C{\equiv}CH$ + $(C_6H_5Y)_2$ $\xrightarrow[60°]{CO,\ Pd(0)}$

Y = S	86%	Z, 100%
Y = Se	80%	Z, 100%

Coupling of acid chlorides with (E)-Bu₃SnCH=CHSnBu₃; 1,4-diketones.[3] The Pd-catalyzed Stille coupling can be extended to coupling of acid chlorides with (E)-1,2-bis(tributylstannyl)ethene. Unexpectedly, a 1,4-dione is formed, evidently by reduction of an intermediate enediketone by Bu₃SnCl. Reaction of an α,β-unsaturated acid chloride also leads to a 1,4-diketone, as does the reaction of an acid chloride with a β-stannyl enone.

Chiral oxazolines; Stille carbonylative coupling.[4] Pd-catalyzed carbonylative coupling of triflates of ketones and phenols with chiral amino alcohols provides β-hydroxy amides, which cyclize to chiral oxazolines when treated with thionyl chloride.

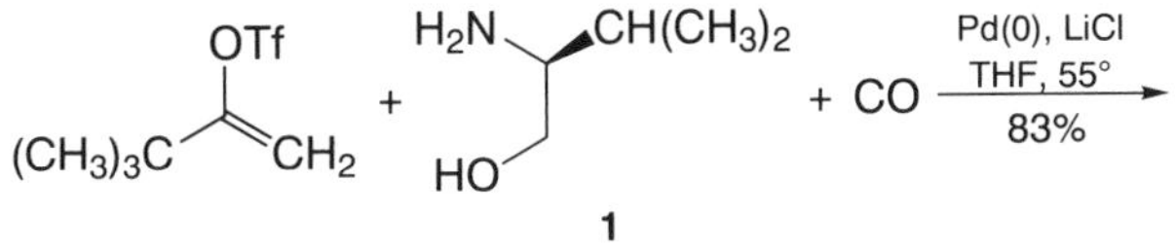

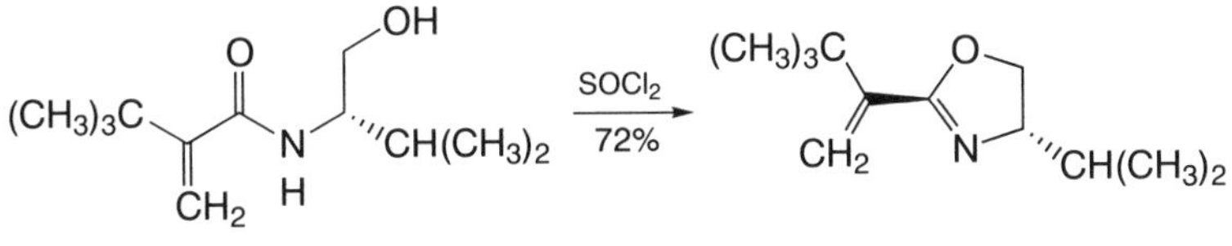

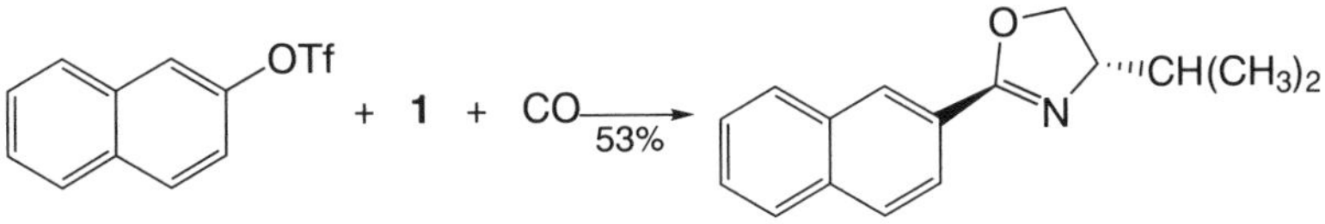

Somewhat more efficient catalysts are Pd(dba)$_2$/P(C$_6$H$_5$) and Pd(OAc)$_2$/dppp.

Cyclization of triynes to benzenes.[5]   The triyne **1** and the bromoalkenylalkyne-alkyne **2** both cyclize to the benzene derivative **3** when treated with this Pd complex (3 mol %) and a base or acid in refluxing CH$_3$CN.

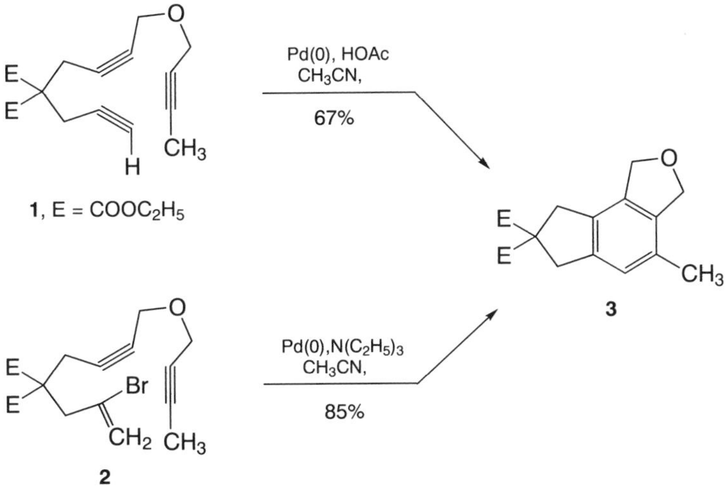

A partially intermolecular verison is possible, but less efficient. Thus the bromoenyne **4** and 3-hexyne cyclize to **5** under the same conditions in 40% isolated yield (equation I). However, cocyclization of (E)-β-bromostyrene and the diyne **6** results in a fulvene derivative **7** (equation II).

Coupling of aryl iodides and arylboronic acids.[6]   This Suzuki cross coupling can be effected with this Pd(0) complex as catalyst and Na_2CO_3 as base. It provides a rapid access to *m*- or *p*-teraryls.

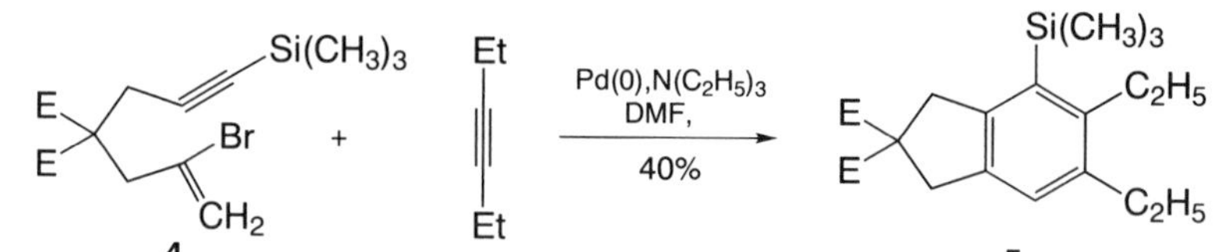

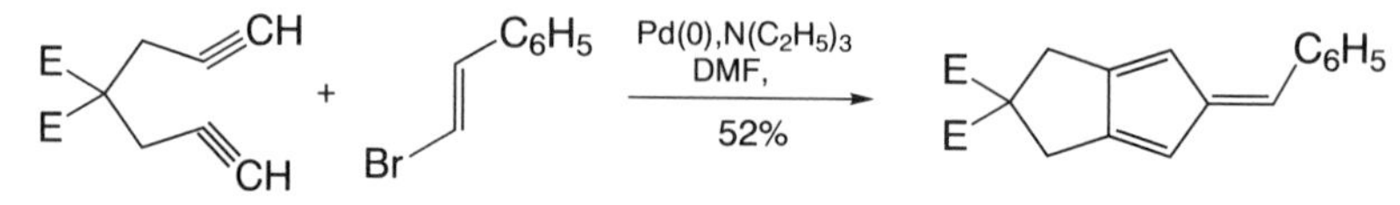

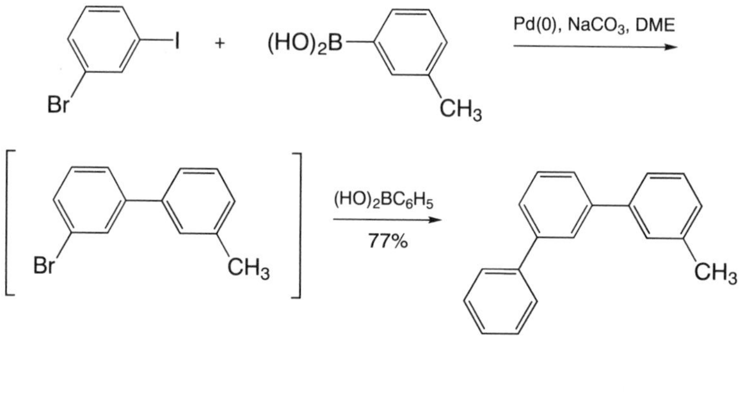

This coupling can be extended to a synthesis of *m*-quinquearyls by using a 1,3-dihalobenzene substituted at C_2 by a directing metalation group: $CON(C_2H_5)_2$, CN OMOM, or NH_2.

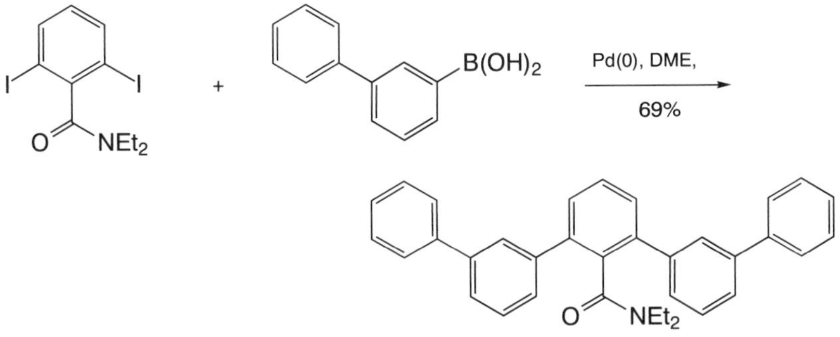

[1] N. Tamayo, A. M. Echavarren, and M. C. Paredes, *J. Org.*, **56**, 6488 (1991).
[2] H. Kuniyasu, A. Ogawa, S.-I. Miyazaki, I. Ryu, N. Kambe, and N. Sonoda, *Am. Soc.*, **113**, 9796 (1991).
[3] M. Peréz, A. M. Castaño, and A. M. Echavarren, *J. Org.*, **57**, 5047 (1992).
[4] A. I. Meyers, A. J. Robichaud, and M. J. McKennon, *Tetrahedron Letters*, **33**, 1181 (1992).
[5] E. Negishi, L. S. Harring, Z. Owczarczyk, M. M. Mohamud, and M. Azy, *ibid.*, **33**, 3253 (1992).
[6] C. M. Unrau, M. G. Campbell, and V. Snieckus, *ibid.*, **33**, 2773 (1992).

Tetramethylammonium triacetoxyborohydride (1, 14, 299–300; 16, 324).

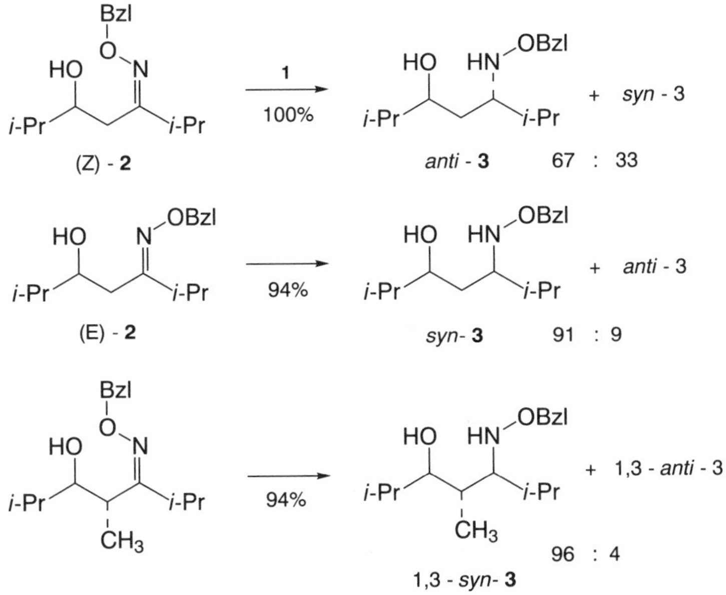

Reduction of α-hydroxy oximino ethers (2).[1] This reduction can be effected by **1** in HOAc/CH₃CN (1:1), −35°. Thus reduction of (Z)-**2** provides mainly the 1,3-*anti* product, whereas (E)-**2** provides mainly the 1,3-*syn* product. Note that an additional chiral center at C_2 can overcome the influence of the oxime geometry (last example).

[1] D. R. Williams and M. H. Osterhout, *Am. Soc.*, **114**, 8750 (1992).

Tetramethylguanidine.

Michael/aldol reaction.[1]   The key step in a synthesis of the alkaloid huperzine A (**3**) involves a Michael/aldol reaction of the β-keto ester **1** with methacrolein to provide the bridged ketol **2**. The usual Michael reaction catalysts (NaOCH$_3$, Bu$_4$NF, ZnCl$_2$) are ineffective, but the desired reaction can be effected in 93% yield with tetramethylguanidine as catalyst. DBU can also catalyze this reaction. Conversion of **2** to **3** is effected by dehydration of the ketol (45% yield), Wittig reaction of the ketone

with ethylidenetriphenylphosporane (73% yield), Curtius rearrangement of the ester to an amine group, and cleavage of the O-methyl ether to a pyridone (iodotrimethylsilane).

[1] A. P. Kozikowski, Y. Xia, E. R. Reddy, W. Tückmantel, I. Hanin, and X. C. Tang, *J. Org.*, **56**, 4636 (1991).

2,2,6,6-Tetramethylpiperidinyl-1-oxyl (TEMPO), 1.

Oxidation of α-amino or α-alkoxy alcohols.[1]   (cf., **14**,302). This oxidation of optically active α-amino or α-alkoxy alcohols with sodium hypochlorite (slight excess) catalyzed by TEMPO and NaBr can provide the corresponding aldehyde with no racemization. Highest yields are obtained with toluene/ethyl acetate/H$_2$O as solvent with 1–2 mol % of TEMPO and 1 equiv. of NaBr. Overoxidation to the acid can be minimized by rapid stirring. Carbamoyl groups (Boc or Cbz) are preferred for protecting groups of the amine. Yields are generally 80–90% and the % ee is >95%.

[1] M. R. Leanna, T. J. Sowin, and H. E. Morton, *Tetrahedron Letters,* **33**, 5029 (1992).

Tetraphenyldistibene, $(C_6H_5)_2SbSb(C_6H_5)_2$ **(1).**

$RCH_2I \rightarrow RCH_2OH$. Irradiation of **1** produces the radical $(C_6H_5)_2Sb\cdot$, which reacts with RCH_2I to form $(C_6H_5)_2SbCH_2R$, which can be oxidized to RCH_2OH. Thus irradiation of a mixture of the iodide **2** and **1** results in the stibene **3** in 80–86% yield. This product is stable to O_2, but is oxidized to **4** in high yield by alkaline H_2O_2. In contrast, the same reaction of **1** with **5** proceeds slowly and in low yield to give **6**,

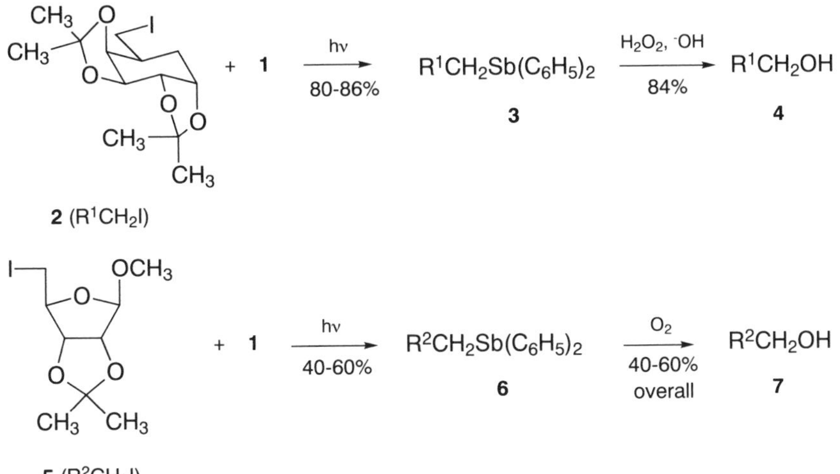

2 (R^1CH_2I)

5 (R^2CH_2I)

which is oxidized in air to the alcohol **7**. The acyclic iodide **8** under these conditions is converted to a stibene that is oxidized by air to **9** in 59% overall yield.

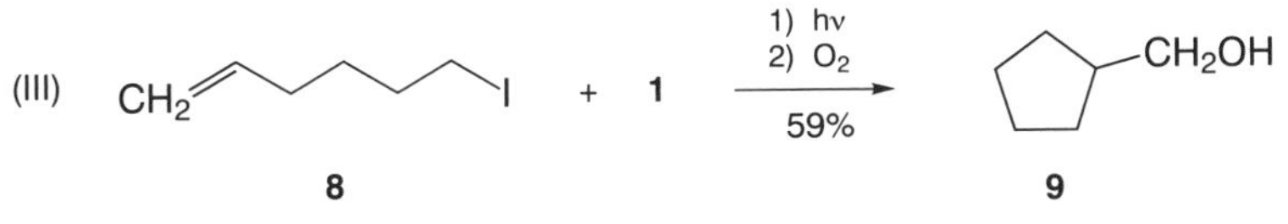

[1] A. G. M. Barrett and L. M. Melcher, *Am. Soc.,* **113**, 8177 (1991).

Tin.

Allylation of —CHO *(13,298).*[1] The addition of allyl bromide to aldehydes promoted by tin in an aqueous/organic solvent has been extended to aldoses. The reaction proceeds with *threo*-selectivity (4–6.5:1). The products can be converted to higher aldoses by ozonolysis. The diastereoselectivity decreases as the chain length decreases and is low in the case of an aldose lacking a hydroxyl group at C_2. No reaction is observed with aldoses with an N-acetyl group at C_2.

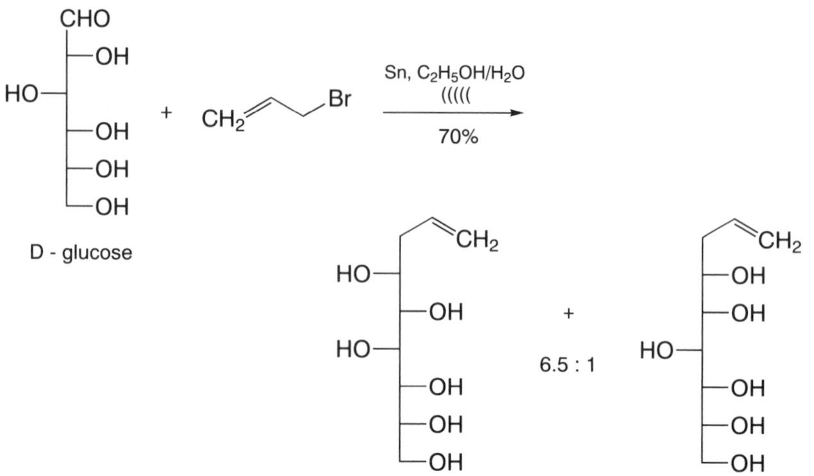

[1] W. Schmid and G. M. Whitesides, *Am. Soc.*, **113**, 6674 (1991).

Tin(II) chloride – Triphenylphospine.

Cleavage of epoxy ketones.[1]   The reaction of benzoyl chloride and an α, β-epoxy ketone proceeds in low yield and regioselectivity. However, addition of $SnCl_2$ complexed with $(C_6H_5)_3P$ results in high regioselectivity. In contrast, use of $Bu_2SnCl_2/P(C_6H_5)_3$ as catalyst results in cleavage at the opposite site.

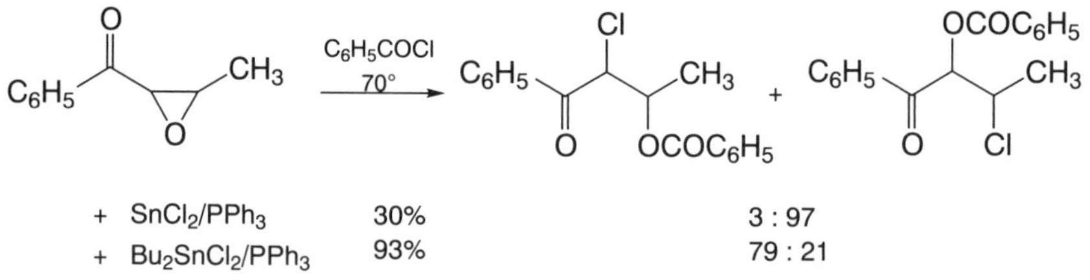

The corresponding tin iodides can also control the reactions of epoxy ketones with TsN=C=O to form 2-oxazolidones.

$$C_6H_5 \xrightarrow{TsN=C=O} $$

(top scheme with structures)

C_6H_5, O, ...C_6H_5 → TsN=C=O

C_6H_5 C_6H_5 ... TsN ... O   + C_6H_5 ... C_6H_5 ... O ... NTs

+ SnI$_2$/ P(Ph)$_3$, 25°	55%	1 : 99
+ BuSnI$_2$ / P(Ph)$_3$, 40°	88%	84 : 16

[1] I. Shibata, N. Yoshimura, A. Baba, and H. Matsuda, *Tetrahedron Letters,* **33**, 7149 (1992).

Tin(IV) chloride.

[4+2]Cycloaddition of nitroalkenes.[1]   Although nitrostyrene reacts as a dienophile in thermal reactions with dienes, it and other nitroalkenes undergo SnCl$_4$-catalyzed reactions with dienes to form *syn*- and *anti*-ring-fused nitronates with some *syn* preference (equation I). Nitronates are also obtained by SnCl$_4$-catalyzed reaction with cycloalkenes (equation II). But in the latter reaction only *anti*-adducts are formed. Nitroalkenes also undergo SnCl$_4$-catalyzed intramolecular [4+2]cycloaddition to form *syn* ring-fused adducts (equation III).

(i) CH_3 , NO_2 , C_6H_4-p-OCH$_3$ + (cyclohexadiene) $\xrightarrow{SnCl_4}$ 53%

CH_3 , p-CH$_3$OC$_6$H$_4$ *syn*   80 : 20   *anti*

(II) CH_3 , NO_2 , C_6H_5 + (cyclopentene) $\xrightarrow[93\%]{SnCl_4, \ CH_2Cl_2, \ 78°}$

CH_3 , C_6H_5 *anti*, 100%

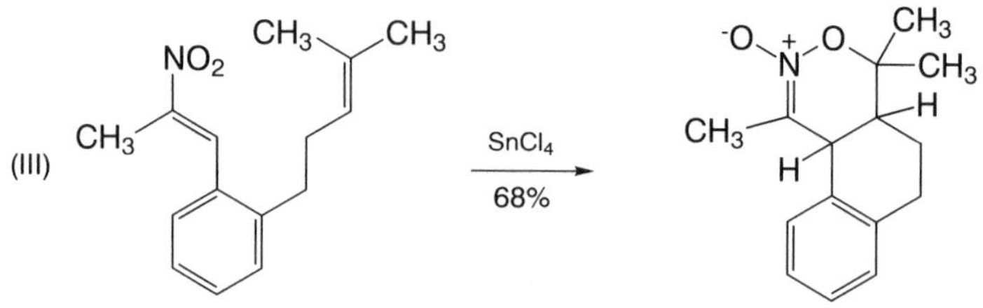

syn

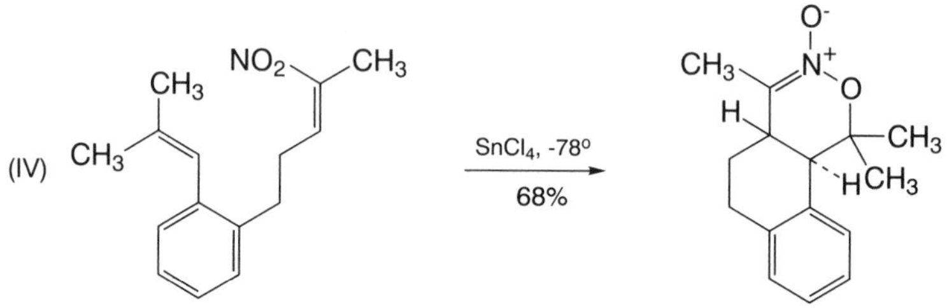

trans / cis = 79 : 21

The intramolecular [4+2]cycloaddition formulated in equation(IV) results in both *anti-* and *syn*-adducts, with some preference for the *anti-* and *trans*-isomer. Note that in all cases the geometry of the dienophile is retained.

Asymmetric Evans aldol reaction (11,379–381).[2] The boron enolate of the Evans imide **1** is widely used for preparation of chiral *syn*-aldols **2**. However, if the boron enolate is generated at −78° and then treated with the aldehydes precomplexed

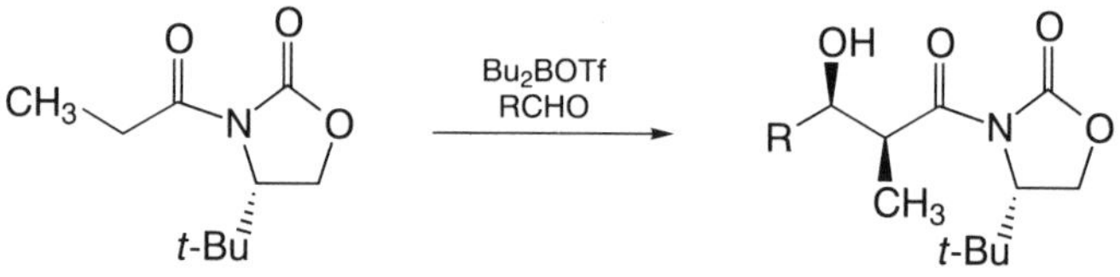

1 *syn* - **2**

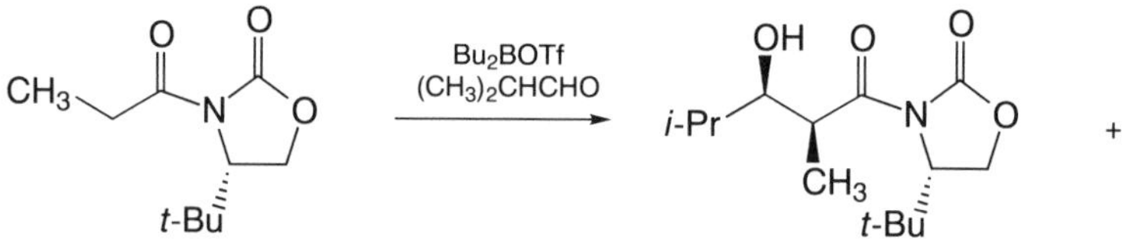

1 *syn* - **2**

anti - 2   +   syn - 3

		syn - 2	anti - 2	syn - 3
+ SnCl₄ (1 equiv.)	51%	0	95	5
+ (C₂H₅)₂AlCl (1 equiv.)	86%	0	86	14
+ TiCl₄ (1 equiv.)	71%	0	17	83

with a Lewis acid at $-78°$, entirely different results obtain, depending on the Lewis acid. Both $SnCl_4$ and diethylaluminum chloride are highly *anti*-selective, giving rise to the *anti*-isomer of the Evans *syn*-aldol. In contrast $TiCl_4$ is *syn*-selective, but gives rise to the chelation-controlled *syn*-aldol. Thus by choice of the Lewis acid and conditions, it is possible to prepare the chelation-controlled *syn*- or *anti*-aldols.

Asymmetric ene reaction of N-sulfinylcarbamates.[3]   The ability of Lewis acids to promote ene reactions (**11**,413,414; **12**,389) is useful for asymmetric reactions. Thus the $SnCl_4$-promoted reaction of chiral N-sulfinylcarbamates (**1**) with alkenes results in thermally unstable adducts (**2**) in 65–91% yield. Use of *trans*-2-phenylcyclohexanol (**13**,244) or 8-phenylmenthol as the source of chirality results in high diastereoselective induction in generation of the new carbon to sulfur bond (usually >95:5). This reaction is applicable to both (E)- and (Z)-alkenes, but the former react more readily. These ene adducts can be transformed into optically active allylic alcohols (**4**) by N-alkylation and conversion to an aryl allylic sulfoxide (**3**), which undergoes rearrangement in the presence of a thiophile (piperidine) to **4**, with retention of configuration at carbon imparted in the ene reaction. The overall process effects enantioselective allylic oxidation of an alkene with retention of the original position of the double bond.

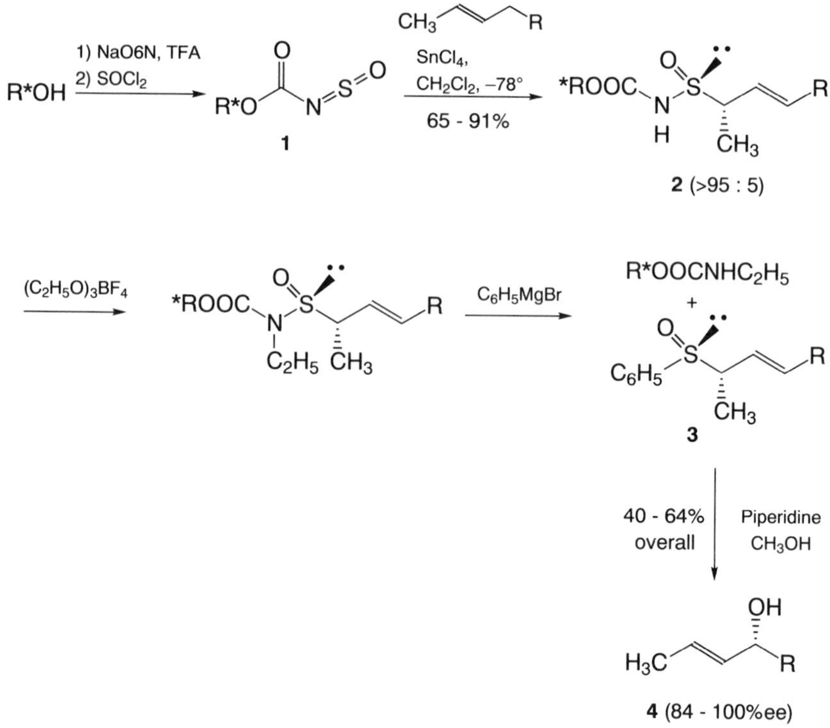

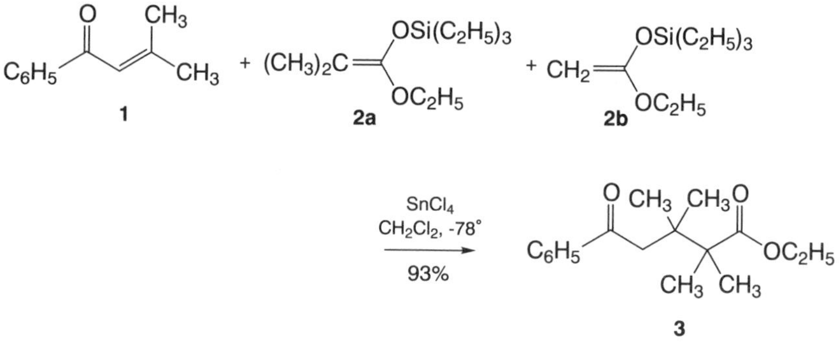

Mukaiyama–Michael reactions.[4]   This reaction involves addition of silyl ketene acetals to enones in the presence of a Lewis acid to form 1,5-keto esters (**13**,306–307; **15**,15). Surprisingly, a β,β-disubstituted enone (**1**) in the presence of various Lewis acid catalysts reacts more rapidly with the disubstituted silyl ketene acetal **2a** than with the unhindered silyl ketene acetal **2b**, to form the hindered adduct **3** with two adjacent quaternary carbon atoms. Similar results obtain with $Bu_2Sn(OTf)_2$ and $TiCl_4$. The result is hardly consistent with a nucleophilic reaction, but suggests that the reaction involves a radical coupling. Otera et al.[4] present evidence that the reaction is initiated by a one-electron oxidation of the silyl ketene acetal followed by electron transfer to the enone. Coupling of the two radicals then regenerates the Lewis acid for further coupling.

Rearrangment of allylic acetals to 3-acyltetrahydrofurans.[5]   A new synthesis of substituted tetrahydrofurans is based on the fact that the acetals (or ketals) formed by condensation of allylic diols with a carbonyl compound rearrange on treatment with a Lewis acid to a tetrahydrofuran. Of various Lewis acids, $SnCl_4$ is the most efficient. The allylic diols are available by reaction of a vinyllithium or vinyl Grignard reagent with α-hydroxy ketones. Of particular interest, this new synthesis shows high stereoselectivity. Thus the acetal **1**, a mixture of four stereoisomers, rearranges to the all-*cis*-3-acylfuran **2**. Even higher stereoselectivity obtains in the rearrangement of **3** to **4**. The reaction was used

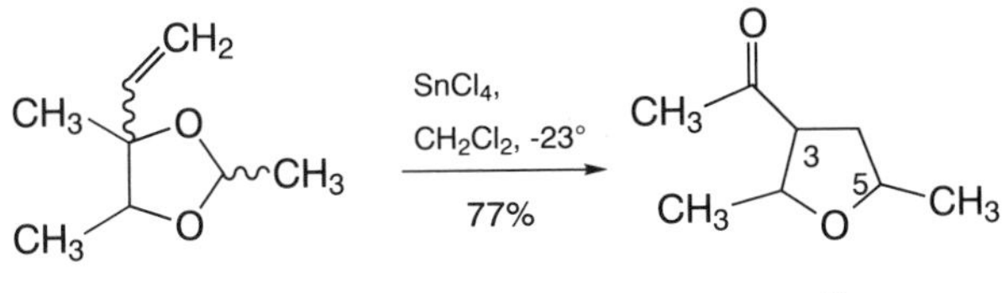

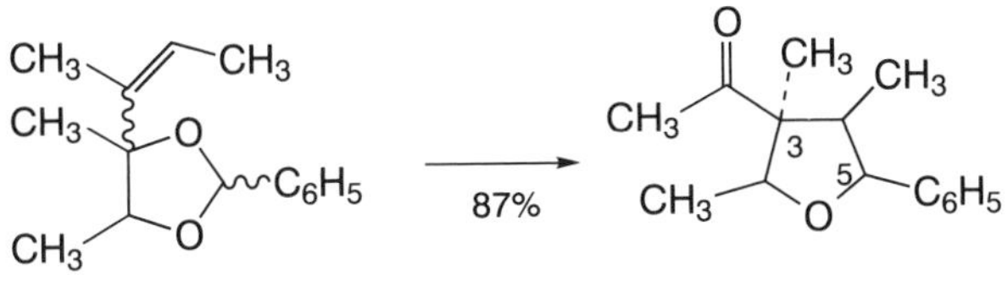

for synthesis of (+)-muscarine (**6**) from (S)-**5**, available from ethyl (+)-lactate (equation I).

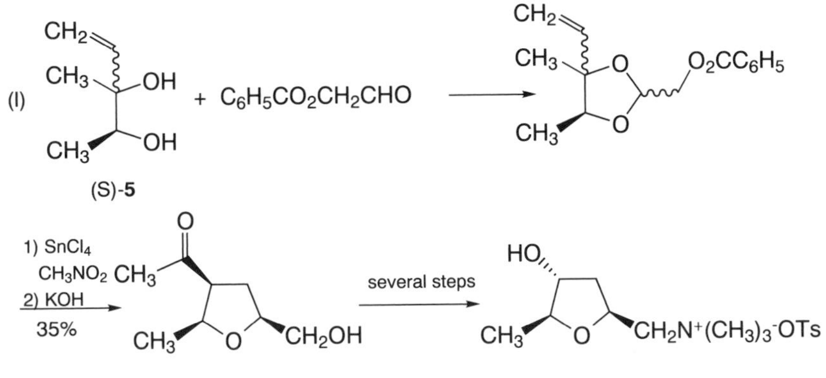

[1] S. E. Denmark, B. S. Kesler, and Y.-C. Moon, *J. Org.*, **57**, 4912 (1992).
[2] M. A. Walker and C. H. Heathcock, *J. Org.*, **56**, 5747 (1991).
[3] J. K. Whitesell, J. F. Carpenter, H. K. Yaser, and T. Machajewski, *Am. Soc.*, **112**, 7653 (1990).
[4] T. Sato, Y. Wakahara, J. Otera, H. Nozaki, and S. Fukuzumi, *Am. Soc.*, **113**, 4028 (1991).
[5] M. H. Hopkins, L. E. Overman, and G. M. Rishton, *Am. Soc.*, **113**, 5354 (1991); L. E. Overman and G. M. Rishton, *Org. Syn.*, **71**, 63 (1993).

Tin(IV) chloride–Trifluoroacetic acid.

Intramolecular ene reactions.[1]   Use of Lewis acid catalysts (particularly $FeCl_3$, **15**,156; **16**,190–101) has greatly extended the usefulness of intramolecular ene cyclization. Thus a new diastereoselective route to corynanthe-type alkaloids involves the ene cyclization of **1** to *trans*-indolo[2,3-*a*]-quinolizidine (**2**), a precursor to methyl corynantheate (**3**) by demethoxycarbonylation. $SnCl_4$ (1 equiv.) is the only common Lewis acid that is useful for this particular ene cyclization, and even so, it also requires the presence of trifluoroacetic acid (1.5 equiv.).

1) TFA, CH_2Cl_2, 20°
2) $SnCl_4$, 20°

42%

2, R = CO_2CH_3
3, R = H

[1] L. F. Tietze and J. Wichmann, *Angew. Chem. Int. Ed.*, **31**, 1079 (1992).

Tin(IV) chloride–Silver perchlorate.

α-Glycosylation.[1]   This combination produces a species such as $SnCl_3^+ ClO_4^-$, which is an effective catalyst for reaction of a protected 1-O-acetylglucose with a silyl ether to form α-glycosides.

$SnCl_3^+ClO_4^-$

ether, 0°

86 – 95%

(α/β = 4:1)   (α/β = 90-95 : 10-5)

[1] T. Mukaiyama, T. Takashima, M. Katsurada, and H. Aizawa, *Chem. Letters*, 533 (1991).

Tin(II) trifluoromethanesulfonate, $Sn(OTf)_2$.

Addition of 1-alkynes to carbonyls.[1]   Alkali metal acetylides are usually used for this addition, but their strong basic properties can cause problems with base-sensitive substrates. 1-Alkynes can add directly if the reaction is carried out in combination with $Sn(OTf)_2$ and a base (1:1). In reactions with aromatic alkynes, DBU is the preferred base, but 1,8-bis(dimethylamino)naphthalne is preferred for reactions with aliphatic alkynes. Silylation of 1-alkynes with R_3SiCl can also be promoted by $Sn(OTf)_2$ and an amine.

$$Ar-C{\equiv}CH \;+\; RCHO \xrightarrow[\text{66 - 87%}]{\substack{Sn(OTf)_2,\ DBU,\\ CH_2Cl_2}} $$

2-Amino-2-deoxy-β-D-glucopyranosides.[2]   In the presence of $Sn(OTf)_2$, 1,3,4,6-tetra-O-acetyl-2-deoxy-(2,2,2-trichloroethoxycarbonylamino)-β-D-glucopyranose (1) reacts with alkyl trimethylsilyl ethers to form the corresponding β-D-glucopyranosides 2 in >90% yield. This procedure is also applicable to preparation of β-D-galactopyranosides.

[1] M. Yamaguchi, A. Hayashi, and T. Minami, *J. Org.*, **56**, 4091 (1991).
[2] T. Mukaiyama and K. Matsubara, *Chem. Letters*, 1755 (1992).

Tin(II) trifluoromethanesulfonate–Chiral diamine.

Asymmetric aldol reaction of silyl enol ethers. (**16**,221–222).[1]   The use of $TiCl_4$ as promoter of aldol condensation of silyl enol ethers with aldehydes, first reported in 1973 (**6**,590–591), has seen wide use, but has the drawbacks that 1 equiv. of the Lewis acid is required and that an asymmetric version requires use of chiral aldehydes or chiral silyl enol ethers. More recently, the combination of a salt and a weak Lewis acid, neither effective catalysts themselves, was found to be effective in catalytic (5–10 mol %) amounts. Further research showed that tin(II) triflate when coordinated with a chiral diamine can effect catalytic asymmetric allylation of aldehydes (**13**,302) and Michael reactions (**15**,313–314), even though this complex cannot promote aldol condensation. Eventually the combination of tin(II) triflate, a chiral diamine, and a tin(IV) salt such as tributyltin fluoride was found to effect catalytic aldol reactions of the silyl enol ether of S-ethyl ethanethioate in good yield and high enantioselectivity (equation I). A number of chiral diamines are effective, but all are substituted deriviates of (S)-1-alkyl-2-methylpyrrolidine. The most effective is (S)-1-methyl-2-[(N-1-naphthylamino)methylpyrrolidine (**1**), as shown in Figure A.

(A)

(I) $(CH_3)_3CCHO$ + ...

$$\text{Sn(OTf)}_2, \text{Bu}_3\text{SnF}$$
$$\textbf{1}, \text{CH}_2\text{Cl}_2, -78°$$
$$90\%$$

>98% ee

(II) RCHO + ...

$$\text{Sn(OTf)}_2, \text{Bu}_2\text{Sn(OAc)}_2$$
$$\textbf{1}, -78°, \text{CH}_2\text{Cl}_2$$
$$85\text{-}96\%$$

(syn/anti = 100:0,
>98% ee)

This chiral catalyst was then found to effect aldol reactions of aldehydes with the silyl enol ether of S-ethyl propanethioate (equation II). In this case dibutyltin diacetate is somewhat superior to tributyltin fluoride as the cocatalyst. With this chiral promotor, chemical yields are high, and only the syn-aldol is formed in >98% ee. This high stereoselectivity obtains with aliphatic and aromatic aldehydes and α, β-enals.

The actual catalyst is believed to be a complex (A) of the three components since it is soluble in CH_2Cl_2 even though none of the components is. The amine-coordinated tin(II) triflate acts as a Lewis acid to activate the aldehyde, and the tin(IV) fluoride or acetate interacts with the silyl enol ether.

Double asymmetric aldol reaction.[2] The asymmetric aldol reaction of aldehydes with silyl enol ethers catalyzed by tin(II) triflate and a chiral diamine such as (S)-**1**, 1-methyl-2-[(N-naphthylamino)methyl]pyrrolidine, (**16**,221–222), has been extended to aldol reactions of a silyl enol ether with chiral aldehydes.

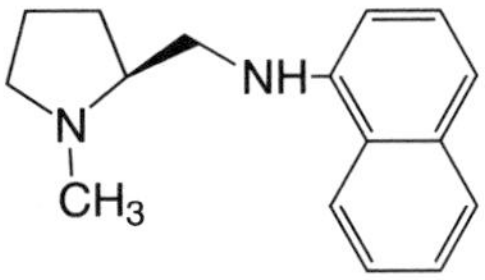

(S)-1

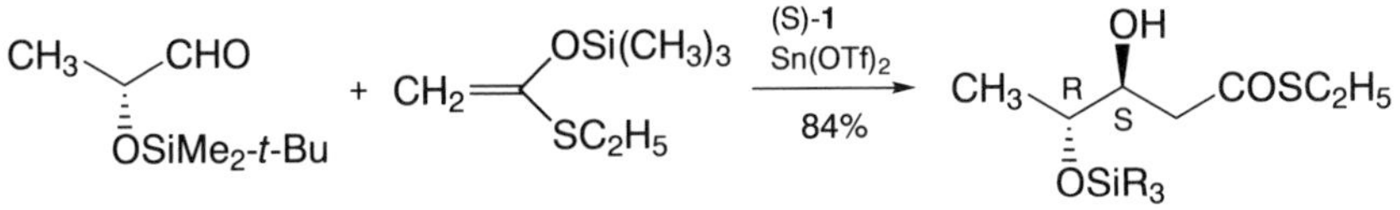

(R)-2   3   (S,R)-anti-4 (94:6)

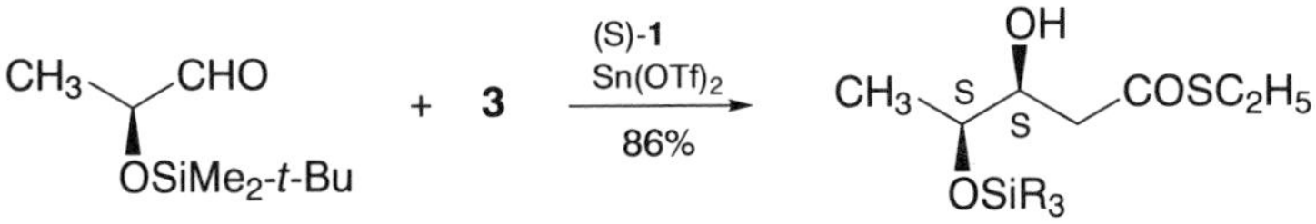

(S)-2   (S,S)-syn-4 (96:4)

The chiral aldehyde, (R)-2, reacts with 1-ethylthio-1-trimethylsilyloxyethene (3) to give *anti*-4 with excellent selectivity (94:6). The enantiomeric aldehyde, (S)-2, undergoes the same reaction to provide *syn*-4. In both reactions the newly formed chiral center has the (S)-configuration, an indication that the stereochemistry is controlled by the chiral amine, and not by the chirality of the aldehyde. Indeed, the same two reactions when catalyzed by (R)-1 and Sn(OTf)₂ provide aldols in which the new chiral center has the (R)-configuration. Thus by a suitable combination of a chiral aldehyde and a chiral

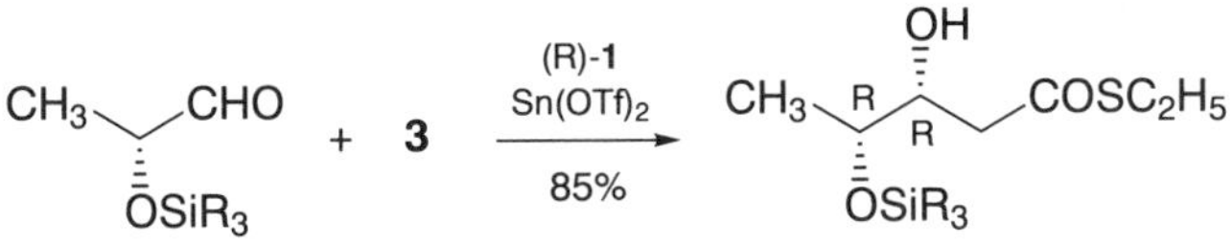

(R)-2   (R,R)-syn-4 (96:4)

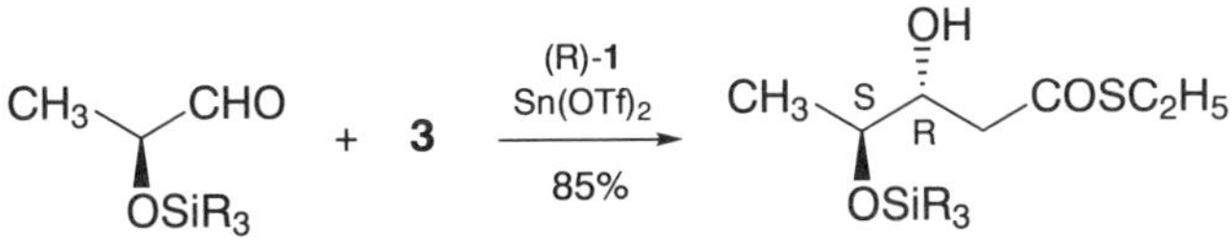

(S)-2   (R,S)-syn-4 (94:6)

catalyst, the four possible diastereomers can be obtained.

Propargylic alcohols.[3]   The highly enantioselective aldol reaction of Mukaiyama can also be used to obtain optically active propargylic alcohols by reaction of acetylenic aldehydes with a silyl enol ether. The most effective chiral catalyst is obtained by a combination of tin(II) triflate with a diamine derived from (S)-proline, (S)-1-methyl-2-[N-naphthylamino)methyl]pyrrolidine (**1**). Thus the reaction of an acetylenic aldehyde

with (Z)-1-ethylthio-1-trimethylsiloxypropene in the presence of $Sn(OTf)_2$ and **1** (20 mol %) provides the trimethylsilyl ethers of propargylic alcohols in 92–97% ee.

[1] S. Kobayashi, H. Uchiro, Y. Fujishita, I. Shiina, and T. Mukaiyama, *Am. Soc.*, **113**, 4247 (1991).
[2] S. Kobayashi, A. Ohtsubo, and T. Mukaiyama, *Chem. Letters*, 831 (1991).
[3] T. Mukaiyama, M. Furuya, A. Ohtsubo, and S. Kobayashi, *ibid.*, 989 (1991).

Titanium(IV) chloride.

Chlorotitanium enolates (**16**,332–334).[1]   The titanium enolate of the N-propionyl-oxazolidone **1**, prepared with $TiCl_4$ and C_2H_5N-i-Pr_2, undergoes highly diastereoselective Michael reactions with ethyl vinyl ketone, methyl acrylate, and acrylonitrile.

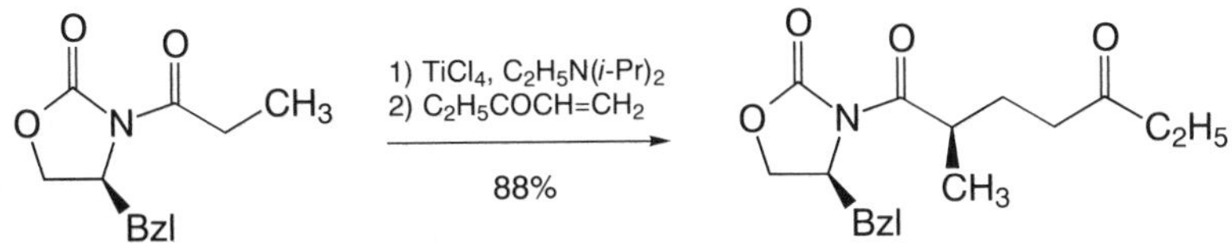

1   **2** (>100 : 1)

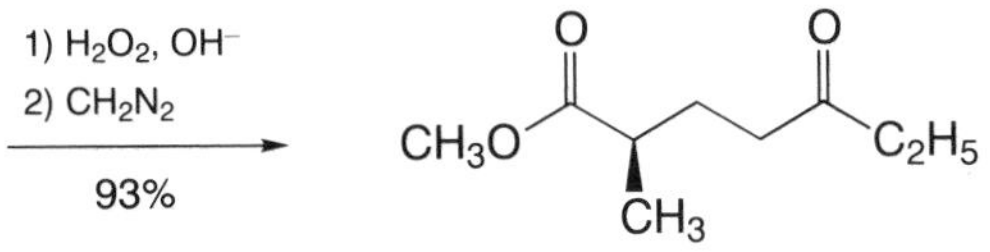

This technique can also effect highly asymmetric intramolecular Michael addition, but in this case i-PrOTiCl$_3$ is superior to TiCl$_4$ and $N(C_2H_5)_3$ is the preferred base (equation I).

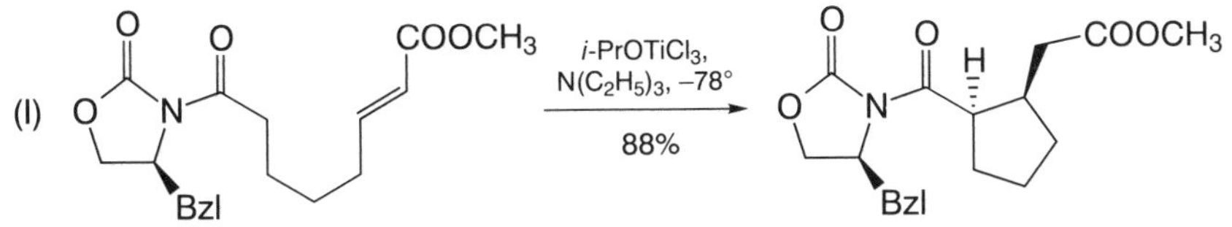

(I)

(93 : 7)

Reaction of RN$_3$ with ketones.[2]   TiCl$_4$ (2.5 equiv.) can effect a Schmidt type reaction of alkyl azides with cyclic ketones to afford N-alkyllactams. An aldol-type reaction can also occur but can be suppressed by use of excess (2 equiv.) of the alkyl azide. Highest yields are obtained with cyclohexanes, but ring expansion products can be obtained in 20–25% yield from cyclopentanone and cyclobutanones.

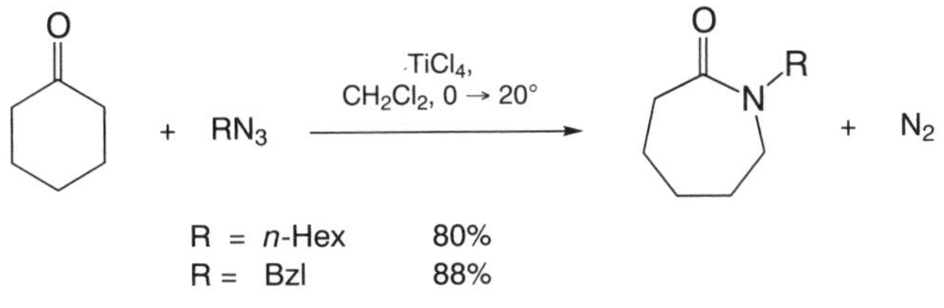

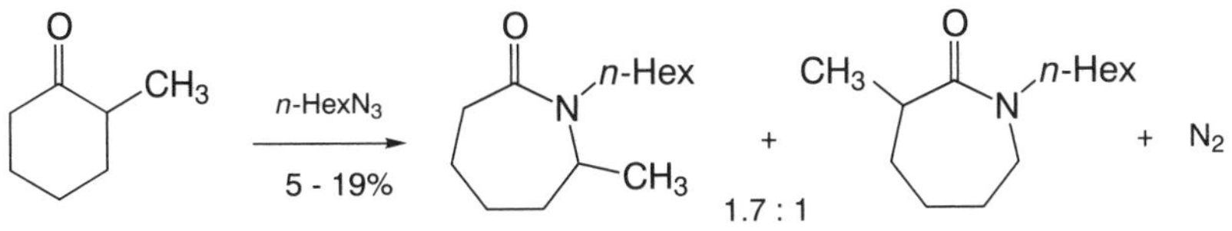

R = n-Hex   80%
R = Bzl   88%

1.7 : 1

In some cases the ring expansion involves the rearrangement of an azidohydrin intermediate (equation I).

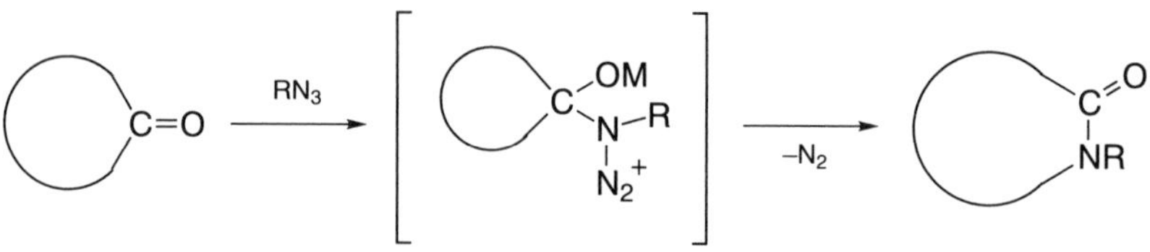

α-Hydroxy β, γ-enoates.[3]   In the presence of TiCl₄, the (E)-vinylsilane (**1a**) reacts with methyl glyoxylate to form an α-hydroxy-β, γ-enoate, (E)-**2a**, in high yield. In contrast, (Z)-**1a** reacts with methyl glyoxylate under the same conditions to give a 1:1 mixture of (Z)-**2a** and the ene product (**3a**). The ene product, **3a** is the exclusive product

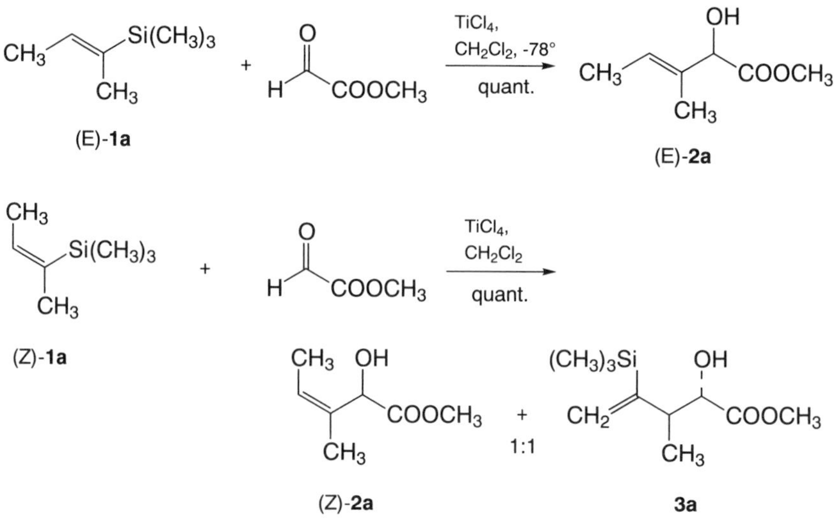

of reaction of (Z)-**1a** with methyl glyoxylate catalyzed by SnCl₄.

The reaction of (E)- or (Z)-vinylsilanes with 8-phenylmenthyl glyoxylate provides a single (2S)-alcohol (equation I).

(>99% ee)

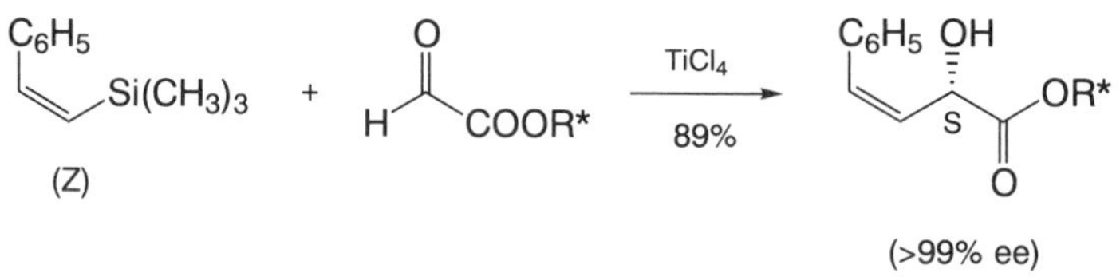

Bicyclic enediynes.[4]   The key step in a synthesis of these enediynes is a TiCl₄-catalyzed Sakurai reaction of a conjugated allylsilane (**1**) with the enone **2** to provide **3** as a single product. This product, after conversion to the ketal, was cyclized to **4** by

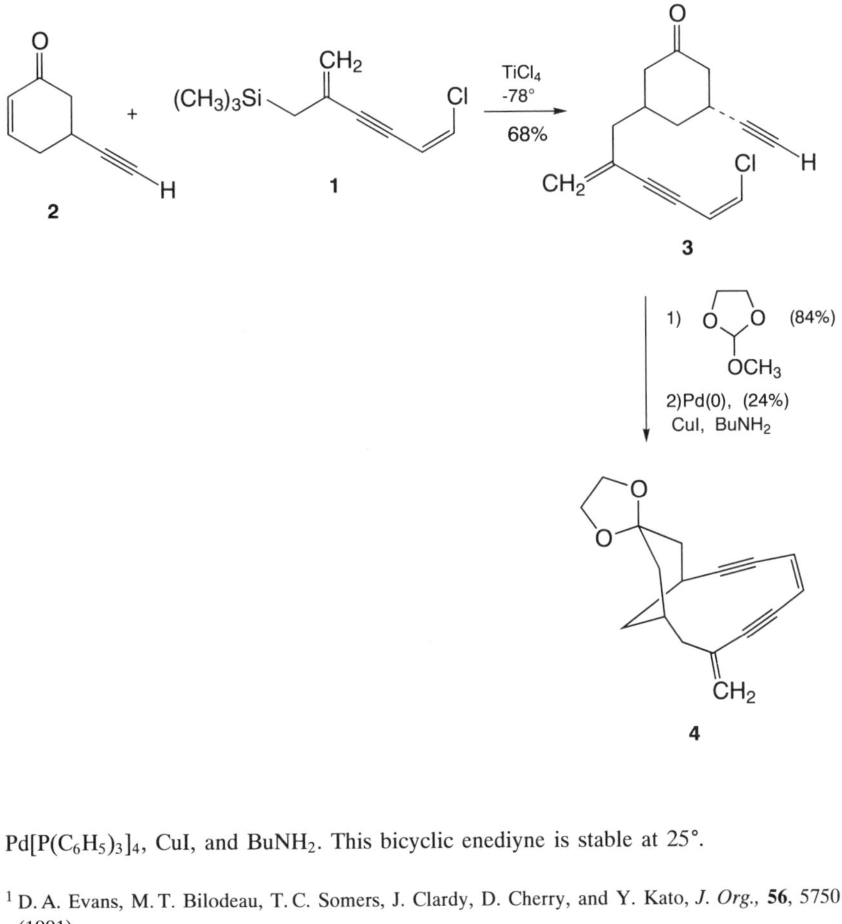

Pd[P(C₆H₅)₃]₄, CuI, and BuNH₂. This bicyclic enediyne is stable at 25°.

[1] D. A. Evans, M. T. Bilodeau, T. C. Somers, J. Clardy, D. Cherry, and Y. Kato, *J. Org.,* **56**, 5750 (1991).

[2] J. Aubé, G. L. Milligan, and C. J. Mossman, *J. Org.,* **57**, 1635 (1992).

[3] K. Mikami, H. Wakabayashi, and T. Nakai, *ibid.,* **56**, 4337 (1991).

[4] D. Schinzer and J. Kabbara, *Synlett,* 766 (1992).

Titanium(IV) isopropoxide.

3-Amino-1,2-diols.[1]   Primary amines react with chiral 2,3-epoxy alcohols in the presence of Ti(O-*i*-Pr)₄ to form 3-amino-1,2-diols with high regioselectivity.

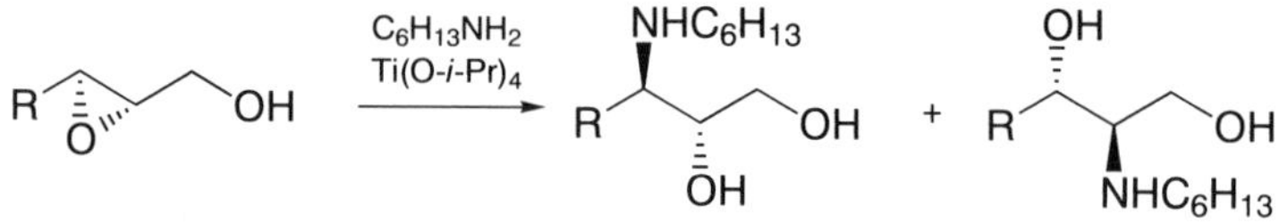

1, R = C_6H_5	83%	100 : 0
= CH_3	96%	93 : 7
= Pr	51%	92 : 8

[1] M. Canas, M. Poch, X. Verdaquner, A. Moyano, M. A. Pericas, and A. Riera, *Tetrahedron Letters,* **32**, 6931 (1991).

N-(*p*-Toluenesulfonyl)iminophenyliodinane, C_6H_5I=NTs, (1).

Preparation.[1]

Aziridination.[2]   Aziridination of alkenes can be effected with this nitrene precursor when catalyzed by various Cu(I) and Cu(II) catalysts, particularly $Cu(CH_3CN)_4ClO_4$ and $Cu(acac)_2$ in yields of 55–95%. Note that aziridination of a silyl enol ether provides an

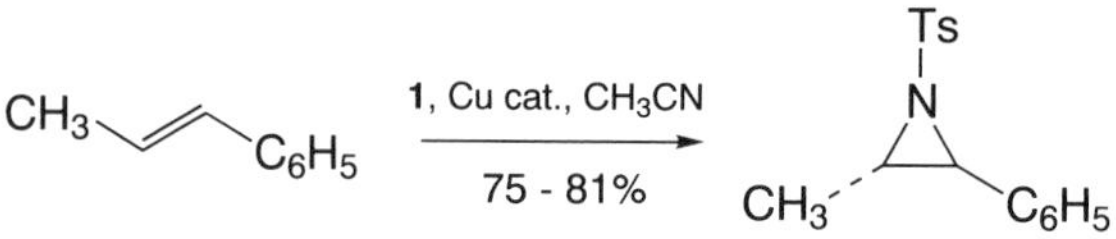

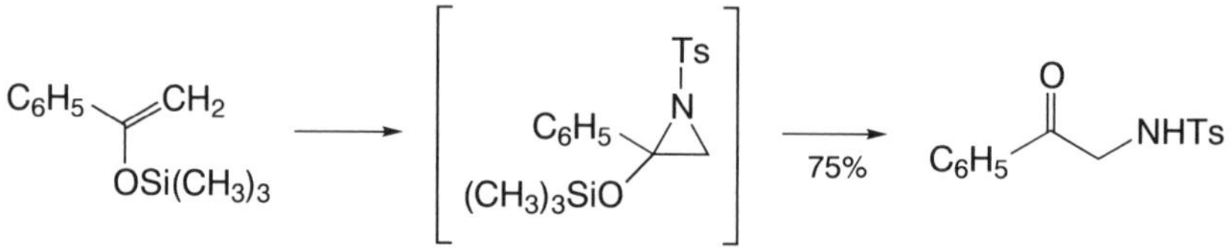

N-substituted α-amino ketone (second example).

[1] Y. Yamada, T. Yamamoto, and M. Okawara, *Chem. Letters,* 361 (1975).
[2] D. A. Evans, M. M. Faul, and M. T. Bilodeau, *J. Org.,* **56**, 6744 (1991).

1-Toluenesulfonyl-3-methylimidazolium triflate (1), prepared by reaction of methyl triflate with *p*-toluenesulfonylimidazole in THF:

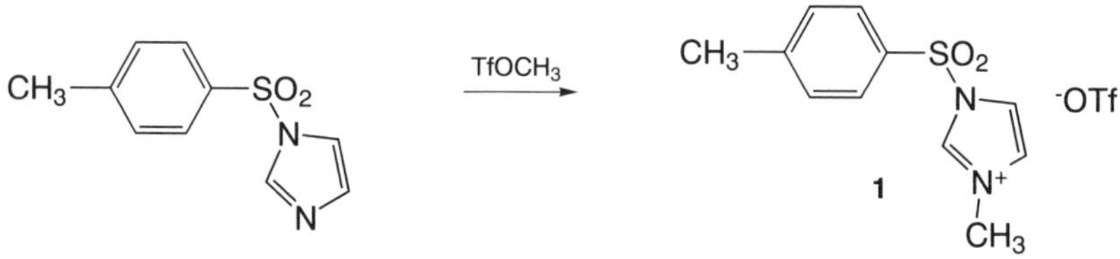

p-Toluenesulfonamides and -sulfonates. The reagent converts primary and secondary amines into the arylsulfonamides in 70–80% yields. A similar reaction with alcohols or phenols provides arylsulfonates. Addition of 1-methylimidazole as an acid scavenger improves the yield.

[1] J. F. O'Connell and H. Rapoport, *J. Org.,* **57,** 4775 (1992).

3-(*p*-Tolylthio)-2-pyrone (1).

Diels–Alder reactions.[1]   2-Pyrone can undergo thermal Diels–Alder reactions but generally requires such high temperatures (100-200°) that the adducts lose CO_2. High-pressure cycloadditions have also been effected, but the products are usually unstable. In contrast, this 3-substituted derivative **1** can undergo thermal Diels–Alder reactions with electrophilic dienophiles at 24–90° to give mainly *endo*-adducts that are stable to chromatography.

EWG		
NO_2	82%	98 : 2
CN	53%	2 : 1
CHO	44%	98 : 2
$COOCH_3$	65%	3 : 1

This Diels–Alder reaction can be used to obtain functionalized bicycloadducts and even an unsaturated spirolactone such as **4** (equation I), which corresponds to the formal product of reaction of α-pyrone itself with α-methylene-γ-butyrolactone without loss of CO_2.

The related 3-(tolylthio)-1-tosyl-2-pyridones (**5**) also undergo thermal Diels–Alder reactions without loss of an isocyanate group (equation II).

(II)

5

[1] G. H. Posner, T. D. Nelson, C. M. Kinter, and N. Johnson, *J. Org.*, **57**, 4083 (1992); G. H. Posner, V. Vinader, and K. Afarinkia, *ibid.*, **57**, 4088 (1992).

Tributyl tin carbamate (1).

$$Bu_3SnOCH_3 \; + \; C_2H_5N{=}C{=}O \longrightarrow Bu_3Sn{-}\overset{\underset{|}{C_2H_5}}{N}{-}COOCH_3$$

Darzens-type reaction.[1]   The Darzens reaction[2] of α-halocarbonyl compounds with an aldehyde requires a base, which can promote side reactions. Darzens reactions can proceed under neutral conditions when effected with this neutral reagent (1 equiv.). Presumably, the reaction proceeds via an organotin(IV) enolate. When the reaction is conducted in THF, the *trans*-epoxide is generally favored, but addition of HMPA can favor *cis*-selectivity. *cis*-Selectivity is also favored in reactions of aliphatic aldehydes. The Darzens-type reaction can also be extended to α-halo esters, particularly in the presence of Bu_4NF or LiBr.

cis/trans = 4 : 96

+ HMPA   40%   = 61 : 39

+ HMPA, 60°   15%   cis/trans = 60 : 40
+ Bu$_4$NF, -78°   81%   = 58 : 42
+ LiBr, 25°   78%   = 8 : 92

This reaction can be extended to a synthesis of disubstituted tetrahydrofurans from γ-halo ketones by use of the stannyl iminocarbamate $Bu_3SnN(C_6H_5)C(OCH_3)=NC_6H_5$ (2), prepared from Bu_3SnOCH_3 and $C_6H_5N=C=NC_6H_5$. In these reactions the more stable *trans*-products predominate.

CH_3—C(=O)—$(CH_2)_3Cl$ + C_6H_5CHO $\xrightarrow[35\%]{\text{2, THF, HMPA, 60°}}$

cis/trans = 38 : 62

C_6H_5—C(=O)—$(CH_2)_4I$ + C_6H_5CHO $\xrightarrow{47\%}$

cis/trans = 21 : 79

[1] I. Shibata, H. Yamasaki, A. Baba, and H. Matsuda, *J. Org.*, **57**, 6909 (1992).
[2] M. S. Newman and B. Magerlein, *Org. React.*, Vol. V, 413 (1949).

Tributyltin hydride.

3,4-Disubstituted tetrahydrofurans.[1] Trimethyltin radicals generated from $(CH_3)_3SnCl$, $NaBH_3CN$, and AIBN (**14**,313–314) add to allyl cinnamyl ethers such as **1** with carbocyclization to **2**. Oxidative cleavage of the $(CH_3)_3Sn$ group with ceric ammonium nitrate provides the aldehyde **3** without attack on the aryl group. This product isomerizes on treatment with DBU to the *trans*-isomer (>23:1). This *trans*-**3** is converted in several steps to the lignan burseran **4**.

$\xrightarrow[92\%]{(CH_3)_3SnCl, \, NaCNBH_3, \, AIBN}$

1

2

1) CAN (58%)
2) DBU (82%)

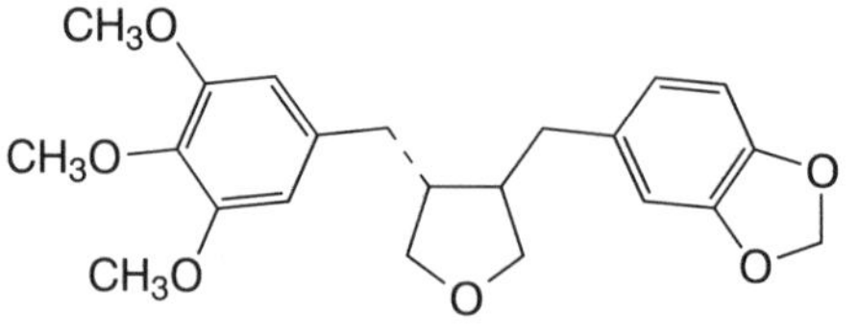

4

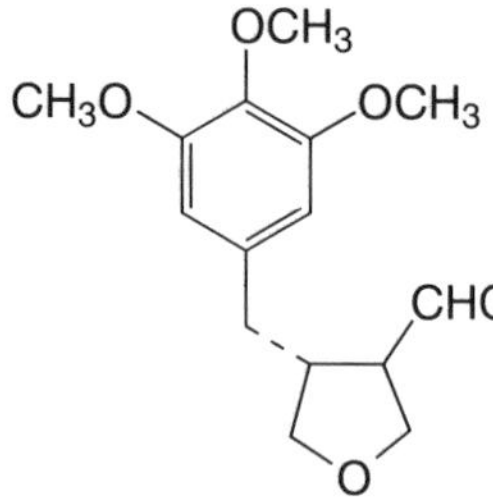

3

Tetrahydrofurans.[2]   A new route to tetrahydrofurans involves cyclization of an alkoxymethyl radical derived from a homoallylic alcohol. The precursor is obtained by conversion of the alcohol (**1**) to the (tributyltin)methyl ether (**9**,475), followed by Sn–Li exchange, and quenching with $(C_6H_5S)_2$ to form an α-phenylselenenylmethyl ether **2**. Treatment of **2** with Bu_3SnH (AIBN) generates an alkoxymethyl radical that cyclizes

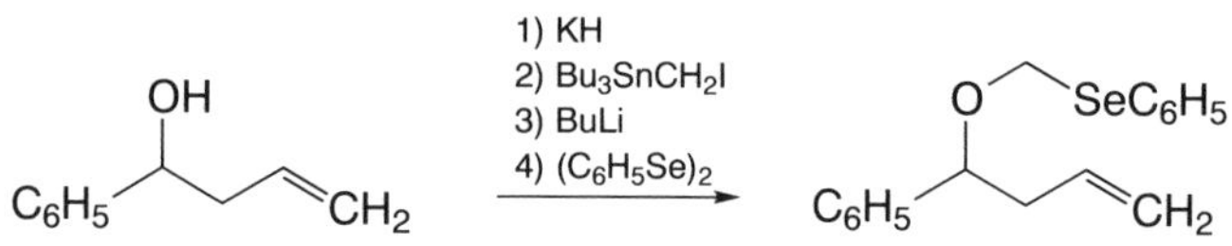

1

1) KH
2) Bu_3SnCH_2I
3) BuLi
4) $(C_6H_5Se)_2$

2

Bu_3SnH, AIBN
C_6H_6, 80°

96%

cis - **3**

+

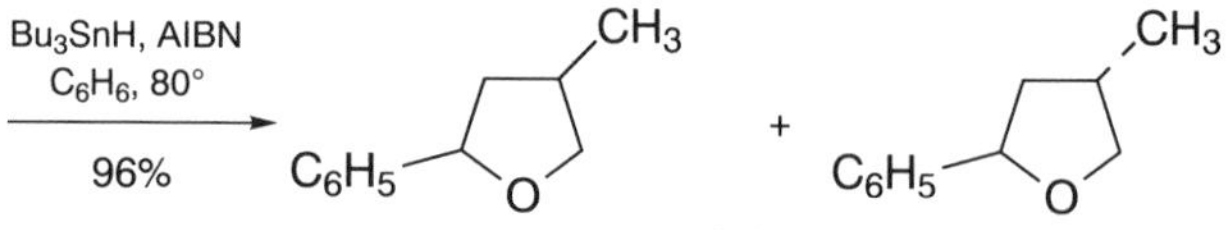

2.6 : 1

trans - **3**

to the tetrahydrofuran with *cis*-selectivity. Only traces of the product of reduction are observed.

This cyclization can be extended to bicyclic tetrahydrofurans and to tetrahydropyrans.

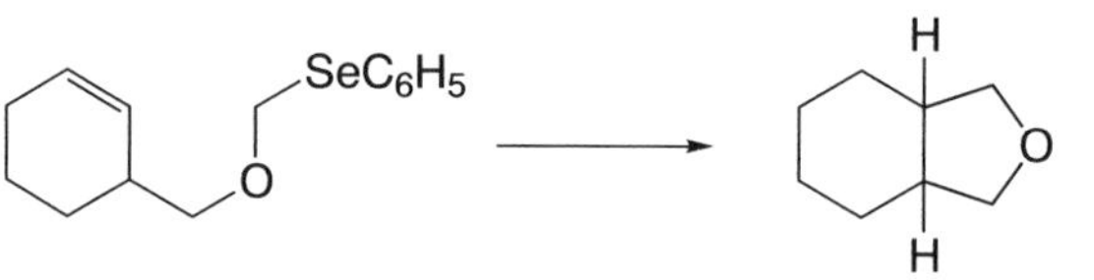

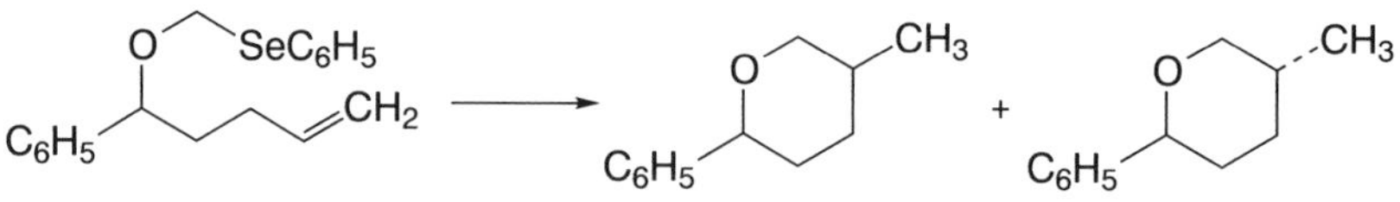

+

8.2 : 1

Radical ring expansion of fused cyclobutanones.[3]   This reaction involves reaction of an ω-bromoalkylchloroketene with an alkene to form a cyclobutanone with an *exo* side chain. Treatment of this adduct with Bu$_3$SnH (AIBN) generates a radical that cyclizes to a ring annelated product because of relief of strain in the four-membered ring. This reaction can be used to append 7-, 8-, or higher-membered rings to appropriate alkenes.

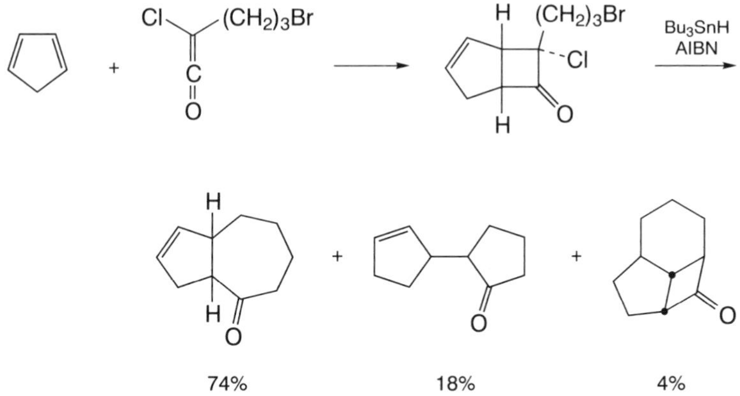

74%   18%   4%

Radical ring expansion of propenylcyclobutanes.[4]   The substrates (**2**) are obtained by alkylation of a spirocyclobutanone **1** with 1,3-dibromopropene. Treatment of **2** with Bu$_3$SnH/AIBN in refluxing benzene provides the tricyclic spiro ketone **3** together with the reduction product **4**.

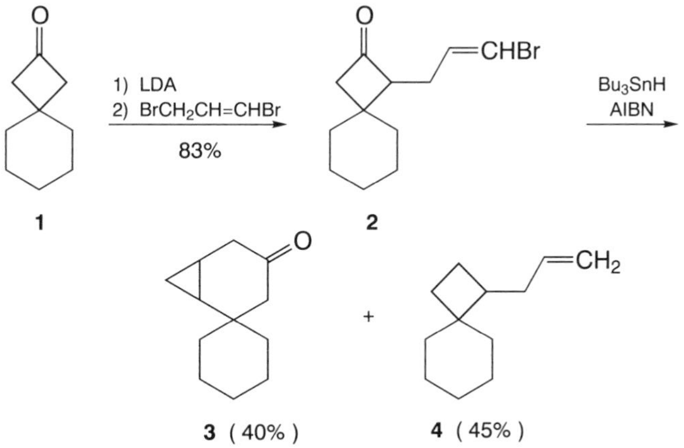

3 (40%)   **4** (45%)

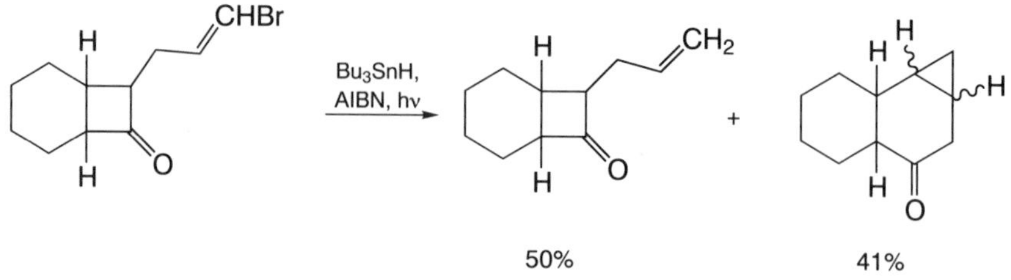

50%   41%

Cyclization and expansion of dichlorocyclobutanones.[5] Reaction of Bu₃SnH/ AIBN with the adduct (**1**) of the reaction of *endo*-6-vinylbicyclo[2.2.1]heptene-2 with dichloroketene results in free radical addition to the free double bond and reduction of the remaining chloride to provide **2**. Reaction of **2** with ISi(CH₃)₃/ZnI₂ results in transformation to a seven-membered cyclic α-iodo ketone, which is converted to enone **3** by DBU.

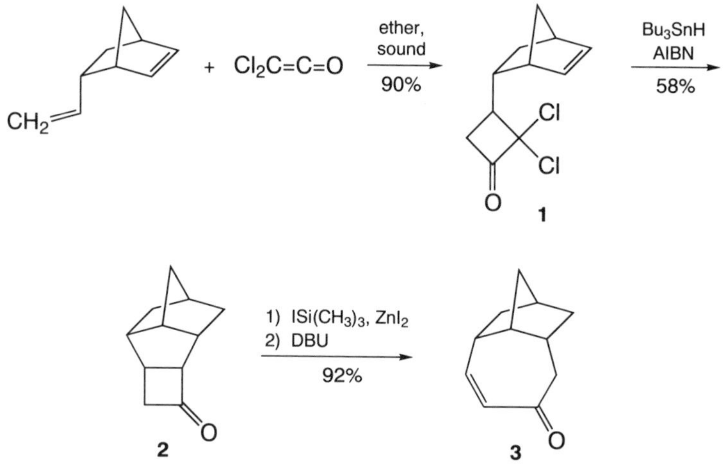

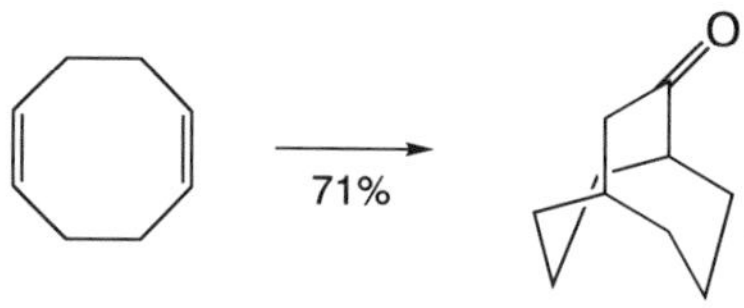

Spiroannelation.[6] A spirocyclobutan-2-one such as **1**, prepared as shown from methylenecyclopentene, when treated with Bu₃SnH and AIBN (2 equiv., slow addition) undergoes radical ring expansion to a spirocycloheptanone (**2**).

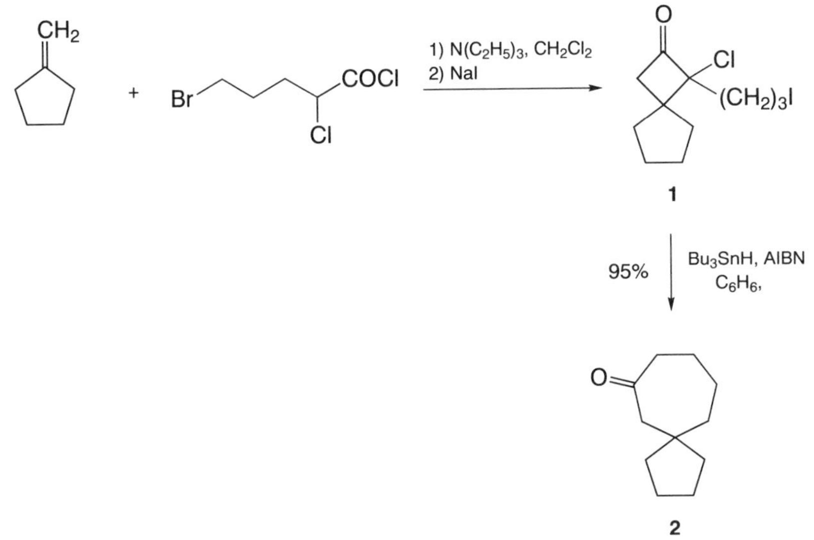

Bu₃SnH insertion into aliphatic chromium carbenes.[7]   Bu₃SnH and pyridine react with aliphatic chromium carbenes (**1**) to form α-alkoxytin compounds with marked diastereoselectivity (4-13:1).

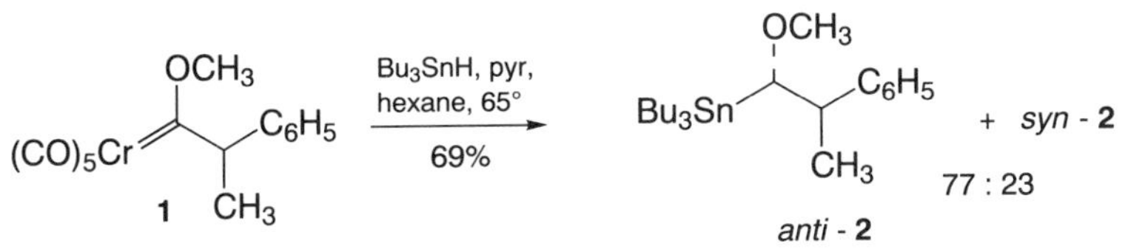

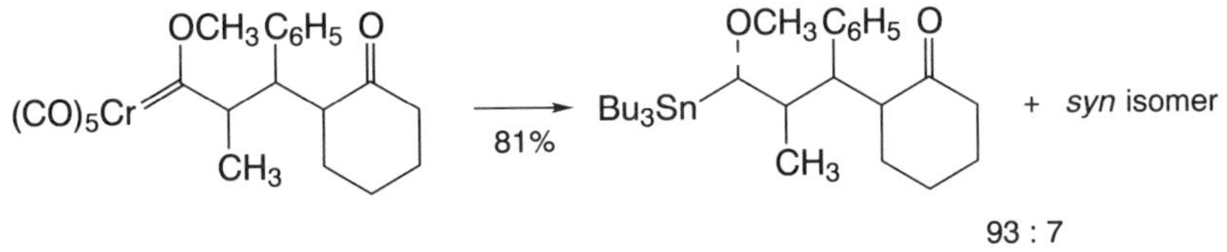

Radical ring expansion of cyclohexanones.[8]   This reaction is a useful route to medium-size ring systems. Thus the precursor **1**, prepared in two steps from cyclohexenone, on treatment with Bu₃SnH and AIBN furnishes the cyclodecenone **2** in 78% yield.

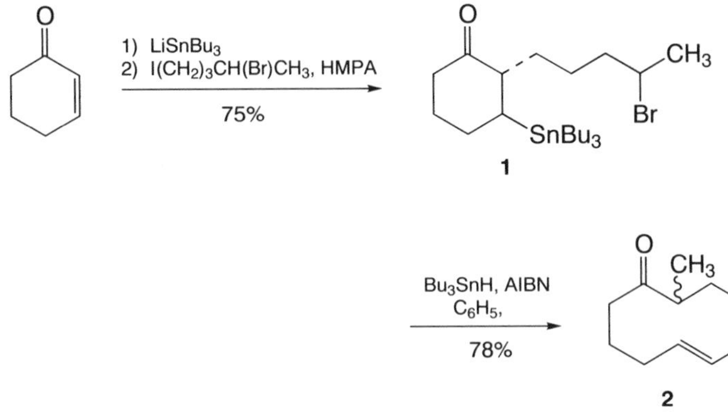

1

2

This ring expansion is applicable to cyclopentenones and cycloheptenones, but the yields are usually somewhat lower. It is not applicable to α, β-unsaturated lactones.

Stereoselective radical reactions. Giese *et al.*[9] have found that radicals can show high 1,2-stereoinduction comparable to ionic reactions, and that the selectivity can be reversed when proceeding from cyclic to acyclic radicals. Thus the Bu$_3$SnH radical addition to the dioxanone **1** provides the only one isomer (**2**) with *trans*-selectivity. The same radical with a corresponding acyclic system (**3**) shows *cis*-selectivity.

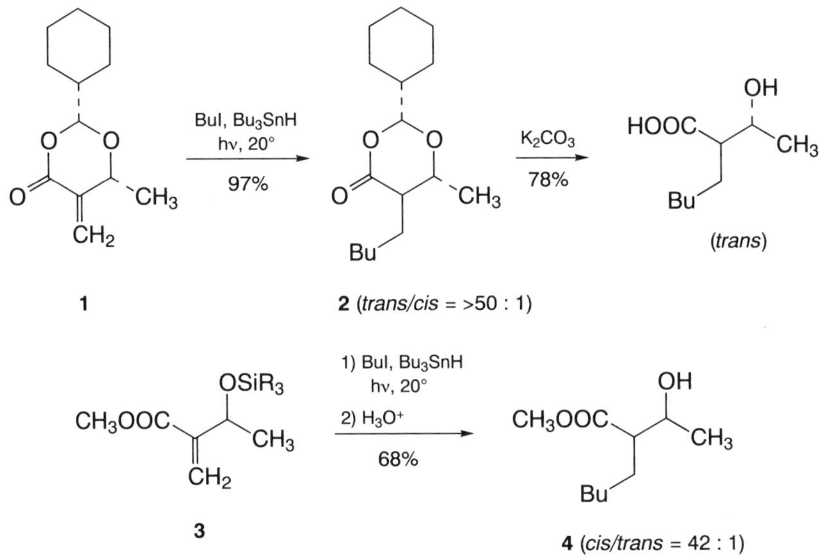

1

2 (*trans/cis* = >50 : 1)

(*trans*)

3

4 (*cis/trans* = 42 : 1)

***trans*-Decalins.**[10]   *trans*-Decalins can be obtained by a radical cyclization of methylenecyclohexanes substituted by an adjacent alkenyl bromide and a carboxylic ester group.

Radical cyclization of acetylenic esters.[11]   Cyclization of ω-iodo-α,β-alkynyl esters **1** with Bu_3SnH/AIBN in refluxing benzene results in the (E)-exocyclic alkene, (E)-**2**, as the major product. In contrast, cyclization of **1** with tris(trimethylsilyl)silane/AIBN provides a mixture of (E)- and (Z)-**2**, with the latter predominating. A different stereoselectivity obtains from cyclization of **1** with Bu_3SnH or $(TMS)_3SiH$ initiated by triethylborane/O_2. Actually the high (E)-selectivity in Bu_3SnH/AIBN is a result of isomerization of (Z)-**2** to (E)-**2**, promoted by Bu_3SnH at high temperatures. In any case, this isomerization is an unexpected example of regiocontrol in radical cyclizations.

Bu_3SnH/AIBN C_6H_6, 80°	82%	98:2
$(TMS)_3SiH$/AIBN	86%	34:66
Bu_3SnH/$(C_2H_5)_3B$	84%	60:40
$(TMS)_3SiH$/$(C_2H_5)_3P$	85%	11:89

Radical reaction of *o*-iodoanilides.[12]   The radical formed by reaction of an *o*-iodoanilide such as **1** is able to translocate to a radical α to the carbonyl group. Thus on treatment with Bu_3SnH (AIBN, 80°), **1** undergoes cyclization to a disubstituted

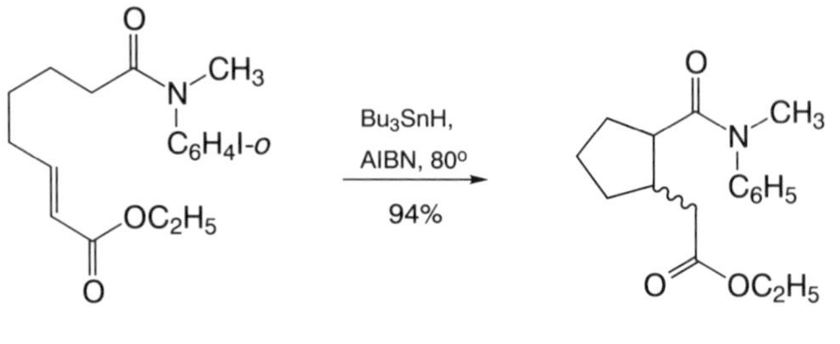

1

2 (2.1:1)

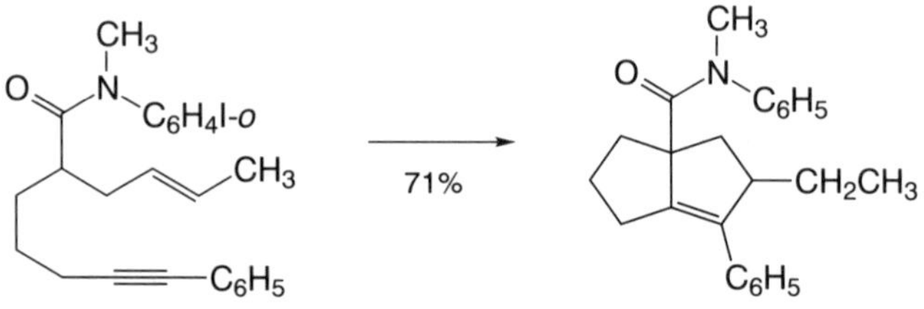

3   71%   **4** (8:1)

cyclopentane **2**. This radical transfer reaction provides a route to bicyclic and even tricyclic products. It can also be used to initiate radical additions to *o*-iodoanilides (equation I).

(I)

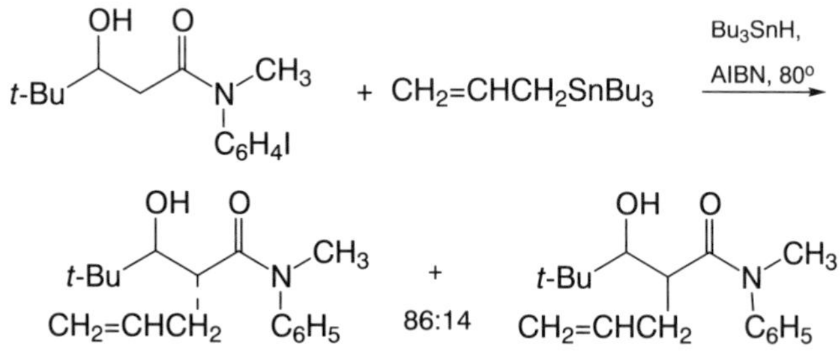

86:14

Fluoromethylenation.[13]   Methods for fluoromethylenation of ketones result in mixtures of (E)- and (Z)-vinyl fluorides, which are difficult to separate. A selective route involves reaction of the ketone with the carbanion of 1-fluoro-1-(phenylsulfonyl)methanephosphonate, prepared from fluoromethyl phenyl sulfones and diethyl chlorophosphate. The reaction provides mixtures of (E)- and (Z)-fluorovinyl sulfones, which can be separated by flash chromatography. Those products cannot be reduced by Al(Hg), but on treatment with tributyltin hydride (AIBN), they are converted into (fluorovinyl)stannanes with retention. Destannylation to vinyl fluorides can be effected with sodium methoxide in methanol or with CsF in methanolic ammonia.[14]

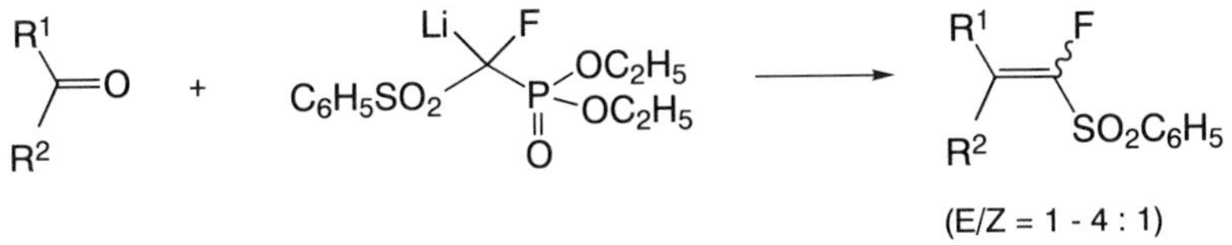

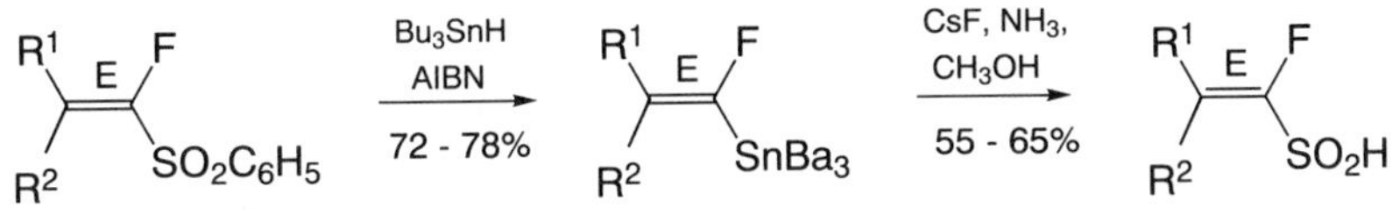

(E/Z = 1 - 4 : 1)

β-Hydroxy ketones from α, β-epoxy ketones.[15] The conversion can be effected by photolysis with Bu_3SnH or by a thermal reaction (98°) with $Bu_3SnH/AIBN$ in C_6H_6. The former reaction is particularly useful for aroyl-substituted epoxy ketones.

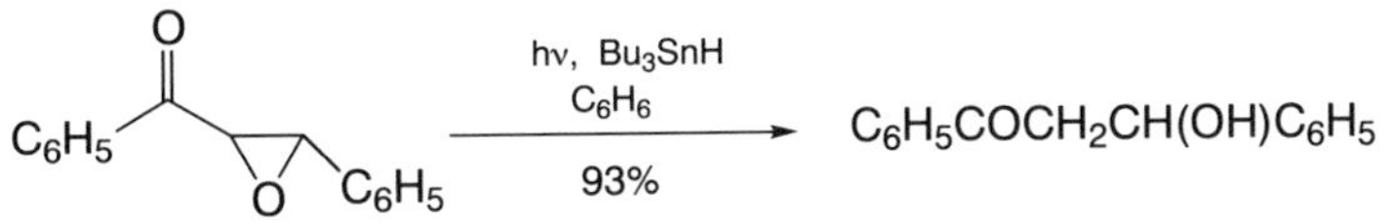

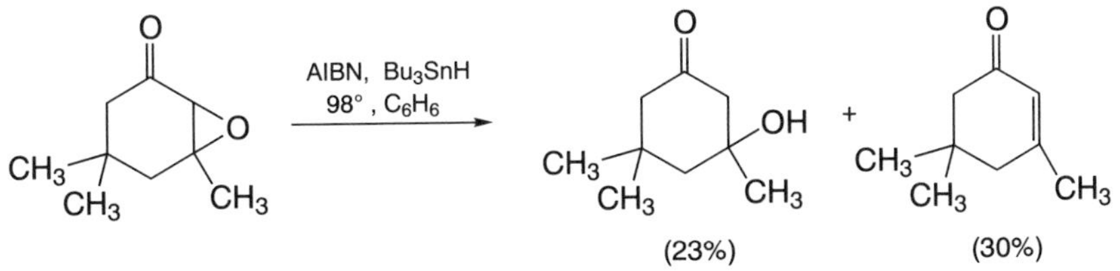

Radical macrocyclization to taxane ring system.[16] Treatment of the iodotrienone **1** with $Bu_3SnH/AIBN$ provides two epimers (**2** and **3**) of the taxane ring system. The ring system **2** corresponds to that present in taxol, the natural alkaloid found in the bark of yew trees.

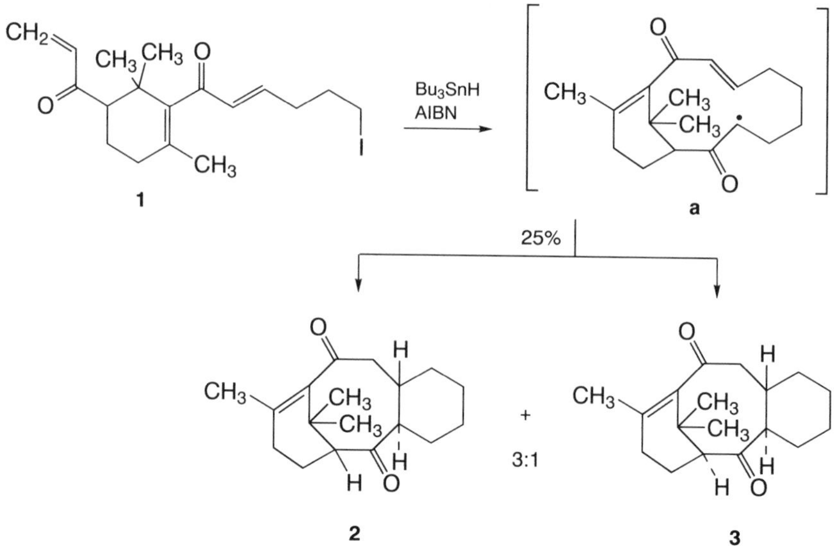

1

a

25%

2   +   3

3:1

Reduction of α-iodo-β-alkoxy esters.[17]   Radical reduction of these estens (1) with Bu₃SnH/AIBN shows high *anti*-selectivity. In contrast, reduction in the presence of MgBr₂ · O(C₂H₅)₂ or MgI₂ (0.25 equiv.) proceeds with high *syn*-selectivity because of chelation (equation I). AlCl₃ is also *syn*-selective, but only when 1 equiv. is present. How-ever, the configuration of the starting material affects the stereochemistry of the reduction.

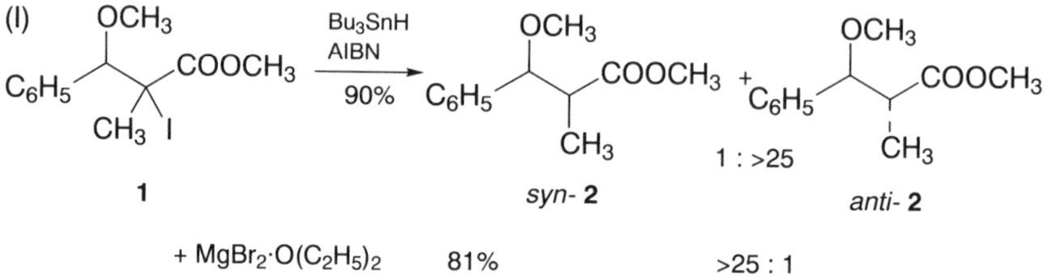

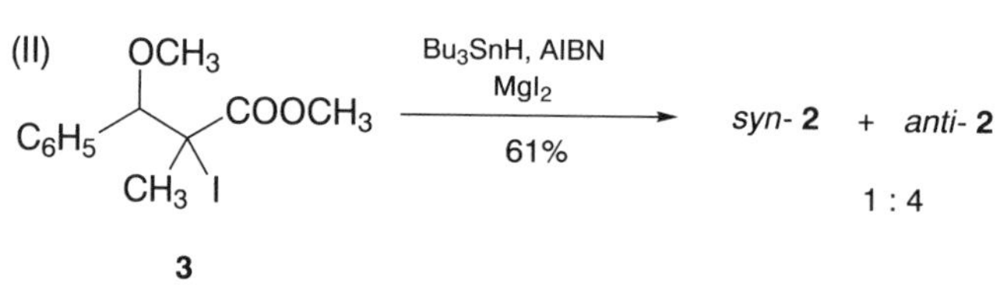

(I)

1   *syn*-2   *anti*-2

1 : >25

+ MgBr₂·O(C₂H₅)₂   81%   >25 : 1

(II)

3

syn-2   +   *anti*-2

1 : 4

Thus the ester **3** in which the iodide is *syn* to the alkoxy group is reduced by Bu₃SnH and MgI₂ with moderate *syn*-selectivity.

Diastereoselective reduction of α-alkoxy ketones.[18]   Bu₃SnH alone reduces these substrates in low yield. The reactivity is enhanced by addition of a base such as HMPA or

Bu_4NCl. Bu_4NF is particularly effective, and this system can show high *syn*-selectivity. In contrast, the combination of Bu_3SnH and Bu_2SnClH^{19} shows good *anti*-selectivity.

Radical deoxygenation of sec.-alcohols, Barton–McCombie reaction.[20]   This reaction proceeds more rapidly with xanthates [$R_2CHOC(S)SCH_3$] than with any of the known phenoxy thiocarbonyl derivatives [$R_2CHOC(S)OC_6H_5$]. Of these the unsubstituted phenoxythiocarbonyl derivative is slightly more reactive than the 2,4,6-trichloro- or the *p*-fluorophenyl derivative. The pentafluorophenyl derivative, $R_2CHOC(S)OC_6F_5$, is the slowest of all known derivatives.

[1] S. Hanessian and R. Léger, *Synlett,* 402 (1992).
[2] V. H. Rawal, S. P. Singh, C. Dufour, and C. Michoud, *J. Org.,* **56**, 5245 (1991).
[3] P. Dowd and W. Zhang, *Am. Soc.,* **113**, 9875 (1991).
[4] W. Zhang and P. Dowd, *Tetrahedron Letters,* **33**, 7307 (1992).
[5] P. Dowd and W. Zhang, *Am. Soc.,* **114**, 10084 (1992).
[6] W. Zhang and P. Dowd, *Tetrahedron Letters,* **33**, 3285 (1992).
[7] E. Nakamura, K. Tanaka, and S. Aoki, *Am. Soc.,* **114**, 9715 (1992).
[8] J. E. Baldwin, R. M. Adlington, and R. Singh, *Tetrahedron,* **48**, 3385 (1992).
[9] M. Bulliard, M. Zehnder, and B. Giese, *Helv.,* **74**, 1600 (1991).
[10] M. Kawaguchi, S. Satoh, M. Mori, and M. Shibaski, *Chem. Letters,* 395 (1992).
[11] T. B. Lowinger and L. Weiler, *J. Org.,* **57**, 6099 (1992).
[12] D. P. Curran, A. C. Abraham, and H. Liu, *ibid.,* **56**, 4335 (1991).
[13] J. R. McCarthy, D. P. Matthews, D. M. Stemerick, E. W. Huber, P. Bey, B. J. Lippert, R. D. Snyder, and P. S. Sunkara, *Am. Soc.,* **113**, 7439 (1991).
[14] J. R. McCarthy, D. P. Matthews, M. L. Edwards, D. M. Stemerick, E. T. Jarvi, *Tetrahedron Letters,* **31**, 5449 (1990).
[15] E. Hasegawa, K. Ishiyama, T. Kato, T. Horaguchi, and T. Shimizu, *J. Org.,* **57**, 5352 (1992).
[16] S. A. Hitchcock and G. Pattenden, *Tetrahedron Letters,* **33**, 4843 (1992).
[17] Y. Guindon, J.-F. Levallée, M. Llinas-Brunet, G. Horner, and J. Rancourt, *Am. Soc.,* **113**, 9701 (1991).
[18] I. Shibata, T. Yoshida, T. Kawakami, A. Baba, and H. Matsuda, *J. Org.,* **57**, 4049 (1992).
[19] W. P. Neumann and J. Pedain, *Tetrahedron Letters,* 2461 (1964).
[20] D. H. R. Barton, J. Dorchak, and J. C. Jaszberenyi, *Tetrahedron,* **48**, 7435 (1992).

Tributyltin hydride–Oxygen.

RX ⟶ ROH (reductive oxygenation).[1] Reaction of an organic halide with Bu_3SnH (2 equiv.) in the presence of dry air in toluene at 0–20° results in conversion to the corresponding alcohol in yields generally >70%. The reaction is particularly useful for conversion of allylic halides to allylic alcohols with retention. It is applicable even to tertiary or hindered halides. The transformation is believed to proceed through a hydroperoxide.

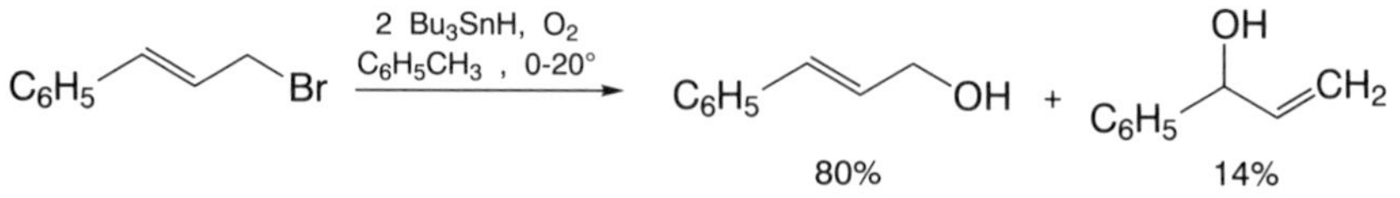
2 Bu3SnH, O2
C6H5CH3 , 0-20°
80%
14%

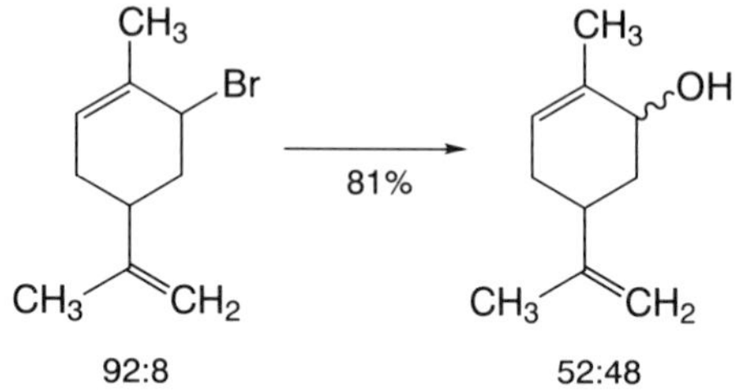
81%
92:8
52:48

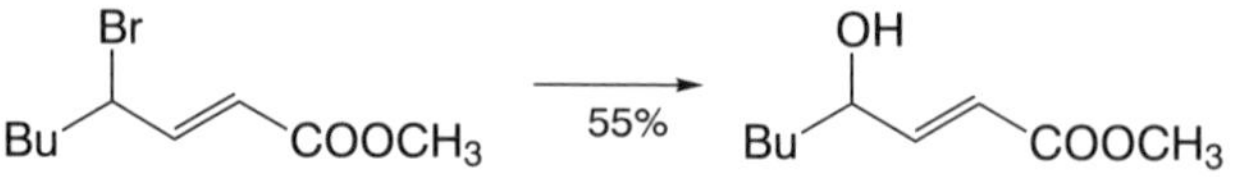
55%

This conversion of RX to ROH can be extended to radical cyclization as shown in equations I and II.

(I)

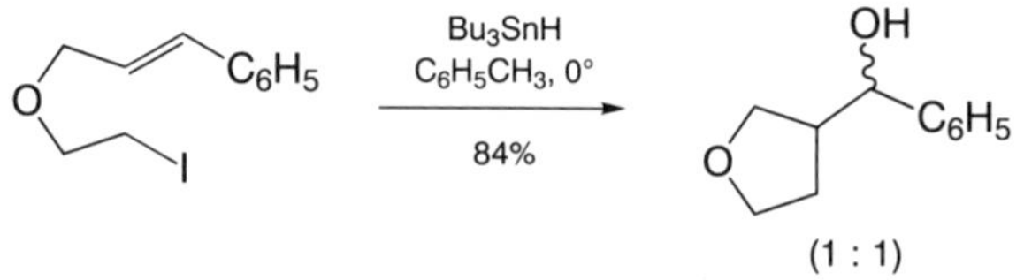
Bu3SnH
C6H5CH3, 0°
84%
(1 : 1)

(II)

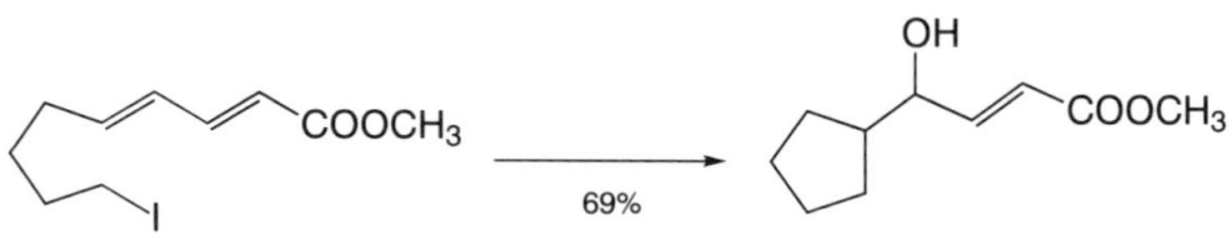
69%

[1] E. Nakamura, T. Inubushi, S. Aoki, and D. Machii, *Am. Soc.,* **113**, 8980 (1991).

Tributyltin hydride–Tetrabutylammonium halide.

Reduction of ketones.[1]   This combination (1:1) can reduce ketones in THF at 25–60° without reduction of halo substituents. *t*-Butylcyclohexanone can be reduced selectively to the more stable *trans*-alcohol (equation I). The reduction is *syn*-selective in the case of α-methoxypropiophenone (equation II). The intermediate tin alkoxide can react with an added electrophile (RX or ROCl) to form ethers or esters (equation III).

(I)   (CH₃)₃C ⟶ (Bu₃SnH - Bu₄NCl, 90%) ⟶ (CH₃)₃C–OH   (79:21)

(II)   C₆H₅–C(=O)–CH(OCH₃)–CH₃ ⟶ (81%) ⟶ C₆H₅–CH(OH)–CH(OCH₃)–CH₃   (100% *syn*)

(III)   cyclohexanone ⟶ [OSnBu₃] ⟶ (C₆H₅COCl, 93%) ⟶ cyclohexyl benzoate

Tributyltin hydride activated by HMPA can reduce aldehydes and some activated ketones such as α-chloro ketones.

[1] I. Shibata, T. Yoshida, A. Baba, and H. Matsuda, *Chem. Letters,* 307 (1991).

Tributyltin hydride–Triethylborane.

Radical cyclization to trans-hydrindanes.[1]   Although 1,5-hexadienyl radicals generally cyclize to a five-membered ring, the radical formed from the vinyl bromide **1** [Bu₃SnH-B(C₂H₅)₃] cyclizes to a *trans*-hydrindane **2** selectively, possibly via radicals **a** and **b**. The presence of an angular methyl group does not prevent a similar

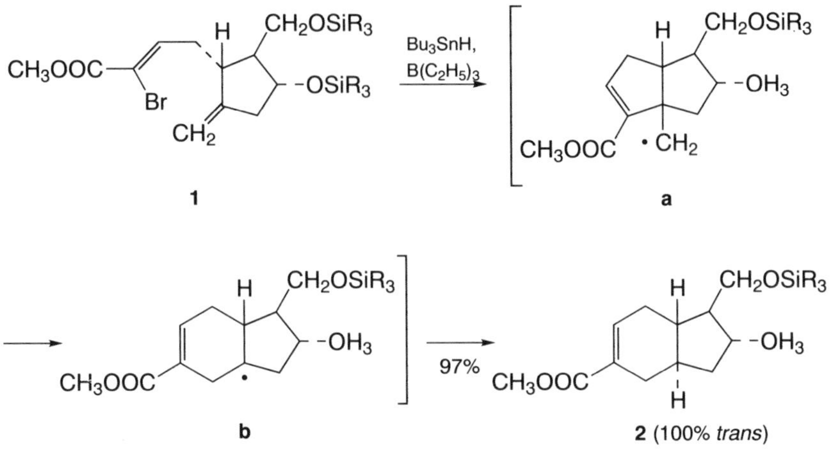

1

a

b

2 (100% *trans*)

6-*endo* cyclization of the bromo ester **3** to **4**.

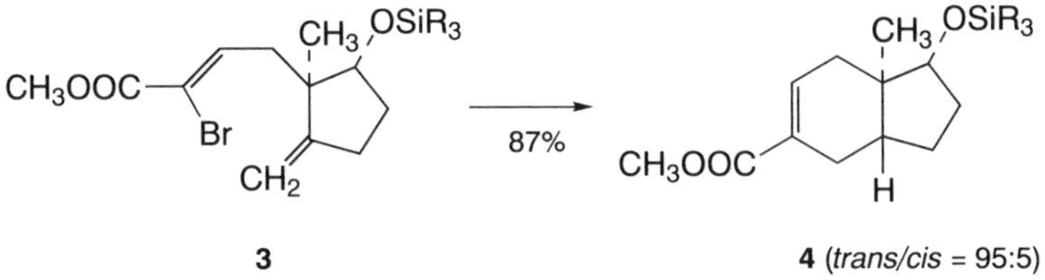

3

4 (*trans/cis* = 95:5)

[1] S. Sato, M. Sodeoka, H. Sasai, and M. Shibasaki, *J. Org.,* **56**, 2278 (1991).

Tributyl(trimethylsilyl) tin, $Bu_3SnSi(CH_3)_3$, **13**, 211–212.

Coupling of vinyl triflates with vinyl halides.[1]   The Pd(0) coupling of vinyl triflates with organotins (**12**, 468–470; **14**, 295–296), first reported by Stille, has wide use in organic synthesis. Stille coupling can be extended to coupling of vinyl triflates with vinyl or aryl halides by addition of $Bu_3SnSi(CH_3)_3$ (1.1 equiv.). Thus the vinyl triflate and the aryl bromide undergo intramolecular coupling to form **3** in the presence of $PdCl_2[P(C_6H_5)_3]_2$ (3 mol %), $Bu_3SnSi(CH_3)_3$ (1 equiv.), Bu_4NBr (3 equiv.), and Li_2CO_3 (1 equiv.).

The report assumes that the key step involves an intermolecular transmetalation of the vinyl triflate to form a vinyltin that undergoes intramolecular cyclization with the aryl bromide.

[1] M. Mori, N. Kaneta, and M. Shibasaki, *J. Org.*, **56**, 3486 (1991).

(Tricarbonyl)chloroiridium, $IrCl(CO)_3$, 1.

Carbonylative hydrosilylation.[1] $IrCl(CO)_3$ [or $Ir_4(CO)_{12}$] catalyzes this reaction with terminal alkenes to form the enol silyl ethers of acylsilanes. Acetal, epoxide, and cyano groups are not affected.

[1] N. Chatani, S. Ikeda, K. Ohe, and S. Murai, *Am. Soc.*, **114**, 9710 (1992).

Trichloromethylsilane–Diphenyl sulfoxide, CH_3SiCl_3-$C_6H_5S(O)C_6H_5$.

Cystine peptides.[1] Conversion of a dicysteine peptide to a cystine peptide is usually effected with air or iodine oxidation. Air oxidation is slow, and iodine oxidation can effect oxidation of some amino acids. The disulfide bond can be formed efficiently with this silyl chloride and a sulfoxide, a system that can also cleave various S-protecting groups of cysteine.

[1] K. Akaji, T. Tatsumi, M. Yoshida, T. Kimura, Y. Fujiwara, and Y. Kiso, *Am. Soc.*, **114**, 4137 (1992).

Triethoxysilane–Rhodium(III) chloride/Copper(II) nitrate.

Reduction of alkynes to (Z)-alkenes.[1]   This reduction can be effected by addition of $(C_2H_5O)_3SiH$ to a mixture of the alkyne, $RhCl_3 \cdot 3H_2O$, and Cu $(NO_3)_2$ in THF/H_2O. Hydrogen is generated by hydrolysis of $(C_2H_5O)_3SiH$, but exogenous H_2 is bubbled through the mixture. Yields of alkenes are 80–96% and the Z/E ratio is 8–20:1. Moreover under standard conditions the following groups are not affected: RNO_2, $ArNO_2$, CbzN, OH, ester, RBr, ArBr, enone. On the other hand ArI, $-C \equiv CH$, and propargylic alcohols are not stable to these conditions.

[1] J. M. Tour, S. L. Pendalwar, C. M. Kafka, and J. P. Cooper, *J. Org.,* **57**, 4786 (1992).

Triethoxysilane–Titanium(IV) isopropoxide.

RCOOR ⟶ *RCH₂OH.*   The combination of this silane and a catalytic amount of $Ti(O-i-Pr)_4$ generates a system that can reduce esters at 40–55°.[1] The product is isolated by addition of THF and NaOH (1N). The reduction requires 4–22 hours, and in some cases addition of $C_6H_5SiH_3$ is required for complete conversion. Yields typically are 70–95%.

[1] S. C. Berk and S. L. Buchwald, *J. Org.,* **57**, 3751 (1992).

Triethylborane.

Stereoselective syn-aldol reaction; syn-3-hydroxy-2-amino esters.[1]   Reaction of the lithium enolate of ethyl N, N-dimethylglycine (**1**) with aldehydes in the presence of $B(C_2H_5)_3$ (1 equiv.) results in *syn*-3-hydroxy-2-amino acid esters in >95% de (equation I). The high diastereoselectivity is explained by the exclusive or predominant

formation of the (Z)-enolate of **1** in the presence of $B(C_2H_5)_3$. Thus the lithium enolate of **1**, when treated with the borane and then a trialkylchlorosilane, is converted exclusively into the (Z)-silyl ketene acetal **3**.

[1] G. I. Georg and E. Akgün, *Tetrahedron Letters,* **32**, 5521 (1991).

Triethylborane $[(C_2H_5)_3B]$; **Triethylaluminum** $[(C_2H_5)_3Al]$.

Ate complexes with ester enolates.[1]   Methylation of the sodium enolate of **1**

provides the α-methyl derivative **2** in high yield, the epimer of the desired β-methyl derivative which is an intermediate to the important 1β-methylcarbapenems. Use of other metal enolates (lithium, zinc) showed no selectivity for β-methylation. However, Merck chemists have reported two diastereoselective routes to the desired β-methyl **2**. The most efficient is epimerization of α-CH_3 **2** to β-CH_3 **2** mediated with triethylborane. Thus reaction of the α-methyl derivative with 1 equiv. of LDA at $-80°$ and then with $(C_2H_5)_3B$ at -75 to $70°$ provides β-methyl **2** in a purity of $>99\%$. A second route to β-CH_3 (**2**) involves methylation of **1** in the presence of $(C_2H_5)_3Al$, which results in formation of both epimers in the ratio 3.5:1 with the β-epimer favored.

[1] D. R. Bender, A. M. DeMarco, D. G. Melillo, S. M. Riseman, and I. Shinkai, *J. Org.*, **57**, 2411 (1992).

Triethylsilane–Benzoyl peroxide.

Deoxygenation of alcohols.[1]   Thionocarbonates or xanthates of primary or secondary alcohols are deoxygenated in 89–100% yield by radical reduction with $(C_2H_5)_3SiH$ and $(C_6H_5CO)_2O_2$.

[1] D. H. R. Barton, D. O. Jang, and J. C. Jaszberenyi, *Tetrahedron Letters*, **32**, 7187 (1991).

Triethylsilane–Titanium (IV) chloride, $(C_2H_5)_3SiH$-$TiCl_4$ (**1**).

Stereoselective reduction of bicyclic ketals.[1]   A new route to 6- and 8-membered cyclic ethers involves stereoselective reduction of the bicyclic ketals such as **2**. Depending on the reducing agent, the bicyclic ketal **2** can be reduced to either the *cis*- or *trans*-cyclic ether **3**. Thus reduction of **2** and **4** with triethylsilane and $TiCl_4$ at $-78°$ yields the *cis*-disubstituted cyclic ethers **3** and **5** with high stereoselectivity. Reduction of **2** and **4**

2		**cis - 3**	
+ **1**	100%	99.93 : 0.07	
+ DIBAH	100%	4 : 96	

4		**cis - 5**	
+ **1**	86%	91 : 9	
+ DIBAH	88%	2 : 98	

to the *trans*-disubstituted ethers is effected with almost as high selectivity with DIBAH.

[1] H. Kotsuki, *Synlett*, 97 (1992).

Triethylsilyl hydrotrioxide, $(C_2H_5)_3SIOOOH$ **(1).**

Preparation by ozonization of triethylsilane, at $-78°$ (**13**, 228).

Oxidative cleavage of alkenes.[1]   Although this reagent (**1**) can cleave terminal and internal alkenes, the yields can be low. However, much higher yields can be obtained if the substrate bears an alkoxy or an ester group even remote from the double bond (equations I and II).

(I) $CH_3(CH_2)_7CH=CH(CH_2)_7R \xrightarrow{\text{1) 1} \atop \text{2) LiAlH}_4} CH_3(CH_2)_7CH_2OH$

R = CH$_3$ 42%
= COOCH$_3$ 83%

(II) $RCH_3CH=CH_2 \longrightarrow RCH_2CH_2OH$

R = C$_6$H$_5$ 54%
= C$_6$H$_5$O   76%

This oxidation can be used to prepare the dioxetane **2** that can rearrange to the 1,2,4-trioxane **3**, a potent antimalarial agent.

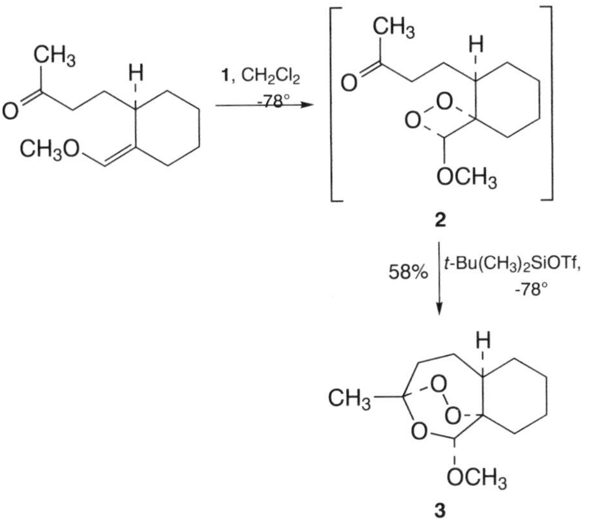

This sequence was used to prepare other related 1,2,4-trioxanes, of which **5** is particularly interesting as an antimalarial against certain strains of *P. falciparum* clones.

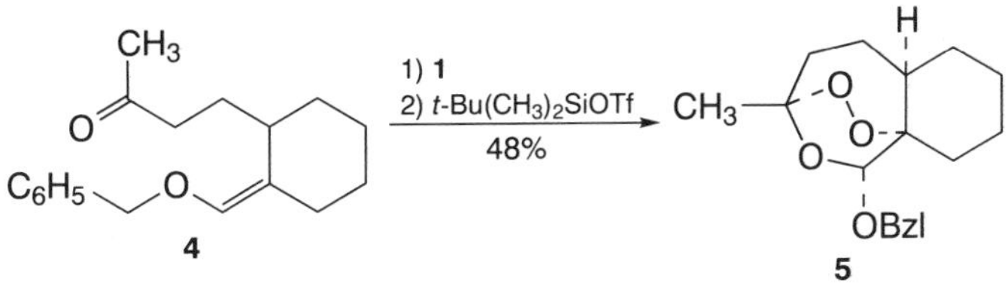

[1] G. H. Posner, C. H. Oh, and W. K. Milhous, *Tetrahedron Letters*, **32**, 4235 (1991).

Trifluoroacetic acid–Trialkylsilane, 5, 695;6, 616.

Intramolecular ionic hydrogenation.[1]   One step in a synthesis of a diterpene, secopseudopterosin A (**4**), from 5-methoxytetralone requires hydrogenation of the trisubstituted double bond of **1**. Hydrogenation catalyzed by Pd/C results in attack from the less hindered face to provide the undesired **3**. Ionic hydrogenation with $(C_2H_5)_3SiH$ and CF_3COOH gives a 2:3 mixture of **3** and the desired **2**. Even higher stereoselectivity can be effected by an intramolecular version of ionic hydrogenation. Thus reaction of **1** with *t*-Bu₂SiHCl and imidazole (DMF, 23°) provides the silyl ether of **1**, which on slow addition to CF₃COOH followed by desilylation (Bu₄NF) gives **2** in 65–75% yield in >95% purity.

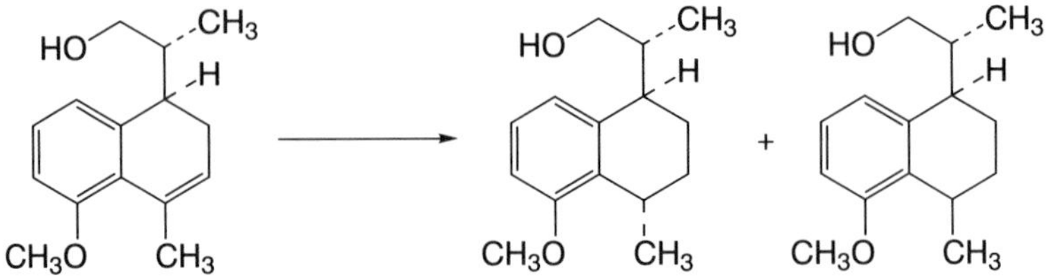

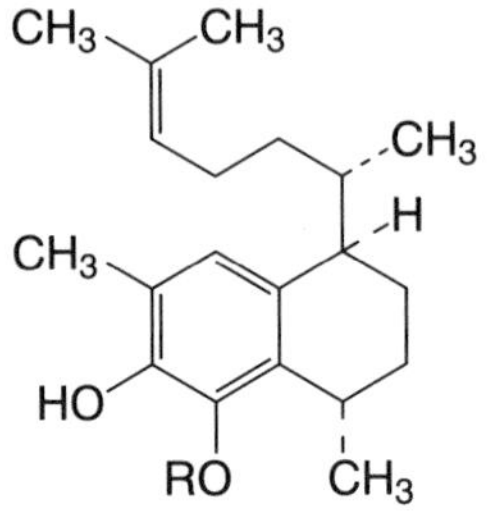

[1] S. W. McCombie, B. Cox, S. Lin, A. K. Ganguly, and A. T. McPhail, *Tetrahedron Letters*, **32**, 2083 (1991).

Trifluoromethanesulfonic acid/Boron(III) trifluoromethanesulfonate, TfOH/B(OTf)₃.

This combination provides the Brønsted superacid TfOH₂⁺B(OTf)₄⁻, **1**.

Stereoselective allylation of alkoxysiladioxanes.[1] Alkoxysiladioxanes **2** are obtained by hydrosilylation of β-hydroxy ketones (**16**, **82**), usually as a 1:1 mixture of *trans/cis* isomers. The allylation is also catalyzed by (CH₃)₃SiOTf, (CH₃)₃SiI, and

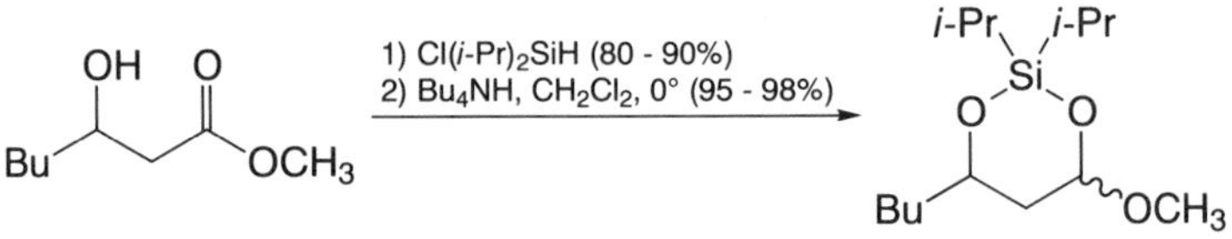

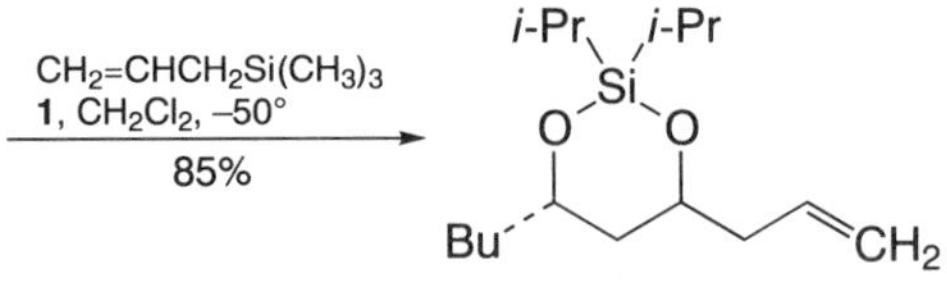

$SnCl_4$, but the highest stereoselectivity obtains with **1**. The products (**3**) can be desilylated to the corresponding 1,3-diols by treatment with HF in CH_3CN.

[1] A. P. Davis and S. C. Hegarty, *Am. Soc.*, **114**, 2745 (1992).

(+)-(S,S,S)-Triisopropanolamine, $\left[\overset{\overset{\displaystyle OH}{|}}{C H_3 C} HCH_2\right]_3 N.$

This amine (NL_3) can be prepared by reaction of (+)-(S)-1-amino-2-propanol with (–)-(S)-propylene oxide.

Enantioselective addition of R_3SiN_3 to meso-epoxides.   A chiral zirconium catalyst **1**, formulated as $(LZrOH)_2$-*t*-BuOH and prepared by reaction of L_3N with $Zr(O$-*t*-Bu$)_4$ followed by addition of H_2O, can effect enantioselective cleavage of epoxides by R_3SiN_3.

[1] W. A. Nugent, *Am. Soc.*, **114**, 2768 (1992).

2, 4, 6-Triisopropylbenzenesulfonyl azide (trisyl azide, 1), 14, 327.

Chiral α-azido arylglycines.[1]   Several glycopeptide antibiotics contain arylglycines in the peptide unit. The original methodology for asymmetric azidation (**14**, 327) to give arylglycines via an intermediate N-sulfonyltriazene, can fail to decompose to an azide group on addition of acetic acid. In such cases, addition of NaI or KOAc at 25° can promote decomposition to an azide.

[1] D. A. Evans, D. A. Evrard, S. D. Rychnovsky, T. Früh, W. G. Wittingham, and K. M. DeVries, *Tetrahedron Letters*, **33**, 1189 (1992).

Triisopropylsilyl enol ethers of cyclohexanones.

Reaction with azidotrimethylsilane.[1]   The triisopropylsilyl enol ether **1** reacts with $(CH_3)_3SiN_3$ and C_6H_5IO (1 equiv.) to form the β-azido adduct **2** as the major product. The azido group of **2** is displaced on reaction with a number of carbon nucleophiles in the presence of a Lewis acid. The overall reaction represents a conjugate addition to cyclohexnone.

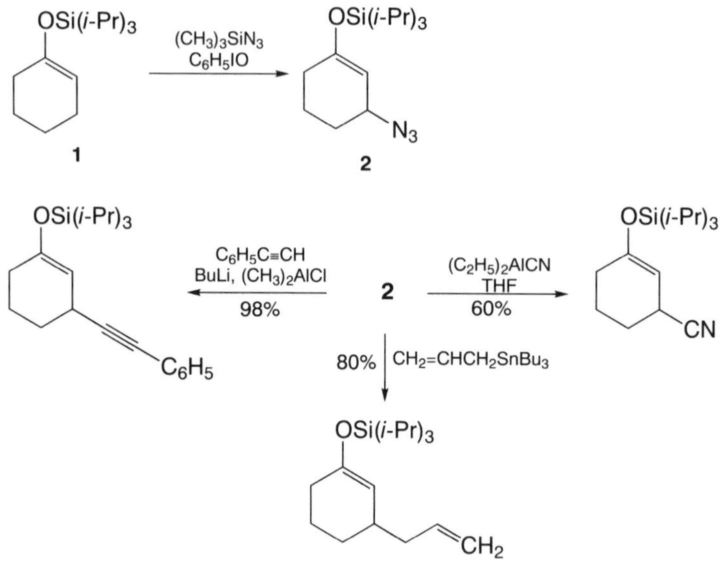

[1] P. Magnus and J. Lacour, *Am. Soc.*, **114**, 3993 (1992).

Trimethylaluminum.

Si-transfer from oxygen to carbon.[1] In the presence of a trialkylaluminum, particularly $(CH_3)_3Al$, silyl ketene acetals rearrange at $-78°$ to $20°$ to esters of trialkylsilylacetic acid. This 1,3-rearrangement of Si from oxygen to carbon is the reverse of the well-known Brook thermal rearrangement[2] of Si from carbon to oxygen. But R_3Al does not rearrange trimethylsilyl enol ethers.

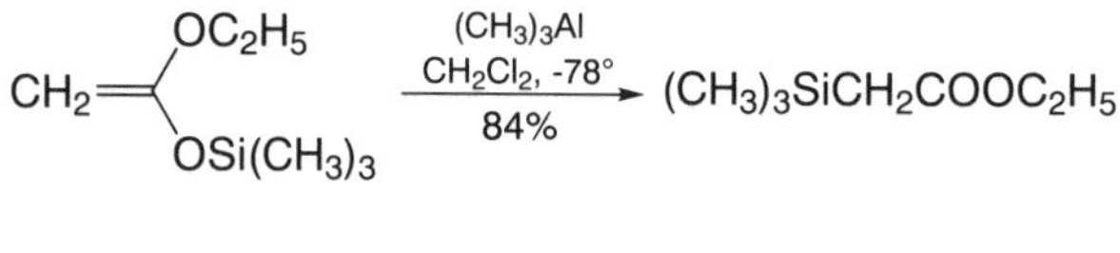

3-Aza-Cope rearrangement of N-alkyl-N-allylenamines.[3] This reaction has seen limited use because it requires elevated temperatures (200–250°). A variety of reagents can catalyze the rearrangement of an enamine of this type, **1**, to the unsaturated

1   **2**

3

imine **2**, which is reduced by LiAlH$_4$ to **3**. Protic acids such as HCl and various Lewis acids (TiCl$_4$, ZnCl$_2$) can effect this rearrangement at 111° even when only 0.03 equiv. is added, but the isolated yields of **3** are in the range 60–85%. In contrast, Al(CH$_3$)$_3$ can effect conversion of **1** to **3** in 99% yield, but it must be used in stoichiometric amounts. Moreover, this reagent is superior to other reagents for a wide variety of substrates including very sensitive enamines.

Methylation of γ,δ-epoxy acrylates.[4]   The reaction of Al(CH$_3$)$_3$ with a simple γ,δ-epoxy acrylate such as **1** results in a mixture of three products (equation I).

However, the same reaction with epoxy acrylates bearing an alkoxy substituent in the ϵ or ξ position results in a single product, especially in reactions conducted with Al(CH$_3$)$_3$ (8–10 equiv.) in CH$_2$Cl$_2$ and water (6–10 equiv.) at -30–$-40°$. Thus the (E)-epoxy acrylate (**2**) is converted exclusively into *anti*-**3** in 96% yield, whereas (Z)-**2** is converted into *syn*-**3** in 93% yield.

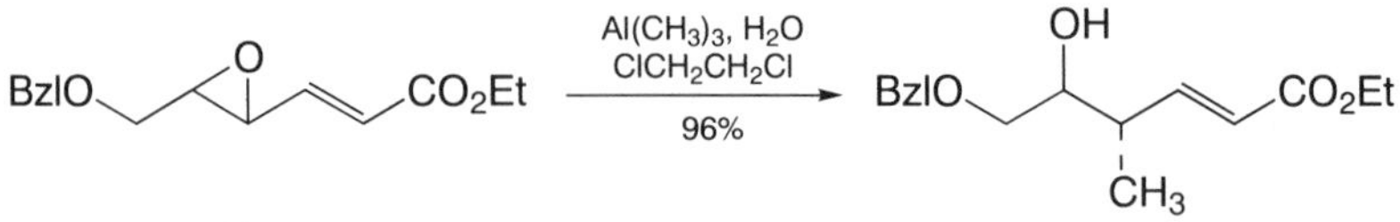

(E) - 2 → anti - 3 (Al(CH₃)₃, H₂O, ClCH₂CH₂Cl, 96%)

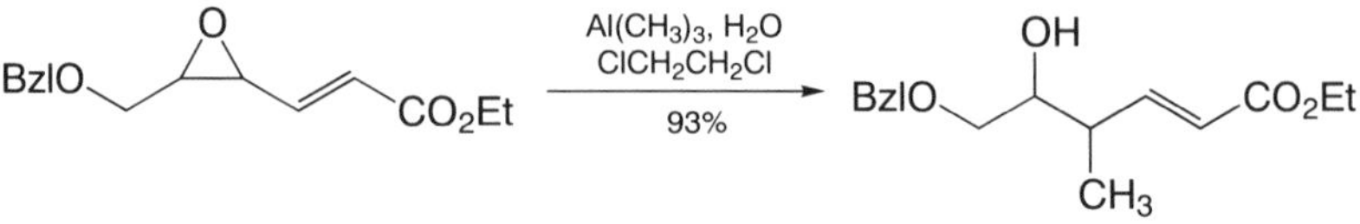

(Z) - 2 → syn - 3 (Al(CH₃)₃, H₂O, ClCH₂CH₂Cl, 93%)

This reaction on extension to **4** provides **5** with five contiguous chiral centers in 95% yield.

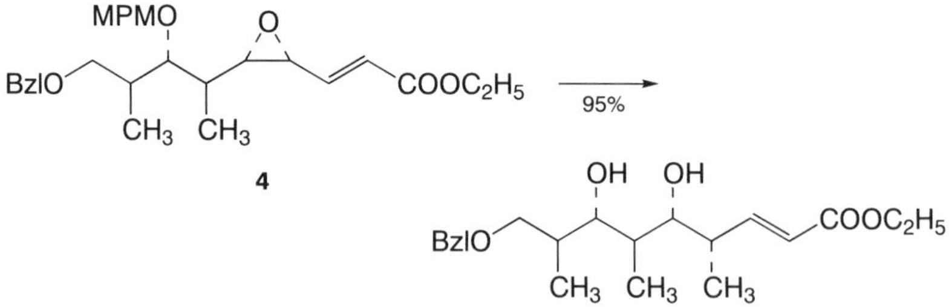

Isopropylidine ketals.[5] The isopropylidine ketals of 1,2-diols on treatment with $Al(CH_3)_3$ (3 equiv.) in CH_2Cl_2 at $-78°$ are cleaved to hydroxy *t*-butyl ethers with high

(I) $R = C_6H_5$, Bu, C_2H_5 ($Al(CH_3)_3$, CH_2Cl_2, $-78°$) 86 - 74%   +   0 - 10%

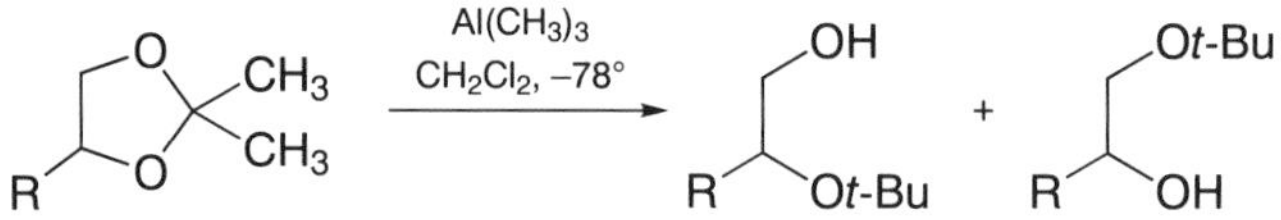

regioselectivity (equation I). This reaction provides a simple method for monoprotection of primary-secondary *vic*-diols. In the case of carbohydrates regioselectivity can be the

(II)

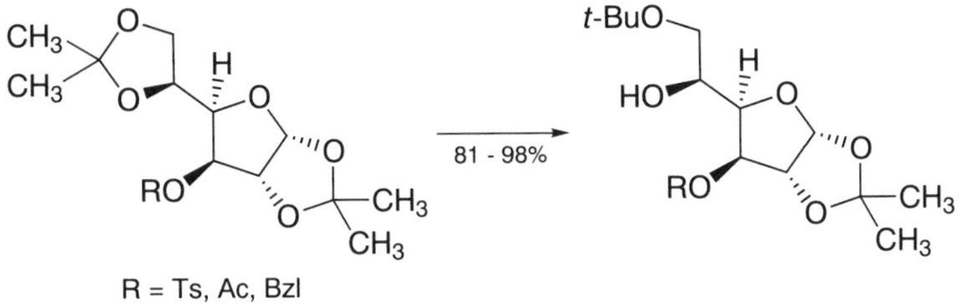

R = Ts, Ac, Bzl

81 - 98%

reverse of that of simple diols (equation II).

[1] K. Maruoka, H. Banno, and H. Yamamoto, *Synlett,* 253 (1991).
[2] A. G. Brook, *Acct. Chem. Res.,* **7**, 77 (1974).
[3] G. R. Cook, N. S. Barta, and J. R. Stille, *J. Org.,* **57**, 461 (1992).
[4] M. Miyashita, M. Hoshino, and A. Yoshikoshi, *ibid.,* **56**, 6483 (1991).
[5] D. H. R. Barton and J. Zhu, *Tetrahedron,* **48**, 8337 (1992).

1,3,5-Trimethyl-1,3,5-cyclohexanetricarboxylic acid (Kemp's triacid), 1.

Supplier: Aldrich. This ester is commercially available but expensive. It can be prepared

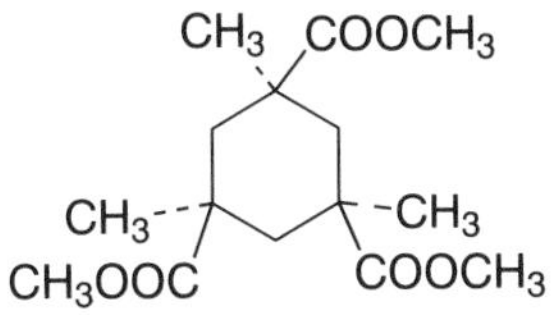

(1)

by trimethylation of 1,3,5-cyclohexanecarboxylic acid.

Chiral auxiliary for asymmetric radical addition and allylation. Curran, Rebek *et al.*[1] have prepared the chiral auxiliary **2** from **1** by manipulation of the carboxylic acid groups to provide (±)-**2**, *endo*-7-(2-benzoxazolyl)-1,5,7-trimethyl-3-azabicyclo[3.3.1]nonan-2-one. The racemic auxiliary is then resolved via its menthyl carbamate to provide both (R)- or (S)-**2**.

1 (±) - 2

(S) - **2**, α_D −7.1° (R) - **2**, α_D +7.4°

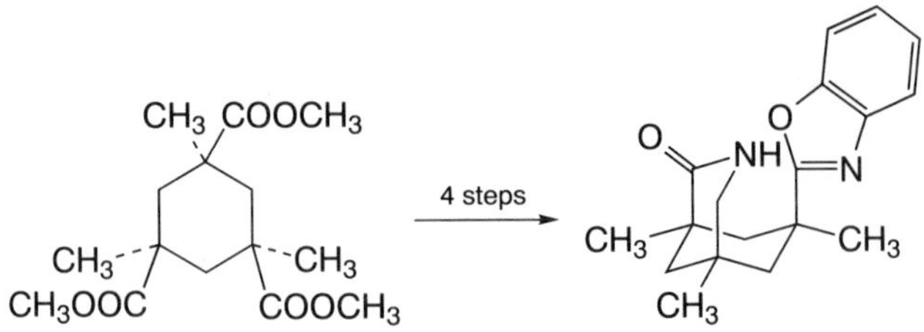

This chiral auxiliary can provide high regio- and β-stereoselective addition to fumarimides equation (I). This is the first chiral auxiliary to control both regio- and

4 : **5** : **6** = 78 : 19 : 3

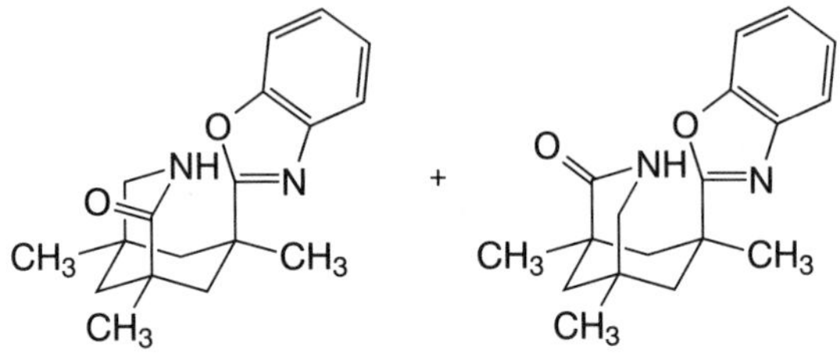

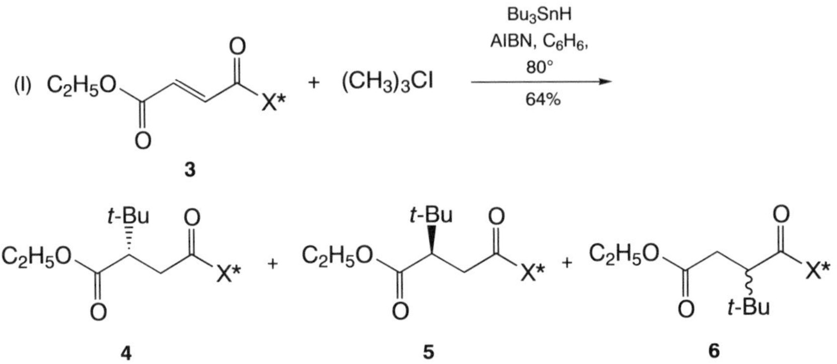

β-stereoselectivity in radical additions. The origin of the regioselectivity is uncertain but may result from inhibition of α-substitution.

This auxiliary can also control the diastereoselectivity of radical allylation (equation II).

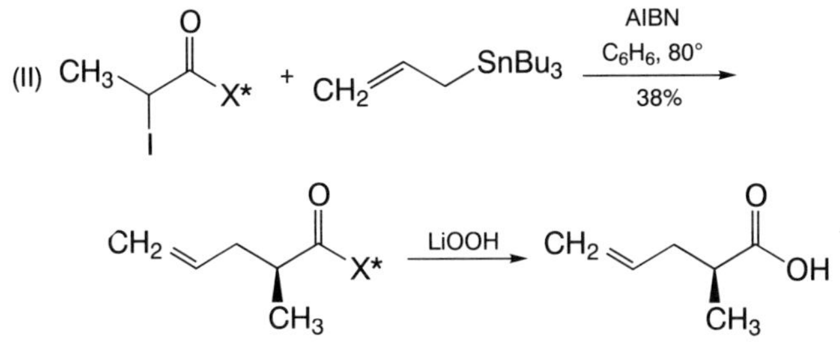

(II)

(S), 96 : 4   (S), α_D +10.5°

[1] J. G. Stack, D. P. Curran, S. V. Geib, J. Rebek, Jr., and P. Ballester, *Am. Soc.*, **114**, 7007 (1992).

Trimethyl orthoacetate, $CH_3C(OCH_3)_3$.

Epoxides from 1,2-diols.[1]   A one-pot conversion of 1,2-diols into epoxides involves conversion to an orthoester **1** by acid-catalyzed reaction with trimethyl orthoacetate. This derivative is then treated with acetyl chloride, bromide, or iodide and $(CH_3)_3SiCl$ with inversion at the halide center formed. Base-catalyzed hydrolysis and cyclization provides the epoxide with a second inversion.

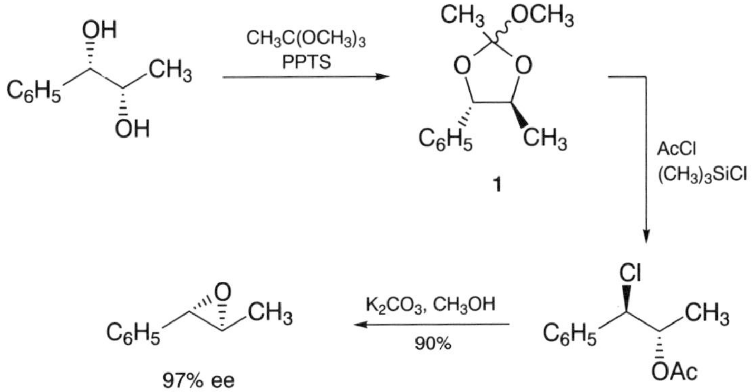

85:15

83% | K$_2$CO$_3$, CH$_3$OH

59% ee

[1] H. C. Kolb and K. B. Sharpless, *Tetrahedron,* 48, 10515 (1992).

Trimethylsilyl azide–Iodosylbenzene. (CH$_3$)$_3$SiN$_3$–C$_6$H$_5$IO (**1**, 2:1).

The reagents in the ratio 2:1 react at −45° to form C$_6$H$_5$I(N$_3$)$_2$, which decomposes at 0° to C$_6$H$_5$I and 3N$_2$.

β-Azido triisopropylsilyl ethers.[1]   Reaction of the triisopropylsilyl enol ether **2** with **1** (2:1) at −45° gives two products, **3** and **4**, as a 1:1 mixture. Addition of 2,6-di-*t*-butyl-4-methylpyridine (BMP) to this reaction gives **3** and **4** in the ratio 1:4. The product **3** can be obtained in >95% yield by reaction of **2** with C$_6$H$_5$I=O and then (CH$_3$)$_3$SiN$_3$ at

−15° → 25°. This unusual β-azido functionalization of triisopropylsilyl ethers is a general reaction, although the best conditions show some variation with the structure of the enol silyl ether.

[1] P. Magnus and J. Lacour, *Am. Soc.,* 114, 767 (1992).

Trimethylsilylboron triflate, (CH$_3$)$_3$SiB(OTf)$_4$.

The reagent is prepared by addition of (CH$_3$)$_3$SiOTf in CH$_2$Cl$_2$ or CHCl$_3$ to freshly prepared B(OTf)$_3$ (obtained from BBr$_3$ and 3HOTf).

Addition of CH₂=CHCH₂Si(CH₃ to RCHO.[1]   This reaction requires fluoride ion, a Lewis acid (1 equiv.), or a superacid. However, this reagent is a particularly effective catalyst, a "supersilylating" reagent.

[1] A. P. Davis and M. Jaspars, *Angew. Chem. Int. Ed.*, **31**, 470 (1992).

[(Trimethylsilyl)ethynyl]-9-BBN (1).
Preparation:

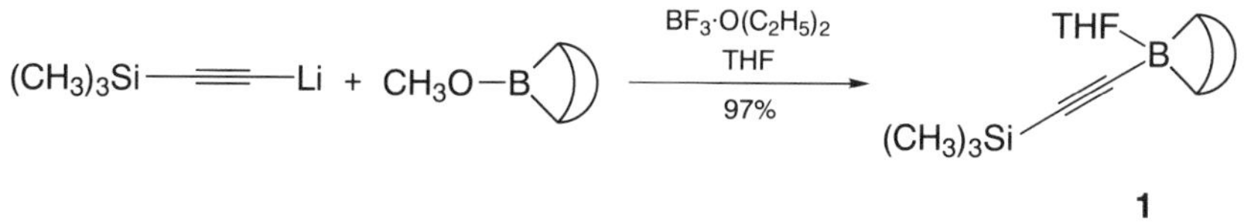

meta-Selective Diels–Alder reactions.[1]   This alkynylborane is less reactive than trimethylsilylvinyl-9-BBN, but it reacts with 1,3-dienes at 100° to form adducts in quantitative yield. The regiochemistry is unusual because the *meta* adducts are formed preferentially over the *para* adducts.

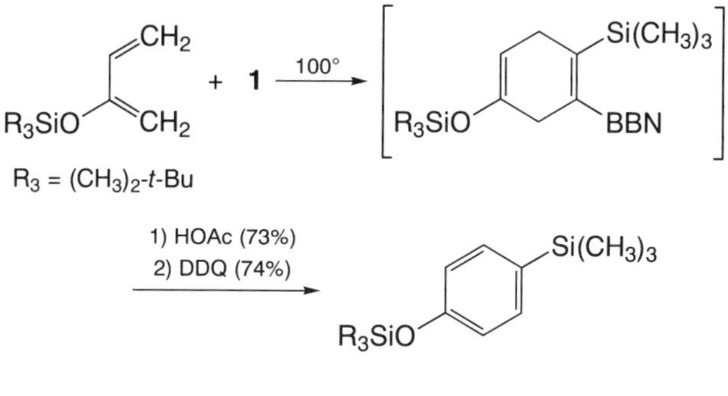

[1] D. A. Singleton and S.-W. Leung, *J. Org.*, **57**, 4796 (1992).

Trimethylsilyl trifluoromethanesulfonate, (CH₃)₃SiOTf (1).

1,2,4-Trioxan-5-ones.[1]   3,6,6-Trisubstituted 1,2,4-trioxan-5-ones (**4**) can be prepared by condensation of trimethylsilyl α-(trimethylsilylperoxy)carboxylic esters **3** with ketones. These precursors are obtained by photooxygenation of the corresponding ketene acetals (**2**). The triflate **1** is an essential catalyst for this condensation and cannot be replaced by Bu$_4$NF. This ring system is rarely encountered in natural products, but is present in artemisinine, a potent antimalarial.

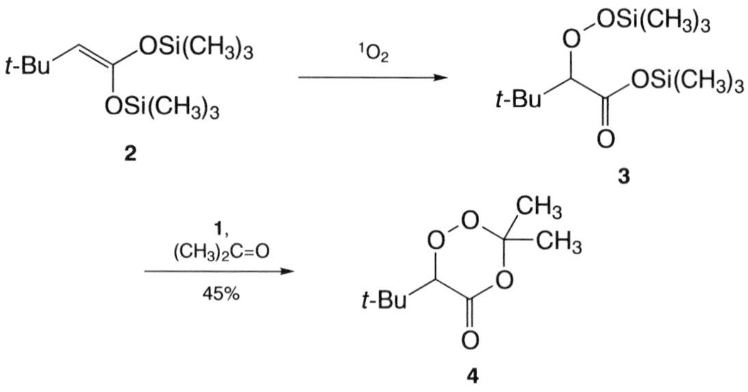

Acetal-β, β-diglucosides.[2]   In the presence of trimethylsilyl triflate, trimethylsilyl-2,3,4,6-tetra-O-acetyl-β-D-glucopyranoside (**1**) reacts with aldehydes at −78° to form acetal-β, β-diglucosides **2** in 30–85% yield. At higher temperatures, mixtures of acetal-β, β-, α, β-, and α, α-diglucosides are obtained. The yields are dependent on the reactivity and steric bulk of the aldehyde. Ketones do not undergo this reaction.

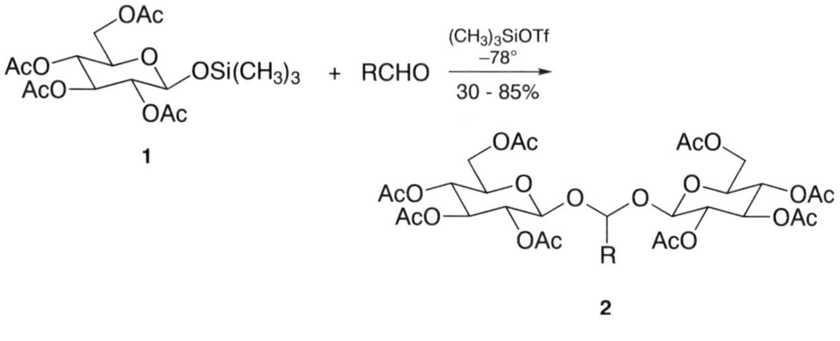

Modified Sakurai reaction.[3]   The original reaction involved the TiCl$_4$-catalyzed addition of allyltrimethylsilane to aldehydes and ketones or the acetals and ketals to form homoallylic alcohols or ethers (**7**,370–371). Markó et al. have extended this reaction to a synthesis of homoallylic ethers by a Lewis acid catalyzed reaction of allyl-trimethylsilane with a carbonyl compound and a trimethylsilyl ether.

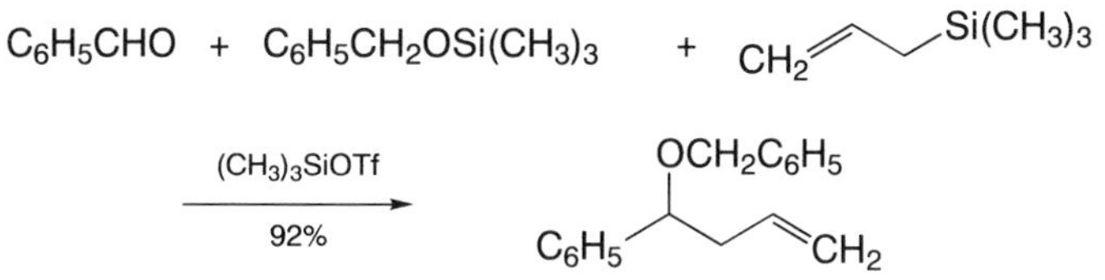

This reaction can show significant diastereoselectivity (equation I).

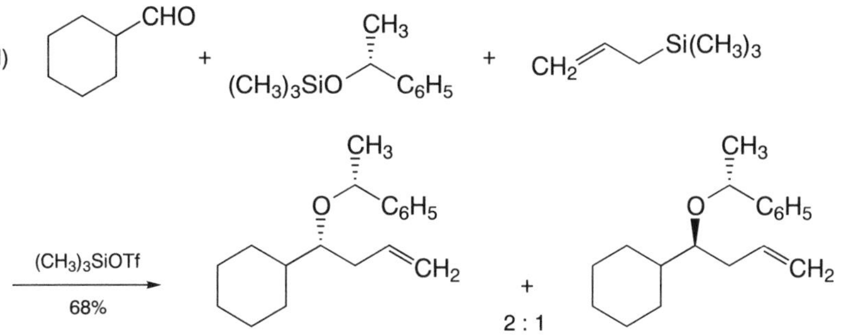

This modified Sakurai reaction has been extended to a synthesis of tetrahydropyrans and -ketals by use of a bis-silylated ether such as **1**.

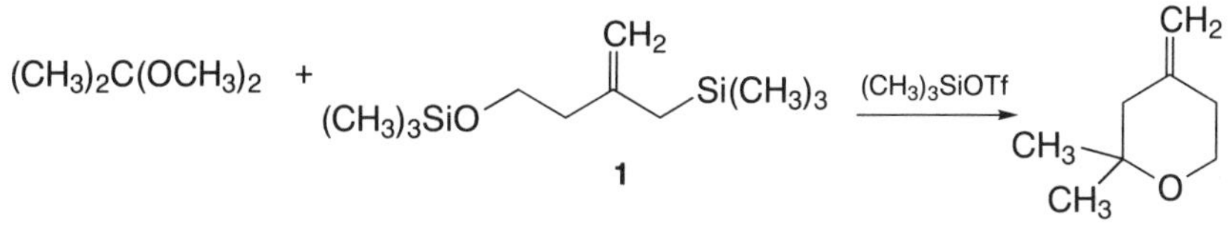

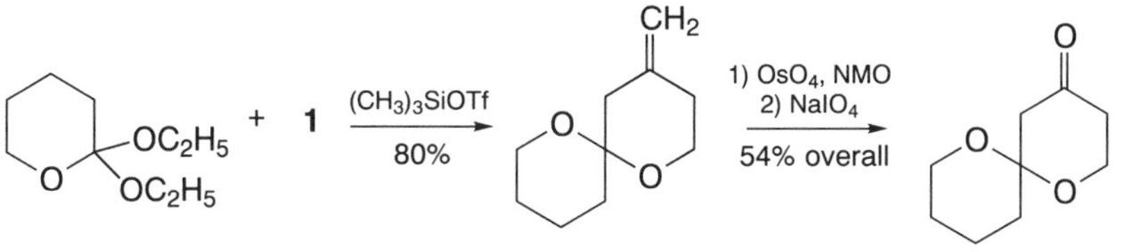

Reaction of crotylsilanes with aryl acetals.[4]   Panek and Yang[1] have extended the Noyori reaction of allylsilanes with acetals catalyzed by trimethylsilyl triflate (**10**,439–440) to optically active (E)-crotylsilanes, such as α-methoxy-β-(dimethylphenyl-silyl)-(E)-hexenoates (**1**). These (E)-crotylsilanes are available by Ireland–Claisen ester rearrangement (**6**,276–277) of optically active vinylsilanes (equation I and II).

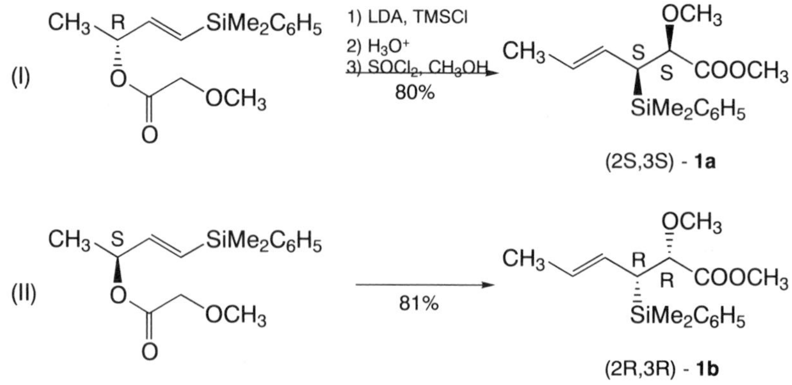

(I)

1) LDA, TMSCl
2) H_3O^+
3) $SOCl_2$, CH_3OH
80%

(2S,3S) - **1a**

(II)

81%

(2R,3R) - **1b**

In the presence of 1 equiv. or less of trimethylsilyl triflate these crotylsilanes react with aryl acetals to form homoallylic ethers with high diastereo- and enantioselectivity. The new C–C bond (C_5C_6) is formed with high *syn*-selectivity (13–40:1).

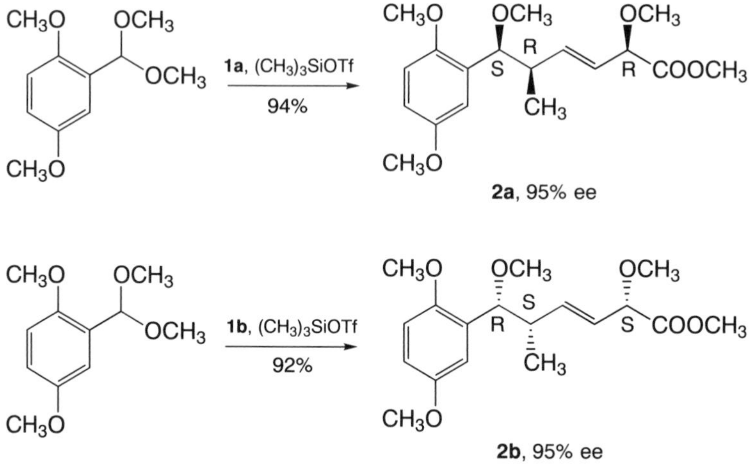

1a, $(CH_3)_3SiOTf$
94%

2a, 95% ee

1b, $(CH_3)_3SiOTf$
92%

2b, 95% ee

Addition of R_2CuLi to epoxy esters.[5]   Addition of $Bu_2CuLi/(CH_3)_3SiOTf$ (1:1) to the *trans*-epoxy ester **1** at −78° followed by quenching at 25° (thermodynamic control) results in a product (**2**) in which the ester group has migrated to the secondary alcohol. The expected product (**3**) can be obtained as the major product if the reaction is quenched at −78°. This unexpected reaction is general as far as the cuprate reagent is

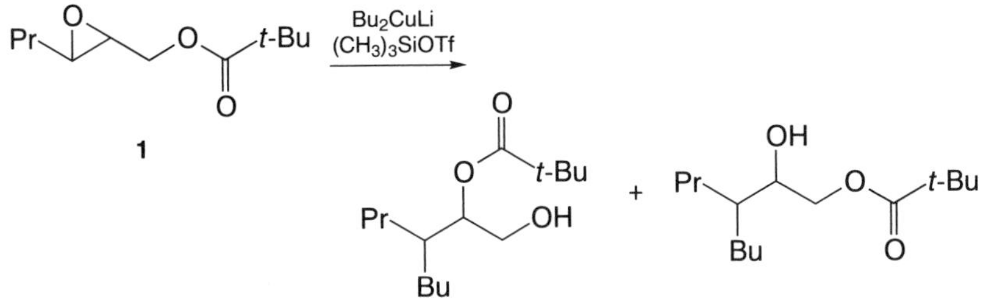

	2	**3**
quench at 25°	73%	5%
quench at −78°	1%	72%

concerned, but has only been reported for pivaloylates. Even so, it provides a convenient route to monoprotected 1,2-diols.

[3+4]Cycloadditions.[6]   In the presence of a Lewis acid, particularly TMSOTf, the bis(trimethylsilyl) enol ether (**2**) of methyl acetoacetate undergoes a [3+4]cycloaddition with 1,4-dicarbonyl compounds (**1**) to form 2-carbalkoxy-8-oxabicyclo[3.2.1]-octan-3-ones (**3**) with high regioselectivity. This Lewis acid ordinarily does not promote reaction of silyl enol ethers with simple aldehydes or ketones, but it is more effective than $TiCl_4$

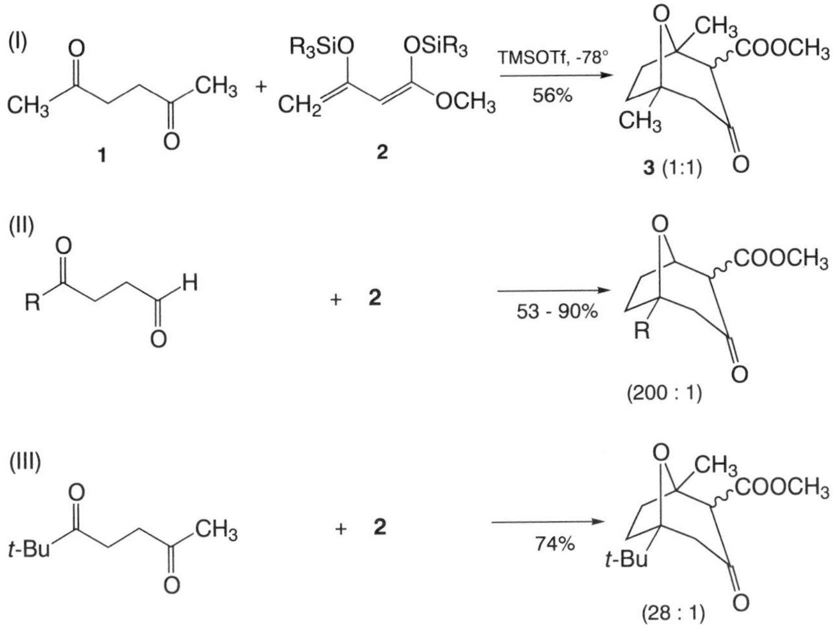

for this [3+4]cycloaddition. Unexpectedly, the reaction of **2** with a 1,4-keto aldehyde involves initial attack with the keto group (equation II). In the case of unsymmetrical diketones, the initial attack occurs with the more hindered ketone (equation III). The reaction is also regioselective in the case of 4-keto aldehydes bearing a substituent at the intervening carbon atoms. The report includes a possible mechanism for this cycloadition involving initial attack of the Lewis acid with the less hindered carbonyl group followed by participation of the more hindered carbonyl to form a cyclic oxonium ion that reacts with the terminal carbon of **2** to form an adduct that then cyclizes to **3**.

The [3+4]cyclization of 4-oxoheptanal (**5**) with 3-substituted bis(trimethylsilyl) enol ethers of β-diketones (**4a**) and of β-keto esters (**4b**) is also regioselective but in the opposite sense (equation IV).

Stereoselective reactions with acetals.[7]   Noyori *et al.* (**10**,438) have used this Lewis acid to promote an aldol-type reaction between enol silyl ethers and acetals and have noted high *syn*-selectivity in this process. Molander and Haar report that reaction of acetals with cyanotrimethylsilane promoted by TMSOTf results in α-alkoxy cyanides and that this reaction can be diastereoselective when the acetal is substituted at the 4-position by an alkoxy group. The diastereoselectivity depends on the nature of the acetal and the 4-alkoxy group. Dimethoxy acetals show slight diastereoselectivity, but diisopropoxy and dibenzyl acetals can show diastereoselectivity of 5–10:1. The diastereoselectivity also depends on the type of 4-substituent. Acetoxy and *t*-butyldimethylsilyloxy groups have no effect on the diastereoselectivity, but methoxy, benzyloxy, and allyloxy groups promote *anti*-selectivity. Since a metal template is not involved, the diastereoselectivity is not a result of chelation, but of neighboring group participation.

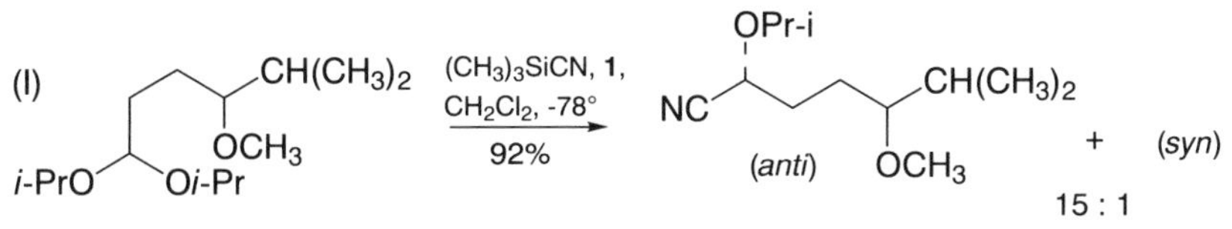

Tandem Sakurai-carbonyl-ene reaction; steroid synthesis.[8] The key step in a short synthesis of the BDC rings of steroids from the chiral aldehyde **1** involves a tandem Sakurai-carbonyl-ene reaction of **2** mediated by trimethylsilyl triflate, which provides **3a** as the only cyclized product in 52% yield. Amazingly, catalysis with $C_2H_5AlCl_2$ leads to two diastereomers, **3b** and **3c** (3:1) in 40% yield. Hydrogenation of **3a** provides **4**, which corresponds to the BCD ring of steroids with an added methyl group in ring B.

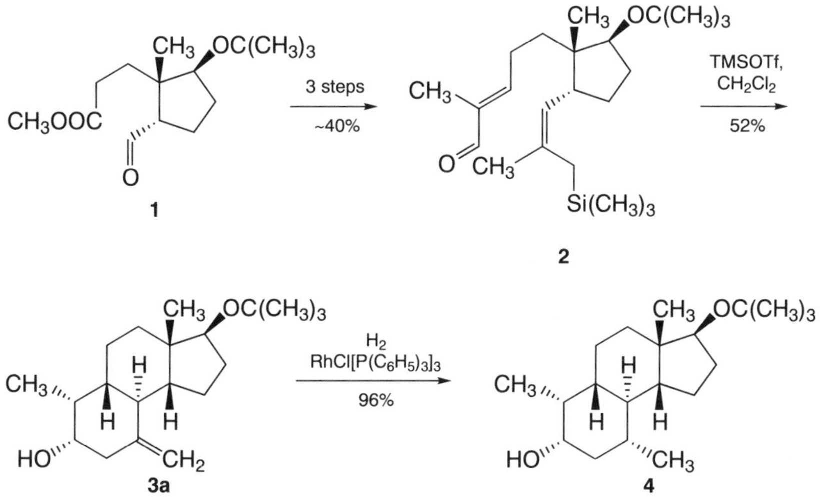

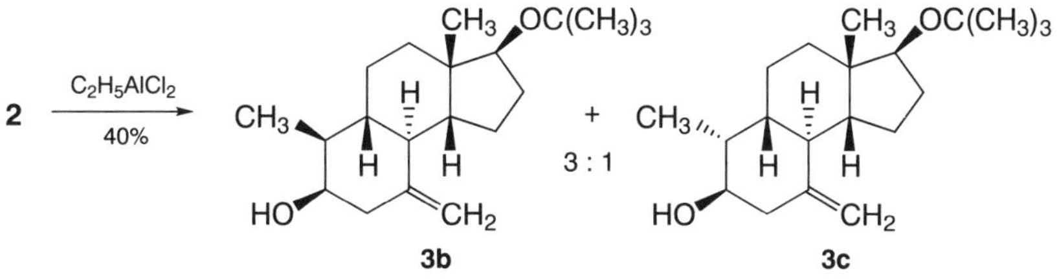

Nazarov cyclization.[9] α, α'-Dienone esters **1**, prepared by acylation/Knoevenagel condensation of various esters, undergo Nazarov cyclization in the presence of trimethylsilyl triflate to form cyclopentenone esters **2**.

Trimethylsilyl triflate (5 equiv.) also converts 3-ethoxycarbonyltetrahydro-γ-pyrones (**3**) into cyclopentenone esters (**4**) by ring opening followed by Nazarov cyclization.

[1] C. W. Jefford, J. Currie, G. D. Richardson, J.-C. Rossier, *Helv.*, **74**, 1239 (1991).
[2] L. F. Tietze and M. Beller, *Angew. Chem. Int. Ed.*, **30**, 868 (1991).
[3] A. Mekhalfia and I. E. Markó, *Tetrahedron Letters*, **32**, 4779, 4783 (1991).
[4] J. S. Panek and M. Yang, *Am. Soc.*, **113**, 6594 (1991).
[5] G. A. Molander and K. L. Bobbitt, *J. Org.*, **57**, 5031 (1992).
[6] G. A. Molander and K. O. Cameron, *ibid.*, **56**, 2617 (1991).
[7] G. A. Molander and J. P. Haar, Jr., *Am. Soc.*, **113**, 3608 (1991).
[8] L. F. Tietze and M. Rischer, *Angew. Chem. Int. Ed.*, **31**, 1221 (1992).
[9] J. F. P. Andrews and A. C. Regan, *Tetrahedron Letters*, **32**, 7731 (1991).

Trimethylsilyl trifluoromethanesulfonate – Silver perchlorate, TMSOTf – AgClO$_4$.

c-Arylglycosylation.[1]   In the presence of a catalyst composed of TMSOTf – AgClO$_4$ (1:1), free sugars or methyl glycosides, unprotected as well as protected, react with α-naphthol to form glycosides in 72–92% yield and with high β-selectivity ($\alpha/\beta = 1:15$–99).

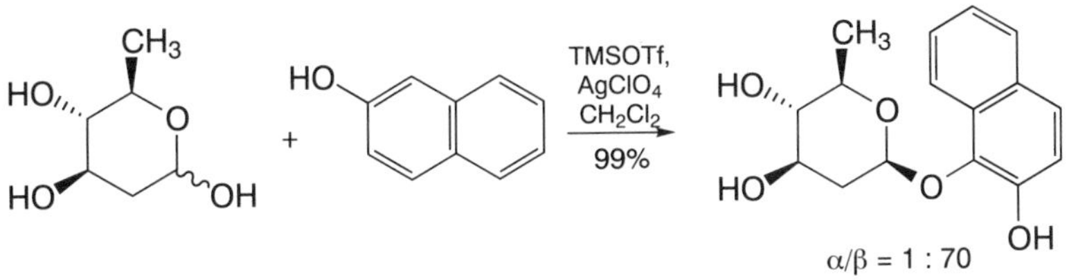

$\alpha/\beta = 1 : 70$

[1] K. Toshima, G. Matsuo, T. Ishizuka, M. Nakata, and M. Kinoshita, *J. C. S. Chem. Comm.,* 1641 (1992).

2-Trimethylsilylvinyl-9-BBN, $(CH_3)_3SiCH{=}CH{-}B$ (1),

air-sensitive, pyrophoric.

The borane is obtained by reaction of trimethylsilylacetylene with 9-BBN.

Diels–Alder reactions.[1]   This reagent is a highly reactive dienophile. Reactions can be effected at 25–85° with marked regioselectivity and high *endo*-selectivity. The adducts on oxidation (H_2O_2, NaOH) afford trimethylsilylcyclohexenols, which on treatment with TsOH at 55° afford 1,4-cyclohexadienes in quantitative yield.

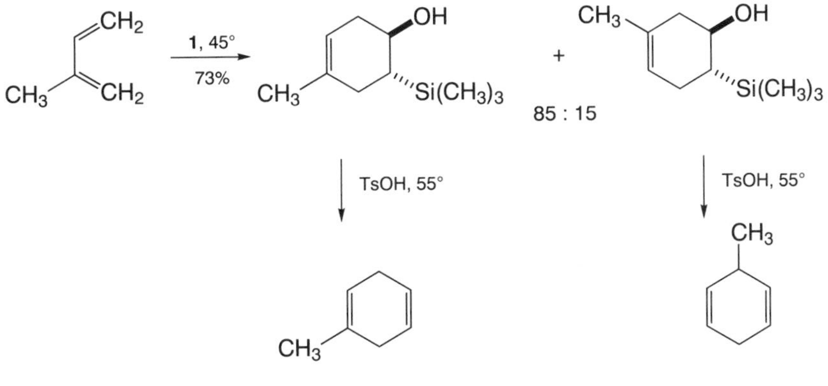

[1] D. A. Singleton and J. P. Martinez, *Tetrahedron Letters,* **32,** 7365 (1991).

(Trimethylsilyl)vinyl phenyl selenide (1).

[2+2]Cycloadditions.[1]   The vinyl selenide **1** undergoes [2+2]cycloaddition with vinyl ketones in the presence of a Lewis acid, usually $SnCl_4$ or $AlCl_3$. If the reaction is quenched with H_2O rather than an amine, acylsilanes are obtained in 37–50% yield.

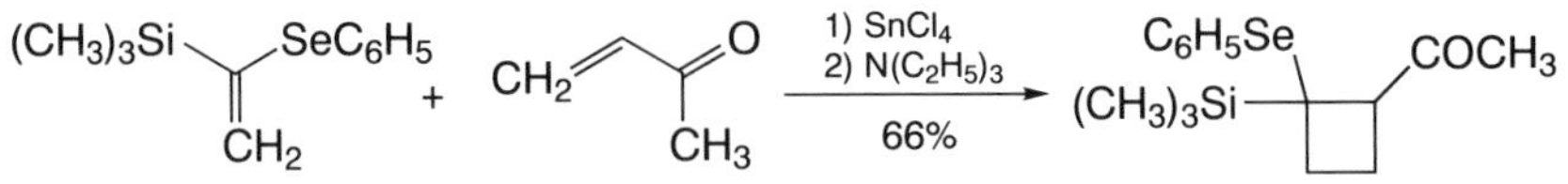

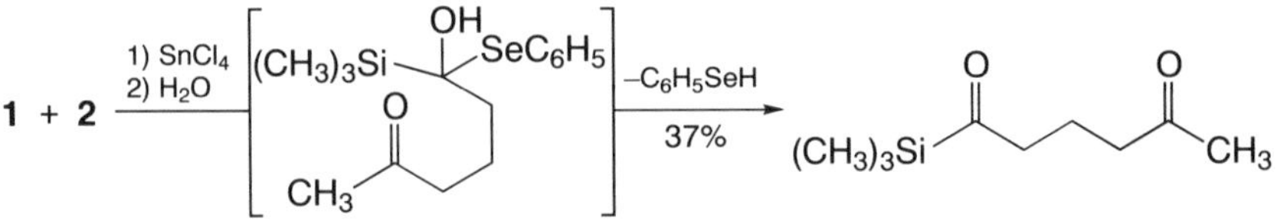

[1] S. Yamazaki, H. Fujitsuka, S. Yamabe, and H. Tamura, *J. Org.*, **57**, 5610 (1992).

Trimethyltin chloride/AIBN/NaBH$_3$CN.

Cyclization of dienes and trienes.[1] Reaction of a diene or triene with at least one terminal double bond with (CH$_3$)$_3$SnCl, AIBN, and NaBH$_3$CN in *t*-BuOH results in addition of the (CH$_3$)$_3$Sn radical to the terminal double bond followed by cyclization to the other double bond. The C–Sn bond of the carbocycle is oxidatively cleaved by CAN in methanol to the dimethyl acetal of an aldehyde. This two-step synthesis provides a useful route to tetrahydrofurans and pyrrolidines as well as 1,6-dioxatriquinanes from allyl ethers.

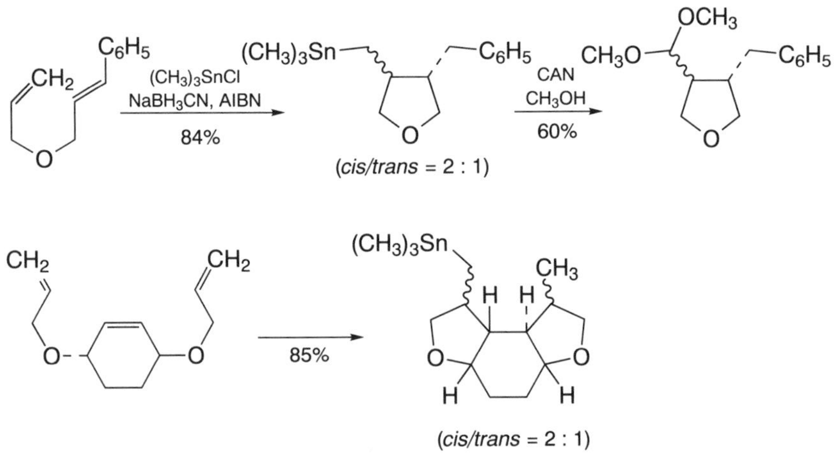

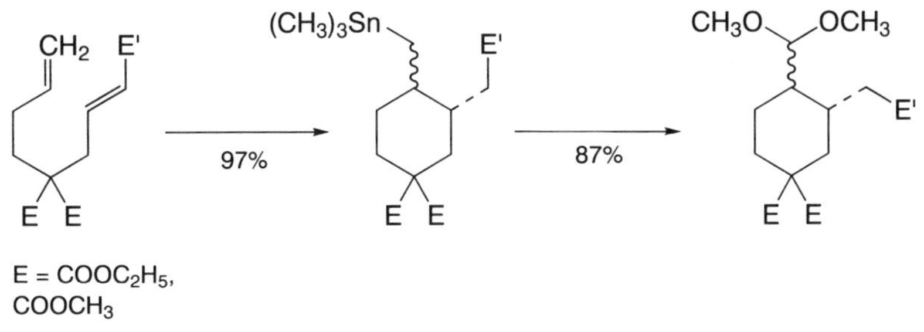

E = COOC$_2$H$_5$,
COOCH$_3$

[1] S. Hanessian and R. Leger, *Am. Soc.*, **114**, 3115 (1992).

Triphenylphosphine – Diethyl azodicarboxylate.

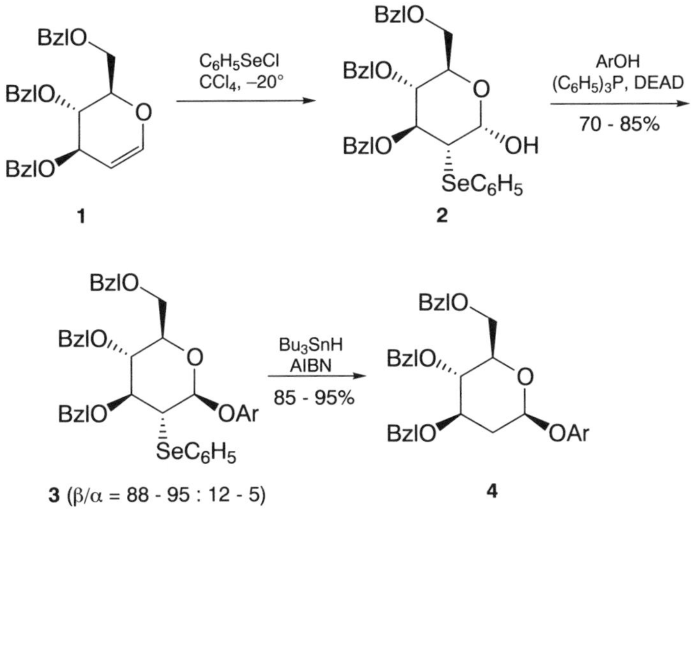

Aryl 2-deoxy-β-D-glycosides.[1]   These glycosides (**4**) can be prepared in high β-selectivity by Mitsunobu coupling of phenols with 2α-(thiophenyl)- or 2α-(seleno-phenyl)-α-D-pyranosnes **2**, prepared by reaction of C$_6$H$_5$SCl or C$_6$H$_5$SeCl with D-glucals (**1**). Reaction of **2** with a phenol under Mitsunobu conditions provides mainly β-glycosides **3**. The thiophenyl or selenophenyl residue of **3** is removed by reaction with Bu₃SnH/AIBN.

Mitsunobu reactions with o-nitrophenylacetonitrile.[2]   The Mitsunobu reaction can be used to form C—C bonds as well as the usual C—O bonds. Thus *o*-nitrophenylacetonitrile (**1**) reacts with various alcohols under Mitsunobu conditions to form the alkylated derivative **2** in 45–70% yield. The products can be converted to indoles (**3**).

R = C₆H₅CH₂   70%   54%
    = CH₃   45%
    = HC≡CCH₂   56%   41%

[1] W. R. Roush and X.-F. Lin, *J. Org.*, **56**, 5740 (1991).
[2] J. E. Macor and J. M. Wehner, *Tetrahedron Letters*, **32**, 7195 (1991).

Triphenylphosphine–Diisopropyl azodicarboxylate.

Mitsunobu reaction of β-hydroxy-α-amino acid peptides.[1]   Reaction of the Cbz-prolyl–threonine N-methyl amide **1** with P(C₆H₅)₃ and DIAD provides the peptide aziridine **2** in 84% yield.

In contrast allothreonine peptides are converted by the reaction into oxazoline derivatives (64–83% yield) with no aziridine formation (equation I). Oxazolines are also

obtained by the Mitsunobu reaction with serine-containing peptides.

[1] P. Wipf and C. P. Miller, *Tetrahedron Letters*, **33**, 6267 (1992).

Tris(4-bromophenyl)aminium hexachloroantimonate, (p-BrC$_6$H$_4$)$_3$N$^+$SbCl$_6^-$ (1).

Radical epoxidation.[1] In the presence of this radical cation, iodosylbenzene can effect epoxidation of stilbene, but only in low yield because of decomposition of the catalyst. Selenium dioxide (SeO$_2$) is a suitable oxidant for this catalyzed epoxidation, but suffers from low solubility in CH$_2$Cl$_2$ at 0°. The most useful oxidant to date is benzeneseleninic anhydride, (C$_6$H$_5$S)$_2$O (BSA). This oxidant, in the presence of **1**, converts *cis*-stilbene to *trans*-stilbene oxide in 85% yield. This radical epoxidation is superior to *m*-chloroperbenzoic for monoepoxidation of 1,3-dienes, such as 1-vinylcyclohexene (equation I). Moreover the endocyclic bond is selectively epoxidized. A similiar selectivity is observed in epoxidation

| | BSA, 1 | 76% | – | – |
| | ClC$_6$H$_4$CO$_3$H | 55% | 21% | 24% |

of 4-isopropenyl-1-vinylcyclohexene (equation II).

[1] N. L. Bauld and G. A. Mirafzal, *Am. Soc.*, **113**, 3613 (1991).

Tris(μ-chloro)hexakis(tetrahydrofuran)divanadium hexachlorodizincate (1), 16,370–372.

Hydroxymethylation with paraformaldehyde.[1] Petersen's pinacol coupling (**16**,371) is applicable to coupling of aldehydes with paraformaldehyde. Even though a large excess of CH$_2$O is necessary for satisfactory yields, there is no evidence of homocoupling to ethylene glycol. Under optimum conditions, the desired 1,2-diols are obtained in yields of >90%. The reaction of chiral aldehydes shows modest to good diastereoselectivity.

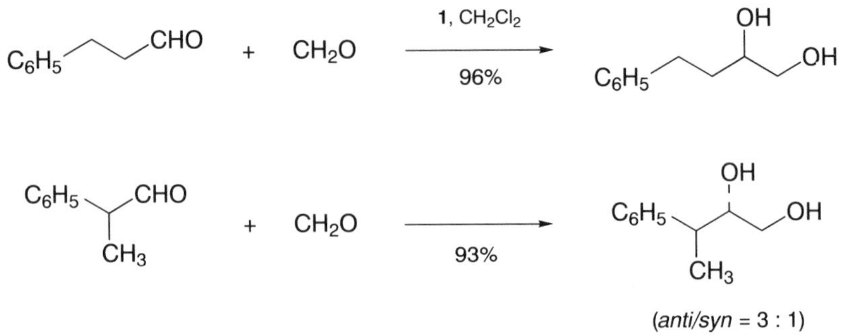

(anti/syn = 3 : 1)

The reagent **1** is generally less effective for pinacol coupling with ketones, but by use of a reagent (**2**) prepared from 2 VCl$_3$(THF)$_3$ and 5 Zn, coupling between ketones and paraformaldehyde proceeds in yields of 80–90%.

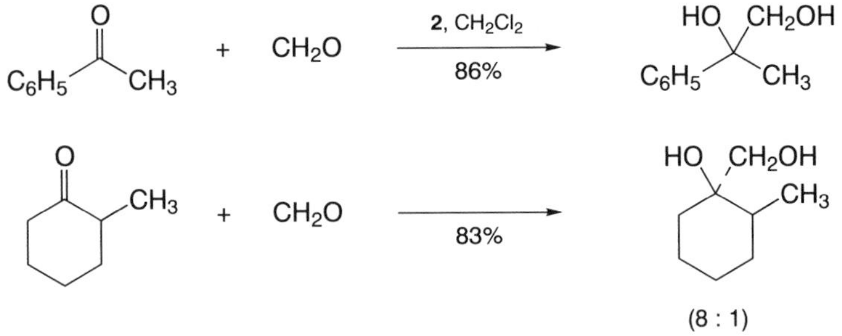

(8 : 1)

Epoxides → olefins.[2]   This reagent (**1**) can convert epoxides to alkenes in 50–98% yield. The reaction requires three equivalents of **1**, and the *cis*-configuration of the substrate is not retained.

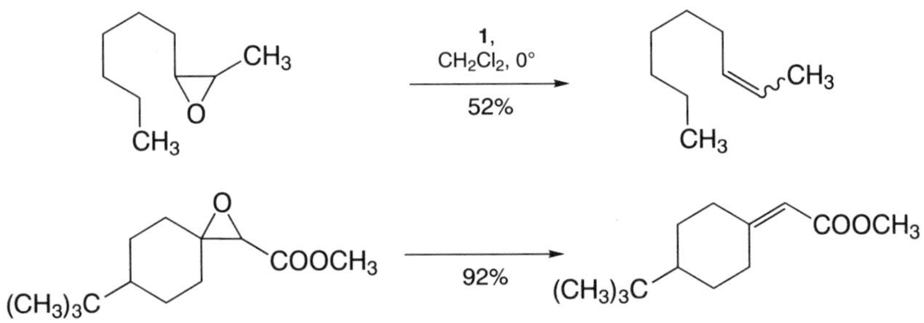

In contrast reaction of the epoxide **2** with VCl$_3$(THF)$_3$ (3 equiv.) affords the halohydrin **3** as the main product.

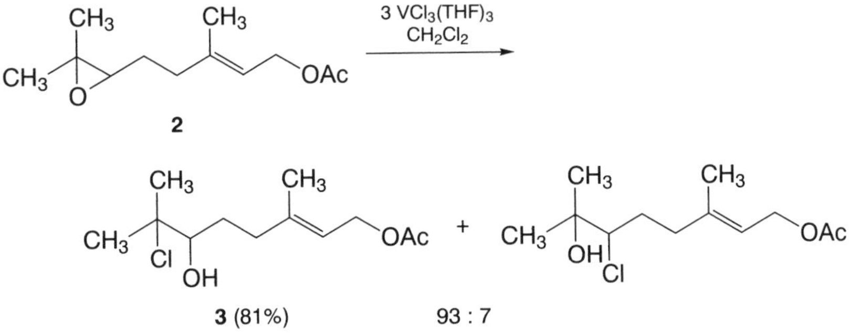

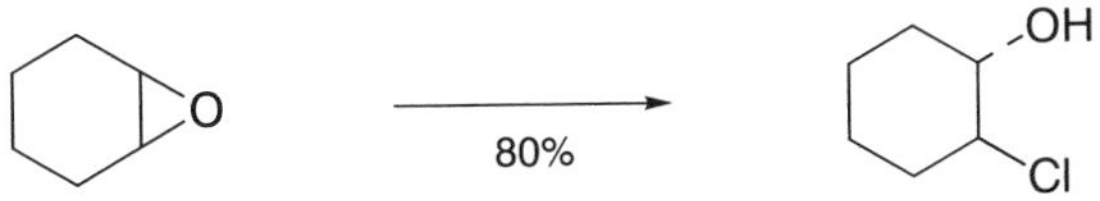

3 (81%)   93 : 7

80%

Asymmetric 1,4-diamino-2,3-diols.[3]   (S)-N-Cbz-amino aldehydes **2** undergo pina-col homocoupling in the presence of **1** to afford (1S,2R,3R,4S)-1,4 bis(Cbz-amino)-2,3-diols (**3**) in good yield and high diastereoselectivity.

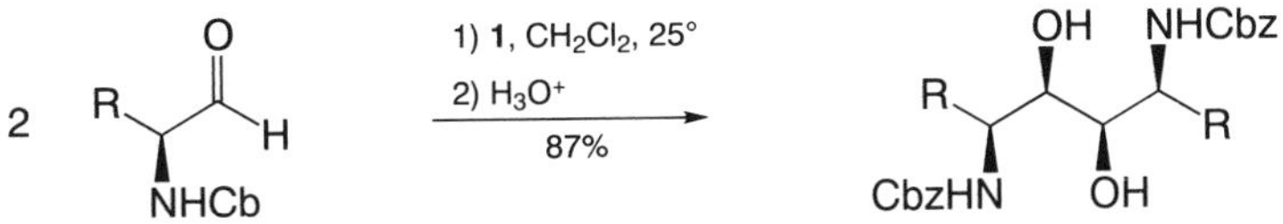

R = i - Pr	89%
R = i - Bu	61%
R = CH₂C₆H₅	76%

[1] J. Park and S. F. Pedersen, *Tetrahedron*, **48**, 2069 (1992).
[2] T. Inokuchi, H. Kawafuchi, and S. Torii, *Synlett*, 510 (1992).
[3] A. W. Konradi and S. F. Pedersen, *J. Org.*, **57**, 28 (1992).

Tris(dibenzylideneacetone)dipalladium, Pd_2dba_3.

Allylic gem.-diacetates.[1] In the presence of this Pd(0) catalyst, propargyl acetates react with acetic acid to form (E)-allylic *gem*-diacetates in 55–79% yield (equation I). Mixed *gem*-dicarboxylates can also be prepared.

An intramolecular version of this reaction can result in macrolides (equation II).

Stille reaction. The preferred catalyst for this coupling of unsaturated halides or triflates with organostannanes has been $Pd[P(C_6H_5)_3]_4$. Much more effective catalysts can be obtained by treatment of Pd_2dba_3 with 4 equiv. of various ligands to form a PdL_4 catalyst.[2] Of a number of ligands, the most effective are tri-2-furylphosphine (TFP)[3] and triphenylarsine $[As(C_6H_5)_3]$.

[1] B. M. Trost, W. Brieden, and K. H. Baringhaus, *Angew. Chem. Int. Ed.,* **31,** 1335 (1992).
[2] V. Farina and B. Krishnan, *Am. Soc.,* **113,** 9585 (1991).
[3] D. W. Allen and B. F. Taylor, *J. C. S. Dalton, 51* (1982).

Tris(phenylthio)methane, $HC(SC_6H_5)_3$ **(1).**

[2+1+2]Cyclopentannelation.[1] This reagent can function as a dilithiomethane equivalent in consecutive conjugate additions to form a dienolate that is oxidized to a cyclopentane. This cyclopentannelation has been used to obtain the triquinane

nucleus **2** as a single *cis, anti, cis*-**2**. This triquinane **2** has been converted into (+)-hirsutene (**3**) by known steps in 18% overall yield from **1**.

[1] K. Ramig, M. A. Kuzemko, K. McNamara, and T. Cohen, *J. Org.*, **57**, 1968 (1992).

Tris(trimethylsilyl)silane (1).

Review.[1]   Chatgilialoglu[1] has reviewed the use of organosilanes, particularly **1**, as radical reducing reagents. One advantage of organosilanes as reducing agents is that the strength of the Si–H can be modulated by a choice of substituents on Si. Thus $[(CH_3)_3Si]_2SiCH_3H$ shows only about 10% of the reactivity of **1** as a hydrogen donor. The silane **1** is useful for reduction of halo, NC, and SeC_6H_5 groups at room temperature in the presence of only small amounts of radical initiators. For this reason the isolation and purification of the products is simplified. Tris(alkylthio)silanes, $(RS)_3SiH$, are also useful for radical reduction and for hydrosilylation of alkenes.

Selective hydrogen atom abstraction by radicals.   Giese and Curran[2] note that radicals such as **4** abstract hydrogen atoms in a ratio that is remarkably similar to that observed by Cram for reduction of ketones by lithium aluminum hydride.

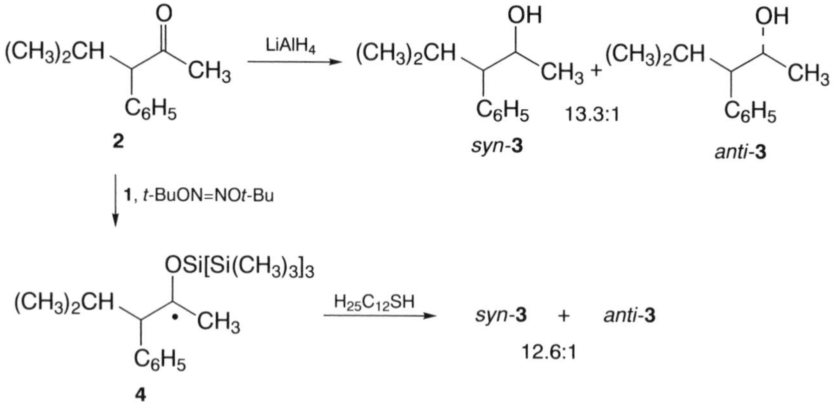

Hydrosilylation of alkenes and alkynes.[3]   This reaction can be effected with **1** and an initiator at 90°. Reaction with monosubstituted and *gem*-disubstituted alkenes shows high *anti*-Markovnikov regioselectivity. *cis*- or *trans*-Disubstituted and trisubstituted alkenes are hydrosilylated in high yield but require longer reaction times.

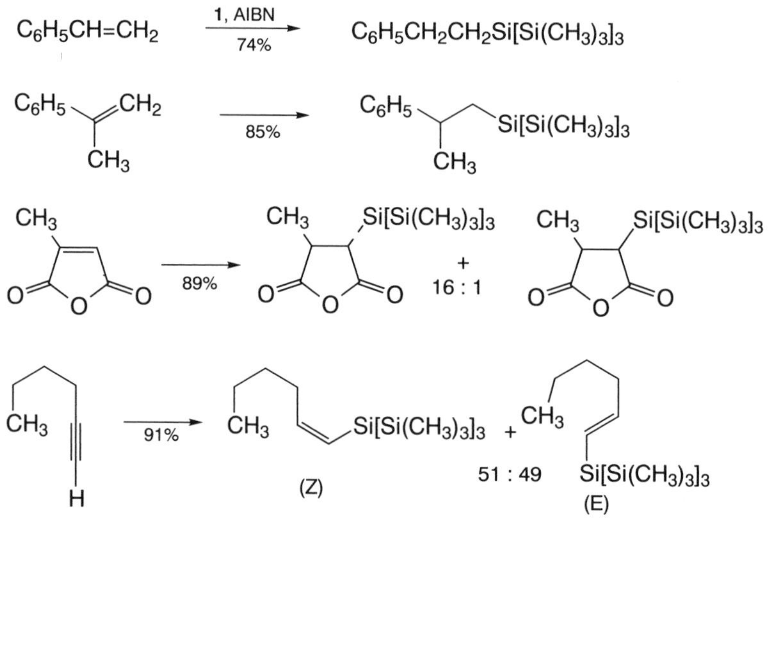

RCOCl → RH.[4]   This reaction can be effected with this silane in combination with AIBN in dodecane or toluene at 80° in high yield, particularly in reactions of *sec-* or *tert-*acid chlorides. Reduction of primary acid chlorides leads also to aldehydes as coproducts.

[1] C. Chatgilialoglu, *Acc. Chem. Res.,* **25**, 188 (1992).
[2] B. Giese, W. Damm, J. Dickhaut, F. Wetterich, S. Sun, and D. P. Curran, *Tetrahedron Letters,* 6097 (1991).
[3] B. Kopping, C. Chatgilialoglu, M. Zehnder, and B. Giese, *J. Org.,* **57**, 3994 (1992).
[4] M. Ballestri and C. Chatgilialoglu, N. Cardi, and A. Sommazzi, *Tetrahedron Letters,* **33**, 1787 (1992).

Tungsten carbene complexes.
   Aminocarbene Diels–Alder reactions.[1]   The known ability of alkenyl alkoxy-chromium carbenes to undergo Diels–Alder reactions even in the absence of a Lewis acid catalyst (**12,**135–136) has prompted an investigation of Diels–Alder reactions of aminocarbene complexes. Unfortunately these complexes are much less reactive than their alkoxy counterparts. But the (E)-amino complex **1** does undergo a Diels–Alder reaction with Danishefsky's diene at 90° to give **2** in 72% yield. Surprisingly, **2** is formed with high *exo*-selectivity (>25:1). However the (Z)-isomer of **1** fails to undergo this reaction.

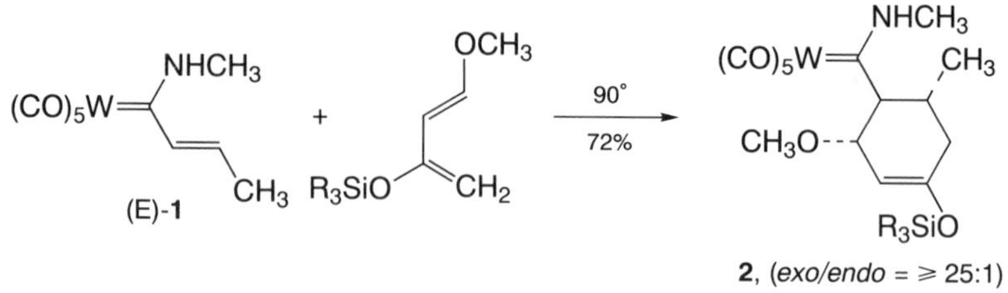

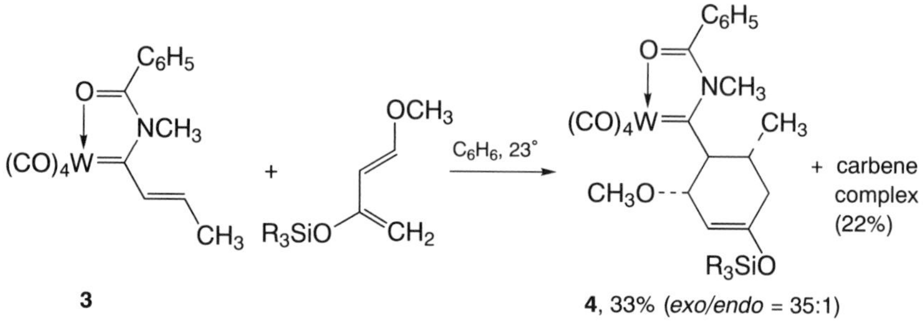

2, (*exo/endo* = ⩾ 25:1)

3

4, 33% (*exo/endo* = 35:1)

The chelated carbene **3** is even more reactive than (E)-**1** and shows even higher *exo*-selectivity, but unfortunately is relatively unstable.

Intramolecular bis-alkyne annelations.[2]   A typical reaction of this type generally leads to a mixture of a phenol and a cyclohexadienone (equation I). However this annelation can be adapted to a construction of the steroid ring system by a

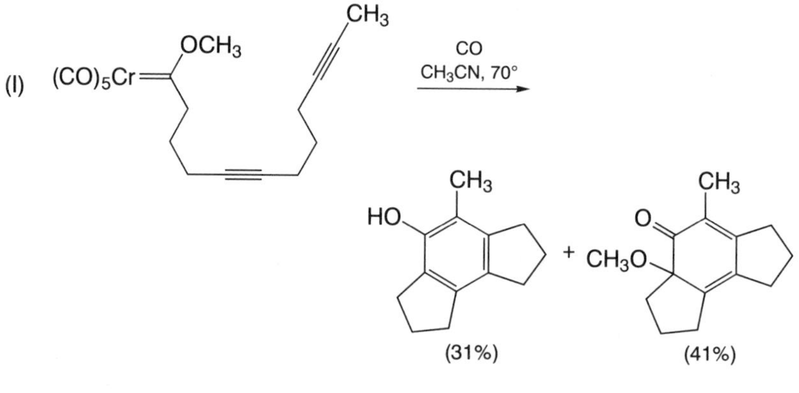

(31%)   (41%)

Diels–Alder reaction of the tungsten carbene complex **1** with Danishefsky's diene (1.5 equiv.) in acetonitrile under CO to give the complex **2** in 62% yield.

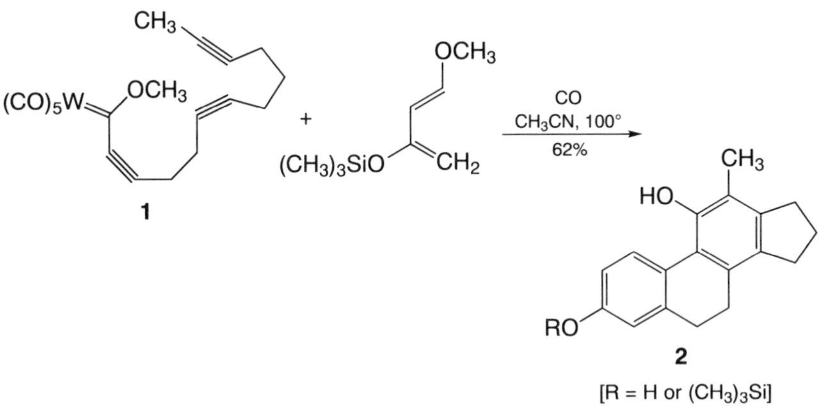

1

2

[R = H or (CH₃)₃Si]

Cycloheptadienones.[3]   The cyclopropylcarbene–tungsten complex **1** reacts with diphenylacetylene at 100° to form a cycloheptadienone **2**, which rearranges to the more

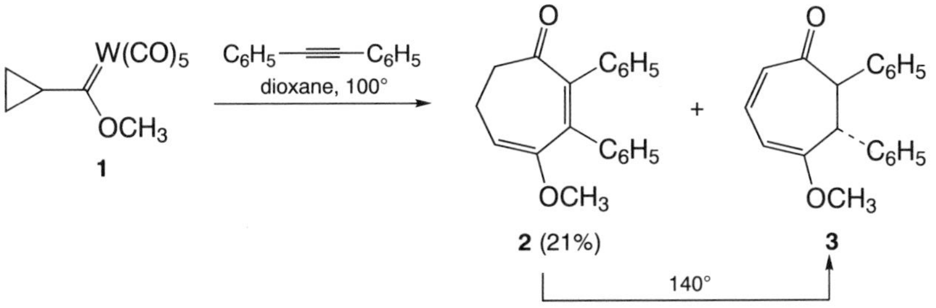

2 (21%)   **3**

stable isomer **3**. Optimal yields of **3** (55%) are obtained in refluxing xylene (140°) in the presence of triarylphosphines. Terminal alkynes do not undergo this reac-

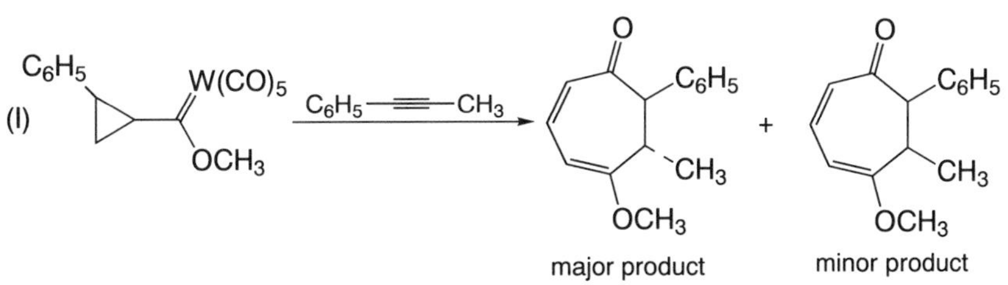

major product   minor product

tion, but unsymmetrical alkynes react with some selectivity (equation I). The complex of a substituted cyclopropane reacts with high regioselectivity (equation II).

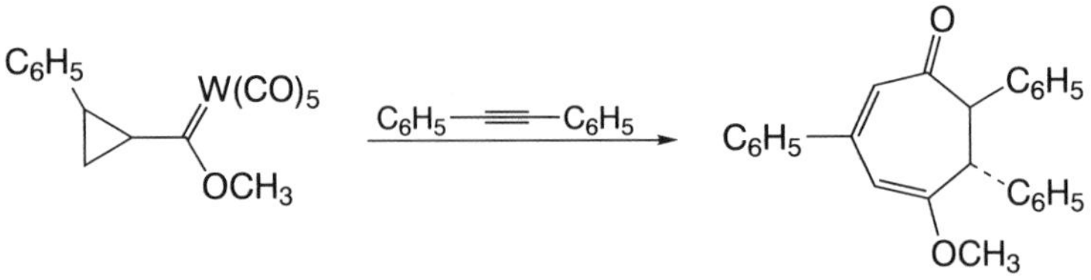

Note that the cyclopropyl chromium carbene corresponding to **1** reacts with acetylenes to form cyclopentenones with loss of ethylene (**14**,93).

[1] B. A. Anderson, W. D. Wulff, T. S. Powers, S. Tribbitt, and A. L. Rheingold, *Am. Soc.,* **114**, 10784 (1992).

[2] J. Bao, V. Dragisich, S. Wenglowsky, and W. D. Wulff, *ibid.,* **113**, 9873 (1991).

[3] J. W. Herndon, G. Chatterjee, P. P. Patel, J. J. Matasi, S. U. Tumer, J. J. Harps, and M. D. Reid, *ibid.,* **113**, 7808 (1991).

U

Urea–Hydrogen peroxide complex, $H_2NCONH_2 \cdot H_2O_2$.

Peroxytrifluoroacetic acid, CF_3COOOH.[1] This peracetic acid is no longer commercially available since 90% H_2O_2 is required for the oxidation of CF_3COOH. However, it can be prepared in the laboratory by oxidation of trifluoroacetic anhydride with the H_2O_2 complex with urea.

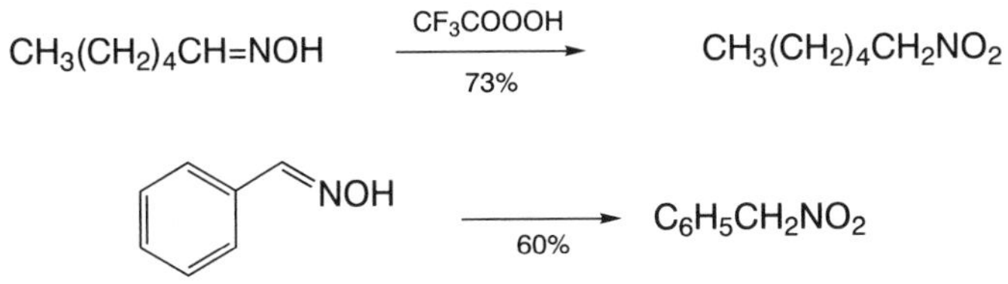

$$CH_3(CH_2)_4CH=NOH \xrightarrow[73\%]{CF_3COOOH} CH_3(CH_2)_4CH_2NO_2$$

$$\xrightarrow[60\%]{} C_6H_5CH_2NO_2$$

[1] R. Ballini, E. Marcantoni, and M. Petrini, *Tetrahedron Letters,* **33**, 4835 (1992).

V

9-Vinyl-9-borabicyclo[3.3.1]nonane (vinyl-9-BBN).
   This reagent is prepared by reaction of B-bromo-9-BBN with vinyltributyltin:

1, b.p. 28-30°/0.25 mm

Diels–Alder reactions.[1]   9-Vinyl-9-BBN is a more reactive dienophile than methyl acrylate, and undergoes Diels–Alder reactions at 25–55°. Although the reaction with isoprenes shows the expected "*para*" selectivity, the reaction with *trans*-piperylene shows an unusual "*meta*"-selectivity, possibly as a result of steric factors. This reaction also can show high *endo*-stereoselectivity (92:8).

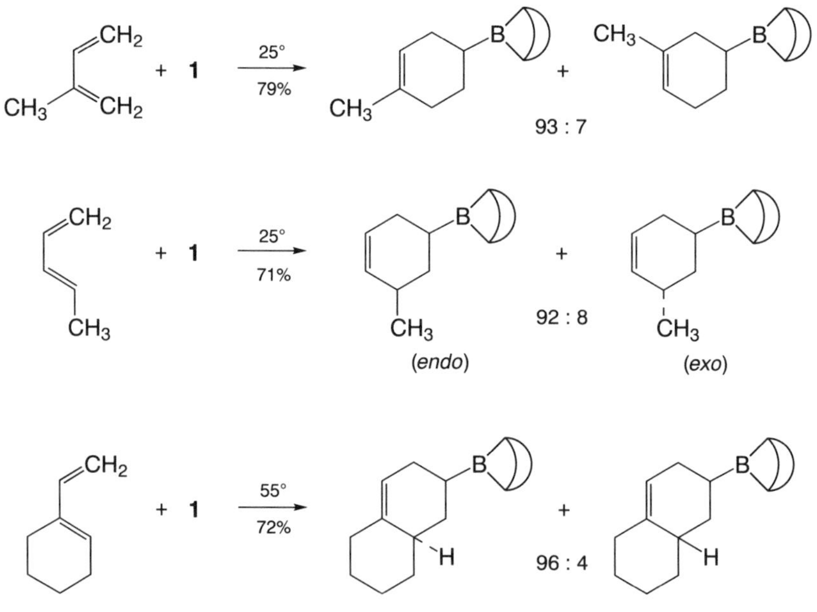

Other vinylboranes.[2]   One disadvantage of vinyl-9-BBN is the conversion at 25° to *trans*- and *cis*-dimers. This reaction can limit yields in Diels–Alder reactions with moderately reactive dienes. This problem is solved by use of 1-vinyl-3,6-dimethylborepane (**2**). Vinyldimethylborane (**3**) is pyrophoric and is best generated *in situ*. Trivinylborane (**4**) has been prepared by reaction of vinylbutyltin with BBr_3. Trivinylborane is the most reactive of these vinylboranes. Vinyldimethylborepane (**2**) is the most stable of the vinylboranes, but shows lower reactivity with dienes; however, the reactions are significantly cleaner and afford better yields, probably because of the absence of dimerization. The most regioselective of these boranes **1–4** is vinyl-9-BBN, probably owing to steric hindrance. All the vinylboranes are *endo*-selective, with the highest selectivity shown by **2**.

$$CH_2=CHB(CH_3)_2 \qquad B(CH=CH_2)_3$$
$$\mathbf{3} \qquad\qquad \mathbf{4}$$

[1] D. A. Singleton and J. P. Martinez, *Am. Soc.,* **112**, 7423 (1990).
[2] D. A. Singleton, J. P. Martinez, J. V. Watson, and G. M. Ndip, *Tetrahedron,* **48**, 5831 (1992).

Vinyldichloroborane.
Diels–Alder reactions.[1]   Vinyldichloroborane reacts with simple dienes in toluene at 110° to form adducts in good yield. These are isolated as the boronic esters of pinacol (equation I). Of greater interest, the C—B bond of the adduct can be replaced by

a C—N bond by reaction with benzyl azide at 25° to give the adduct of a secondary amine (equation II).

[1] N. Noiret, A. Youssofi, B. Carboni, and M. Vaultier, *J. C. S. Chem. Comm.*, 1105 (1992).

Vinylsilanes.

Regioselective Diels–Alder reactions.[1]   The intramolecular [4+2]cycloaddition of **2**, prepared by reaction of a vinylsilane with the diunsaturated alcohol **1**, provides an adduct (**3**), which on oxidative desilylation provides the cyclohexenes **4** and **5** with marked preference for the *cis*-1,3-diol **4**.

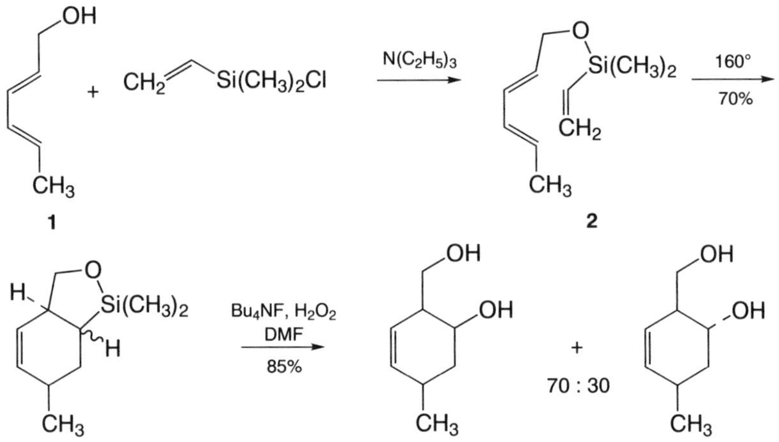

The advantage of a silicon tether is illustrated further by reaction of **1b** with (E)-(chlorodimethylsilyl)acrylate **6** to provide **7**. This cyclizes to a single, *endo*-adduct **8**, which can be converted to **9** and **10** as shown.

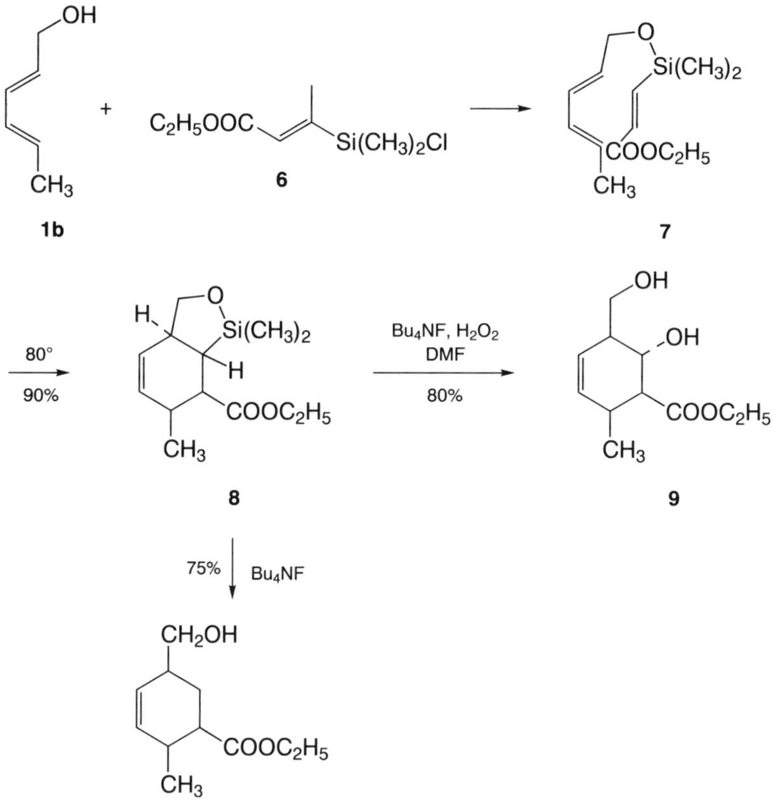

[1] G. Stork, T. Y. Chan, and G. A. Breault, *Am. Soc.,* **114**, 7578 (1992).

Z

Zinc.

(1R,2R)-Diarylethylenediamines.[1]   A novel route to these useful diamines involves esterification of an N-protected valine with 1,3-propanediol followed by N-deprotection to give the (R,R)-diamines **1**. This product is converted to the diimine **2** by reaction with C_6H_5CHO. Reduction of **2** with Zn in CH_3OH/THF at 0° results in intramolecular coupling to provide (R,R)-**3** and (R,S)-**3** in the ratio 91:9. Hydrolysis of (R,R)-3 followed by oxidation, $Pb(OAc)_4$, and N-protection yields R,R-**4** in 72% yield.

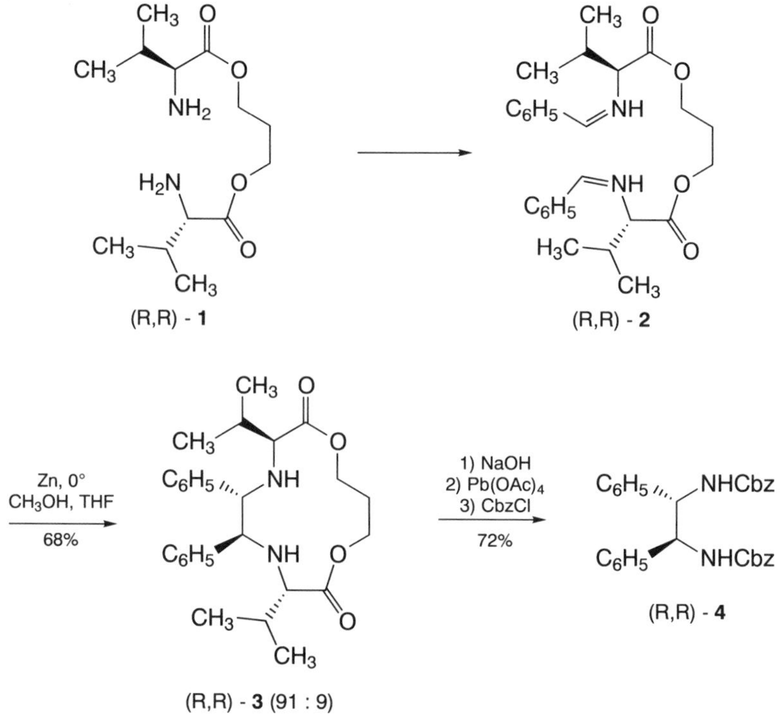

(R,R) - **1**

(R,R) - **2**

(R,R) - **3** (91 : 9)

(R,R) - **4**

Note that zinc dust reduction of the phenylimine derived from (S)-valine methyl ester itself provides a mixture of (R,R)-, (R,S)-, and (S,S)-diphenylethylenediamines in the ratio 63:30:7.

Stereoselective addition of t-BuBr to α,β-enoates.[2]   Reaction of t-BuBr with diethyl mesaconate with reduction by zinc with ultrasound results in syn-**1** and anti-**1** in the ratio 85:15. Similar stereoselectivity obtains in radical reactions promoted by vitamin B_{12} or with t-BrHgCl, $NaBH_4$ in CH_2Cl_2. Even higher stereoselectivity obtains when different amines are added.

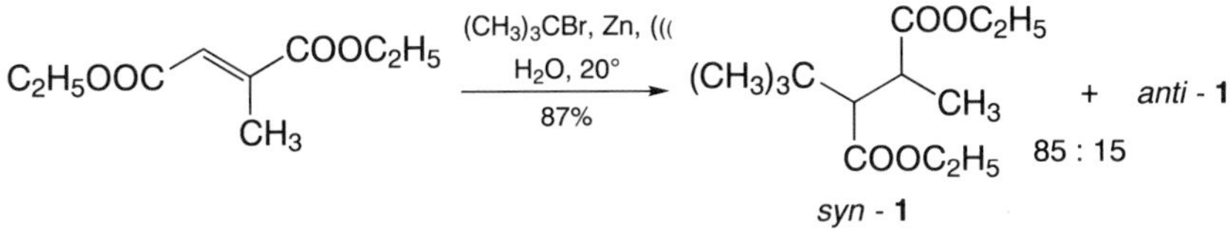

[1] T. Shono, N. Kise, H. Oike, M. Yoshimoto, and E. Okazaki, *Tetrahedron Letters,* **33**, 5559 (1992).
[2] P. Erdmann, J. Schäfer, R. Springer, H.-G. Zeitz, and B. Giese, *Helv.,* **75**, 638 (1992).

Zinc–Copper couple.

Cleavage of 2,3-epoxy halides.[1] A zinc–copper couple, prepared by sonication of zinc powder and CuI in aqueous ethanol, cleaves epoxy halides to a radical that rearranges to an allylic alcohol. Since the epoxy halide is prepared by epoxidation of an allylic alcohol (*m*-chloroperbenzoic acid) followed by reaction with $P(C_6H_5)_3$ and CBr_4, the reaction effects 1,3-transposition of the hydroxyl group.

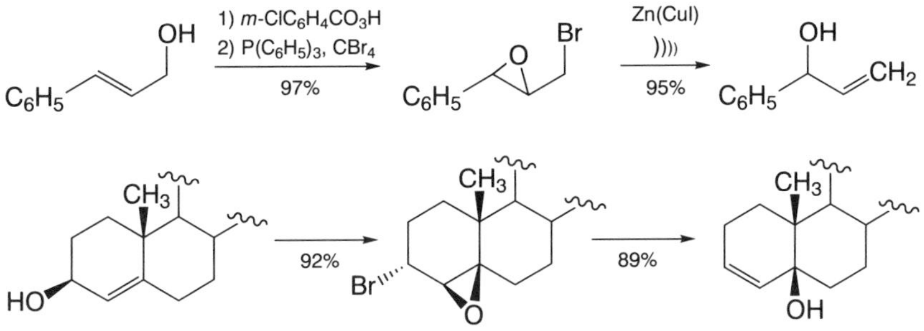

[1] L. A. Sarandeses, A. Mouriño, and J.-L. Luche, *J. C. S. Chem. Comm.,* 818 (1991).

Zirconium bisamides, $Cp_2Zr(NHR)_2$.

Preparation:

$$Cp_2ZrCl_2 + 2LiNHR \xrightarrow[-\,2LiCl]{THF,\ 25^\circ} Cp_2Zr(NHR)_2$$

$$Cp_2Zr(CH_3)_2 + H_2NR \xrightarrow{C_6H_5CH_3,\ 140^\circ} Cp_2Zr(NHR)_2$$

Hydroamination of alkynes.[1]   Zirconium bisamides can react with internal alkynes at 85–120° to form azametallacyclobutenes, which can dissociate to $Cp_2Zr=NR$ and an enamine (equation I).

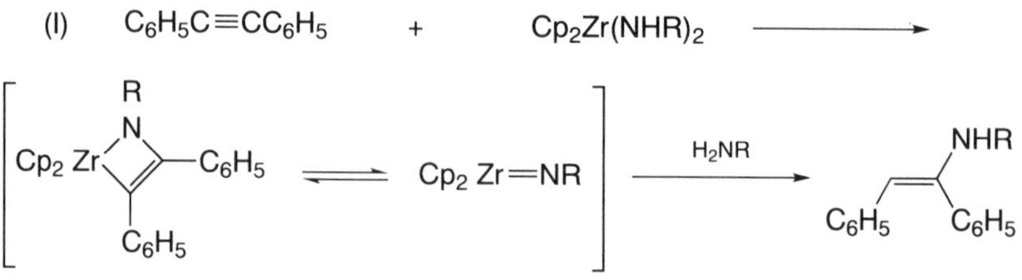

(I) $C_6H_5C\equiv CC_6H_5$ + $Cp_2Zr(NHR)_2$ ⟶

$$\left[\underset{\displaystyle C_6H_5}{Cp_2\,Zr} \overset{\displaystyle \overset{R}{\underset{}{N}}}{\diagup} C_6H_5 \rightleftharpoons Cp_2\,Zr=NR \right] \xrightarrow{H_2NR} \underset{C_6H_5 \quad C_6H_5}{\overset{NHR}{\diagup}}$$

Of more practical value zirconium bisamides can serve as catalysts for reaction of primary amines with alkynes to form enamines. This reaction cannot be extended to alkenes, but allenes undergo this hydroamination to form imines.

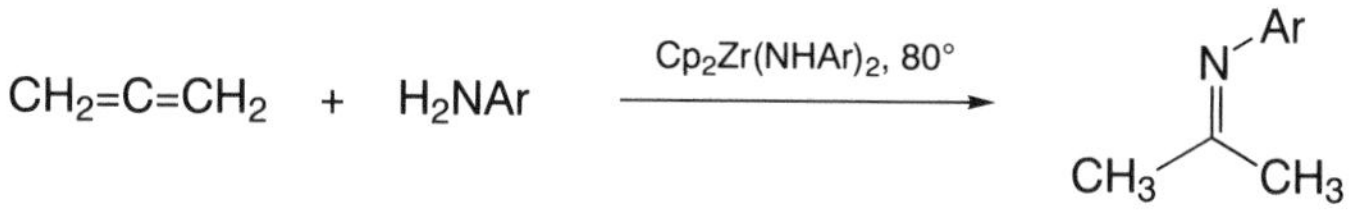

$$CH_2{=}C{=}CH_2 \;+\; H_2NAr \xrightarrow{Cp_2Zr(NHAr)_2,\ 80°} \underset{CH_3 \quad CH_3}{\overset{N^{\diagdown Ar}}{\diagup}}$$

[1] P. J. Walsh, A. M. Baranger, and R. G. Bergman, *Am. Soc.,* **114**, 1708 (1992).

AUTHOR INDEX

SUBJECT INDEX